The Halal Food Handbook

The Halal Food Handbook

Edited by

Yunes Ramadan Al-Teinaz
Independent Public Health & Environment Consultant,
London, UK

Stuart Spear
Freelance Journalist, London, UK

Ibrahim H. A. Abd El-Rahim
Department of Environmental and Health Research,
the Custodian of the Two Holy Mosques Institute for Hajj
and Umrah Research, Umm Al-Qura University,
Makkah Al-Mukaramah, Saudi Arabia

Infectious Diseases, Department of Animal Medicine,
Faculty of Veterinary Medicine, Assiut University,
Assiut, Egypt

This edition first published 2020
© 2020 John Wiley & Sons Ltd

The right of Yunes Ramadan Al-Teinaz, Stuart Spear, and Ibrahim H. A. Abd El-Rahim to be identified as the authors of the editorial material in this work has been asserted in accordance with law.

Registered Offices
John Wiley & Sons, Inc., 111 River Street, Hoboken, NJ 07030, USA
John Wiley & Sons Ltd, The Atrium, Southern Gate, Chichester, West Sussex, PO19 8SQ, UK

Editorial Office
The Atrium, Southern Gate, Chichester, West Sussex, PO19 8SQ, UK

For details of our global editorial offices, customer services, and more information about Wiley products visit us at www.wiley.com.

Wiley also publishes its books in a variety of electronic formats and by print-on-demand. Some content that appears in standard print versions of this book may not be available in other formats.

Library of Congress Cataloging-in-Publication Data
Names: Al-Teinaz, Yunes Ramadan, 1952– editor. | Spear, Stuart, editor. | Abd
    El-Rahim, Ibrahim H. A., editor.
Title: The halal food handbook / edited by Yunes Ramadan Al-Teinaz, Stuart
    Spear, and Ibrahim H.A. Abd El-Rahim.
Description: Hoboken, NJ : John Wiley & Sons, Inc., 2019. | Includes
    bibliographical references and index. |
Identifiers: LCCN 2019014356 (print) | LCCN 2019015538 (ebook) | ISBN 9781118823118
    (Adobe PDF) | ISBN 9781118823101 (ePub) | ISBN 9781118823125 (hardback)
Subjects: LCSH: Halal food–Handbooks, manuals, etc. | Muslims–Dietary
    laws–Handbooks, manuals, etc. | Halal food industry–Handbooks, manuals,
    etc. | Food–Religious aspects–Islam–Handbooks, manuals, etc.
Classification: LCC BP184.9.D5 (ebook) | LCC BP184.9.D5 H325 2019 (print) |
    DDC 297.5/76–dc23
LC record available at https://lccn.loc.gov/2019014356

Cover Design: Wiley
Cover Image: © hallimsib99/Shutterstock

Set in 9.5/12.5pt STIXTwoText by SPi Global, Pondicherry, India
Printed and bound in Singapore by Markono Print Media Pte Ltd

10 9 8 7 6 5 4 3 2 1

This book is dedicated to those who cherish the prevention of cruelty to animals.

Contents

About the Editors

Prof. Dr Ibrahim H. A. Abd El-Rahim

Ibrahim is a Professor of Infectious Diseases & Epizootiology in the Department of Environmental and Health Research at The Custodian of the Two Holy Mosques Institute for Hajj & Umrah Research, Umm Al-Qura University, Makkah Al-Mukaramah, Saudi Arabia. He is also a Professor of Infectious Diseases in the Department of Animal Medicine at the Faculty of Veterinary Medicine, Assiut University, Assiut, EGYPT. He was a Consultant at the Veterinary Laboratory Department in the Ministry of Agriculture, Riyadh, Saudi Arabia, from 2003 to 2010. He worked as Professor of Infectious Diseases and Clinical Medicine at the College of Agriculture and Veterinary Medicine, Qassem University, Saudi Arabia, from 2010 to 2011.

He is a member of the German Academic Exchange Services (DAAD), the International Society of Cattle Diseases, and the Egyptian Society of Biochemistry and Molecular Biology. He received a DAAD scholarship for PhD research at the cattle clinic, Justus-Liebig University, Giessen, Germany, from 1993 to 1996. He was also awarded a post-doctoral Danish scholarship as a researcher at the Danish Veterinary Institute for Virus Research, Lindholm, Kalvehave, Denmark in 1999. In the same year, he received a post-doctoral DAAD scholarship as a researcher at the Faculty of Veterinary Medicine, Free University, Berlin, Germany.

He received two awards for Scientific Excellence in the field of Veterinary Medicine in Egypt in 2000 and 2002. He also obtained an ALMARAI award for a Distinguished Researcher in the field of veterinary medicine in Saudi Arabia in 2012. He has registered three patents in the United States Patent and Trademark Office (USPTO) (US 2014/0193307A1, US 2014/0190201A1 and US 2016/0273428A1). He has published 50 articles in local and international scientific journals in the field of veterinary science and has participated as a speaker in international scientific conferences in the field of veterinary medicine in Germany, Austria, Italy, Denmark, Hungary, Egypt, Saudi Arabia, Tunisia, and Syria. In addition, he has participated as a speaker in international scientific conferences in the field of halal food in Saudi Arabia, Kuwait, and the United Arab Emirates.

Stuart Spear

Stuart Spear is a journalist who has been working with the Islamic community for around 20 years to combat illegal practices in the halal industry and the sale of unfit food to ethnic communities in the UK. As editor of *Environmental Health Practitioner*, the UK's national magazine for environmental health officers, he spearheaded a number of UK-wide campaigns to stop criminal gangs exploiting UK Muslims. In particular, he headed up the Stamp it Out campaign, which focused on illegal slaughter, false halal declaration, and the smuggling of bush meat along with other food safety issues impacting the Muslim community. In recognition for his support for UK Muslims he received the International Pioneer and Diversity Award and an award for excellence from the International Health Foundation at the London Central Mosque.

He has written extensively about halal practices and has over the years campaigned for the introduction of a common, easily auditable standard for all halal produce. He has also written about all aspects of food safety and how criminal gangs exploit weaknesses in the food chain for high reward and low risk. Over the last few years he has been editing an on-line news magazine aimed at the environmental health profession, covering food safety issues, and now works as a freelance writer on food safety and public health.

Dr Yunes Ramadan Al-Teinaz
A British citizen of Libyan origin, Yunes is an Independent Public Health and Environment Consultant, who holds a PhD from Liverpool University, a Masters in Public Health Medical School from Dundee University, a BSc in Environmental Science from Kings College London, a Certificate in Tropical Community Medicine and Health from the Liverpool School of Tropical Medicine. He is a Chartered Environmental Health Practitioner, a Fellow of the Royal Society Public Health, a Fellow of the Royal Society of Medicine, a former Head of Environmental Health in the London Borough of Hackney, and has been a consultant/expert for several organizations, including the Metropolitan Police Wildlife Crime Unit, British meat, Agriculture Development Board (AHDB), European Food Standards Agency (EFSA), Royal Society for the Prevention of Cruelty to Animals (RSPCA), Department for Environment, Food & Rural Affairs (DEFRA), the Food Standards Agency and the European Commission. He is a member of the British Standards Institute Halal Standards Committee.

Dr Al-Teinaz was Advisor to the London Central Mosque Trust and the Islamic Cultural Centre (ICC), representing the organization on aspects of food and halal issues in the Food Standards Agency's Muslim Organization Working Group. He has overseen the ICC response to FAWC'S report published in 2003. Ha is a consultant to food businesses and other agencies, for example he is an advisor to the EU for the SSA project DIALREL (2006–2010), which published *Recommendations to Good Animal Welfare Practices During Religious Slaughter*. He served as a Vice-Chair of the Muslim Council of Britain Health and Food Standards Committee and as Trustee for the Foundation for Women's Health, Research and Development (FORWARD), an African diaspora women's campaign and support charity (registered in the UK). For years Yunes was at the forefront of a campaign to stamp out the illegal trade food fraud and mis-description of halal meat in the UK. He has trained many environmental health officers in the UK to deal with food frauds and crimes of major public health significance, such as the illegal trade in unfit meat, and has many public health publications.

Yunes has investigated many food frauds, illegal meat crimes, and food poisoning incidents, and taken numerous successful enforcement legal actions at the Magistrates and Crown Courts in the UK relating to the illegal meat trade, food fraud, public health safety, and mis-description of food as halal. He has given papers and presentations at international and national conferences on food safety and food fraud, the illegal meat trade, and halal food meat fraud at the invitation of the House of Commons and the House of Lords, the Food Standards Agency, the Department for Environment, Food and Rural Affairs, the Chartered Institute of Environmental Health, Cardiff University, and local authorities.

Yunes is a widely experienced broadcaster on radio and television, and acts as technical adviser and contributor to television and radio programmes on food safety issues, including the BBC's 'Life of Grime', 'Worst Restaurant', and 'Newsnight', Channel 4's 'Dispatches', Al Jazeera, and Radio 4. Yunes won the Best Arabic Scientist in the West Award in 2012.

Notes on Contributors

Dr Majed Alhariri

Dr Majed Alhariri is a food scientist who holds a PhD in Food Science and Technology from Cairo University. He carries out research on natural antioxidants and anticancer agents, focusing on halal auditing, halal program training, and halal assurance systems. Dr Majed has developed a program of halal assurance system training to increase understanding of this area. He is currently working in GIMDES as R&D Coordinator and Lead Auditor. He has many published scientific articles and has given scientific presentations on halal topics worldwide at international conferences.

Dr Hani Mansour M. Al-Mazeedi

Dr Hani Mansour M. Al-Mazeedi is a Kuwaiti scientist who specializes in halal requirements for food and non-food products and food safety management systems (HACCP/prerequisite programs and ISO22000). Dr Mazeedi obtained his BSc from USA and his PhD from UK and is the author of a number of books, such as *Concepts on Food Hygiene* (1998), *Practical Guide to Food Safety* (2002), and a six-book series, *My Food*, to be published in 2019. He also published a four-part book series entitled *Index of Official Papers Related to Food and Slaughter According to Islamic Rites* over the period of 1979–2017. Since 2011, through the Kuwait government (Ministry of Awqaf and Islamic Affairs and Kuwait Institute for Scientific Research) and the Gulf Standard Organization of the GCC Cooperation Council for the Arab States of the Gulf, he has organized several Gulf conferences on the halal industry and its services in Kuwait. Being an international speaker, Dr Mazeedi has over 35 years of experience of delivering presentations on the halal industry and its services and food safety systems in countries including South Africa, Thailand, Malaysia, India, Russia, Denmark, France, Turkey, Syria, Yemen, the United Arab Emirates, Saudi Arabia, the USA, and Brazil. Since 1981 he has represented Kuwait when visiting halal certification bodies in major meat exporting countries, such as New Zealand, Australia, Malaysia, Russia, France, Denmark, Belgium, Turkey, the USA, and Brazil. Dr Mazeedi won the Halal Journal Award (Malaysia) in 2009.

Dr Mehmet Haluk Anil

Mehmet is a former Senior Research Fellow of the Veterinary School, University of Bristol, UK, and Professor of Medical Physiology at the Higher Education Council, Turkey. He has been a consultant/expert scientist for several organizations, including Cardiff University,

Agriculture Development Board (AHDB), European Food Standards Agency (EFSA), Royal Society for the Prevention of Cruelty to Animals (RSPCA), UK. He has been a research leader for projects funded by the UK Department for the Environment, Food and Rural Affairs, the Food Standards Agency and the European Commission. His research interests include farm animal physiology, animal welfare with particular reference to stunning and slaughter, and food safety. He is currently an assessor for food control and safety, an assessor for halal production and certification for the Dubai Municipality, a tutor for DG-SANTE courses and an lRSPCA/FAO consultant.

Professor Dr Faqir Muhammad Anjum

Professor Dr Faqir Muhammad Anjum is a renowned and well-respected expert in food science and technology not only in Pakistan but also worldwide. He holds two PhD degrees, one in Grain Science from Kansas State University, USA and other in Food Technology from the University of Agriculture, Faisalabad, Pakistan. He has a solid scientific/academic food research background, particularly in the field of halal science. He possess 38 years of teaching and research experience. He is presently working as Vice Chancellor (President) of the University of the Gambia, The Gambia. He is food scientist with international publications and research projects.

Dr Muhammad Sajid Arshad

Dr Muhammad Sajid Arshad is currently working as Assistant Professor at the Institute of Home and Food Sciences, Government College University, Faisalabad, Pakistan. He served as visiting research scholar at the University of Illinois, Urbana Champaign, USA for a period of six months. He received his doctoral degree from the University of Agriculture, Faisalabad, Pakistan in 2013. From 2016 to 2017 he worked as a post-doctoral researcher at Kyungpook National University, South Korea. He is the author of about 35 publications and has five book chapters to his credit. His area of research is food science, particularly meat science and halal foods.

Dr Rossella Bottoni

Rossella Bottoni specializes in law and religion. She wrote her PhD dissertation on the origins of secularism in Turkey (1839–1938), and it received the 2007 Arturo Carlo Jemolo Award for the best PhD dissertation in law and religion-related disciplines. She is currently Researcher and Adjunct Professor of Religious Factor and New Constitutions: Europe and Africa in the Faculty of Political and Social Sciences at the Catholic University of Milan, and Adjunct Professor of Law and Religion in the Department of Political and Legal Sciences and International Studies at the University of Padua, Italy. Her research interests include the legal regulation of religious slaughter in the EU and EU member states, the accommodation of religious and ethical dietary rules in secular legal systems, secularism, Islam and religious minorities in Turkey, the legal condition of Islam in European countries, and legal pluralism.

Mary Critchley

After obtaining a first degree at London University, Mary Critchley worked as a teacher in British state schools. A concern for all aspects of animal health led her, at the beginning of the 2001 UK foot and mouth disaster, to set up the public information website warmwell.

com. Updated daily from 2001 to her retirement in 2016, the website provided a platform for experts in veterinary science, diagnostics, virology, and public health, as well as for concerned and informed members of the public, to contribute to the debate on governmental livestock disease policies.

Magfirah Dahlan

Magfirah Dahlan receives her PhD from Virginia Tech. She specializes in the ethics and politics of religious food practices. She is a faculty in Philosophy, Relegous Studies and Political Science at Craven Community College. Her research interests include religious ethics, food ethics, food politics, animal ethics, posthumanism, and postliberalism. Her current project is a monograph on the religious, ethical, and political aspects of modern meat.

Dr Adrian Evans

Adrian is a Senior Research Fellow at the Centre for Agro ecology Water and Resilience (CAWR) at Coventry University. His current research seeks to understand food and drink consumption practices, especially in relation to exploring the ethics, sustainability, and resilience of different types of food consumption. He has a particular interest in understanding the embodied nature of food consumption practices. He also researches the cultural and political embeddedness of different markets for food and drink, and explores the interconnections between food consumption and systems of provision, distribution, and ownership.

Professor Dr M. Diaa El-Din H. Farag

Professor Farag is Chairman of the Industrial Irradiation Division, Atomic Energy Authority, Nasr City, Egypt. He was Head of the Food Irradiation Research Department, Former National Center for Radiation Research and Technology (NCRRT).

Professor Dr Beniamino Cenci Goga

Beniamino is currently Head of the Food Bacteriology Laboratory at the Faculty of Veterinary Medicine, University of Perugia. He has a background in private practice, research, and academia, and consults for slaughterhouses and private companies. His research interests lie in veterinary public health, community-oriented veterinary extension, livestock and animal welfare, and veterinary regulation. He is the Italian primary investigator of the EU-funded project DIALREL (FP6–2005-FOOD-4-C) on animal welfare and religious slaughter. He is an expert for the European Commission (Research Directorate-General) and for the European Food Safety Authority, as well as Deputy for the ERASMUS scheme of the School of Veterinary Medicine, Perugia, Italy and Lead Auditor for Quality Management Systems (ISO 19011).

Professor Dr Temple Grandin

Professor Grandin is a Professor of Animal Science at Colorado State University and the owner of Grandin Livestock Handling Systems, Inc. Her entire career has been dedicated to animal welfare, including major contributions to the improvement of animal welfare during the religious slaughter of animals. She also works to help those with autism. Dr Grandin is generally considered the foremost animal welfare expert in the world. She was listed as one of the 100 most influential people by Time Magazine in 2010 and a movie about her early life won the Emmy in 2010 as the best documentary movie along with six other Emmy awards.

Dr Shahzad Hussain

Dr Shahzad Hussain is currently working as Associate Professor in the Food Science and Nutrition Department, College of Food and Agricultural Sciences, King Saud University Riyadh, Saudi Arabia. He also served as a Trained Sensory Panellist at the Department of Food Science and Nutrition, University of Minnesota, USA for a period of six months in 2008. He received a University Merit Scholarship for the duration of his academic carrier. He has published a number of research and review articles in journals of high repute. His particular area of research is cereal science and functional foods. He also works for the betterment of food safety, particularly halal food issues.

Dr Mah Hussain-Gambles, MBE

Mah founded and directed Saaf Pure Skincare in 2004, the world's first known organic and halal certified skincare range, which is alcohol- and animal-free and thus permissible and lawful under Shariah or Islamic law delivered in accordance with an eco-ethical business model.

She is a chartered chemist and is recognized as a pioneer in applying chemistry to the halal world. She has travelled extensively, speaking at international halal conferences, writing articles on halal ingredients, and engaging with scientists around the world, culminating in developing the first halal standard for the cosmetic and pharmaceutical industry.

Mah has also worked with large multi-national companies to independently verify their ingredients as halal compliant, as well as with international non-governmental organizations to develop halal certification standards for cosmetics and pharmaceuticals. She is considered an expert in this field and has been officially recognized for her contribution to the halal industry by the Malaysian government. Mah was also awarded an MBE in the Queen's Birthday Honours List (2010) for her contribution to international trade and the beauty industry.

Mah has a degree in pharmacology from Sunderland University and a Masters and a PhD from the School of Medicine, University of Leeds.

Dr A. Majid Katme

Dr A. Majid Katme is a qualified medical doctor (Bachelor of Medicine MBBCh, Cairo University, 1969–1970) and a qualified psychiatrist (DPM: Diploma in Psychological medicine, London). He has worked in many psychiatric hospitals in the UK. He is ex-President of the Islamic Medical Association, UK, a Muslim speaker on medical Ethics, former Director of Muslim Welfare House (Finsbury Park, London), Co-Founder/Ex-Chairman of Palmers Green Mosque (Enfield, North London). He is Long Muslim Campaigner and speaker on halal meat and food and healthy diet. He is Campaigner for the prophetic method of DHABH (religious slaughter/without stunning). He published scientific medical paper entitled "Comparison between the Religious Prophetic method of animal slaughter without stunning and the western method of animal slaughter with stunning". Also, he published "Assesment of The Muslim Method of Slaughtering" - Azhar Halal Foods. www.azhar.jp/info/halal-eng/halal5.html
He has got his own website on Halal meat: www.halaltayyibmeat.com

Dr Mariam Abdul Latif

Prof. Mariam is a staff member at the Faculty of Food Science and Nutrition, University of Malaysia Sabah, 88400 Kota Kinabalu, Sabah, Malaysia. Prof. Mariam has had a much-decorated three-decade career in public service where she served for 17 years in the Ministry of Health (MOH), Malaysia. She was directly involved in the international food standard development, Codex Alimentarius Commission (CAC), while she was the Codex Regional Representative for Asia from 1996 till 1998. This led to the development of the Draft General Guidelines for Use of the Term "Halal" before it was adopted in 1997 at the CAC. She went on to an illustrious career at the Department of Islamic Development Malaysia (JAKIM) and the Halal Industry Development Corporation (HDC). At JAKIM, she established the technical audit and the Malaysian halal certification system. She also published the Handbook of Halal Ingredients, translated the first and second JAKIM Halal Manual Procedure and drafted the National Halal Laboratory concept paper. Prof. Mariam's stint with HDC proved to be as noteworthy as she led the Integrity Department in developing the Halal Training Programs, Halal Standard Unit, Halal Certification Unit, Halal Reference Unit and Halal Consultancy. She also initiated and organized the first World Halal Research Summit in 2007. She had presented more than 200 papers on Halal and Halal industry in Malaysia as well as abroad (China, Australia, France, Netherlands, Thailand, Singapore, South Africa, Taiwan, Kuwait, Indonesia, Philippines, Turkey, Korea, Russia, New Zealand, Iran and Spain). She was a Fellow Researcher at the Halal Product Research Institute (HPRI), University Putra Malaysia; Panel Expert at the Institute of Halal Research and Management (IHRAM), University Sains Islam Malaysia; Panel Assessor of the Malaysian Qualifications Agency (MQA) and is a Board Member of Labuan Halal Hub, Malaysia. She was retired from the Government in 2011 and currently on contract with University Malaysia Sabah. She headed the Food Safety and Quality Unit under the Faculty of Food Science and Nutrition for four years besides lecturing Halal Food, Food Laws and Human Nutrition. She conducts trainings and consultancy on food safety and halal food for the industry, the regulators and the public. Her research areas are in the Prophet's Diet, Islamic eating practices, Halal nutrition, Halal standards, Halal certification, Halal industry development and halal consumerism.

Dr John Lever

John is a Senior Lecturer in Sustainability at the University of Huddersfield Business School. He has conducted research and published on many aspects of the food industry and is particularly interested in farm animal welfare, local food production, and halal and kosher meat markets.

Professor Dr Mara Miele

Dr Mara Miele is Professor of Human Geography in the School of Geography and Planning at Cardiff University. Her research focuses on the fields of Science and Technology Studies (STS) and geographies of science (animal welfare science) as well as food. Her research interests include animal welfare science, food taste, and the ethics of eating animal food. She coordinated the EU SSA project DIALREL (2006–2010), and was member of the Steering Committee and Management Team of the EU VI Framework Project Welfare Quality (2004–2009) and EUWelnet (2012).

Umar Moghul

Umar is a corporate and finance attorney based in New York whose practice has involved various aspects of Islamic law. He is also an adjunct faculty member at the University of Connecticut School of Law, Hartford, CT. For nearly 20 years, Umar has advised on a variety corporate and finance matters. Much of his work has been on behalf of ethical and responsible sponsors, investors, and business owners seeking to create a positive impact. Clients call on him to design socially responsible and environmentally conscious legal and business terms to help create more responsible transactions, markets, and economies. Umar is adjunct faculty at the University of Connecticut School of Law, Michigan State Law School, and Hartford Seminary. He teaches courses in Islamic law, business ethics, Islamic finance and investment, and halal food. He is author of *A Socially Responsible Islamic Finance: Character and the Common Good* (Palgrave-MacMillan) and has spoken at numerous forums.

John Pointing

John Pointing is a barrister with over 25 years experience in environmental law. His practice includes advising local authorities, central government, commercial clients and private individuals. He has experience of prosecuting cases involving food fraud, bush meat, 'smokies', and other food safety offences. John has lectured in law and provided professional training in environmental health for many years. He is the Legal Partner of *Statutory Nuisance Solutions*. He is co-author (with Rosalind Malcolm) of *Statutory Nuisance: Law and Practice* (OUP, 2nd edn 2011) and *Food Safety Enforcement* (Chadwick House Publishing, 2005). His most recent book is S. Battersby and J. Pointing, *Statutory Nuisance and Residential Property* (Routledge, 2019). Since 2003 he has been interested in issues affecting the Muslim community, particularly those to do with fake halal food and food crime. Besides writing papers on such matters, he has presented at several conferences, including the *World Halal – Europe* conference, held in London in 2010, and the *First Gulf Conference on Halal Industry and its Services*, held in Kuwait in 2011.

Joe M. Regenstein

Professor Regenstein is Professor Emeritus of Food Science in the Department of Food Science at Cornell University and the Head of the Cornell Kosher and Halal Food Initiative. He is also the editor-in-chief of *Food Bioscience*, the first peer-reviewed food science journal sponsored by China. Dr Regenstein is currently involved in activities with the Muslim Council of Britain, the Islamic Society of North America, and the Islamic Food and Nutritional Council of America. He still teaches kosher and halal food regulations at Cornell and also a distance learning course at Kansas State University.

Sol Unsdorfer

The grandson of a rabbi killed in the holocaust, Sol Unsdorfer was born in London and has been an active member of the orthodox Jewish community for over 60 years. His interests are his profession as a management surveyor and his advocacy for Israel. He and his family keep a kosher home and observe the strictures of the sabbath, the Jewish festivals, and fast days. He firmly believes that such strictures make for a better life and a better person. The sabbath laws include that he must not touch his mobile phone from sunset on Friday until nightfall Saturday. No TV, no travel, no football games, no money or commerce. How many people these days could keep that discipline one day a week? The same applies to keeping kosher. Self-control puts us above the animals we eat.

Mufti Mohammed Zubair Butt

Mufti Mohammed Zubair Butt memorized the Holy Qur'an at the age of 15. He concluded the Shahāda al-ʿĀlimiyyah programme at Darul Ulum, Karachi, where he also received post-graduate training in Islamic legal edicts. Mufti Zubair is now Senior Advisor on Islamic law at the Institute of Islamic Jurisprudence, Bradford. He is a Sharia Advisor to the MCB, the Halal Monitoring Committee, the Gardens of Peace Cemetery, Ilford and the National Burial Council. He is also a serving Imam at the Masjid Ibraheem Centre, Bradford and a lecturer on Sahih al-Bukhari at Madrasah Madania Tahfeezul Quran, Bradford. He is chair of the Al-Qalam Sharia Scholars Panel. Mufti Zubair has served as a hospital chaplain at the Leeds Teaching Hospitals for over 17 years and is the Chair of the Muslim Health Care Chaplains Network. He lectures on Islamic medical ethics as part of the MA in Muslim chaplaincy at the Markfield Institute of Higher Education and in the Diploma in Contextual Islamic Studies and Leadership at the Cambridge Muslim College.

Acknowledgements

Completion of this book provides an opportunity to recognize the advice and help provided by a number of very important individuals. We particularly want to express our heartfelt gratitude to Dr Ahmad Al Dubayan, Director General, The London Central Mosque Trust and the Islamic Cultural Centre, Dr Hani Mansour Al-Mazeedi, Kuwait Institute for Scientific Research, State of Kuwait; Professor Dr Haluk, University of Bristol, Dr A. Majid Katme Medical Doctor, London, and Professor Dr Joe Regenstein of Cornell University, USA, for their guidance and exhaustive suggestions in the preparation of the manuscript. Their combination of sage advice, the capacity to listen and discuss, the ability to challenge us to expand our vision, and a never-ending willingness to share their time will always be greatly valued. They were all extremely helpful in providing the most accurate information and critique throughout this work. We also wish to express our thanks and gratitude to our colleagues, Mary Critchley, animals welfare activist, and John Pointing, English barrister, for his help with the final editing and organization, and, more importantly, for motivating us in writing this book. Thanks are also due to all our authors from throughout the world: their contributions have given this book comprehensiveness and coherence. Finally, we wish to thank our friends, family members, and colleagues, who provided help in numerous ways.

Yunes Ramadan Al-Teinaz
Stuart Spear
Ibrahim H. A. Abd El-Rahim

In the name of Allah the Beneficent the Merciful

Foreword

The London Central Mosque Trust and the Islamic Cultural Centre (ICC) have great pleasure in presenting *The Halal Food Handbook*, edited by our former advisor Dr Yunes Ramadan Al-Teinaz together with Stuart Spear and Professor Dr Ibrahim H.A. Abd El-Rahim. We are grateful too for the advice and assistance provided by John Pointing and Professor Dr Joe M. Regenstein.

Dr Al-Teinaz has for many years given his time as an advisor to the Trust. He is an eminent public health and environmental consultant who has published widely on various public health, halal and food fraud issues. His publications are used as reference works throughout the English-speaking world. During his career as a public servant, culminating as a local authority Chief Environmental Health Officer, he highlighted many public health concerns, including food fraud and the false description of food as halal. He was the first Chartered Environmental Health Practitioner in the UK to take on a local authority prosecution with respect to food falsely described as halal in an east London market. His enforcement work was so effective that food criminals offered his employer £20 000 to dismiss him from his job and they even made threats on his life. For many years he did his best gratefully in many ways to create awareness in society and among communities about food and health issues.

Stuart Spear is a freelance journalist specializing in public health and food safety. For 10 years he edited the UK's leading environmental health magazine, *Environmental Health Practitioner*. He worked on the national meat crime campaign 'Stamp it out' and campaigned with Dr Al-Teinaz to stop illegal meat gangs from selling unfit meat to ethnic communities. He also campaigned with Dr Al-Teinaz for the proper certification of halal meat and for an end to illegal practices in the halal industry. He has contributed to a number of other books on the food industry, mental wellbeing, and the history of public health.

Professor Dr Ibrahim H. A. Abd El-Rahim is a Professor of Infectious Diseases and Epizootiology, and the Custodian of the Two Holy Mosques Institute for Hajj and Umrah Research at Umm Al-Qura University, Makkah, Saudi Arabia.

John Pointing is an English barrister specializing in environmental health law. He is the legal partner of Statutory Nuisance Solutions, which provides advice, consultancy, and training to government, local authorities, and industry. Over the last 25 years he has lectured and written books, papers, and articles on environmental health matters, including food crime and food safety. He has worked with Dr Al-Teinaz for many years on food issues, including halal. Together they have prosecuted food criminals, written articles, and made presentations on food crime and halal issues to a wide range of audiences, both in the UK and internationally.

Professor Dr Joe M. Regenstein is a Professor Emeritus of Food Science in the Department of Food Science at Cornell University and the Head of the Cornell Kosher and Halal Food Initiative. He is also the co-editor-in-chief of *Food Bioscience*, the first peer-reviewed food science journal sponsored by China. Dr Regenstein is currently involved in activities with the Muslim Council of Britain, the Islamic Society of North America, and the Islamic Food and Nutritional Council of America. He still teaches kosher and halal food regulations at Cornell and also as a distance learning course at Kansas State University.

The public at large is not well informed about halal and remains confused about such important matters as the stunning of animals prior to slaughter, the use of food additives, and food fraud. Such confusion is perhaps not surprising as what constitutes halal now varies between cultures and regions as well as among the followers of different schools of Islamic thought.

Islamic dietary laws determine which foods are permitted for Muslims: halal means permitted, whereas haram means prohibited. Several foods are considered harmful for humans to consume and are forbidden. This includes the prohibition of the consumption of pork, blood, alcohol, carrion, and meat that has not been slaughtered according to Islamic prescriptions. Meat is the most strictly regulated food. The animal (of a permitted species) must be slaughtered by a sane adult Muslim by cutting the throat quickly with a sharp knife. The name of Allah must be invoked while cutting. The halal method is the Muslims' religious-humane method of slaughtering animals and birds for food. All other methods, including stunning pre- or post-slaughter, even if done in conjunction with halal, renders the meat haram, that is it is non-halal according to most authorities and thus forbidden for Muslims. A ban on this halal method of slaughter or imposition of other methods would mean that Muslims would be unable to buy and consume acceptable meat, poultry or meat products as required by Muslim law.

The subject of halal food, including meat and meat products, is thus of central importance to Muslims. There are many issues involved that demand the community's greater attention, including a growing focus on the ethical aspects of meat production, the position of stunning animals prior to slaughter, the importance of food safety and proper sanitation, and the welfare of animals. However, there is more to the halal industry than simply food and meat production: halal issues touch many areas of Muslim life. The range of halal regulation extends to foods, beverages, drugs, cosmetics, chemical products, biological products, genetically modified products, and consumer goods. Industry is driven by the market realities of supply and demand, but effective regulation, ethical production, and a better understanding of Islamic principles are also vital.

Exploring these questions in detail is the goal of this book, which is a unique work comprising 25 chapters with about 30 authors involved, including leading scientists, religious experts, and academics from around the world. The fundamental principles of halal are examined in the context of the Holy Quran. This book also offers practical advice to Muslim scholars, those managing and working in the food industry, local authorities and food enforcement officers, government departments, halal certifiers, and academics. It also seeks to advance the consistency of halal standards and provides practical guidance on the entire food chain from farm to fork. We can look forward to the introduction of specific halal food laws in the EU and the UK to protect Muslim consumers and regulate the halal industry, as is the case today in some states in the USA and in many other countries world-wide. This book provides an important step forward towards such a future.

Finally, I have to mention the main Islamic principle, which determine wasting water, food, forest, is totally haram and not permitted, since natural resources must be protected for the present and future generations.

Dr Ahmad Al Dubayan
Director General
The London Central Mosque Trust and the Islamic Cultural Centre

Introduction

John Pointing

Barrister, London, UK

Halal regulation of food sits within a context comprising secular as well as religious elements. Both elements are complex – as are the relationships between them – and this forms a theme which the authors of this book have sought to elucidate. Institutions also have to engage with this complexity, but do so with varying levels of awareness, and sometimes those speaking for secular institutions seem obtuse about recognizing the religious norms of others.

An example of this is the campaign against the religious slaughter of animals for food that has been run over many years by the British Veterinary Association (BVA). A crucial element has been lobbying by the BVA to repeal the exemption provided for Muslim and Jewish religious slaughter from the effect of European laws which require pre-stunning to be carried out in the slaughtering of animals for food. The point is that the BVA shows no appreciation that religious norms have validity, and its own position is based on the supremacy of its secular norms which it believes must also govern the export trade in halal meat. As reported in *The Times* on 12 June 2018:

> The British Veterinary Association (BVA) said that killing animals in this way for the export market was against the spirit of a derogation from EU laws that was designed to meet the needs of religious communities within member states. It said that UK abattoirs were producing much more non-stun meat than appeared to be required by the combined Muslim and Jewish population.
>
> John Fishwick, the BVA president, said: 'BVA will continue to push for an end to non-stun slaughter in the interests of animal welfare. With Brexit on the horizon and in the light of announcements about export deals with non-EU countries, there is a pressing need for clarity on the quantities and destinations of exports of non-stun meat. While not illegal, if meat from non-stun religious slaughter is exported we consider this to be outside the spirit of the legislation.'

Moving on to consider the wider regulatory picture, it is clear that the European Commission has expanded its scope of influence in regulating food safety across the EU over recent years. There are various reasons for this, such as the globalization of trade in food products and the length and complexity of food chains from producers to consumers.

The Halal Food Handbook, First Edition. Edited by Yunes Ramadan Al-Teinaz, Stuart Spear, and Ibrahim H. A. Abd El-Rahim.
© 2020 John Wiley & Sons Ltd. Published 2020 by John Wiley & Sons Ltd.

Poor practices throughout the food industry, weak supervision by public regulators, and the undermining of public trust in food safety governance have formed the backdrop to this expansion (Caduff and Bernauer 2006). Major threats to public health, having the potential to cross national boundaries, have introduced a dimension of panic, giving a further impetus to change. The paradigm case of this was the bovine spongiform encephalopathy (BSE) crisis in the early-1990s (Zwanenberg and Millstone 2002). The manner in which this was handled by public authorities demonstrated panic and confusion, but the dimensions of the problem also ensured that food safety was placed at the forefront of policy priorities for the European Commission.

The regulatory framework that governs food law in most of the European continent is provided mainly by EU regulations, which are directly applicable for Member States of the EU. The General Food Law: EC Regulation 178/2002 provides the primary legislative framework within which more detailed regulations are connected. These regulations deal with the hygiene of foodstuffs; specific hygiene rules for food of animal origin, and specific rules for the organization of official controls on products of animal origin intended for human consumption (Malcolm and Pointing 2005). A sophisticated and comprehensive body of law has been created by European institutions in order to regulate a very large, complex, and diverse industry. It is likely that should the UK leave the EU much of this body of law will remain intact.

The food industry is not regulated exclusively by officials employed by Member States. Self-regulation from within the food industry has become increasingly important during a period of change, as 'command and control' methods of regulation have receded (Hawkins 2002). Self-regulation often develops having little engagement with state institutions, and there may be a lack of transparency in the way it functions (Martinez et al. 2013). It also forms a major plank of deregulation initiatives, both in the UK and in Europe generally (Pointing 2009). A fundamental problem with all forms of self-regulation is that where the regulatory culture embedded in parts of an industry or a sector is weak, lacks transparency or is otherwise defective, it becomes very difficult to change and bring up to acceptable standards (Eccles and Pointing 2013). Recourse to prosecution and criminal sanctions can play an important but nevertheless limited role in changing the culture of the industry.

The engagement of religious law with food production, preparation, processing, trade, and consumption provides an additional level of regulation. State law and religious law may be integrated in Muslim countries (Ahmad et al. 2018), but in the majority of states – where Muslims form a minority of the population and in which secular laws are sovereign – religious norms have to be given 'normative space' in order to allow religious law to have effect. This results in a complex architecture of regulation in which considerable tensions may be induced (Pointing 2014). However, a process of accommodation can also take place, and many of the contributions to this book demonstrate this. Accommodation is not unusual and religious law (whether Islamic, Jewish or other religions) 'often has its own independent institutional existence' and is 'often intertwined with and incorporated into the state legal system' (Tamanaha 2001).

It has taken several years for this book to reach publication and many leading experts in their different fields have contributed. The principal driving force behind the project has been Dr Yunes Ramadan Al-Teinaz, the internationally renowned public health and environmental consultant. The range of issues surrounding halal and the different areas of expertise drawn from several different cultures and nations have induced complexity and difficulty. There is a diversity of views and norms as well as many nuances in religious interpretation. Just as there is no worldwide consensus on halal regulation, there is no seamless weaving of a common web in the contributions to this book. One of its aims, however, is to provide a benchmark which may set an authoritative standard in halal regulation. The intention has been to inform, discuss, and suggest ways forward in respect of such regulation, not to try and impose a particular formulation. Integrating halal regulation with state regulation is another major objective and a very important one for those working in the food industry, as well as for regulators, legislators, policy makers, and consumers. This has been an extremely challenging project fusing together both religious and scientific knowledge to produce an illuminating text on such a complex subject. At times translations from the Arabic, when dealing with religious texts, are not always in the most common English vernacular; the focus, however, has been on accuracy of meaning. In parts of the book there are references to often long and complex appendices. Rather than reproducing these a web address has been given where they can be easily accessed.

The meaning of halal – its relationship to Shariah law and the examination of its importance in relation to food – is considered by Yunes Ramadan Al-Teinaz in Chapter 1. The fundamental importance of halal rules for regulating food at every stage from farm to fork is examined, as is the normative basis for rules that disallow the pre-slaughter stunning of animals. A detailed examination of halal and Shariah law is provided in Chapter 2 by Mufti Mohammed Zubair Butt. He demonstrates that halal is a much wider concept than one pertaining solely to food and beverages, and shows how practices regarding marriage, divorce, raiment and adornment, financial matters, and devotion are affected.

A comprehensive and wide-ranging analysis of halal issues affecting animals and, in particular, the situation for Muslims regarding the use of stunning during the process of slaughter is provided in Chapters 3–8. The implications raised for halal ethics by the consumption of animals for food form the subject of Chapter 3, written by Magfirah Dahlan-Taylor. Animal welfare considerations that follow from the holding and restraining of animals prior to slaughter and those pertaining to the different slaughtering methods are analysed by Temple Grandin in Chapter 4. She concludes that halal methods of slaughter without stunning can be performed within acceptable welfare parameters, but adds that prescribed methods must be scrupulously adhered to if animal welfare objectives are to be achieved. The humane treatment of animals, in compliance with halal requirements at the pre-slaughter stages of production and during slaughter, is a theme further explored in Chapter 5 by Mehmet Haluk Anil. Additional examination of halal requirements regarding the slaughter process is provided in the following chapter by Mehmet Haluk Anil and Yunes Ramadan Al-Teinaz. They also consider methods involving some pre-stunning and compare these with those where stunning is prohibited. The impact of slaughtering methods on food hygiene and meat quality is analysed in Chapter 7 by Ibrahim H.A. Abd El-Rahim. He confirms the findings of Temple Grandin, highlighted earlier, that methods of slaughter that eschew stunning – when carried out correctly – cause minimal pain and suffering to the animals, but adds that the benefits include the production of better quality meat.

Some of the wider implications of the religious slaughter of animals are discussed by Joe M. Regenstein in Chapter 8. The long traditions of religious laws regulating the production and consumption of foods to be found in both the Muslim and Jewish communities are examined, together with how they relate to animal welfare and modern methods of slaughter. Professor Regenstein argues that the two communities need to work together to make improvements to the methods of slaughtering animals and to resist attacks from the secular world.

Halal ethics and factory farming form the subject of the next chapter, provided by Faqir Muhammad Anjum, Muhammad Sajid Arshad, and Shahzad Hussain. They make a distinction between the unchanging core beliefs of Islam and their application, in which tools and methods change over time. The authors emphasize that halal ethics have a central importance for food hygiene practices, animal husbandry, and the conditions prevailing in slaughterhouses. The need to ensure that food ingredients conform to halal requirements and that halal foods are labelled to include all processing ingredients and additives is strongly argued by Yunes Ramadan Al-Teinaz in Chapter 10. He emphasizes the importance of embedding strict controls at all stages in the food chain, based on effective monitoring and backed by rigorous certification.

Halal issues go beyond the preparation and consumption of foodstuffs. In Chapter 11, Majed Alhariri discusses the Islamic perspective on genetically modified organisms. He objects to genetic modification because it interferes with divine work, and because traditional forms of breeding and organic farming are superior. He argues that genetic modification is inconsistent with halal and tayyib principles because of the risks, hazards, and threats it poses to humans, animals, and the environment. In the next chapter, Mah Hussain-Gambles examines the halal implications that arise with the ingredients used in the manufacturing of personal care products and cosmetics. She highlights the importance of establishing rigorous certification systems in order to better protect consumers.

The need to improve standards, systems of control, and certification procedures is explored in more detail in the following three chapters. Halal and hazard analysis and critical control point (HACCP) regulation is discussed in Chapter 13 by Hani Mansour Mosa Al-Mazeedi, Yunes Ramadan Al-Teinaz, and John Pointing. The authors maintain that halal requirements are compatible with HACCP and provide an additional layer of regulation and protection for consumers. The major topic of international standards and certification for halal products and services is taken up by Mariam Abdul Latif. She argues, in Chapter 14, that the lack of harmonized international standards means that the integrity of the supply chain has been compromised. She adds that this problem is exacerbated by the inconsistencies and regulatory weaknesses of the various halal certification bodies operating in different countries. Yunes Ramadan Al-Teinaz and Hani Mansour Mosa Al-Mazeedi provide in the following chapter an analysis of the Halal Certification Model. Their purpose is to establish standard guidelines for the halal industry and services in respect of: the preparation, handling, and storage of food and drinks, medicines, cosmetics, skincare, and healthcare products.

Rossella Bottoni, in Chapter 16, examines the legal aspects of halal slaughter and certification. An overview of EU rules applying to halal is provided and it is noted that secular authorities lack competence in regulating halal matters. Chapter 17, by John Pointing, examines the legal framework of the General Food Law of the EU. The law governing the stunning of animals prior to slaughter is examined in terms of the relationship between religious law and state law. It is argued that the 'legislative space' created by the exemption from stunning requirements for Muslim and Jewish religious slaughter indicates compatibility between religious and state law.

The scientific detection of adulteration in halal foods is discussed by M. Diaa El-Din H. Farag in Chapter 18. Techniques used for identifying food that is processed in accordance with halal rules but is subsequently adulterated with haram ingredients are considered. Also discussed are DNA-based methods for identifying different species, which are particularly important for detecting when cheaper meats are passed off as lamb, or when pork ingredients are used in food purporting to be halal. Fraudulent undermining of the integrity of food production occurs at every stage of the food chain, for both halal and non-halal foods generally. Besides crimes involving misdescription, false labelling, and passing off, fraud also has implications for food safety and food quality. Food fraud extends beyond regulatory breaches and slides into mainstream criminal activity carried out by organized criminals, and this makes it particularly difficult for enforcement agencies to counteract. These wider aspects of food fraud form the subject of Chapter 19, written by John Pointing, Yunes Ramadan Al-Teinaz, John Lever, Mary Critchley, and Stuart Spear.

Establishing a dialogue between secular and religious norms is an important aspect of democratic societies and essential for giving effect to human rights to practise religious beliefs freely. Chapter 20, written by Yunes Ramadan Al-Teinaz, Joe M. Regenstein, John Lever, Dr A. Majid Katme, and Sol Unsdorfer, provides a comparison between kosher and halal dietary and slaughtering rules as well as considering their foundations in religious laws. The Dialrel Project (Encouraging Dialogue on Issues of Religious Slaughter) was set up with EU funding in 2006 to address issues pertaining to Jewish and Muslim religious slaughter in Member States. The project explicitly sought to encourage dialogue between religious, scientific, animal welfare, and legal professionals. An account of the Dialrel findings is provided in Chapter 21 by Mara Miele, John Lever, and Adrian Evans.

Variation between countries – in their rules and practices regarding religious requirements – was a key reason for setting up the Dialrel project. This theme is explored further in the remaining four chapters of this book, which focus on halal practices in different countries. Chapter 22, by Majed Alhariri and Hani Mansour Mosa Al-Mazeedi, examines halal food production in the Arab world in general. In Chapter 23, M. Diaa El-Din H. Farag discusses the situation in Egypt, where 90% of the population is Muslim. The author argues that with Egyptian government agencies in control of halal regulation, high standards of food safety and food quality are maintained. It is also asserted that animal slaughtering methods fully conform with Shariah law in Egypt. By contrast, in the USA halal certifiers are commercial organizations operating independently of government, as is outlined by Joe M. Regenstein and Umar Moghul in Chapter 24. The authors state that whilst religious slaughter has been declared humane in the USA, the halal meat industry is being challenged to improve on animal welfare considerations and operate consistently with the guidelines produced by the North American Meat Institute. Variations in practices and inconsistencies in the rules governing religious slaughter are issues considered in Chapter 25 by Beniamino Cenci Goga. Empirical research carried out by the author into religious slaughter in Italy has established that most halal (and all kosher) slaughter occurs without stunning.

References

Ahmad, A., Abidin, U., Othman, M., and Rahman, R. (2018). Overview of the halal food control system in Malaysia. *Food Control* 90: 352–363.

Caduff, L. and Bernauer, T. (2006). Managing risk and regulation in European food safety governance. *Review of Policy Research* 23 (1): 153–168.

Eccles, T. and Pointing, J. (2013). Smart regulation, shifting architectures and changes in governance. *International Journal of Law in the Built Environment* 5: 71–88.

Hawkins, K. (2002). *Law as Last Resort: Prosecution Decision-Making in a Regulatory Agency*. Oxford: Oxford University Press.

Malcolm, R. and Pointing, J. (2005). *Food Safety Enforcement*. London: Chadwick House Publishing.

Martinez, M., Verbruggen, P., and Fearne, A. (2013). Risk based approaches to food safety regulation: what role for co-regulation? *Journal of Risk Research* 16: 1–21.

Pointing, J. (2009). Food law and the strange case of the missing regulation. *Journal of Business Law* 6: 592–605.

Pointing, J. (2014). Strict liability food law and halal slaughter. *Journal of Criminal Law* 78 (5).

Tamanaha, B. (2001). *A General Jurisprudence of Law and Society*. Oxford: Oxford University Press.

Zwanenberg, P. and Millstone, E. (2002). "Mad cow disease" 1980s–2000: how reassurances undermined precaution. In: *The Precautionary Principle in the 20th Century: Late Lessons from Early Warnings* (eds. P. Harremoees, D. Gee, M. MacGarvin, et al.), 170–184. London: Earthscan.

Part I

What is Halal

1

What is Halal Food?

Yunes Ramadan Al-Teinaz

Independent Public Health & Environment Consultant, London, UK

1.1   Introduction

1.1.1   Basic Terms

The editors appreciate this book is aimed at a wide cross-section of readers, from the devout Muslim practitioner to those who are interested in learning about halal but may know little about the religious context behind it. To aid readers who may be non-Muslim or non-Arabic speakers, we first lay out a few basic Arabic terms:

- Quran: means 'recitation' in Arabic and is the literal word of God recited to the Prophet Mohammed (peace be upon him "pbuh") (pbuh) in Arabic by the Angel Gabriel.
- Sura: a chapter of the Glorious Quran
- Hadith: means 'traditions' in Arabic and is a written record of Mohammed's (pbuh) life and thoughts.
- Shariah: means 'legislation' in Arabic and provides the moral code and religious law for Muslims.
- Halal: means 'lawful, allowed or permitted' in Arabic.
- Haram: means 'prohibited and unlawful' in Arabic.
- Makrooh: means 'disapproved, disliked, hated or detested' in Arabic.
- Mushbooh: means 'doubtful or questionable' in Arabic.
- Tayyab: means 'wholesome or fit for consumption' n Arabic.

1.1.2   What is Halal?

To understand halal and its importance to Muslims, one must first understand something about Islam. Every day a quarter of the earth's population is called to prayer five times a day. From Indonesia to Bangladesh, from Nigeria to Morocco, from Egypt to the USA, no matter where in the world Muslims live, whether alone or with others, they are daily united with fellow believers in this common experience through which each Muslim is required to express devotion to God. These five obligatory prayers take place at daybreak, midday, mid-afternoon, evening, and at sunset while facing Makkah, the holiest city of Islam. Prayer is always done in Arabic regardless of the worshiper's native tongue.

Much as Islam prescribes the time and nature of worship, it also provides a set of standards by which Muslims are required to live their lives. At the core of these standards are the five pillars of Islam, which are:

1) Shahadah, the declaration of faith
2) Salah, the five daily prayers
3) Zakah, an obligation to give 2.5% of your savings to the poor each year
4) Sawm, fasting during the ninth Muslim month, Ramadan
5) Hajj, the pilgrimage that must be made once in every Muslim's lifetime to the Holy City of Makkah in Saudi Arabia. This takes place during the twelfth Muslim month.

A Muslim who believes in God and accepts his works as revealed by the Prophet Mohammed (pbuh) is also required to carry out a set of duties and obligations that impact on every aspect of life. It is only by adhering to this set of instructions that have been clearly laid out by God that you can practice the faith or call yourself a Muslim.

In Arabic, the word halal means lawful or permissible. To the non-Muslim, it is a word that is often exclusively associated with the foods that Muslims are allowed to eat, but in reality it is a term that describes everything that it is permissible for a Muslim to do, both in deed and thought. Halal impacts every aspect of a Muslim's life, from the clothes that can be worn to attitudes towards work, from relations between men and women to the treatment of children, from the way business is carried out to the treatment of a fellow Muslim, the principal of halal must be applied. Financial products, holidays, sports, films, even how you play a game of chess can be either halal, permissible, or the opposite, haram, unlawful. Haram covers everything that is prohibited for a Muslim. Haram in this sense is just as important as the principles of halal. Its importance to Muslims is due to the Islamic belief that everything put on this earth by God is here for our benefit

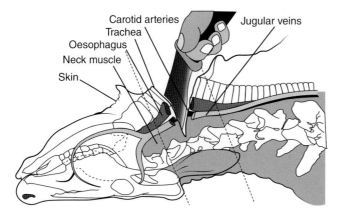

Figure 1.1 Slaughtering of cattle showing proper swift cut. *Source:* The London Central Mosque Trust and the Islamic Cultural Centre.

unless he or she has specified otherwise. In practical terms, this means that everything is halal or permissible unless God has specifically stated that it is not, which is when it becomes haram.

But sometimes things are not that clear-cut. In addition to halal and haram, there are two other terms that are used to describe whether a deed or thought should be permitted. Makrooh is a term meaning disapproved of, disliked, hated or detested. While a lesser sanction than haram, which is something that is prohibited, it is used to describe an action or deed that is described negatively. The other term that is commonly used is mushbooh, which is an action or deed that is doubtful or questionable. In relation to food, something is mushbooh when there is no consensus about whether it is halal or haram. In such cases, it may be wise to avoid the food item in question.

Understanding what is haram and what is halal, and that which is in between, lies at the core of a Muslim's faith. The reason Muslims believe so profoundly in these distinctions is that they have been laid down in Arabic in the Holy Quran, which for a Muslim is the literal word of God.

1.1.3 Halal and the Holy Quran

The first place a Muslim turns to understand what is halal and what is haram is the Holy Quran. In Arabic Quran means 'recitation'. Over a period of 23 years from Mohammed's (pbuh) 40th year to his death in 632, the angel Gabriel visited Mohammed (pbuh) and recited in Arabic the word of God. These recitals were later written down in Arabic to form the Holy Quran we know today. This is why Arabic is the language of prayer for Muslims regardless of their mother tongue.

The Holy Quran is about four-fifths the size of the Christian New Testament and is made up of 114 chapters of varying length, each known as a sura. The suras are not ordered thematically so they start with the longest first and end with the shortest. It means that a Muslim can open the Holy Quran at any page and start reciting at the start of any paragraph, as each represents a

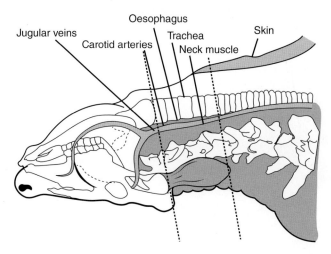

Figure 1.2 The proper site for slaughtering in cattle and arteries, veins and organs to be cut. *Source:* The London Central Mosque Trust and the Islamic Cultural Centre.

lesson to be learned and reflected upon. The Holy Quran covers such issues as the existence of God, the early prophet's historical events during Mohammed's (pbuh) time, as well as moral and ethical lessons. It also describes what actions are right and what are wrong, or what is halal and what is haram, and it is these exhortations that we will be focusing on in this book.

Islam is not just a religion; it is also a source of law and a guide to social behaviour that sets out a standard by which the good Muslim should live their life. Islam is an Arabic word that means 'submission' or 'submission to the word of God'. Muslim is Arabic for 'one who submits'. A Muslim submits to the word of God through the expression of the Islamic faith, which involves accepting the ethical standards and practices that are laid out in the Holy Quran. Understanding and adhering to what the Holy Quran defines as halal and haram is part of this process of acceptance. Because these laws are laid out in the Holy Quran, and so it is the literal word of God as described to the Prophet Mohammed (pbuh), it follows that all these rules of belief and conduct are the rules that God requires you to follow to be a good Muslim.

Food and its derivatives are mentioned in the Holy Quran 49 times. The two chapters or suras that provide the clearest instruction on what can and cannot be eaten by Muslims are Al-Ma'ida, which can be translated as the Table or the Feast, and Al-Baqara, translated as the Cow or Cattle. But it is not just as an expression of faith that Muslims adhere to the laws of halal. Islam teaches of a day of judgement. At a time only known to God, a day will come when everyone is judged along with the dead, who will be resurrected for judgement. The reward comes in the form of passing into heaven, punishment involves passing into hell. The judgement itself is carried out by an omnipotent God who will have witnessed all your deeds throughout your life and so be able to judge your fitness to enter heaven. A Muslim believes God to be ever-present and knowing of every deed and thought, and so is aware of who follows his command and who does not. An individual's adherence to the rules of the faith, including halal, is part of the judgement that will determine their status in the afterlife.

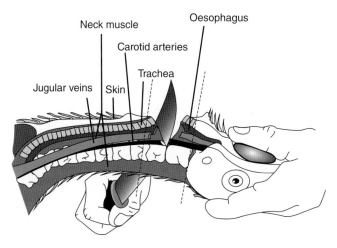

Figure 1.3 Slaughtering of poultry showing proper swift cut. *Source:* The London Central Mosque Trust and the Islamic Cultural Centre.

1.1.4 Other Sources of Halal Instruction

While the Holy Quran is the literal word of God and so the primary source of all instruction, Muslims also look to two other sources for guidance. The first is the Hadith, often translated as the prophetic 'traditions', which is also known as the Sunnah, which is translated as the 'clear path' or 'trodden path'. Muslims see Mohammed (pbuh) as not only the (pbuh) but also as a very human figure. Unlike a Christian's relationship with Jesus, he is not seen in a spiritual light, but rather as a man who set the best possible example for Muslims to follow when striving to do God's will on earth. Muslims look to the words and deeds of Mohammed (pbuh) as providing guidance in all aspects, of life including what to eat and drink.

Because it is believed that the actions and words of Mohammed (pbuh) provide a living example of the meaning of the Holy Quran, early Muslims were keen to record as much of what Mohammed (pbuh) said and did as possible. These oral stories about his life and the example of the path that he trod (Sunnah) are known as the Hadith (the traditions). The Hadith were collected and written down over several 100 years and it is to these written texts that Muslims turn for guidance.

The other source of guidance for Muslims is Shariah law (see Chapter 2). For a Muslim, there is only one authority and that is the authority of God and being a good Muslim means submitting to this one authority. A Muslim does not drink alcohol not because this is a law laid down by the state, but because it is the will of God that he or she should not do so. For a Muslim, there should ideally be no need for a secular legal system because there should be no difference between your duty to God and your duty to the state. To help in this process, religious scholars drew up a set of rules that instruct Muslims on what is the right thing to

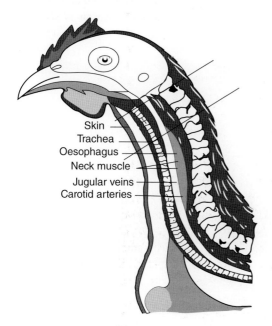

Skin
Trachea
Oesophagus
Neck muscle
Jugular veins
Carotid arteries

Figure 1.4 Method of slaughtering in chicken. *Source:* The London Central Mosque Trust and the Islamic Cultural Centre.

do in response to the different circumstances that life throws at them. These laws cover everything from religious rituals to setting standards for international law, politics, and economics as well as more personal rules around such things as marriage, divorce, diet, hygiene, and prayer. Islamic law or Shariah law provides a set of standards that unite all Muslims in their struggle to obey the will of God. However, it needs to be noted that legal systems have been developed differently in different countries and so Shariah law is not always consistent. Different countries have interpreted the Quran and Sunnah differently.

The two primary sources for Shariah law are the precepts laid out in the Holy Quran and the Sunnah or example set by Mohammed (pbuh). However, there may be occasions when neither provides an adequate answer to the question 'What is the right course of action?' It then becomes the role of Islamic judges (*qadis*) along with religious scholars (*ulama*) to look at similar situations or principles (*qiyas*) that may be used to gain insight into God's will. When making their decisions religious scholars will also apply the principle of consensus, translated as '*ijma*' in Arabic, to guide them. In the Sunnah Mohammed (pbuh) is reported to have said that 'my community will never agree on an error'. This is taken to mean that if a consensus view can be formed it will ultimately be the expression of the will of God.

1.2 What is Halal Food?

1.2.1 Halal Foods in Islam

Food has a great status in Islam and one of the fundamental principles of halal is that unless it is specifically stated that something is haram or unlawful in the Holy Quran or in the Hadith then it is has been put on earth by God for human benefit. To cause unnecessary hardship by unjustifiably prohibiting something and so causing privation is seen by

Figure 1.5 Installing the animal before slaughtering in the rotation slaughter box: Weinberg model.

Muslims as going against the will of God. It is not up to humans to second-guess what should be permitted and what should not. When Mohammed (pbuh) was asked about whether animal fat or cheese could be eaten he replied:

'The lawful is that which Allah has made lawful in His Book and the prohibited is that which He has prohibited in His Book, and that concerning which He is silent He has permitted as a favour to you.'

When it comes to food, Islamic dietary law also requires that food is not only halal but also that it is wholesome and fit for human consumption (*tayyab*). If something is not fit for consumption and wholesome it automatically becomes haram. This means that all foods have to be produced in hygienic conditions to ensure that no food pathogens are allowed to contaminate the food.

While the consumption of meat is permitted, the choice of animals that are allowed (halal) is restricted and there is a further requirement to follow the prescribed method of slaughter (Al Dabah/Al Zabah) and to ensure that the principles of animal welfare have been adhered to throughout the life of the animal, including its slaughter.

So for a food to be halal it must first of all not be haram, it must be wholesome and fit for consumption, in the case of meat it must be slaughtered in the prescribed manner and good animal welfare must have been practiced throughout the life of the animal and its slaughter.

1.2.2 Halal Food Defined

Food made from the following substances is halal unless it contains or comes into contact with a haram substance:

- All plant and their products.
- Halal slaughtered meat, poultry, game birds and halal animal ingredients which include sheep, lamb, goats, cattle, buffalo, camel, rabbit, and grasshoppers. Wild animals that are non-predatory, e.g. deer, big horn sheep, gaurs, and the antelope. Non-predatory birds, e.g. chicken fowl, quails, turkey, hens, geese, pigeons, sparrows, partridges, ostriches, and ducks.

Figure 1.6 The head holder in cattle (Dr.Temple Grandin)

- All water creatures, fish, crustaceans, and mollusks. There is no prescribed method of killing them. Dying fish must not be made to suffer or cut open while alive, and shall not be cooked alive.
- Eggs can only come from acceptable birds.
- Non-animal rennet (NAR, culture) or rennet from halal slaughtered calves.
- Gelatine produced from halal beef bones or skins.
- What has been slaughtered under non-normal conditions of the animal, e.g. a battered or a strangled animal about to die but still alive.

Where a Muslim is forced to eat what is not permitted, to avoid the risk of dying, then he or she can eat only an amount sufficient to stay alive.

1.2.3 Haram Food Defined

- Pork/swine and its by-products.
- Animals improperly slaughtered or dead before slaughtering,
- Carrion or dead animals.
- Animals killed in the name of anyone other than Allah (God), and lawful animals not slaughtered according to Islamic rites. (Fish is exempt from slaughtering rules.)
- Carnivorous animals and animals with fangs such as tigers, lions, cats etc.
- Birds that have talons with which they catch their prey such as owls, eagles, etc.
- Land animals without external ears.
- Animals which Islam encourages people to kill such as scorpions, centipedes, rats etc.
- Animals which Islam forbids people to kill such as bees etc.
- Animals which have toxins or poisons or produce ill-effects when eaten, such as some fish.
- Amphibious animals such as crocodiles, turtles, frogs etc.
- Blood and blood by-products, faces and urine, and placental tissue.
- Almost all reptiles and insects, which are considered ugly or filthy, such as worms, lice, flies, etc.
- Wine, ethyl alcohol, spirits, and intoxicants such as poisonous and intoxicating plants.
- Foods contaminated with any of the above products.

Figure 1.7 Imperfect bleeding due to delaying of the slaughter cut of the neck (few minutes) after stunning.

- Foods not free from contamination while prepared or processed with anything considered *najiis* (filthy).
- Foods processed, made, produced, manufactured, and/or stored using utensils, equipment, and/or machinery that have been not cleansed according to Islamic Shariah law.

1.2.4 Fit and Wholesome Food

Tayyab means in Arabic to be wholesome and fit for consumption and it is a requirement of halal food that is should be tayyab. If it is not it cannot be described as halal and so it becomes haram. Of course, regardless of our faith, we all wish to eat food that is fit for human consumption, so countries have their own food hygiene laws that both producers and retailers are legally bound by. Any food that is produced outside these food hygiene regulations is by definition unfit for human consumption so as a starting point all food sold as halal has to be legally produced and has to have met that particular country's food hygiene standards.

In the case of meat in the UK that would mean the animal has been slaughtered in a licensed abattoir where procedures are monitored by the Food Standards Agency, that the meat has been transported in suitable vehicles, and that any further processing has been done under the appropriate food hygiene regulations.

Where problems often arise with halal foods produced in non-Muslim countries is where halal and haram foods are processed in the same food plant. While halal and haram foods can be stored together, if they come into contact with each other then the halal food is rendered haram.

A production line that produces both non-halal foods and halal foods at different times will often struggle to clean equipment to a standard where cross-contamination does not

Stunning in cattle

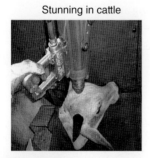

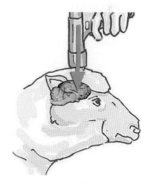

Electrical stunning tools

Captive bolt shooting in
sheep
(Their ideal shooting position
in sheep)

Figure 1.8 Some kinds of animals stunning.

occur. This can be extremely problematic where haram foods such as pork is being processed along with halal foods. There have been many instances where small quantities of pork have been detected in foods that are labelled halal and these have caused quite understandable distress. This particularly came to light during the 2011 horsemeat scandal in the UK and Europe.

1.3 Animal Welfare and Halal Slaughter

1.3.1 Animal Welfare in Islam

In Islam the mistreatment of an animal is considered a sin. Animals are not considered as a merely a resource for humans but as creatures that are dependent on God and that are organized into social groups. They are seen as having their own lives and purpose that is valuable to both themselves and God. In Islam a good deed done to an animal is equivalent to a good deed done to a human. Equally an act of cruelty to an animal is the equivalent of an act of cruelty to a human. Most importantly, Muslims believe that animals engage in the active worship of God. In one Hadith Mohammed (pbuh) tells of a past prophet who ordered an ants' nest to be burned after being bitten by an ant. Mohammed (pbuh) recounts how God reprimanded the prophet for destroying a community that glorified him.

The message that humans should show kindness to animals can be found in the Holy Quran and the Hadith. The Prophet Mohammed (pbuh) provided many examples of his concern for animals. He chastised anyone who mistreatesd animals while giving praise to those who show kindness. It is forbidden to strike or beat an animal as well as to brand an animal or mark it on the face. Mohammed (pbuh) also introduced, at the time radical, prohibitions against the practice of cutting off the tails and humps of living animals for food.

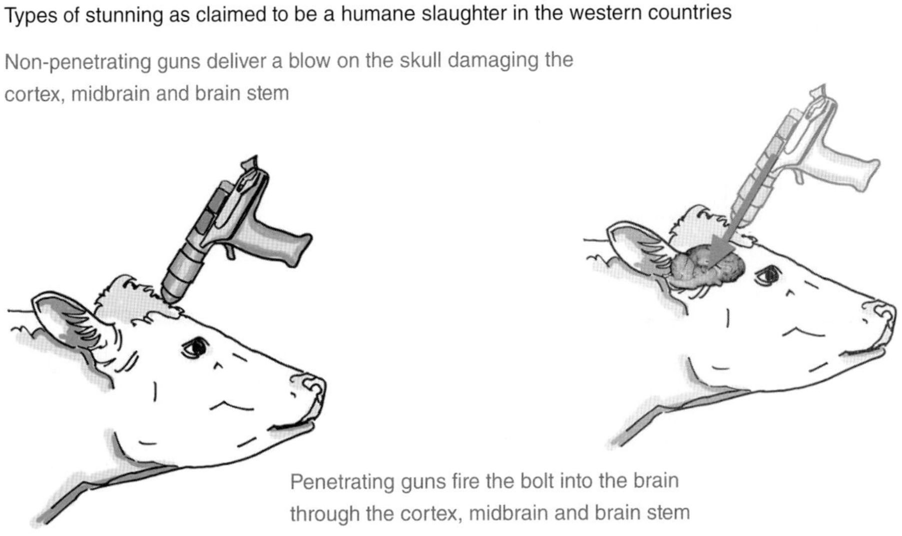

Types of stunning as claimed to be a humane slaughter in the western countries

Non-penetrating guns deliver a blow on the skull damaging the cortex, midbrain and brain stem

Penetrating guns fire the bolt into the brain through the cortex, midbrain and brain stem

Figure 1.9 Penetrating guns fire the bolt into the brain in cattle.

Mental cruelty to animals is also forbidden. In one instance described in the Hadith Mohammed (pbuh) orders his companions on a journey to return two young birds they had taken from a nest after he saw the mother's distress.

An animal that has been poorly treated is not halal. In our modern food chain this should begin at the primary point of production, where best husbandry practice should be applied. In Chapter 5 on animal welfare Mehmet Haluk Anil goes into more detail to explain how this can be achieved throughout the whole food chain, including loading, transportation, and slaughter. In Chapter 4 Temple Grandin explains how our knowledge of animal behaviour can instruct us on reducing distress to animals before and during slaughter.

In addition to animal husbandry, the welfare of animals in the Islamic faith extends to the keeping of pets and hunting for sport. A Muslim who chooses to keep a pet must take on the responsibility of the animal's care and well-being. This means providing the animal with appropriate food, water, and shelter. The Prophet Mohammed (pbuh) described how a woman who had cruelly confined her pet cat as well as failing to properly feed it was punished by being forced to enter the fire after death.

Equally, the hunting of animals for sport is prohibited in the Muslim faith. Muslims are only allowed to hunt for food. In the seventh century, when the Prophet (pbuh) was alive, hunting for sport was common practice and Mohammed (pbuh) took every opportunity to condemn the practice as being cruel to animals.

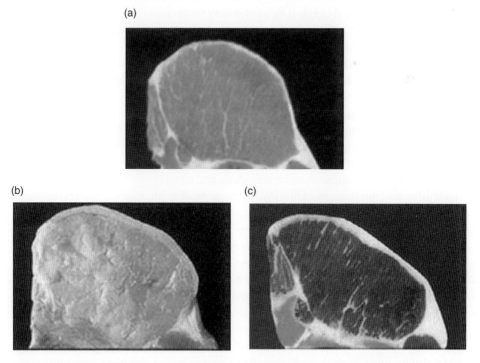

(a)

(b)   (c)

Figure 1.10 Meat texture and colour (a)Normal meat; (b) Pale Soft and Exudative (PSE) meat; (c) Dark Firm and Dry (DFD) meat chambers. *Source:* Chambers, P.G. and Grandin, T. (2001): Guidelines for humane handling, transport and slaughter of livestock. G. Heinz and T. Srisuvan (Eds.).

1.3.2 Halal Slaughter

The slaughtering (*dhabh*) rules for halal slaughter are based on the Holy Quran, the Hadiths and Shariah law. In Chapter 5 Mehmet Haluk Anil goes into greater detail on halal slaughter good practice but here we outline the Hadith and Shariah laws as follows:

- The abattoir or factory must be under the close and constant supervision of an Islamic religious organization.
- Animals should have a preslaughter rest, and be well fed and well looked after at the point of slaughter.
- Animals that are slaughtered should be securely restrained, particularly the head and neck, before the throat is cut.
- The premises, equipment, and machinery must be classed according to Islamic Shariah law before any production takes place.
- Muslim men must be trained to slaughter animals in a licensed slaughterhouse that implements all hygiene and animal welfare regulations.
- The slaughterman must be a mature and pious Muslim of sound mind who understands fully the fundamentals and conditions relating to halal slaughter and is approved by the religious authorities and the meat hygiene services.
- The animal/bird must have been allowed to feed and grow up on a natural vegetarian diet.
- The animal/bird must be alive, healthy, and free from any disease or injury at the time of slaughter, as certified and checked by a veterinary surgeon.
- The animal skin or fur and bird feathers must be cleaned prior to slaughter and free from faeces, dirt or other unhygienic substances.
- The animal must be fed and not be hungry or thirsty before slaughter.
- The animal must not be slaughtered in front of other animals and should not see any blood.
- The animal must be handled gently and individually, and the knife should not be sharpened in front of any animal before slaughter.
- No stress or discomfort should be caused to any animal.
- No stunning is allowed before slaughter.
- The knife used for slaughter must be very sharp and clean.
- The Muslim slaughterman must first say, 'In the name of Allah, Allah is greater (Bismillah, Allah Akbar)'.
- The cut must be made in the correct anatomical site in the neck by cutting the two carotids, the two jugulars, the windpipe, and the gullet, but without gutting the spinal cord.
- All blood should be allowed to flow from the carcass.
- Animals should not be shackled and hoisted before bleeding.
- Hoisting should be done only after the animal has lost consciousness.
- Restraining equipment should be comfortable for the animal.
- A specific time should be allowed till the animal ceases any movement.
- De-feathering, de-skinning, and evisceration can be done after slaughter.
- Any unlawful meat, such as pork, should not contaminate halal meat. Separate knives, equipment, and utensils should be used for halal meat.

1.4   The Holy Quran and the Hadith

1.4.1   Verses that Explain Why Muslims Can Only Eat Halal Foods

Consuming halal is an order of Allah and an essential part of the Islamic faith. Allah has repeatedly emphasized the consumption of halal in His Book. The following are some examples of such verses from the Holy Quran:

> O Messengers, eat from the pure foods and work righteousness.
>
> *Holy Quran 23:51*

> O you who have believed, eat from the pure things which we have provided you.
>
> *Holy Holy Quran 2:172*

> O mankind, eat from whatever is on earth (that is) lawful and pure.
>
> *Holy Quran 2:168*

> and eat of the lawful and good (things) that Allah has given you, and be careful of (your duty to) Allah, in whom you believe.
>
> *Holy Quran 5:88*

> Therefore eat of what Allah has given you, lawful and good (things), and give thanks for Allah's favor if him do you serve.
>
> *Holy Quran 16:114*

> O men! Eat the lawful and good things out of what is in the earth, and do not follow the footsteps of the shaitan; surely he is your open enemy.
>
> *Holy Quran 2:168*

> O you who believe! Do not forbid (yourselves) the good things which Allah has made lawful for you and do not exceed the limits; surely Allah does not love those who exceed the limits.
>
> *Holy Quran 5:87*

> O you who believe! Intoxicants and games of chance and (sacrificing to) stones set up and (dividing by) arrows are only an uncleanness, the shaitan's work; shun it therefore that you may be successful.
>
> *Holy Quran 5:90*

> O messengers! Eat of the good things and do good; surely I know what you do.
>
> *Holy Quran 23:51*

> O children of Adam! Attend to your embellishments at every time of prayer, and eat and drink and be not extravagant; surely he does not love the extravagant.
>
> *Holy Quran 7:31*

1.4.2 Verses That Explain Why Only Allah Can Prescribe What is and is not Halal

And, for what your tongues describe, do not utter the lie, (saying) this is lawful and this is unlawful, in order to forge a lie against Allah; surely those who forge the lie against Allah shall not prosper.

Holy Quran 16:116

And what reason have you that you should not eat of that on which Allah's name has been mentioned, and he has already made plain to you what he has forbidden to you – excepting what you are compelled to; and most surely many would lead (people) astray by their low desires out of ignorance; surely your lord – he best knows those who exceed the limits.

Holy Quran 6:119

Say: tell me what Allah has sent down for you of sustenance, then you make (a part) of it unlawful and (a part) lawful. say: has Allah commanded you, or do you forge a lie against Allah?

Holy Quran 10:59

Oh, ye messengers! Eat of the good things [tayyibat] and do righteous deeds. Surely, I know what you do.

Holy Quran 23:51

Oh believers! Eat what we have provided for you of lawful and good things, and give thanks for Allah's favour, if it is He whom you serve.

Holy Quran 2:172, 16:114

1.4.3 Verses That Explain What is Halal and What is Haram

Haram to you (for food) is carrion; blood; the flesh of swine; that which any name other than Allah has been invoked; that which has been killed by strangling; and beat to death, or dead through falling from a height; and killed by the goring of horns; and that which was eaten by wild beasts, unless you are able to perform dhabiha; and that which is sacrificed to idols.

Al-Quran Surah Al-Maidah

I do not find in that which has been revealed to me anything forbidden for an eater to eat of except that it be what has died of itself, or blood poured forth, or flesh of swine – for that surely is unclean – or that which is a transgression, other than (the name of) Allah having been invoked on it; but whoever is driven to necessity, not desiring nor exceeding the limit, then surely your lord is forgiving, merciful.

Holy Quran 6:145

He has only forbidden you what dies of itself and blood and flesh of swine and that over which any other name than that of Allah has been invoked, but whoever is driven to necessity, not desiring nor exceeding the limit, then surely Allah is forgiving, merciful.

Holy Quran 16:115

Forbidden to you is that which dies of itself, and blood, and flesh of swine, and that on which any other name than that of Allah has been invoked, and the strangled (animal) and that beaten to death, and that killed by a fall and that killed by being smitten with the horn, and that which wild beasts have eaten, except what you slaughter, and what is sacrificed on stones set up (for idols) and that you divide by the arrows; that is a transgression. this day have those who disbelieve despaired of your religion, so fear them not, and fear me. this day have I perfected for you your religion and completed my favor on you and chosen for you Islam as a religion; but whoever is compelled by hunger, not inclining willfully to sin, then surely Allah is forgiving, merciful.

Holy Quran 5:3

They ask you as to what is allowed to them. say: the good things are allowed to you, and what you have taught the beasts and birds of prey, training them to hunt – you teach them of what Allah has taught you – so eat of that which they catch for you and mention the name of Allah over it; and be careful of (your duty to) Allah; surely Allah is swift in reckoning.

Holy Quran 5:4

This day (all) the good things are allowed to you; and the food of those who have been given the book is lawful for you and your food is lawful for them; and the chaste from among the believing women and the chaste from among those who have been given the book before you (are lawful for you); when you have given them their dowries, taking (them) in marriage, not fornicating nor taking them for para mours in secret; and whoever denies faith, his work indeed is of no account, and in the hereafter he shall be one of the losers.

Holy Quran 5:5

Lawful to you is the game of the sea and its food, a provision for you and for the travellers, and the game of the land is forbidden to you so long as you are on pilgrimage, and be careful of (your duty to) Allah, to whom you shall be gathered.

Holy Quran 5:96

And he it is who has made the sea subservient that you may eat fresh flesh from it and bring forth from it ornaments which you wear, and you see the ships cleaving through it, and that you might seek of his bounty and that you may give thanks.

Holy Quran 16:14

Allah is he who made the cattle for you that you may ride on some of them, and some of them you eat.

Holy Quran 40:79

1.4.4   Examples from the Hadith (traditions) Covering Animal Welfare

One Hadith quotes Prophet Mohammed (*pbuh*) as saying:

> A good deed done to an animal is as meritorious as a good deed done to a human being, while an act of cruelty to an animal is as bad as an act of cruelty to a human being.
> There is a reward (ajr) for helping any living creature.
> *Bukhari and Muslim*
> It is a great sin for man to imprison those animals which are in his power.
> *Muslim*
> The worst of shepherds is the ungentle, who causes the beasts to crush or bruise one another.
> *Muslim*
> You will not have secure faith until you love one another and have mercy on those who live upon the earth.
> *Bukhari, Muslim, and Abu Dawud*
> Fear God in these mute animals, and ride them when they are fit to be ridden, and let them go free when... they [need to] rest.
> *Abu Dawud*
> There is no man who kills a sparrow or anything beyond that, without its deserving it, but God will ask him about it.
> *Ahmad and al-Nasai*
> The grievous things are: shirk (polytheism); disobedience to parents; the killing of breathing beings ...
> *Bukhari and Muslim*
> May god curse anyone who maims animals.
> *Ibn al-Athir*
> It is related from Abdullah ibn Umar that the Messenger of Allah, may Allah bless him and grant him peace, said:
> 'A woman was once punished after death because of a cat which she had kept confined until it died, and because of this she entered the Fire. She had neither given it food or drink while confining it, nor had she let it free to eat the creatures of the earth.'
> *Muslim*

The Prophet Mohammed (pbuh) often chastised his Companions who mistreated animals, and spoke to them about the need for mercy and kindness. Here are several examples from the Hadith instructing Muslims about how to treat animals.

Reward for mercy: It is related from Abu Umama that the Messenger of Allah, may Allah bless him and grant him peace, said, 'Whoever is merciful even to a sparrow, Allah will be merciful to him on the Day of Judgment.'

Animals are like humans: 'A good deed done to an animal is like a good deed done to a human being, while an act of cruelty to an animal is as bad as cruelty to a human being.'

Animals cannot speak up for themselves: It is related from Sahl ibn Al-Handhaliyya that the Messenger of Allah, may Allah bless him and grant him peace, once passed by a camel that was so emaciated that its back had almost reached its stomach. He said, 'Fear Allah in these beasts who cannot speak.'

Abu Dawud

Mental cruelty is also forbidden: It is related from AbdulRahman bin Abdullah that a group of Companions were once on a journey with the Prophet, may Allah bless him and grant him peace, and he left them for a while. During his absence, they saw a bird with its two young, and they took the young ones from the nest. The mother bird was circling above in the air, beating its wings in grief, when the (pbuh) came back. He said, 'Who has hurt the feelings of this bird by taking its young? Return them to her.'

Muslim

Forgiveness of sins: It is related from Abu Hurayra, from the Prophet, may Allah bless him and grant him peace, that a prostitute once saw a dog on a very hot day going round and round a well, lolling its tongue because of its thirst. She drew some water for it using her shoe, and for this action all her sins were forgiven her.

Muslim

Mistreatment is a sin: It is related from Jabir that the Messenger of Allah once saw a donkey which had been branded on its face and he said, 'May Allah curse the one who branded it.'

Muslim

Give rest to beasts of burden: It is related from Abu Hurayra that the Prophet, may Allah bless him and grant him peace, said, 'Do not use the backs of your animals as chairs. Allah has made them subject to you, so that by them you can reach places that you would not otherwise be able to reach except with great fatigue.'

Abu Dawud

1.4.5 Verses from the Hadith Concerning the Slaughter of Animals

So eat of that (meat) upon which Allah's name has been mentioned, if you are believers in His verses.

Holy Quran 6:118

And do not eat that upon which the name of Allah has not been mentioned, for indeed it is a grave disobedience.

Holy Quran 6:121

The humane slaughter of animals is strongly supported in The Islamic tradition.

Sahih Muslim (Book 21, Chapter 11, Number 4810) records Mohammed (*pbuh*) saying, 'Verily Allah has enjoined goodness to everything; so when you kill, kill in a good way and when you slaughter, slaughter in a good way. So every one of you should sharpen his knife, and let the slaughtered animal die comfortably.'

Prophet Mohammed (*pbuh*) has also said, 'When one of you slaughters, let him complete it', meaning that one should sharpen the knife well and feed, water, and soothe the animal before killing it.

He also said, 'Do you intend inflicting death on the animal twice – once by sharpening the knife within its sight, and once by cutting its throat?'

The Holy Quran is explicit with regard to using animals for human purposes. A closer look at the teachings of the Holy Quran and tradition reveals teachings of kindness and concern for animals. Nonetheless, the Holy Quran clearly supports the use of animals, including for food.

> If you kill, kill well.
>
> *Imam Nawawi 40:1*

> And cattle He has created for you (men); from them ye derive warmth and numerous benefits, and of their (meat) ye eat.
>
> *Surrah An-Nahl 16:5*

> There is not a moving (living) creature on earth, nor a bird that flies with its two wings, but are communities like you. We have neglected nothing in the Book, then unto their Lord they (all) shall be gathered.
>
> *Surrah Al-Anam 6:38*

> And they carry your heavy loads to lands that ye could not (otherwise) reach except with souls distressed: for your Lord is indeed Most Kind, Most Merciful.
>
> *Surrah An-Nahl 16:7*

> We have made animals subject to you, that ye may be grateful.
>
> *Surrah Al Haj 22:36*

> And (He has created) horses, mules, and donkeys, for you to ride and as an adornment; And he has created other things of which ye have no knowledge.
>
> *Surrah An-Nahl 16:8*

> Seest thou not that it is Allah Whose praise all beings in the heavens and on earth do celebrate, and the birds (of the air) with wings outspread? Each one knows its own (mode of) prayer and praise, and Allah knows well all that they do.
>
> *Surrah An-Noor 24:41.*

> There is not an animal that lives on the earth, nor a being that flies on its wings, but they form communities like you. Nothing have we omitted from the Book, and they all shall be gathered to their Lord in the end"
>
> *Holy Quran 6:38*

Although animals do not have free will, they follow their natural, God-given instincts, and in that sense they submit to God's will, which is Islam.

> Seest thou not that it is Allah Whose praise all beings in the heavens and on earth do celebrate, and the birds (of the air) with wings outspread? Each one knows its own (mode of) prayer and praise, and Allah knows well all that they do.
>
> *Holy Quran 24:41*

For the most up-to-date Codex Alimentarius guidelines for the use of the term "Halal" please visit http://www.fao.org/docrep/005/Y2770E/y2770e08.htm

2

Halal and Shariah Law

Mufti Mohammed Zubair Butt

Senior Advisor on Islamic law at the Institute of Islamic Jurisprudence, Bradford, UK

2.1 Introduction

Halal is a term that Muslims are well acquainted with and most will appreciate its wide connotations. It is a term that is also now familiar to many western non-Muslims due to the public discourse surrounding halal food and the halal signs that are a consistent feature of Muslim retail outlets selling fast food, fresh meat, and other groceries. However, the wider application of the term is much less appreciated and is often considered to be related exclusively to the food and beverage sector. This chapter will seek to demonstrate that halal is a much wider concept than simply that related to food and beverages.

2.2 Lexical Definition

Halal, also hilal (Al-Ferozabādī 1999), is derived from the root *h-l-l* and is defined by its opposite, haram (Ibn Manẓūr 1994). It takes metaphorical significance from *ḥall al-ʿuqdah* ('untying of the knot') and refers to that which is not forbidden. It therefore includes what is disapproved and what is not disapproved. According to some, it is that for which one is not punishable (Al-Zabīdī 1994; Lane 2003). According to Ibn Manẓūr, everything that Allah has permitted is halal and that which He has prohibited is haram (Ibn Manẓūr 1994).

2.3 Legal Definition

Muslim jurists have used various expressions to define halal. According to al-Jurjānī, it is all that which one is not punished for in its use, and that which the Sharia has allowed (Al-Jurjānī 1997). According to al-Thānawī, it is that which the Book and the Sunna have permitted on account of a permissible cause (Al-Thānawi 1998). However, this definition is arguably deficient as it does not allow for recommended, reprehensible or mandatory

The Halal Food Handbook, First Edition. Edited by Yunes Ramadan Al-Teinaz, Stuart Spear, and Ibrahim H. A. Abd El-Rahim.
© 2020 John Wiley & Sons Ltd. Published 2020 by John Wiley & Sons Ltd.

causes that are considered to be separate from simple permissibility. In view of this, some have defined halal as that which is permitted in Sharia. This definition includes the recommended, the permissible, and the reprehensible in that it is permissible to perform and is not proscribed in Sharia. Performance of the recommended enjoys preponderance as opposed to abstention, whilst in the permissible both performance and abstention are of equal status. Abstention from the reprehensible enjoys preponderance as opposed to performance (Al-Mawsû'ah 2006). However, owing to the different nuances in definitions, the Ḥanafīs restrict the scope of the reprehensible to that which is non-blameworthy. Al-Qardawi has defined it as the permissible from which the knot of prohibition has been opened and the lawgiver has allowed its performance (Al-Qardawi 1997).

2.4 Halal and the Values of Islamic Law

Communication from the Lawgiver regarding the conduct of one who is the locus of obligation may consist of a demand, an option or an enactment. A demand is usually communicated in the form of a command or a prohibition and may be either binding or non-binding (Kamali 2003). The majority of Muslim jurists accept five values of law: *wājib* (obligatory), *mandūb* (recommended), *mubāḥ* (permissible), *makrūh* (abominable), and *ḥarām* (forbidden). *Wājib* represents a binding demand with respect to a performance which leads to reward, whilst omission leads to punishment. *Mandūb* represents a non-binding demand with respect to a performance which leads to reward, but the omission of which does not lead to punishment. *Mubāḥ* represents communication that gives the locus of obligation the option to perform or not perform something. It leads neither to reward nor to punishment. *Makrūh* represents a demand that requires the locus of obligation to abstain from something, but not in strictly prohibitory terms. Performance does not lead to punishment nor incur moral blame. *Ḥarām* represents a binding demand with respect to abstention from something which leads to reward, whilst performance leads to punishment.

The Ḥanafī School recognizes seven values of law. First, it recognizes *farḍ* (obligatory in the first degree) as a distinct value from *wājib* (obligatory in the second degree), which is based on the definitiveness of evidence. If the text that conveys a binding demand of performance is definitive in terms of both meaning and authenticity then this is referred to as *farḍ*. If either meaning or authenticity or both are not definitive, then a binding demand of performance results in a *wājib* (Kamali 2003). In terms of performance, both are the same: obligatory. However, in terms of belief, denial of a *farḍ* leads to disbelief whilst denial of a *wājib* leads to punishment only (Al-Bukhārī 1890). In contrast, the majority of jurists make no distinction between *farḍ* and *wājib*. Similarly, the Ḥanafī School requires that the text of a binding demand of abstention is definitive in terms of both meaning and authenticity to effect a value of *ḥarām*. If either the meaning or authenticity or both are not definitive, then a binding demand of abstention results in a *makrūh taḥrīmī* (prohibitive abomination). This is the second of the two additional values of law. The value of *makrūh* recognized by the majority of jurists is qualified as *makrūh tanzīhī* (blameless abomination).

All of the values of Islamic law except *ḥarām*, and according to the Ḥanafī School *makrūh taḥrīmī* too, fall within the parameters of halal (Ḥamd 2004; 'Uthmān 2002). This is because the first four of the five values of Islamic law or, according to the Ḥanafī School, the first five of the seven values of Islamic law, are not prohibitive. Only a binding demand of abstention is prohibitive in nature and falls outside the parameters of halal.

2.5 Halal and the Original Norm

Muslim jurists have discussed at length whether the normative ruling in cases where the evidentiary texts are silent is permission, prohibition or suspension. This discussion stems from the notion of whether an event can exist without a ruling from the Lawgiver. Some accept the possibility whilst others reject it. Others still recognize the theoretical possibility but aver a legal impossibility. The latter two positions assert that the event must be governed by a ruling of binding performance, binding abstention or a choice between the two (Al-Duwaihī 2007:93). However, this notion is only applicable to cases where legal permission or prohibition is a possibility. Where this does not exist, such as disbelief and rejection of monotheism in the case of permission and the consciousness of God and His oneness in the case of prohibition, it is beyond the scope of this discussion. The discussion is restricted instead to elective sayings and deeds concerning an individual's livelihood, food, drink, raiment, and all transactions and activities. Involuntary actions, such as breathing, fall out of the scope of this discussion (Al-Duwaihī 2007:97). The majority opinion amongst Muslim jurists is that the normative ruling in such cases is permission. This is the apparent position of the founder of the Shāfiʿī School. There are similar indications from the founder of the Ḥanbalī School. It is the position of the majority within the Ḥanafī School, the majority within the Shāfiʿī School (Al-Suyūṭī 1983:99; Ibn Nujeym 1998:1:209), the majority within the Ḥanbalī School, and the position of the Literalist School. Individuals within the Mālikī School also maintain this position (Al-Suyūṭī 1983:84; Ibn Nujeym 1998:1:99). The proponents of this position derive their justification from textual evidence of the Holy Quran and prophetic sayings as well as from the sayings of the Prophet's companions. A few examples are given here:

1) 'It is He Who has created for you all things that are on the earth' (Quran, 2:29). Here God mentions His creation as a favour from which humankind benefits. Naturally, this necessitates permission.
2) 'Say: Who has forbidden the beauty of Allah that He has produced for His servants, and the clean and pure sustenance?' (Quran, 7:32)
   There is reproach in this verse for one who forbids the beauty of Allah and clean and pure sustenance. This indicates that Allah has created things with a normative ruling of permission.
3) 'The lawful is that which Allah made lawful in His Book, the unlawful is that which Allah made unlawful in His Book, and what He was silent about; then it is among that for which He has pardoned' (Al-Tirmidhi 2007; Al-Hakim 1990). This statement of the Holy (pbuh) is possibly the most express proof for the proponents of this position.
4) 'Indeed Allah has set boundaries so do not cross them; and He has obligated for you obligations so do not waste them; and He has made unlawful for you things so do not violate them; and He has left things without forgetfulness from your Lord but rather as

mercy from Him to you, so accept it and do not pursue it' (Al-Hakim 1990). This statement of the Holy (pbuh) is also very clear in supporting this position.

This rule may also be expressed as follows: the normative ruling in cases where the evidentiary texts are silent is lawfulness (*ḥilla*) as all that which is not prohibited, and so does not attract punishment, is halal. This latitude in Islamic law allows for the different experiences of individuals of varying places, eras, and cultural norms.

However, there are some major exceptions to this normative ruling which Muslim jurists express as legal maxims. These exceptions do not render this rule null but rather restrict the scope within which it remains valid.

1) The normative rule in harms is prohibition (*al-aṣl fī al-maḍārr al-taḥrīm*) (Badr Al-Din 2000:4:322; Al-Razi 1997). This effectively qualifies the normative ruling of permissibility in that it requires that a thing must be void of harm to be permissible. The harm and benefit here is that which is ultimately regarded as harmful and beneficial by Sharia and not harm and benefit per se, for wine has benefit despite being haram whilst medication may carry a degree of harmful side-effects but still be halal (Al-Duwaihī 2007:145).

2) The normative rule in devotional practices is suspension (*al-aṣl fī al-ʿibādāt al-tawqīf*). According to this rule it is not permitted to adopt a worship practice that does not find sanction in revelation, viz. the Holy Quran or the words of the Messenger. Worship is a right that Allah enjoys over His subjects; a right that He has decreed, determined, and legislated. It is thus necessary to adhere and restrict oneself to revelationary sources in this regard without allowing reason to decide in the matter. As far as a subject is concerned, once a devotional practice is established by revelation, reason and underlying wisdom are irrelevant (Al-Shāṭibī 2004:2:211). Ibn al-Qayyim goes one step further and states that the normative rule in devotional practices is nullity until proof of an imperative is established (*al-aṣl fī al-ʿibādāt al-buṭlān ḥattā yaqūma dalīlun ʿalā al-amr*) (Ibn Al-Qayyim 1996), i.e. instead of suspension until otherwise established, the normative rule is nullity until otherwise established.

3) The normative rule in sexual activity is prohibition (*al-aṣl fī al-abḍāʿ al-taḥrīm*) (Al-Suyūṭī 1983:84; Ibn Nujeym 1998:1:210; Badr Al-Din 2000:4:325). This rule has also been expressed as: the normative rule in pudenda is prohibition except that which Allah and His Messenger have decreed as lawful (*al-aṣl fī al-furūj al-taḥrīm illā mā abāḥahu Allah wa rasūluhū*) (Ibn Al-Qayyim 1997). Similarly, the normative rule in marriage is prohibition; it has been permitted on account of need (*al-aṣl fī al-nikāḥ al-ḥazar, wa ubīḥa li al-ḍarūra*) (Ibn Nujeym 1998:1:210). The purport of this rule is that Islam has granted honour and protection to the pudenda of both genders by legislating prohibition as the original norm. Permission for sexual activity is achieved by only that which the Sharia has allowed, principally lawful marriage. This serves to ensure the protection of honour and progeny.

4) The normative rule in persons and limbs is prohibition (*al-aṣl fī al-anfus wa al-aṭrāf al-taḥrīm*) (Al-Duwaihī 2007:145), i.e. persons and limbs are exempt from the normative rule of permission as they are protected in law. Only with the permission of the law can a life or limb be taken. Thus, legal protection for an individual of suspect protection status remains as the original norm and it is not permissible to take the life of such an individual.

5) The normative rule in properties is prohibition (*al-aṣl fī al-amwāl al-taḥrīm*) (Badr Al-Din 2000:4:325). This rule is based on, amongst other evidence, the Quranic verse, 'O you who have believed, do not consume one another's wealth unjustly but only [in lawful] business by mutual consent' (Quran, 3:29) and the saying of the Prophet Mohammed (pbuh), 'The wealth of a Muslim is not lawful without his wilful consent' (Al-Dāraquṭnī). The founder of the Shāfi'ī School states when commenting on unlawful trades, 'And that is because the property of every individual is unlawful to another except by that which it becomes lawful. And lawful trades are those that have not been prohibited by the Messenger of Allah' (Al-Shāfi'ī 1938).

6) The normative rule in meat is prohibition (*al-aṣl fī al-luḥūm al-ḥurma*) (Al-Sa'dī 2007; Ibn al'Arabī 1967). This is the preponderant opinion across the different schools. Permission to eat meat is only achieved once the requirements of slaughter laid down by Islamic law are known to have been met (Al-Kāsānī 1910; Al-Ḥaṣkafī 2002; Al-Shāṭibī 2004:1:181; Ibn al'Arabī 1967; Al-Khaṭṭābī 1996; Al-Ghazālī 1996; Ibn Qudāma 1968). For captive slaughter, these are that the animal (i) itself is of the lawful category, (ii) is slaughtered by cutting the appropriate vessels, (iii) is slaughtered by a qualifying individual, and (iv) is slaughtered whilst invoking the name of God. If a credible (Al-Ghazālī 1996) uncertainty remains that any one or more of these four requirements have not been met the meat will retain its normative ruling of prohibition.

2.6 Halal in Different Spheres

2.6.1 Food and Beverages

The most recognized of associations of the terms halal and haram is in the food and beverage sector. It is also the most oft-repeated association in the evidentiary texts. Food and beverages fall under the normative rule of permission with three notable exceptions: meat and that which is harmful or repugnant. Effectively, all foods and beverages are halal as the original norm except for meat and harmful or repugnant substances. In relation to meat, only that which is expressly permitted in the evidentiary texts, either specifically or under a stated rule, is halal whilst harmful and repugnant items are eternally haram. Consequently, the range of halal foods and beverages is extensive and so the evidentiary texts, on the whole, identify what is haram.

The Holy Quran exhorts mankind in general, not only Muslims, to eat what is lawful and good:

> O you people! Eat of what is on earth lawful [halal] and good; and do not follow the footsteps of Satan for he is for you an avowed enemy.
>
> *(Quran, 2:168)*

Permission to eat from the halal and good things on the earth is mentioned here as a favour to mankind in general. Ibn Kathīr has interpreted 'good' as that which is deemed intrinsically good and is not harmful to the body or intellect (Ibn Kathīr 1998:1:277). This is followed by an interdiction of walking in the footsteps of Satan, which Ibn Kathīr interprets as Satan's ways and methods by which he misguides his followers, such as prohibiting certain camels that were dedicated to idols as well as other similar practices that Satan had embellished for them during the era of ignorance. Ibn Kathīr then relates a narration recorded by Muslim from 'Iyāḍ bin Ḥimār from the Messenger of Allah that, 'Allah the Exalted says, "Everything that I have endowed My servants with is lawful [halal] for them" ... "And, indeed, I have created My servants professors of the truth, but the devils came to them and led them astray from their (true) religion and prohibited them from what I had made lawful [halal] for them"' (Ibn Kathīr 1998:1:227).

Four verses later, the Holy Quran specifically exhorts Muslims to eat what is good:

> O you who have believed! Eat of the good things that We have provided for you as sustenance, and be grateful to Allah, if it is He that you worship.
>
> *(Quran, 2:172)*

Here too, the provision of good is mentioned as a favour for which gratitude is due, which is a form of worship in itself. This is then followed by a summary discussion of the major categories of prohibited articles of food:

> He has only forbidden you carrion, and blood, and the flesh of swine, and that on which any other name has been invoked besides that of Allah ...
>
> *(Quran, 2:173)*

Blood and carrion are naturally repugnant to any cultured individual whereas the flesh of swine is repeatedly declared to be prohibited (haram). Food dedicated to idols or false gods is clearly improper for those that profess belief in the one deity. A second verse provides further elaboration on the four types of prohibited articles of food mentioned here:

> Forbidden to you [for food] are carrion, blood, the flesh of swine, and that on which any other name has been invoked besides that of Allah, that which has been killed by strangling, or by a violent blow, or by a headlong fall, or by being gored to death, and that which has been eaten by a predatory animal unless you are able to slaughter it, and that which is sacrificed on stone altars, ...
>
> *(Quran, 5:3)*

This verse mentions 10 prohibited articles of food of which animals killed by strangling, a violent blow, a headlong fall, being gored to death, or from which a predatory animal has partially eaten are essentially types of carrion. Animals sacrificed on stone altars are synonymous with those on which any other name has been invoked besides that of Allah. That is why a third verse identifies the prohibited articles as no more than four in the Revelation:

> Say: 'I find not in that which has been revealed to me anything [meat] forbidden to be eaten by one who wishes to eat it unless it be carrion, or blood poured forth, or the flesh of swine, for it is an abomination, or what is impious on which a name other than Allah's has been invoked, ...'
>
> *(Quran, 6:145)*

Verse 5:3 comprises an effective negative imperative defining what is not lawful for food, namely, articles that are gross or naturally repugnant, or dedicated to superstition and false gods. The succeeding verse positively defines what is lawful, namely, all things good and pure:

> They ask you what has been made lawful [halal] to them [as food]. Say: 'Lawful for you are [all] things good and pure, ...'
>
> *(Quran, 5:4)*

Declaring lawful (halal) what is good and forbidding what is repugnant is mentioned as an attribute demonstrated by the Messenger and unlettered Prophet:

> Those who follow the Messenger, the unlettered Prophet, ... he declares lawful [halal] for them what is good and he declares unlawful [haram] to them what is repugnant ...
>
> *(Quran, 7:157)*

Chapter 5 of the Holy Quran is notable for its repeated references to halal in relation to food. Indeed the very first verse contains one such reference:

> Lawful [halal] for you have been made all the beast of cattle ...
>
> *(Quran, 5:1)*

Then, in addition to verse 5:3 mentioned above, the following verses of the same chapter also make such references:

> This day [all] things good have been made lawful [halal] for you. The food of the people of the Book is lawful [halal] for you and your food is lawful [halal] for them ...
>
> *(Quran, 5:5)*

Ibn Kathīr quotes several authorities who interpret 'food' to mean slaughtered animals and then states, 'This is a consensus position amongst the scholars that their slaughtered animals are lawful for Muslims, as they believe that slaughtering for other than Allah is prohibited and they do not mention other than the name of Allah upon their animals, even if they hold beliefs concerning Allah from which He is free' (Ibn Kathīr 1998:2:28).

> O you who believe! Make not unlawful the good things which Allah has made lawful [halal] for you, but commit not excess: for Allah loves not those given to excess. And eat of that which Allah has provided for you lawful [halal] and good ...
>
> *(Quran, 5:87–88)*

> Lawful [halal] has been made for you the pursuit of water-game ...
>
> *(Quran, 5:96)*

Further references to halal in relation to food from other chapters include:

> All food was lawful [halal] to the children of Israel, except what Israel made unlawful for itself ...
>
> *(Quran, 3:93)*

For the iniquity of the Jews we made unlawful to them certain [foods] good which had been made lawful [halal] for them ...

(Quran, 4:160)

So now eat of that which you have received as booty, lawful [halal] and good ...

(Quran, 8:69)

Say: 'Do you see what Allah has sent down to you for sustenance and you make of it lawful [halal] and unlawful? ...'

(Quran, 10:59)

So eat of the sustenance which Allah has provided for you, lawful [halal] and good, ...

(Quran, 16:114)

But say not for any false thing that your tongues may put forth, 'This is lawful [halal], and this is unlawful' ...

(Quran, 16:116)

... and lawful have been made for you cattle except those mentioned to you ...

(Quran, 22:30)

O Prophet! Why do you make unlawful that which Allah has made lawful [halal] for you? ...

(Quran, 66:1)

Declaring lawful (halal) what is good and forbidding what is repugnant is mentioned as an attribute demonstrated by the Messenger and unlettered Prophet:

Those who follow the Messenger, the unlettered Prophet, ... he declares lawful [halal] for them what is good and he declares unlawful [haram] to them what is repugnant ...

(Quran, 7:157)

Examples of the statement of the (pbuh) include:

'Every beverage that intoxicates, so it is unlawful [haram]' (Al-Bukhārī 1961:1:38).
'Every intoxicant is unlawful [haram]' (Al-Naysābūrī 1956).
'Every intoxicant is wine and every intoxicant is unlawful [haram]' (Al-Naysābūrī 1956).

2.7 Marriage and Divorce

A second area where evidentiary texts frequently employ the terms halal and haram and their derivatives is in expounding the rules relating to marital unions and divorce. The Holy Quran goes into some detail when explaining a number of marital impediments and then declares the remaining women to be lawful for marriage:

... and lawful [halal] have been made for you all other women besides these ...

(Quran, 4:24)

References related specifically to the (pbuh) in this regard include:

O Prophet! We have made lawful [halal] for you your wives to whom you have paid their dowers ...

(Quran, 33:50)

It is not lawful [halal] for you [to marry more] women after this, ...

(Quran, 33:52)

Other references to permission to marry include:

So if he divorces her [a third time] she will not be lawful [halal] for him after that until she marries a husband besides him. ...

(Quran, 2:230)

... They [believing women refugees] are not lawful [halal] for them [pagan husbands] and nor are they [pagan husbands] lawful [halal] for them [believing women refugees]. ...

(Quran, 60:10)

If the husband and wife are unable to maintain the limits set by Allah and separation between them both becomes inevitable:

... It is not lawful [halal] for you that you should take [back] anything from what you have given them [wives], except when both parties fear that they would be unable to maintain the limits ordained by Allah. ...

(Quran, 2:229)

Once divorced, the wife must observe a waiting period of three monthly menstrual cycles or the delivery of her child, if pregnant. The wife must not conceal what is in her womb, whether child or menstruation, to ensure that her waiting period is known.

... And it is not lawful [halal] for them [divorcees] to hide what Allah has created in their wombs, ...

(Quran, 2:228)

The fasts of Ramadhan comprise an abstention from food, drink and conjugal relations from dawn to sunset. Initially, these restrictions extended to after sunset if one had fallen asleep without breaking the fast. This ruling was later repealed and permission to engage in conjugal relations even if one had fallen asleep without breaking the fast was given in the following verse:

Lawful [halal] has been made for you on the night of the fasts the approach to your wives, ...

(Quran, 2:187)

2.7.1 Raiment and Adornment

A few examples of the use of evidentiary texts in relation to clothing and adornment are as follows:

> Say: 'Who has made unlawful [haram] the adornment of Allah which he has produced for His servants, and the good and pure sustenance?'
>
> *(Quran, 7:32)*

> Silk clothes and gold have been made unlawful [haram] to the males of my community and has been made lawful [halal] for their women (Al-Tirmidhī 1996)
>
> It is reported from 'Ali that the Prophet of Allah (may Allah bless him and grant him peace) took some silk and put it in his right hand and took some gold and put it in his left hand and then said: 'Indeed, these two are unlawful to the males of my community' (Al-Sajistānī 2009).

2.7.2 Financial Matters

In order to further the interests of economic justice, Islam encourages enterprise but imposes restrictions on all detrimental forms of earning. In addition to prohibiting the trade of unlawful commodities and services, it also proscribes those structures that concentrate wealth with the few to the detriment of the many. Foremost in this regard is the prohibition of usury:

> ... but Allah has made lawful [halal] trade and forbidden usury. ...
>
> *(Quran, 2:275)*

The Holy (pbuh) forewarned of a time when the ethics of how one earns will receive scant attention whilst the profit motive will solely dictate whether one engages in an economic enterprise.

> A time will most definitely come upon the people that a person will care not how he earns wealth, whether from lawful [halal] means or from unlawful means (Al-Bukhārī 1961:1:279).

The Holy Quran also makes a reference to halal in a matter of inheritance:

> O you who believe! It is not lawful [halal] for you to inherit women against their will. ...
>
> *(Quran, 4:19)*

2.7.3 Devotional Practices

Halal also features in matters that are purely devotional. The Holy Quran contains a number of such references:

O you who believe! Do not violate the sanctity of [make halal] the symbols of Allah, nor of the sacred month, ...

(Quran, 5:2)

'the symbols of Allah' refer to the rites of the Hajj, the mountains of al-Ṣafā and al-Marwa in Makkah and the sacrificial animals or to proscribed practices in general. Violation here is expressed as deeming the unlawful to be lawful (halal). A second reference is as follows:

... and when you are clear of the state of pilgrimage [become halal] you may hunt, ...

(Quran, 5:2)

There was a long established custom amongst the Pagan Arabs of observing four months in which fighting was prohibited. However, they changed the months about or added or deducted months when it suited them to get an unfair advantage over the enemy. A prohibited month became an ordinary month and the ordinary month would be observed as a prohibited month. The Holy Quran (9:36) first advises that the number of months, as ordained by Allah the day He created the heavens and the earth, is twelve of which four are sacred, and then cautions against meddling with the old-established custom of close time for warfare during these months. Violation of the sanctity of these months is expressed as deeming them to be lawful (halal).

... They make it lawful [halal] one year and unlawful another year, in order that they may conform [only] to the number of what Allah has sanctified, and so make lawful [halal] what Allah has made unlawful. ...

(Quran, 9:37)

2.7.4 General

The Holy Quran relates a conversation of Jesus with the children of Israel wherein Jesus also states:

... and to make lawful [halal] for you part of what was forbidden to you, ...

(Quran, 3:50)

The above is a reference to relaxation brought by Jesus in some of the restrictive laws of the Torah. The relaxation is referred to as making lawful (halal).

2.8 Conclusions

'Halal' is a much wider concept than simply that related to food and beverages. Whilst it is true to say that the most oft-repeated association in the evidentiary texts is with food and beverages, it is most certainly not exclusively so. On the contrary, it has an all encompassing application including, marriage, divorce, raiment and adornment, financial matters, devotional practices, and all matters in general.

References

Al-Bukhārī, A.A. (1890). *Kashf al-Asrār Sharḥ Uṣūl al-Bazdawī*, vol. 2, 548. Karachi: Qadimi Kutub Khana.

Al-Bukhārī, M. (1961). *Ṣaḥīḥ al-Bukhāri*. Karachi: Qadimi Kutub Khana.

Al-Duwaihī, A. (2007). *Qāʿida al-Aṣl fī al-Ashyāʾ al-Ibāḥa*. Ministry of Higher Education, Imam Muhammad ibn Saud Islamic University.

Al-Ferozabādī, M. (1999). *Al-Qāmūs al-Muḥīṭ*, vol. 3, 493. Beirut: Dar al-Kotob al-Ilmiyah.

Al-Ghazālī, M. (1996). *Iḥyāʾ ʿUlūm al-Dīn*, vol. 2, 112. Beirut: Dar al-Kotob al-Ilmiyah.

Al-Hakim, M. (1990). *Al-Mustadrak ʿalā al-Ṣaḥīḥain, Hadith No. 44/7115*, vol. 4, 129. Beirut: Dar al-Kotob al-Ilmiyah.

Al-Ḥaṣkafī, ʿ.A. (2002). *Al-Durr al-Mukhtār*, vol. 6, 294. Karachi: H.M. Saeed Company.

Al-Jurjānī, A. (1997). *Kitāb al-Taʿrīfāt*, 66. Beirut: Dar al-Fikr.

Al-Kāsānī, ʿ.A. (1910). *Badāʾiʿ al-Ṣanāʾiʿ fī Tartīb al-Sharāʾiʿ*, vol. 5, 40. Quetta: Maktaba Rashidiya.

Al-Khaṭṭābī, H. (1996). *Maʿālim al-Sunan*, vol. 4, 262. Beirut: Dar al-Kotob al-Ilmiyah.

Al-Mawsūʿah (2006). *Al-Fiqhyyiah Al-Kuwaitiyyah*, vol. 10, 18:74. Kuwait: Ministry of Waqf and Islamic Affairs. (2006-1984/1427-1404), Kuwait.

Al-Naysābūrī, M. (1956). *Ṣaḥīḥ Muslim*, vol. 2, 167. Karachi: Qadimi Kutub Khana.

Al-Qardawi, Y. (1997). *Al-Ḥalāl wa al-Ḥarām fī al-Islām*, 15. Cairo: Maktaba Wahba.

Al-Razi, F.M. (1997). *al-Maḥṣūl fī ʿIlm Uṣūl al-Fiqh*, vol. 6, 105. Beirut: Muʾassasa al-Risāla.

Al-Saʿdī, A. (2007). *Al-Qawāʿid al-Fiqhiyya: al-Manẓuma wa Sharḥuhā*, 128. Kuwait: Ministry of Endowments.

Al-Sajistānī, S. (2009). *Sunan Abū Dāwūd*, Hadith No. 4057, vol. 6, 165. Damascus: Al-Resalah Al-Aʾlamiah Ltd Publishers.

Al-Shāfiʿī, M. (1938). *Al-Risāla*, 348. Egypt: Maktaba Muṣṭafā al-Bājī al-Ḥalabī.

Al-Shāṭibī, I. (2004). *Al-Muwāfaqāt fī Uṣul al-Aḥkām*. Beirut: Dar al-Fikr.

Al-Suyūṭī, A. (1983). *Al-Ashbāh wa al-Naẓāʾir fī Qawāʿid wa Furūʿ Fiqh al-Shāfiʿiyya*, 2e. Beirut: Dar al-Fikr.

Al-Thānawi, M.A. (1998). *Kash-shāf Iṣṭilāḥat al-Funūn*, vol. 1, 475. Beirut: Dar al-Kotob al-Ilmiyah.

Al-Tirmidhi, M. (2007). *English Translation of Jāmiʿ al-Tirmidhī*, vol. 3, 456. Riyadh: Darussalam.

Al-Tirmidhī, M. (1996). *Al-Jāmiʿ al-Kabīr Sunan al-Tirmidhī*, Hadith No. 1817, vol. 3, 515. Damascus: Al-Resalah Al-Aʾlamiah Ltd Publishers.

Al-Zabīdī, M.M. (1994). *Tāj al-ʿArūs min Jawāhir al-Qāmūs*, vol. 14, 162. Beirut: Dar al-Fikr.

Badr Al-Din, M. (2000). *Al-Baḥr al-Muḥīṭ fī Uṣūl al-Fiqh*. Beirut: Dar al-Kotob al-Ilmiyah.

Ḥamd, Y.M. (2004). *Al-Qawāʿid al-Fiqhiyya fī Ijtimāʿ al-Ḥalāl wa al-Ḥarām wa Taṭbīqātihā al-Muʿāṣara*, 25. Jordan: Maktaba Jāmiʿa al-Urduniyya.

Ibn alʿArabī, M. (1967). *Aḥkam al-Quran*, vol. 2, 548. Beirut: Dar al-Jil.

Ibn Al-Qayyim, M. (1996). *Iʿlām al-Muwaqqiʿin ʿan Rabb al-ʿĀlamīn*, vol. 1, 259. Beirut: Dal al-Kotob al-Ilmiyah.

Ibn Al-Qayyim, M. (1997). *Aḥkām Ahl al-Dhimma*, vol. 2, 715. Dammam: Ramadi Publications.

Ibn Kathīr, I. (1998). *Tafsīr al-Quran al-ʿAẓīm*. Damascus: Maktaba Dār al-Fīḥāʾ.

Ibn Manẓūr, M. (1994). *Lisān al-ʿArab*, vol. 11, 167. Beirut: Dar Sadir.

Ibn Nujeym, Z. (1998). *Al-Ashbāh wa al-Naẓāʾir*. Karachi: Idaratul Quran wal Ulum al-Islamiyya.

Ibn Qudāma, ʿ.A. (1968). *Al-Mughnī*, vol. 13, 271. Riyadh: Dar ʿĀlam al-Kutub.

Kamali, M.H. (2003). *Principles of Islamic Jurisprudence*, 410. Cambridge: The Islamic Texts Society.

Lane, E.W. (2003). *An Arabic English Lexicon*, vol. 1, 621. Lahore, Pakistan: Suhail Academy.

ʿUthmān, M.H. (2002). *Al-Qāmūs al-Mubīn fī Iṣṭilāḥāt al-Uṣūliyyīn*, 146. Riyadh: Dar al-Zāḥim.

Part II

Animal Welfare and Slaughter

3

Animals in Islam and Halal Ethics

Magfirah Dahlan

Faculty in Religious Studies, Philosophy and Political Science at Craven Community College, North Carolina, USA.

3.1   Introduction

The discourse on halal food is inseparable from the issue surrounding the animals that are being consumed. In addition to the explicitly prohibited animal products, such as pork, the way the animals are slaughtered is central to the practice of keeping halal. For Muslims around the world, eating halal is almost synonymous with eating meat from animals slaughtered by or in the presence of a Muslim pronouncing the name of Allah. In places where Muslims are a minority group, many Muslims would consume only meat obtained from specified halal butchers, or meat products bearing the labels of halal certification. The concerns for the particularity of the slaughter are partly related to the ethical concern for animal welfare. All Muslims believe that the method of slaughter required by Islamic teachings and tradition minimizes the suffering experienced by the animals at the time of slaughter.

A growing number of Muslims, however, have argued that halal food discourse on animals needs to include other ethical concerns for animal welfare beyond minimizing the suffering of the animals at the time of slaughter. This need is of growing importance especially given the current conditions of animal farming systems and meat production prevalent in industrial countries and spreading onto industrializing countries. For these Muslims, it is important to ensure that the meat products they consume come from a production system in which the animals have been treated humanely. They base their view on numerous Prophetic sayings and traditions that speak to the need for Muslims to practice kindness and compassion to animals long before the invention of modern industrial farming. Furthermore, there are also those who argue that Muslims should refrain from eating meat altogether. Those who argue for Islamic vegetarianism believe that fundamental values in Islam, including the kindness and compassion to animals modelled by the (pbuh), are best realized in the practice of vegetarianism. This chapter will explore how this currently minority position fundamentally questions not only the permissibility of consuming animals as food but also human–animal relations in general.

This chapter addresses the different debates on animals in Islam relating to the understanding and practice of halal food. It begins with a discussion of concerns for the animal at the time of slaughter, especially given the context of mechanical slaughter. Muslims disagree on whether the mechanical slaughter procedure that includes stunning animals prior to their slaughter is considered permissible. There continues to be an ongoing debate on the importance of having animals conscious at the time of slaughter and which method (stunning vs non-stunning) causes more pain to the animals. The second part of the chapter addresses the concerns that Muslims have for animals beyond the time of slaughter, particularly in the context of industrial farming. Many Muslims who consider the slaughter requirement important but not sufficient to determine whether the animals consumed are permissible refer to the importance of purity or wholesomeness in addition to permissibility. Finally, the third part of the chapter discusses the argument that challenges the basic assumption that Muslims are permitted to consume animals as food. Those with this view argue for a non-anthropocentric reading of Islamic teachings and traditions, one that radically questions the underlying assumptions of current human–animal relations.

3.2 Halal in the Era of Mechanical Slaughter

As previously mentioned, for many Muslims eating halal is almost synonymous with eating meat from animals slaughtered by or in the presence of a Muslim pronouncing the name of Allah. While we can still find animals consumed as food being hand-slaughtered in many parts of the world, most animals in industrialized countries are raised on intensive farms and slaughtered using mechanical procedures. This poses some challenges for Muslims living in those countries, one of which is the use of electrical shock or stunning on the animals prior to their slaughter by a machine. The procedure for adapting the requirements of halal slaughter to mechanical slaughter is described by Riaz and Chaudry (2004), who stated that animals or birds must be of halal species and alive at the time of slaughter. Slaughtering must be done by a mature Muslim of sound mind, trained in the slaughtering method for the type and size of the animal to be slaughtered. The name of Allah (Bismillah Allahu Akbar) must be verbally invoked by the Muslim slaughter person while slaying the animal. Slaughtering must be carried out on the neck from the front, cutting the oesophagus, trachea, jugular veins, and carotid arteries, without cutting the spinal cord beyond the neck muscle. Slaughtering must be carried out by a sharp knife in a swift sweep so that the animal does not feel the pain of slaying. Blood must be drained out thoroughly and the animal must die of bleeding rather than any other injury, inflicted or accidental.

Concerning the mechanical slaughtering of birds, Riaz and Chaudry (2004) also describe the *adaptation* of the typical halal slaughter to the industrial procedure for mechanical slaughter. They mentioned that the birds must be of halal species (chicken, ducks, or turkey). The slaughter person, while pronouncing Bismillahi Allahu Akbar, starts the machine. The birds are hung on the conveyor railing one at a time without agitating them. The birds are passed over electrified water, which touches the beak to shock them unconscious. A Muslim slaughter person is positioned behind the machine and bleeds the birds missed by the machine while continuously invoking the name of God (two Muslim workers might be required, depending on the line speed). Halal birds are completely segregated (from birds slaughtered the non-halal way) throughout the process.

Many Muslims, however, believe that stunning is not permissible because halal slaughter requires the animal to still be conscious at the time of slaughter. For some, the importance of having the animal stay conscious is more than simply following the tradition or the letter of the religious dietary law. They argue that stunning animals prior to slaughter does not actually decrease, and can in some cases increase, the suffering experienced by them. While the debate continues on whether the stunning or the non-stunning method is more humane for the animals, the concerns for the humane treatment of the animals at the time of slaughter is central to the current halal discourse. Discussions on such concern can be found, for example, in the works of Mukerjee (2014) and Bergeraud-Blackler (2007).

There are other concerns regarding animal welfare apart from whether stunning lessens animal suffering. Islamic teachings contain numerous Prophetic sayings and orders that speak to the different aspects of ensuring humane slaughter, which include both the physical and psychological suffering of the animals. These include, for example, the prohibition of sharpening one's slaughtering knife in front of the animals, slaughtering an animal in the presence of other animals, and beating the animals and causing physical harm prior to the slaughter, as cited in Masri (2009). The conditions of mechanical slaughterhouses, where animals are subject to physical and psychological suffering, are problematic for Muslims who seriously consider these Prophetic sayings and traditions.

3.3 Halal in the Era of Industrial Farming

For some Muslims, the concerns for the well-being of the animals they consume extend beyond the time of slaughter. According to this view, Muslims should be critical of the way animals are treated in industrial farming. In his work *Animal Welfare in Islam*, Masri (2009) argues that the conditions in industrial animal farming are far from conforming to the teachings of Islam. In industrial farms, animals are inhumanely confined in cages and pens, and subjected to unnatural diets, hormones, and antibiotics. Unlike in their natural settings, industrial farming does not allow animals to fulfil their social and behavioural needs, such as the need for chickens to dust-bathe. These practices are contrary to Islamic teachings that speak to the importance of recognizing and respecting the animals' capacity to suffer physical and mental capacities. Numerous Prophetic sayings that speak to the importance of being kind to animals include the example of the (pbuh) reminding his wife, A'ishah, against rough handling of her camel, his condemnation of branding cattle on their face as it is a sensitive part, and his order to his companions to return young birds to their mothers because separation causes mother–bird emotional distress.

Muslims who argue for the need to pay attention to the welfare of animals, in general, believe that, as Muslims, it is important to ensure not only that the animals are slaughtered appropriately, but also that they have been raised in a respectful and humane environment. Some of them have used terms such as 'eco-halal', 'organic halal', or 'green *zabiha*' to capture their concerns about the welfare of animals prior to the time of slaughter (Arumugam 2009; Power 2014). Those who believe in this view utilize the concept of *tayyib* (wholesome or pure) to distinguish it from conventional halal that focuses on the way the animals are slaughtered. For them, Islamic dietary laws require Muslims to consume food that is not only permissible but also wholesome.

Amongst their concerns for animal welfare is the importance of ensuring that the animals have been given natural diet and living conditions. In one farm based on the ethos of tayyib, for example, the owner describes his belief that it is as important to ensure that the chickens he raised 'are fed on a purely natural diet, allowed to grow at a healthy rate, and given bug-filled pastures to explore', as it is to say Bismallah Allah-u-Akbar at the time of slaughter. For him, it is problematic for Muslims to consume chickens raised in industrial farming whose feed legally contains animal by-products, such as animal meal or dehydrated blood (blood meal), including from non-halal animal species such as pigs (Arumugam 2009). For many Muslims with these concerns, the way to ensure that the animals they consume have been raised in humane conditions is to have a close and transparent relationship with the farms on which the animals are raised. They tend to support local small farms, either individually or as a community-supported agriculture group, sometimes called a 'halal-CSA' (Arumugam 2009).

What is commonly shared by Muslims with the differing views that have been discussed thus far is the underlying assumption that it is lawful for Muslims to consume animals. Although those who belong to this position rarely describe themselves as such, those who are critical of the position have described this assumption as anthropocentric speciesism in nature. The next section will discuss a radically different view, which starts by questioning this assumption and employing a fundamentally different reading of Islamic teachings and traditions with regards to human–animal relations.

3.4   Islamic Vegetarianism and Alternative Views of Animals

To be clear, most Muslims are not vegetarians. Some Muslims even consider vegetarianism to be potentially 'un-Islamic'. This is because they view the prohibition of meat-eating as contrary to fundamental islamic teachings. This view is based on Qur'anic passages that specify the lawfulness of food that are not explicitly forbidden, as well as passages that warn against forbidding good things that God has made lawful to Muslims (Qur'an 2:172–173; 5:87–88; see, for example, the discussion in Foltz 2006: 30). While they may accept that Muslims are not obligated to eat meat, they do not think that Muslims should be prohibited from consuming animals as food altogether. In other words, for most Muslims concerns for animal welfare do not override the permissibility of consuming animals as food.

There are those, however, who argue for a radically different reading and interpretation of Islamic teachings and traditions concerning animals. Some of them focus on the value of compassion, which serves as the foundation for such concern. This can be seen, for example, in the argument for Islamic vegetarianism by Foltz (2006) that directly challenges the permissibility of meat consumption widely accepted by Muslims. He argues in his work *Animals in Islamic Tradition and Muslim Cultures* that the principle of compassion that underlies the slaughter practices observed by Muslims can be used to derive an argument for doing away with slaughtering animals altogether. He mentions that the Qur'an and the Sunna have been shown to enjoin Muslims to treat animals with compassion. This is clearly reflected in the established procedure for halal slaughter. It should be obvious, however, that not slaughtering the animals at all would be even more compassionate.

Furthermore, he argues that the historical permissibility of consuming animals in Islamic teachings and traditions must be understood within the context of when the laws originated in order to find the appropriate application of the laws in the current context. While meat-eating was essential in the context of seventh-century Arabia, when the Islamic dietary laws were revealed, he argues that the original intention of the dietary laws is best fulfilled through vegetarianism, considering the current context of modern industrial meat production. He stated that factory farms did not exist in seventh-century Arabia. Traditional Arab pastoralists needed animal products in order to survive, yet their practices did not result in the destruction of the entire ecosystem. For the most part, the early community lacked the vast dietary alternatives available to most Muslims today.

In essence, this argument suggests that given the vast difference between the original contexts of the laws and the current modern context, there is a need to redraw the boundaries between the permissible and the forbidden with regards to consuming animals as food in order to retain the intention or the purpose of the laws.

Another argument that is critical of the mainstream position focuses on the assumption of human superiority and dominion over animals (Tlili 2012). According to this view, the mainstream position that allows animals to be used for human needs, including for food, is anthropocentric and speciesist. Anthropocentrism refers to the belief of the centrality and superiority of humans above non-humans, including animals. Speciesism is the belief that non-human species are different from humans with regard to their status for moral consideration. According to this alternative view, the pervasiveness of anthropocentrism and speciesism in our society makes it difficult for us to see animals as having any value other than to serve our needs and interests. Furthermore, according to this view, Islamic teachings themselves are not anthropocentric or speciesist; rather, Muslims have interpreted the teachings through the anthropocentric or speciesist lens. The challenge is to read the teachings through a radically different lens in order to achieve a better understanding and practice of the religion.

Those who argue for this view emphasize the similarities and shared experiences and characteristics between humans and animals. These similarities and shared characteristics include the complexity of animals' psychology, spirituality, and morality. Those who argue for a non-anthropocentric view place humans as part of the animal kingdom, rather than separate from and superior to other animals. In her work *Animals in the Qur'an*, Tlili (2012) challenges the special status of humans above other creations by arguing that the speciesism that underlies the interpretation of *Khalifa* as vicegerents, giving humans special status above non-human animals, contradicts the principles of God's oneness (*tawhid*) and justice. She critically analyses Qur'anic verses that give permission to use domestic animals and argues that although the Qur'an permits the instrumental use of domestic animals, this permission does not necessarily mean that animals are inferior to humans. She argues that rather than being inherent in the Qur'an, the prevailing belief amongst Muslims that humans are superior to animals is more likely a result of the anthropocentric reading of the Qur'an. She provides an alternative, non-anthropocentric reading of the concepts of subordination (*tadhlil*), servitude (*taskhir*), and vicegerency (*istikhlaf*), and demonstrates that while an anthropocentric reading would interpret the concepts of subordination, servitude, and vicegerency as an indication of human superiority of and dominion over animals, a non-anthropocentric reading would interpret the same concepts in terms of the superiority of God and His dominion over all creation (Tlili 2012).

Ali (2015) discussed the criticism against the hierarchical relationship between human and animals that justifies of the permissibility of consuming animals as food. In her article *Muslims and Meat-Eating: Vegetarianism, Gender, and Identity*, she uses feminist moral philosophy to draw a parallel between the inequality and injustices that characterize the current relationships between humans and animals, and those that characterize the relationships between genders in a patriarchy. She argues that the assumption of hierarchy between humans and animals is closely related to the assumption of hierarchy between genders. With regard to the mainstream view of the permissibility of meat-eating, she stated that this insistence on the potential for meat eating, even if one is choosing to abstain, reflects an attachment to a hierarchical cosmology that subordinates women. It is similar to the model of marriage frequently advocated where husband and wife are expected to typically arrive at agreements on matters after consultation; however, the husband's authority to impose a unilateral decision remains and conditions all prior negotiations. She also argues that, just as some Muslims have begun and continued to use feminism to challenge the patriarchal assumption in Islamic traditions, it is also important for Muslims to practice vegetarianism as part of their stand against the dominant form of injustice.

References

Ali, K. (2015). Muslims and meat-eating: vegetarianism, gender, and identity. *Journal of Religious Ethics* 43 (2): 268–288.

Arumugam, N. (2009). The Eco-Halal Revolution in https://www.patheos.com/blogs/altmuslim/2009/12/the_eco-halal_revolution/ retrieved June 2019.

Bergeraud-Blackler, R. (2007). New challenges for Islamic ritual slaughter: a European perspective. *Journal of Ethnic and Migration Studies* 33 (6): 965–980.

Foltz, R. (2006). *Animals in Islamic Tradition and Muslim Cultures*. Oxford: Oneworld.

Masri, A.B.A. (2009). *Animal Welfare in Islam*. Leicestershire: The Islamic Foundation.

Mukerjee, S.R. (2014). Global halal: meat, money, and religion. *Religions* 5: 22–75. https://doi.org/10.3390/rel5010022.

Power, C. (2014). Ethical, organic, safe: the other side of halal food. *The Guardian* (18 May 2014). http://www.theguardian.com/lifeandstyle/2014/may/18/halal-food-uk-ethical-organic-safe retrieved June 2019.

Riaz, M.N. and Chaudry, M.M. (2004). *Halal Food Production*. Boca Raton, FL: CRC Press.

Tlili, S. (2012). Animals in Islam and Halal Ethics. In: *Animals in the Qur'an* (ed. M. Dahlan-Taylor). New York, NY: Cambridge University Press.

4

Animal Behaviour and Restraint in Halal Slaughter

Temple Grandin

Department of Animal Science, Colorado State University, USA

4.1   Introduction

There are two issues concerning animal welfare and halal slaughter without stunning: welfare concerns about the throat cut and how the live animal is restrained and handled. In chapter 4, pre-slaughter handling, restraint methods to position the cattle, sheep, or goats for slaughter, and best commercial practices will be discussed. Other chapters contain a review of scientific literature on welfare issues concerning throat cutting without prior stunning.

In some countries, highly stressful methods of restraint are used to hold the animal in a position such as shacking and hoisting or shackling and dragging. Westervelt et al. (1976) and Giger et al. (1977) found that suspension of calves by their rear leg was more stressful than being held in an upright position. The World Organisation for Animal Health or Office International des Epizootics (OIE) (2007) states in their slaughter guidelines that methods of restraint that cause pain and stress should not be used on conscious animals. These methods are:

- suspending or hoisting animals (other than poultry) by the feet or legs
- mechanical clamping of the animal's legs (applies to mammals)
- use of electro-immobilization to restrain conscious animals (Electro-immobilization must not be confused with proper electric stunning, which produces unconsciousness by inducing a grand mal seizure (Croft 1956; Croft and Hume 1956; Warrington 1974; Lambooy 1982). Electro-immobilization that does not induce unconsciousness is detrimental to welfare and should not be used (Grandin et al. 1986; Pascoe 1986; Lambooy, 1985).
- cutting leg tendons or blinding animals in order to immobilize them.

This chapter will discuss the proper operation of restraining devices to hold cattle and sheep in position for halal slaughter. Common names for these restraint devices are a box, pen, head holder, rotating box, American Society Prevention of Cruelty to Animals

(ASPCA) pen or upright box. The design of handling and restraint equipment for conventional slaughter with stunning is covered in many other publications (Grandin 1992, 2003, 2007, 2013). This chapter will concentrate on the operation and behavioural principles of low-stress restraint devices for religious slaughter.

4.2   Pre-slaughter Restraining Stress

A question that often gets asked is 'Do animals know they are going to get slaughtered?' During a long career working with slaughter plants all around the world, I have observed that the behaviour of cattle waiting in line in a race at a slaughter plant is usually the same as that of cattle waiting in line to go into the veterinary chute. A review of the literature on physiological stress levels in both cattle and sheep during on-farm handling and during pre-slaughter handling indicates that the cortisol levels in both places are approximately in the same range (Mitchell et al. 1988; Grandin 1997, 2013). The highest and lowest cortisol levels in cattle after conventional slaughter and during on-farm handling ranged from 20 to 60 ng/ml (Grandin 1997).

Another indicator of pre-slaughter stress is vocalization (moos and bellows) in cattle during driving into a restraint device or while being held in a restraint device. Increased vocalization in cattle during restraint was associated with higher cortisol levels (Dunn 1990; Hemworth et al. 2011). Grandin (1998) reported that 99% of cattle vocalizations were associated with an obvious aversive event such as electric goads, excessive pressure from a restraint device or slipping on the floor. A French study also showed that excessive pressure applied by a restraint device was associated with 25% of cattle vocalization (Bourquet et al. 2012). In one plant, reducing the pressure applied by a head restraint reduced the percentage of cattle that vocalized from 23% to 0% (Grandin 2001).

In a well-run slaughter plant, the percentage of cattle that vocalize while held in a restraint device for religious slaughter is 5% or less (Grandin 2012; www.grandin.com). The cattle were scored on a per animal basis as either silent or vocal. Stress from handling during pre-slaughter may be due partially to the novelty of the new environment. The results of two studies indicated that the cattle and sheep that are most reactive to a novel stimulus on the farm have higher cortisol levels after slaughter (Deiss et al. 2009; Bourquet et al. 2010). Sheep do not vocalize in response to painful or frightening events. This may be due to them being a prey species animal whose only defence against predators is flocking. Sheep, however, will vocalize (baa baa) when a lamb is separated from the flock. They respond vocally to separation from the flock and usually remain silent when they are frightened or in pain. Cattle will often vocalize loudly when a painful stimulus, such as excessive pressure from a restraint device, is applied (Grandin 1998). Obviously, cattle cannot vocalize after the throat cut as it separates the larynx from the lungs.

4.3   Benefits of Reduced Pre-slaughter Restraining Stress

Reducing pre-slaughter stress helps to improve both meat quality and animal welfare. Multiple application of electric prods a few minutes before the slaughter of cattle resulted in tougher beef (Warner et al. 2007). Problems with severe bruising from either a poorly designed restraint devices or rough handling have been observed. Animals can be bruised any time before they are bled. Bruising can occur in cattle after they are stunned with a captive bolt (Meischke and Horder 1976). Some of the most common problems with restraint boxes that may cause bruising are rear pusher gates with sharp edges or that apply excessive pressure. Keeping animals calm will also help to reduce petechial haemorrhages in the meat (Grandin and Regenstein 1994). Another benefit of reducing pre-slaughter restraining stress is that animals will be less fearful and more willing to walk into a restraint device that is covered with blood. Blood from cattle that had prolonged stress for 10 or 15 minutes was avoided by other cattle (Boissy et al. 1998). Cattle will willingly enter a restrainer that is covered with blood if the previous cattle held in it had remained calm (Grandin 1992). There is evidence that there is an alarm pheromone in blood from stressed animals (Stevens and Saplikoski 1973). Observations during restraint equipment start-ups indicate that blood, urine, and saliva from cattle that become highly agitated for 10–15 minutes due to equipment malfunctions caused the next cattle in line to baulk and refuse to enter until the equipment was completely washed. Blood from cattle that remained calm did not cause this problem. In one plant, it was observed that cattle often put their heads down and hesitated to walk over a silver portion of the floor where the blood was rubbed off. This behaviour may be similar to the behaviour of animals when they stop where the flooring surfaces changes. Shadows, metal strips, and reflections will often cause animals to baulk (Grandin 1996, 2007).

4.4 Design Requirements for Animal Handling and Restraint Equipment

The design requirements for handling and restraint equipment vary depending on whether or not the cattle or sheep are wild with a high flight zone or tame. When animals are tame and accustomed to close contact with people, simpler, less expensive facilities may work effectively. A system that would be appropriate for cattle trained to lead may be terrible if used with high flight zone extensively reared animals that have seldom been around people. Designs for races and lairages are given in Grandin and Deesing (2008), Grandin (2007), and www.grandin.com. These systems will work for both tame and wild cattle with high flight zones.

4.5 Improving Animal Movement

It is essential that races, pens, and the entrances of restraint devices have non-slip flooring. Animals are often difficult to move if they are slipping. They may also refuse to move if they see distractions. Differences in flooring type, lighting, and shadows can slow down animal movement and increase baulking (Kilgour 1971; Grandin 1996, 2001, 2007; Grandin and Deesing 2008; Klingimair et al. 2011). Removing the distractions will improve movement. If a distraction is impossible to remove, then the animal should be given an opportunity to look at it for a few seconds before attempting to drive it forward. Calm cattle and sheep will show you where distractions are located by orienting both their eyes and ears towards them.

Distractions that slow down animal movement and methods for improving animal move-ment into a restrainer include the following:

- Seeing people walking by the front of the restrainer. Install a metal shield or instruct people to stand where the approaching animals will not see them.
- Reflections on wet floors or shiny metal. Experiment with adding, moving, or changing lights to eliminate reflections.
- Race entrance is too dark. Experiment with installing a lamp at the entrance. Animals do not like entering dark places.
- Animals can see vehicles or moving equipment. Install solid shields to block the animals' vision.
- Air blowing in the faces of approaching animals will cause baulking. To correct this, change the direction of the airflow.
- Air hissing and metal clanging frighten animals. Silence these sounds.

4.6   Use of Driving Aids from Moving Animals

Animal handlers need to be trained to understand the behavioural principles of animal handling. Animals that are trained to lead can be led. More extensively raised animals will have a flight zone. If a handler gets too close to an extensively raised animal while it is standing in a race, it may rear up. The animal will stop rearing if the handler backs away. When a handler moves an animal forward in a race, he should stand behind the point of balance at the shoulder (Figure 4.1). The handler should avoid the common mistake of

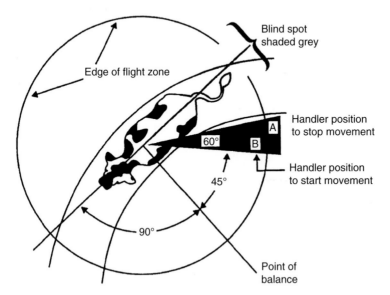

Figure 4.1   Handlers need to understand the animal's flight zone and point of balance at the shoulder. To move an animal forward, the handler must stand behind the point of balance. Quickly walking from the animal's head towards its tail will often induce the animal to move forward when the point of balance is crossed. Diagram courtesy of Temple Grandin.

standing at the head of an animal and poking it on the rear. Animals will often move forward when the handler walks back past the shoulder in the opposite direction of the desired movement.

An electric goad must never be a person's primary driving tool. If an electric goad is required to move a stubborn animal, it should be picked up, used once, and then put away. Small flags or plastic bags are more suitable livestock-driving aids. Handlers should remain quiet. Yelling at cattle is very stressful (Waynert et al. 1999; Hemworth et al. 2011).

4.7 Design of Restraint Devices

There are two basic types of restraint devices for halal slaughter that provide an acceptable level of welfare for large cattle. They are a box with a head holder where the animal stands in an upright position or a rotating box that inverts an animal onto its side or back. Inversion onto the back for 90 seconds resulted in higher cortisol levels than use of an upright restraint (Dunn 1990). Velarde et al. (2014) reported that cattle vocalized less in an upright box compared to a rotating box. Struggling was greater in the upright box. Unfortunately, Velarde et al. (2014) did not differentiate between struggling before and after the loss of consciousness. Struggling is a welfare concern before the loss of consciousness and not a welfare concern after the animal loses consciousness and the ability to stand. In another study, the relative aversiveness of different handling treatments was tested in a Y-maze. Sheep avoided the race that led to a device that inverted them (Rushen 1986). This is why it is important to perform the cut immediately after inversion. Another major advantage of a box where the animal stands or is held in an upright position is low cost. An economical upright box which requires no pneumatic or hydraulic cylinders is being used successfully in Indonesia for tame cattle. When an upright box is equipped with hydraulic or pneumatic cylinders it is still much more economical than a rotating box. The mechanism for full inversion of a box for holding adult cattle is very costly. When a rotating box is used, it must fully support the body with an adjustable side. This will prevent the animal from slipping and becoming agitated during rotation. In large sheep plants, a V-conveyor restrainer is often used. In this system, the sheep are held between two conveyors that form a V. For slaughter without stunning, the conveyor is stopped for each animal. After the throat is cut, it is ejected onto another moving conveyor. When the chin lift is used on a conveyor system it parts into two pieces so the animal can pass through it (Grandin 1993; Derouin 2003). The two biparting chin lift pieces are attached to two sliding panels. These sliding panels open like biparting sliding doors used in supermarkets or other stores. A similar system has also been used in the USA for sheep. For small plants, in a well-designed sheep system, the sheep straddles a metal frame which fully supports the body. Sheep can also be easily restrained by a person straddling the animal's body and holding up the head for the throat cut; Velarde et al. (2014) reported that hoisting sheep and goats on a shackle was a common practice. Sheep hanging on the shackle took longer to lose the righting reflex (hypertonic).

Further details of holders and lifts can be found at http://www.grandin.com/index.html.

4.8 Basic Restraint Principles

There are basic principles that apply to the design and operation of all types of restraint equipment. Restraint devices should be designed to minimize vocalization and struggling (OIE 2007). If animals vocalize or struggle before losing consciousness while they are restrained, there is a problem that needs to be corrected. Causes of vocalization or struggling are listed below. Problems that cause animals to vocalize (moo or bellow) or struggle while still conscious when they are held in a restraint device should be eliminated.

- *Avoid excessive pressure.* To correct this, install pressure-limiting regulators on hydraulic or pneumatic powered equipment. The restraint device should automatically stop before it applies excessive pressure (Grandin 1992). Different parts of a restraint device require different amounts of pressure. Cylinders that operate the head holder require much less hydraulic or pneumatic pressure than systems that rotate the box, or open and close heavy gates. On many systems, three different pressure regulators will be required to prevent excessive pressure from being applied to the animal. The head holder should be set to the lightest maximum pressure. Gates set at a medium pressure and rotation or tilt mechanisms may need higher settings.
- *Remove pinch points.* When the animal's skin is pinched, it is likely to struggle or vocalize. Find the pinch point and eliminate it.
- *Remove sharp edges.* A very small sharp edge may cause pain and cause cattle to bellow. Remove sharp edges with a grinder or change the design. Head holders should be constructed from smooth, round rod or pipe.
- *Avoid excessive neck bending.* Bending the neck too much may cause pain. The best head holders will hold the throat tight without bending the neck.
- *Prevent the body of the animal from slipping during rotation.* When the box is rotated, the animal's body should not slip or slide. The box should have an adjustable side and a backrest to support the body.
- *Reduce jerky sudden motion.* Sudden motion frightens animals. On hydraulic or pneumatic powered equipment, flow (speed) controls should be installed to prevent sudden rapid movement of the head holder or other parts of the restrainer. The best hydraulic or pneumatic systems provide positive mid-stroke position control of the cylinders that control the head holder and other parts of the restrainer. Simple hand-operated hydraulic or pneumatic valves often work better than electric solenoid valves. The best hand-operated valves enable the box operator to smoothly control the speed of the head holder similar to a car's accelerator (Grandin 1992). Hand-operated valves also provide better mid-stroke position control of the cylinders that move the various parts of the restrainer. When the operator has mid-stroke position control, is easier for the operator to avoid applying excessive pressure. The restrainer should still be equipped with pressure-limiting devices to prevent a careless operator from applying excessive pressure.
- *Prevent bruises.* If the carcasses of many animals from many different origins have bruises in the same location, it is likely the bruise is happening at the plant. A common cause of bruises is a poorly designed rear pusher gate. Fresh bruises that occurred on the same day of slaughter will have a fresh red appearance. Older bruises that are several days old will often

have yellow mucous. Animals can have severe bruises and the hide may appear normal and undamaged. If the restraint device is causing bruises, the problem should be corrected. Bruises can occur up until the animal's blood pressure is reduced by the loss of blood.

4.9 Best Commercial Practices

Good equipment is essential to have an acceptable level of animal welfare but it must be accompanied by good management. Below is a summary of best practices.

- *Keep animals calm.* A calm animal will usually lose sensibility faster.
- *Record time to collapse or eye rollback.* Animals become unconscious when they lose the ability to stand (Benson et al. 2012). Another term for this is a loss of posture. When slaughter without stunning is correctly performed in an upright restraint box over 90% of the cattle will lose the ability to stand within 30 seconds (Grandin 2010a). Gregory et al. (2010) reported that 90% collapsed in 34 seconds. Continuous monitoring of collapse times will enable operators to continuously improve and shorten the collapse times. In rotating boxes, time to eye roll back can be used. Sheep will become unconscious more quickly than cattle due to differences in blood vessel anatomy (Baldwin and Bell 1963a,b,c).
- *Vocalization scores.* If over 5% of the cattle vocalize in the restraint box, or while entering it, there is a problem that needs to be corrected (Grandin 2012). Vocalization is a score on a yes/no per animal basis. Each bovine is either silent or vocalizes. Continuous measurement of the percentage of cattle vocalizing will enable continuous improvements to reduce vocalization. Do not use vocalization scoring for sheep. Vocalization scoring criteria for goats need to be developed.
- *Slaughter promptly.* To reduce stress on animals of all species, the throat cut should be performed promptly after the head is restrained. In rotating boxes, slaughter should be performed immediately after inversion.
- *Long sharp knife.* To prevent gouging of the wound; use a knife that is sufficiently long so that the throat can be cut without the tip of the knife entering the wound (OIE 2007). The knife should be sharp enough to pass a paper test. A paper test is performed by dangling a single sheet of standard A4 printer paper by one corner in one hand. A dry knife held in the other hand should easily slice through the paper. The author has observed problems with halal slaughter being performed with either dull knives or knives that are too short. Cattle and sheep may struggle violently if the wound closes back over the knife or the end of the knife gouges in the wound. Issues concerning painfulness of the cut will be addressed in other chapters. Field observations by the author indicate that when a razor-sharp long knife is used, cattle have a less behavioural reaction to the cut compared to a person deeply invading their flight zone by waving their hands at the animal's head (Grandin 1994; Grandin and Regenstein 1994).
- *Cut in the C1 position.* Researchers have learned that cutting close to cervical vertebrae 1 (C1) will help reduce problems with arteries sealing off in cattle (Gregory et al. 2011). Another advantage of cutting in the C1 position is that it will cut a sensory nerve, which helps to prevent aversive sensations from the aspiration of blood into the trachea

after the throat cut. Some cattle aspirate blood after slaughter without stunning (Gregory et al. 2009).

- *Guidance for plants that choose to use pre- or post-cut stunning.* If either pre- or post-cut stunning is used, good maintenance of equipment is essential (Nakyinsige et al., 2013). There are stunning methods that are acceptable from an Islamic perspective. Poor maintenance is a major cause of captive bolt gun failure (Grandin 1998). When the captive bolt is used, the heart will continue to beat for several minutes (Vimini et al. 1983). Properly applied electrical stunning induces unconsciousness by passing an electric current through the brain. The current must have sufficient amperage to induce a grand mal epileptic seizure (AVMA 2013) (Croft 1952). When head-only electric stunning is used, the period of unconsciousness is temporary (Blackmore and Newhook 1982). After head-only stunning the heart still keeps beating (Gilbert and Devine 1982; Weaver and Wotton 2008). If slaughter is delayed, the animal will fully recover. Head-only electric stunning is used for the halal slaughter of cattle and sheep in many countries.
- *Use upright restraint.* The use of a system that holds the animal in a comfortable upright position is strongly recommended to help reduce stress on the animal. Upright restraint systems also have the advantage of being much less expensive.
- *Restrict or eliminate electric prods (goads).* There is a problem that needs to be corrected if electric prods need to be used on a high percentage of animals. Causes of high electric prod use are either untrained employees or animals that refuse to move forward and repeatedly baulk. If electric prods are used, they should be battery operated and only applied to the hindquarters of the animal. The OIE (2007) states that electric prods should not be used on sheep and small calves.
- *No acts of abuse.* People moving cattle into the restraint box must never beat animals or poke them in sensitive locations such as the eyes, anus, nose, ears, genitals, or udder. Dragging of fully conscious animals is forbidden in most codes of practice.
- *Prevent falling.* If more than 1% of the animals fall down in any part of the yard, unloading ramp, race, or during entry into the restraint box there is a problem that should be corrected.

4.10 Auditing Animal Handling and Slaughter

People manage the things they measure. It is important to keep measuring and assessing handling and slaughter practices to maintain high standards (Grandin 2005, 2010b). People tend to slip back into old bad practices unless they are continuously evaluated. It is just like traffic laws. The police have to keep measuring speeding and giving out fines to keep cars moving at a reasonable speed. The author has worked with numerous restaurant companies in implementing animal welfare auditing programmes. These programmes have resulted in big improvements (Grandin 2000, 2010b). The best auditing systems have three components: internal audits done by the company welfare officers, third-party independent audits done by an outside auditing company, and additional audits done by a major meat buyer. To solve the problem of people 'acting good' when they are being watched, two major US meat companies use video auditing. A third-party auditing company watches the video on the internet. The following criteria should be numerically scored:

1) *Stunning.* If pre- or post-stunning is used, the percentage of animals rendered unconscious with one application of the stunner should be tabulated. In a well-run plant, a score of 95% or more of the animals rendered unconscious with a single shot can be easily attained. For electric stunning, 99% correct electrode placement is easily attainable. The electrodes must be placed so that the electric current passes through the brain and induces an epileptic seizure (Lambooy and Spanjaard 1972; Lambooy 1982; Gregory and Wotton 1984; Cook et al. 1991).
2) *Vocalization.* Vocalization should be present in 5% or less of the cattle in the restraint box and while they are entering it. Do not use vocalization scoring for sheep. Vocalization targets for goats need to be established (Grandin 2012).
3) *Falling.* 1% or less of the animals anywhere in the entire facility.
4) *Electric goad use.* The target is either 0 or 5% or less of the animals to be moved with an electric prod. Twenty-five percent is the absolute maximum. Score as yes/no: touched with electric goad or not touched. Every touch with the electric goad should be counted because it is impossible to determine if the goad was energized.
5) *Time to collapse or eye rollback.* Ninety percent or more of the cattle collapse or eye roll back within 30 seconds. Sheep and goats should collapse within 15 seconds.
6) *Knife passes paper test.* Described in this chapter.
7) *No acts of abuse.* Described in this chapter
8) *Unconsciousness.* All animals should be fully unconscious and insensible before invasive dressing procedures are started. All the animals must be 100% insensible and unconscious before skinning, leg removal or other invasive procedures are started. Information on determining insensibility is in AVMA (2013) and (Grandin 2010b). The corneal reflex must be absent before dressing procedures start. The corneal reflex may be present in some animals that are unconscious (Vogel et al. 2011).

4.11 Conclusions

Halal slaughter without stunning can be performed with an acceptable level of welfare but it requires more attention to detail in the procedure than conventional slaughter with pre-slaughter stunning. Cattle present greater welfare concern compared to sheep or goats. This is because they are large and more difficult to restrain and take longer to lose sensibility after slaughter without stunning compared to sheep. Observations by the author in many plants indicate that the use of pre-slaughter stunning will improve animal welfare.

References

AVMA (2013). *Guidelines for the Euthanasia of Animals, 2013 Edition*. Schaumberg, IL: American Veterinary Medical Association.

Baldwin, B.A. and Bell, F.R. (1963a). The effect of temporary reduction in cephalic blood flow on the EEG of sheep and calf. *Electroencephalography and Clinical Neurophysiology* 15: 465–473.

Baldwin, B.A. and Bell, F.R. (1963b). The anatomy of the cerebral circulation of sheep and ox; the dynamic distribution of the blood supplied by the carotid and vertebral arteries to cranial regions. *Journal of Anatomy London.* 97: 203–215.

Baldwin, B.A. and Bell, F.R. (1963c). Blood flow in the carotid and vertebral arteries of the sheep and calf. *Journal of Physiology* 167: 448–462.

Benson, E.R., Alphin, R.L., Rankin, M.K. et al. (2012). Evaluation of EEG based determination of unconsciousness vs. loss of posture in broilers. *Research in Veterinarian Science* 93: 960–964.

Blackman, N., Cheetham, K., and Blackmore, D.K. (1986). Differences in blood supply to the cerebral cortex between sheep and calves. *Research in Veterinary Science* 40: 252–254.

Blackmore, D.K. and Newhook, J.C. (1982). Electroencephalographic studies of stunning and slaughter of sheep and calves. Part 3: The duration of insensibility induced by electrical stunning in sheep and calves. *Meat Science* 7: 19–28.

Boissy, A., Terlow, C., and LeNeindre, P. (1998). Presence of pheromones from stressed conspecifics increases reactivity to aversive events in cattle, evidence for the existence of alarm substances in urine. *Physiology and Behavior* 4: 489–495.

Bourquet, C., Deiss, V., Gobert, M. et al. (2010). Characterizing emotional reactivity of cows to understand and predict their stress reactions to slaughter procedures. *Applied Animal Behaviour Science* 125: 9–21.

Bourquet, C., Deiss, V., Tannugi, C.C., and Terlouw, E.M. (2012). Behavioral and physiological reactions of cattle in a commercial abattoir, relationships with organizational aspects of the abattoir and animal characteristics. *Meat Science* 68: 158–168.

Cook, C.J., Devine, C.E., Gilbert, C.E. et al. (1991). Electroencephalograms and electrocardiograms in young bulls following upper cervical vertebrae to brisket stunning. *New Zealand Veterinary Journal* 39: 121–125.

Croft, P.S. (1952). Problems with electric stunning. *Vet Rec* 64: 255–258.

Croft, P.S. (1956). Problems with electric stunning. *Veterinary Record* 64: 255–258.

Croft, P.G. and Hume, C.W. (1956). Electric stunning of sheep. *Veterinary Record* 68: 318–321.

Deiss, V., Temple, D., Ligout, S. et al. (2009). Can emotional reactivity predict stress responses at slaughter in sheep? *Applied Animal Behaviour Science* 119: 193–202.

Derouin, K.L. (2003). Headgate assembly for restraining livestock. US Patent 6,537,145 B1, US Patent Office, Washington, DC.

Dunn, C.S. (1990). Stress reactions of cattle undergoing ritual slaughter using two methods of restraint. *Veterinary Record* 126: 522–535.

Giger, W., Prince, R.P., Westervelt, R.G., and Kinsman, D.M. (1977). Equipment for low stress animal slaughter. *Transactions of the American Society of Agricultural Engineers* 20: 571–578.

Gilbert, K.V. and Devine, C.E. (1982). Effect of electric stunning methods ion petechial hemorrhages and on blood pressure of lambs. *Meat Science* 7: 197–207.

Grandin, T. (1988). Double rail restrainer conveyor for livestock handling. *Journal of Agricultural Engineering Research* 41: 327–338.

Grandin, T. (1992). Observations of cattle restraint devices for stunning and slaughtering. *Animal Welfare* 1: 85–91.

Grandin, T. (1993). Handling and welfare of livestock in slaughter plants. In: *Livestock Handling and Transport* (ed. T. Grandin), 289–311. Wallingford: CABI Publishing.

Grandin, T. (1994). Euthanasia and slaughter of livestock. *Journal of the American Veterinary Medical Association* 204: 1354–1360.

Grandin, T. (1996). Factors that impede animal movement at slaughter plants. *Journal of the American Veterinary Medical Association* 209: 757–759.

Grandin, T. (1997). Assessment of stress during handling and transport. *Journal of Animal Science* 75: 249–257.

Grandin, T. (1998). The feasibility of using vocalization scoring as an indicator of poor welfare during slaughter. *Applied Animal Behaviour Science* 56: 121–128.

Grandin, T. (1998). Objective scoring of animal handling and stunning practices in slaughter plants. *Journal of the American Veterinary Medical Association* 212: 36–93.

Grandin, T. (2000). Effect of animal welfare audits of slaughter plants by a major fast food company on cattle handling and stunning practices. *Journal of the American Veterinary Medical Association* 216: 848–851.

Grandin, T. (2001). Cattle vocalizations are associated with handling and equipment problems in slaughter plants. *Applied Animal Behavior Science* 71: 191–201.

Grandin, T. (2003). Transferring results from behavioral research to industry to improve animal welfare on the farm, ranch and slaughter plants. *Applied Animal Behaviour Science* 81: 216–228.

Grandin, T. (2007). Handling and welfare of livestock in slaughter plants. In: *Livestock Handling and Transport* (ed. T. Grandin), 329–353. Wallingford: CABI Publishing.

Grandin, T. (2010a). Improving livestock, poultry, and fish welfare in slaughter plants with auditing programs. In: *Improving Animal Welfare: A Practical Approach* (ed. T. Grandin), 160–185. Wallingford: CABI Publishing.

Grandin, T. (2010b). Auditing animal welfare at slaughter plants. *Meat Science* 86: 56–65.

Grandin, T. (2012). Developing measures to audit welfare of cattle and pigs at slaughter. *Animal Welfare* 21: 351–356.

Grandin, T. (2013). Making slaughter houses more humane for cattle, pigs, and sheep. *Annual Reviews of Animal and Veterinary Sciences* 2: 491–512.

Grandin, T. and Deesing, M. (2008). *Humane Livestock Handling*. North Adams, MA: Storey Publishing.

Grandin, T. and Regenstein, J.M. (1994). Religious slaughter and animal welfare: a discussion for meat scientists. In: *Meat Focus International*, 115–123. Wallingford: CAB International.

Grandin, T., Curtis, S.E., and Widowski, T.M. (1986). Electro-immobilization versus mechanical restraint in an avoid-avoid choice test. *Journal of Animal Science* 62: 1469–1480.

Gregory, N.G. and Wotton, S.B. (1984). Sheep slaughtering procedures, III head to back stunning. *Meat Science* 140: 570–575.

Gregory, N.G., Von Wenziawowcz, M., and Von Holleben, K. (2009). Blood in the respiratory trait during slaughter with and without stunning of cattle. *Meat Science* 82: 13–16.

Gregory, N.G., Fielding, H.R., Von Wenzlawowicz, M., and Von Holleben, K. (2010). Time to collapse following slaughter without stunning of cattle. *Meat Science* 85: 66–69.

Gregory, N.G., Schuster, P., Mirabito, L. et al. (2011). Arrested blood flow during false aneurysm formation in the carotid arteries of cattle slaughtered with and without stunning. *Meat Science* 90: 368–372.

Hemworth, P.H., Ric, M., Karlen, M.G. et al. (2011). Human–animal interactions at the abattoir, relationships between handling and animal stress in sheep and cattle. *Applied Animal Behaviour Science* 135: 24–33.

Kilgour, R. (1971). Animal handling in works: pertinent behavior studies. *Proceedings of the 13th Meat Industry Research Conference*, Hamilton, New Zealand, pp. 9–12.

Klingimair, K., Stevens, K.B., and Gregory, N.G. (2011). Luminaire and glare in indoor handling facilities. *Animal Welfare* 20: 263–269.

Lambooy, E. (1982). Electric stunning of sheep. *Meat Science* 6: 123–135.

Lambooy, E. (1985). Electro-anesthesia or electro-immobilization of calves, sheep and pigs by Feenix Stuckstill. *Veterinary Quarterly* 7: 120–126.

Lambooy, E. and Spanjaard, W. (1972). Electric stunning of veal calves. *Meat Science* 6: 15–25.

Meischke, H.R.C. and Horder, J.C. (1976). A knocking box effect on bruising in cattle. *Food Technnology in Australia* 28: 369–371.

Mitchell, G., Hattingh, J., and Ganhao, M. (1988). Stress in cattle assessed after handling, transport, and slaughter. *Veterinary Record* 123: 201–205.

Nakyinsige, K., Cha Man, Y.B., Agywan, Z.A. et al. (2013). Stunning and animal welfare Islamic and scientific perspectives. *Meat Science* 95: 352–361.

Pascoe, P.J. (1986). Humaneness of electro-immobilization unit for cattle. *American Journal of Veterinary Research* 10: 2252–2256.

Rushen, J. (1986). Aversion of sheep for handling treatments – paired choice studies. *Applied Animal Behaviour Science* 18: 363–370.

Stevens, D.A. and Saplikoski, N.J. (1973). Rats reactions to conspecific muscle and blood evidence for alarm substances. *Behavioral Biology* 8: 75–82.

Velarde, A., Rodriguez, P., Dalmau, A. et al. (2014). Religious slaughter: evaluation of current practices in selected countries. *Meat Science* 96: 278–287.

Vimini, R.J., Field, R.A., Riley, M.L., and Varnell, T.R. (1983). Effect of delayed bleeding after captive bolt stunning and heart activity and blood removal in lambs. *Meat Science* 7: 197–207.

Vogel, K.D., Badtram, G., Claus, J.R. et al. (2011). Head only stunning followed by cardiac arrest electrical stunning is an effective alternative to head only stunning in pigs. *Meat Science* 89: 1412–1418.

Warner, R.D., Ferguson, D.M., Cottrell, J.J., and Knees, B.W. (2007). Acute stress induced by preslaughter use of electric prodders causes tougher meat. *Australian Journal of Experimental Agriculture* 47: 782–788.

Warrington, P.D. (1974). Electrical stunning: a review of the literature. *Veterinary Bulletin* 44: 617–633.

Waynert, D.E., Stookey, J.M., Schartzkopf-Genswein, J.M. et al. (1999). Response of beef cattle to noise during handling. *Applied Animal Behavior Science* 62: 27–42.

Weaver, A.L. and Wotton, S.B. (2008). Jarvis beef stunner, effects of prototype chest electrode. *Meat Science* 81: 51–56.

Westervelt, R.G., Kinsman, D., Prince, R.P., and Giger, W. (1976). Physiological stress measurement during slaughter of calves and lambs. *Journal of Animal Science* 42: 831–834.

World Organisation for Animal Health (OIE) (2007). *Terrestrial Animal Health Code: Guidelines for the Slaughter of Animals for Human Consumption*. Paris: Organization mundial de la Sante Animals (World Organization and Animal Health).

5

A Practical Guide to Animal Welfare during Halal Slaughter

Mehmet Haluk Anil

Formerly at Bristol University, based Bristol, UK

5.1 Animal Welfare During Primary Production and Transportation

There is evidence in available religious recommendations (Hadith) for the requirements to humanely treat animals in Islam. To meet halal rules, ensuring good animal welfare must surely begin at the time of primary production and optimum husbandry practices should prevail. This should involve good facilities, housing, appropriate stocking density, and humane handling during production as well as loading, transport, and unloading. These days much emphasis is placed on these issues and legislation requires good animal welfare maintenance. The following principles can be observed to maintain good animal welfare (welfare quality):

1) Animals should not suffer from prolonged hunger, e.g. they should have a suitable and appropriate diet.
2) Animals should not suffer from prolonged thirst, i.e. they should have a sufficient and accessible water supply.
3) Animals should have comfort when they are resting.
4) Animals should have thermal comfort, e.g. they should neither be too hot nor too cold.
5) Animals should have enough space to be able to move around freely.
6) Animals should be free of injuries, e.g. skin damage and musculoskeletal disorders (e.g. lameness).
7) Animals should be free from disease
8) Animals should not suffer pain induced by inappropriate management, handling, slaughter, or surgical procedures (e.g. castration, dehorning).
9) Animals should be able to express normal, non-harmful, social behaviours (e.g. grooming, preening).
10) Animals should be able to express other normal behaviours, e.g. species-specific natural behaviours such as foraging.

11) Animals should be handled well in all situations, e.g. handlers should promote good human–animal relationships.
12) Negative emotions such as fear, distress, frustration, and apathy should be avoided, and positive emotions such as security and contentment should be promoted.

Regarding poultry, modern production facilities in the UK are fully integrated, with production from the breeder flocks, hatcheries, growing units, and food mill through to transport, slaughter, and processing controlled by the producers. This has resulted in more accountability at all levels of poultry production and measurements taken during production can be readily passed on up the production chain. Subsequently processing information can likewise be fed back down the integrated chain. The following conditions are causes of welfare problems in poultry: daily mortality, gait abnormalities due to abnormal bone formations, broken wings and bones, hock burn and foot pad pododermatitis caused by poor litter, and bruising in broilers, and high stocking density, feather pecking, old and new broken bones, poor lighting, and depopulation damage in hens.

Welfare problems during transport can include dead on arrival (DOA), high stocking density, metabolic exhaustion, dehydration, emotional stress, and temperature stress.

5.2 Pre-slaughter Handling

On arrival at a slaughter house welfare assessment must be carried out. The scoring system shown in Table 5.1 can be used.

The following areas and procedures should be checked/observed:

- unloading bay and animal transfer from lorries
- passage to lairage pens
- layout/construction design/thermal environment
- lairage pens
- isolation pen
- passage to slaughter area/race
- stunning/slaughter pen/area
- stunning/slaughter operation
- office set up
- standard operating procedures (SOPs) (new requirement by law)
- records of welfare problems/actions.

Table 5.1 Welfare assessment scores.

1	2	3	4	5
No welfare problems	Minimal welfare problems without causing distress	Considerable welfare compromise causing discomfort and distress	Significant welfare problems that would cause quality problems and need addressing	Unacceptable welfare conditions and practices

The smooth transfer of animals from lorries to lairage pens must be carried out by experienced trained handlers. Any injured animals need prompt attention. Hitting animals is not allowed and any form of coercion must be minimized. Different groups of animals should not be mixed.

5.3 Restraint During Slaughter

Whether for conventional or religious slaughter, animals need to be transferred from the lairage pens either directly or through a race into the stunning and slaughter area (European Community 1993; Anil and Lambooij 2009). In order to facilitate slaughter, and also to protect the operatives, some form of restraint is necessary. Restraint should allow the correct application of stunning equipment, if used, facilitate access for neck cutting and rapid bleeding, protect animal welfare, and protect the operatives, especially from large animals. Restraint could be achieved by the following methods:

- *Manual restraint in an open pen.* This is usually done by manually handling the free-standing animal in an open area or a pen. Animals can enter the pen either directly from holding areas or through raceways. Electrical or captive bolt stunning in sheep and religious slaughter can be carried out this way. However, safety and welfare issues pose problems with this method, especially with cattle.
- *Restraint in a squeeze/crush pen.* This involves holding the animal by lateral pressure from both sides. Usually one side moves. This method is not commonly used.
- *V-type restrainer.* Animals are held in a funnel-shaped apparatus which usually has a conveyor system. It seems to work well for sheep that can be electrically stunned, either head-only or head-to-back, at the end of the conveyor either manually or automatically. The same system or a stationary restrainer can be used for halal slaughter of sheep without stunning.
- *Monorail restrainer.* This system holds the animal in a straddle position over a rail. Combined with a conveyor system, animals are moved to the point of stunning with possibly less stress than with a V-type restrainer. This method is commonly used for cattle and calves in North America.
- *Cattle stunning pen.* Cattle often present with more handling, safety, and welfare problems before and during slaughter than other species. Different designs of cattle restraint pens exist so that stunning and slaughter can be carried out effectively and safely. Animals usually enter the pen after going through a race. Pens must have gates to close after entry. Races should have smooth curved sides if long and have sufficient light. The use of prods should be minimized. Cattle pens aim to incorporate facilities to present the head for correct stunning at the front. Some cattle pens are specially constructed for captive bolt, electrical stunning, and/or religious slaughter for neck cutting without stunning. Upright and Facomia pen designs have special functions for extra restraint such as belly lift, back push, and chin-lift. A Facomia pen (Figure 5.1) rotates the animal around 45°. Old-fashioned rotary pens that turn the animal 180° are regarded as being more stressful and are banned in the UK and the USA. However, a recent report for the European Commission concluded that more scientific evidence is needed to substantiate the claims, therefore rotary pens are still in use in the rest of Europe.

Figure 5.1 Facomia rotating pen (45°).

The recent European Council Regulation ((EC) No 1099/2009) (European Community 2009) now requires mechanical restraint of ruminants. Regarding the choice of pens, a commissioned study of cattle restraint systems is expected to be submitted and likely to confirm banning of complete rotation. Some existing rotating restraint apparatus employed for cattle could have inherent undue stress factors. In particular, periods of restraint before and after a neck cut can be long in some rotary systems that turn 180°.

Regarding the restraint of sheep for religious slaughter, either a cradle or a V-type restraining device can be employed. In the former a mechanical cradle is used, allowing individual lifting and placement of animals in a horizontal position before neck cutting. Following the neck cut the animal has to be held until the end of the legally prescribed period (see below) before release. This is a permissible method in the UK. However, disadvantages include the stress of pre-slaughter handling and potential carcass damage as well as slow operation.

Although more in-depth detail regarding halal slaughter is presented in the next section, slaughter/bleed-out method need to be referred to in relation to restraint. Neck cutting can be made within a V-type restraining conveyor or at the exit point. If stunning is employed it is usually applied at the exit, followed by neck cutting in the horizontally positioned animal on a moving conveyor before shackling and hoisting. In regard to neck cutting without stunning, both EU and UK regulations require a time period during which no manipulation is applied to the animals. In the former regulation (WASK 1995), this interval is 20 seconds for sheep (30 seconds for cattle) to ensure no signs of recovery before death are present, whereas the 1099/2009 EC regulation does not specify a figure and instead regular checks have to be made.

Poultry slaughter is usually carried out on shackled birds (Figure 5.2). If stunning is used, birds are lowered into a bath where an electric current is passed through the head and the shackle. Otherwise neck cutting can be performed on the shackled bird (Jewish poultry slaughter is carried out on hand-held birds).

5.4 Religious Slaughter Methods: Halal Method

Religious slaughter has been a controversial issue and has received much attention in recent decades (Anil 2012). Increases in the Muslim populations in European countries

(a)

(b)

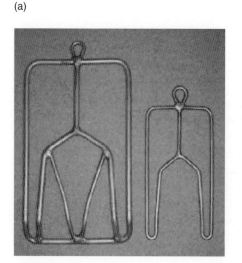 

Figure 5.2 (a) Poultry slaughter restraint equipment and (b) shackled birds on-line.

have increased demands from consumers for authentic halal products as required by religious rules. In the meantime, the debate on aspects of halal slaughter has intensified, especially in regard to animal welfare, involving both secular and Muslim groups. Objections and demands for changes to current practices and legislation have also become more frequent. Most religious slaughter in Europe and Western countries, where it is allowed by law, is carried out mostly by Muslim/halal methods and to a lesser extent by Jewish (*shechita*) methods. The EC funded a project, called Dialrel, which involved consultations with interested parties, collecting relevant information whilst stimulating a debate about religious slaughter (www.dialrel.eu).

Legislation in most European countries requires pre-slaughter stunning, rendering the animal unconscious prior to slaughter by exsanguination (bleeding) (Velarde et al. 2003). There are exemptions for religious slaughter methods (Ferrari and Bottoni 2010). Several countries in Europe (in the EU and others) do not allow slaughter without stunning (e.g. Sweden, Denmark, Norway, and Switzerland). In the UK, where this exemption exists, the Farm Animal Welfare Council published a report on religious slaughter methods in 2003 (FAWC 2003) on the welfare of livestock when slaughtered by religious methods. This report recommended that the UK government should repeal the exemption, but this was rejected. Current legislation (WASK 1995; to be replaced by Welfare of Animals at the Time of Killing [WATOK]) allows these practices with or without pre-slaughter stunning. It is likely that this exemption will continue in the near future.

This chapter focuses on animal welfare aspects of halal slaughter, but inevitably includes some references to shechita (Jewish slaughter) because of common legislation and their similarities and differences.

Concerns and discussions about religious slaughter relate to three aspects:

i) the stress of handling and restraining procedures prior to religious slaughter (Dunn 1990; Grandin 1994; Grandin and Regenstein 1994)

ii) potential pain during the neck cut and/or immediately afterwards (Gibson et al. 2009a,b)

iii) delayed loss of sensibility and consciousness following exsanguination (bleeding after neck cutting/sticking) (Daly et al. 1988; Kalweit et al. 1989; Grandin and Regenstein 1994; Anil et al. 1995a,b; Rosen 2004).

In regard to handling, as no strict specific religious requirements exist to adhere to, the first question also applies to all other conventional methods of slaughter. There are, however, anecdotes of traditional practices, such as tying and shackling the legs of live sheep and cattle before religious slaughter, that are of concern. Aspects of pre-slaughter restraint and handling were addressed in the previous section.

The second and third aspects need to be addressed together as they are interrelated. Scientific methods to measure 'pain' used to have limitations and no 'proof' was obtained to answer the second question conclusively until recently. However, some past studies involving measurement of evoked responses and brain activity reported early loss of sensibility (10–20 seconds) following incision although others claimed delays lasting up to 2 minutes. Objections to these findings have been made on the grounds that possible sensations did not necessarily mean pain. A scientific team in New Zealand has recently developed a new technique to study pain in slaughtered animals. Their publications reported analysis of EEG patterns in calves following neck cutting (Gibson et al. 2009a,b) and concluded that ventral neck cutting results in responses to painful stimuli, in particular when blood vessels are cut. In spite of the evidence results are disputed. Grandin and Regenstein (1994) reported that they noticed no visible reaction from the body and legs of cattle to the neck cut, provided that animals were restrained without stress in upright pens, but only a slight flinch where the blade made contact (Figure 5.3). However, reactions may be masked by the position of the animal, restraint, haemorrhagic shock, or severance of trachea and inability to vocalize. Therefore little or no reaction does not necessarily indicate absence of pain (EFSA 2004). Rosen (2004) argued that a shechita cut is painless due to rapid physiological changes and the extreme sharpness of the shechita knife (*chalaf*), and the smooth incision prevents activation of the pain pathways, comparing it to surgeons who sometimes cut themselves only to notice it after an operation. However, it must be borne in mind that a neck cut would involve a large area and also pain is not merely related to the quality of the cut.

The possibility of potential pain after neck cutting has implications, especially for cattle slaughter without stunning. There is evidence to suggest restrictions to blood flow and loss may occur in calves and adult cattle because the cut carotid ends develop clots (Anil et al. 1995a,b). The development of these occlusions could be due to inadequate sharpness of the knife as well as other reasons because the same problem has also been reported following shechita slaughter with a razor-sharp knife (Anil, personal observations; Levinger (1976). Gregory et al. (2011) observed an incidence of 10% carotid occlusions (aneurysm) in cattle slaughtered by halal and shechita methods, and suggested an alternative neck-cutting position higher up in the neck. In cattle, there is an extra link (anastomosis) between the carotid arteries and the base of the brain. Therefore, during restriction to flow by aneurysms blood may still be pumped to the brain. Anil et al. (1995a) found that carotid occlusion delayed the time to isoelectric electrocorticogram (ECoG) in calves. In the same study, when carotid occlusions occurred, vertebral artery blood flow was maintained at about 30% of its initial

Figure 5.3 Upright pen with chin lift.

level for up to three minutes and in some animals it increased substantially following sticking.

The potential problem of sensibility during and after neck cutting could be minimized by the use of pre-slaughter electrical stunning, provided that it is correctly applied and is acceptable. If stunning does not stop the heart or kill before exsanguination, it is accepted by certain Muslim groups as part of the slaughter process. Other Muslim groups object, preferring the authentic method of neck cutting without stunning. In some countries in Europe, the Far East, and invariably in New Zealand and Australia, stunning is used before halal slaughter for export to countries in the Middle and Far East, with the approval of the appropriate religious authorities. Most poultry processors also use water bath stunning for halal slaughter, albeit in low voltages and currents. However, objections to stunning during halal slaughter have been increasing in recent years in Europe. Reasons for this trend include potential welfare problems during stunning, perceived uncertainty regarding the effects of stunning on heart function, and other myths about stunning and reluctance to move away from tradition (see section 5.5).

Contentious issues include captive bolt stunning, which fractures the head and fails to stun. More difficult and current problems relate to water bath electrical stunning of poultry. Failure to induce a stun and/or achieve unconsciousness need to be addressed, in particular with low currents and high frequencies that deliver only electric shocks. This is encountered during halal slaughter if very low electric currents are applied. Some business operators maintain that they may be allowed to use low currents due to the exemption.

5.5 Background on Halal Slaughter and Rules

The Muslim method of slaughter, known as the halal method, varies in practice. The variations may be because of differences in the interpretation of the Holy Quran and the Hadith (the sayings of the prophet Mohammed, (pbuh)), different traditions, lack of trained slaughtermen, interested individuals, and certifiers. This is in contrast to shechita organizations, which have stricter and more consistent practices.

The act of slaughter (*al-dhabh*) is regarded as a privilege, hence pronouncing the name of Allah coincides with slaughter. Following restraint, slaughter is carried out by severing (ideally a single cut) vessels in the neck to achieve instant and copious exsanguination with a sharp knife. The usual type of incision is a transverse cut in the retrograde fashion.

Muslims believe rapid and maximum blood loss is crucial during and after halal slaughter because consumption of blood is forbidden. Effective exsanguination can be impeded by occlusions in carotid arteries causing a delay in loss of consciousness (Anil et al. 1995a,b). It has also been claimed that stunning methods could impede blood loss. However, comparative studies in sheep and cattle have shown no significant difference between stunned and non-stunned sheep (Anil et al. 2004) and cattle (Anil et al. 2006).

Differences in the interpretation of rules have caused confusion and controversy regarding halal slaughter. The Dialrel project undertook consultations and listed the following (Dialrel 2009):

- Islam is comprehensive; Sharia looks after everything for good. Allah u Teala provided rules.
- All food, fish, nuts, grains, vegetables, and fruits are good for us. Haram things are unlawful. Animals are lawful and must be killed according to Islamic rules.
- Haram (unlawful) animals include pigs, dogs, donkeys, carnivores, reptiles, insects, animals killed by strangulation or blow/clubbing or that died from natural causes, beasts with fangs, and birds of prey. Fish are regarded as halal as the Prophet (pbuh) allowed things that come from sea as lawful, therefore seafood is not carrion. Animals, if not slaughtered according to Islamic rules, and those killed for gods other than Allah are haram.
- To avoid certain diseases blood must be cleared out of the animal's body. Blood should not be retained in the veins and congeal, for hygiene reasons. Good flow of blood is required.
- Animals must only die from slaughter, no dressing while alive is allowed.
- Anything can be eaten during necessity.
- Muslims or people of the books (Christian and Jews), male or female can slaughter animals.
- *Besmele/tasmiyyah*, verbal citing of god's name, is a religious requirement for halal slaughter.
- Facing Kible (Makkah) is recommended, but not required (agreed by the majority of scholars)
- An animal's head must not be removed during slaughter.
- Runaway animals that are out of control can be shot.
- In regard to stunning, if suffering occurs, or if the animal dies before slaughter, or if the blood is congealed and retained, then that would be haram. Otherwise, stunning is acceptable if the following are observed: (i) tasmiyyah, (ii) no suffering, and (iii) good flow of blood.

It is understood that there are three main requirements:

1) mercifulness to animal
2) the slaughtered animal must be healthy
3) death must be the result of blood loss.

New technological methods are fine as long as suffering is minimized and sufficient blood flows out to protect consumers.

The Organisation of Islamic Countries (OIC) has also issued guidelines for halal slaughter. These make references to slaughter with and without stunning. Although they do not necessarily recommend stunning, if used the permissible methods and currents are described in Table 5.2.

References Used by OIC

1) General Guidelines for the use of the term halal CAC/GL 24–1997, The Codex ISO 22005 : 2007 Traceability in the feed and food chain – General principles and basic requirements for system design and implementation.
2) ISO 9001 : 2005, Quality management systems – Requirements.
3) MS 1500 : 2004 Halal Food Production, Preparation, Handling, and Storage – General Guidelines, Malaysia.

In conclusion, it seems there is a two-tier system of halal slaughter: neck cutting without prior stunning and slaughter with reversible stunning.

Table 5.2 Guideline parameters for electrical stunning issued by the OIC (provided by Dr Hamid Ahmed).

Type of animal	Current (Ampere)	Duration (Second)
Chicken[a]	0.25–0.50	3.00–5.00
Lamb	0.50–0.90	2.00–3.00
Goat	0.70–1.00	2.00–3.00
Sheep	0.70–1.20	2.00–3.00
Calf	0.50–1.50	3.00
Steer	1.50–2.50	2.00–3.00
Cow	2.00–3.00	2.50–3.50
Bull	2.50–3.50	3.00–4.00
Buffalo	2.50–3.50	3.00–4.00
Ostrich	0.75	10.00

Note: Electrical current and duration shall be validated and determined by the organization, taking into account the type and weight of the animal and other varying factors.
[a] Dialrel note: This table should be amended. For example, the high currents given for poultry if used with 50 Hz in the UK would kill the birds before slaughter.

There are three views in regard to stunning:

i) those who accept reversible stunning (Al-Masri 1989)
ii) those who reject the idea of stunning completely as they think stunning is not necessary, is against religious rules, and is inhumane (Katme 2017)
iii) those who are not sure or need assurances in both cases.

The Dialrel project found that consumer trust in halal products is low in Europe. The following are problem/contentious areas highlighted by Dr. Yunes Al Teinaz:

- halal certifiers: authenticity, lack of standards, insufficient checks/documentation
- illegal slaughter and unfit meat sale
- lack of auditing standards (from stable to the table)
- mechanical killing of animals before slaughter
- training of slaughtering personnel
- recorded tasmiya (Islamic prayer) during poultry slaughter
- animal welfare compromises not being recorded
- labelling: stun/non-stun (new European law may require labelling of non-stunned animals in future)
- Islamic rules: interpretation unclear
- lack of training for Muslim slaughtering personnel and certifiers
- almost all attention of halal on fresh meat
- hygiene standards questionable.

5.6 Physiological Effects of Neck Cutting

Neck cutting is one of two slaughter methods used to achieve exsanguinations in both cattle and sheep. The other cut, chest sticking, is carried out to sever large vessels inside the thoracic cavity for rapid blood loss. However, during religious slaughter a transverse neck cut is invariably used to sever tissues and blood vessels in the neck, except the spinal cord (because severance of the cord impedes blood loss). In regard to religious slaughter without stunning, the efficiency of the cut is of utmost importance otherwise delays in blood loss and consciousness can lead to welfare problems such as pain and prolonged sensibility before death.

Under optimum slaughter conditions, it could be argued that if the incision is performed by a highly skilled slaughterer, using a sharp knife, the least amount of pain will be inflicted though not totally eliminated (Woolf 2004; Brooks and Tracey 2005). Deviation from this scenario will probably worsen the severity of pain in an exponential manner. The greater the damage to tissues in the neck, the more nociceptors will be activated than after good cuts, thereby firing fibres and relaying signals to the brain (EFSA 2004). Therefore, neck cutting during halal slaughter should ensure the following:

- clear access to the neck must be provided before the cut is made
- a cut in a restrainer must be made without delay
- knives must be maintained and kept sharp (1099/2009 reg. requirement)

- cuts should only be made by skilled and trained slaughter personnel (1099/2009 reg. requirement)
- ideally the incision should be single, ensuring severance of both carotid arteries completely
- development of carotid occlusions should be observed and prevented in cattle by taking appropriate action
- if bleeding is impeded corrective action (stun and kill) must be taken and the problem investigated (even if it means rejecting the carcass)
- systematic checks to ensure no signs of life are present before moving the carcass need to be carried out by law.

The following reflexes can be used as tools to confirm loss of consciousness:

- no eye reflex or blinking (palpebral or corneal)
- widened pupils
- fixed eye
- no response to threatening movements
- absence of breathing activity
- floppy head and relaxed tongue.

The following reflexes may indicate residual consciousness:

- rhythmic breathing
- vocalization
- kicking/struggling movements (except typical convulsions during epileptiform activity)
- righting
- attempts to stand up and escape.

5.7 Exsanguination and Loss of Consciousness

Of the circulating blood volume (8% of body weight), 18% of cardiac output supplies the brain (EFSA 2004). Following effective cuts, 40–60% of blood volume is lost in similar patterns and rates in the different species (Warriss and Wilkins 1987). This rapid loss should result in a dramatic drop in blood pressure. This leads to inadequate perfusion of tissues, and a state of shock and failure of the system's compensatory mechanism (Gregory 2004).

In cattle, following exsanguination it takes a certain amount of time for the blood loss to reach critical levels. It is estimated that 50% of total blood volume is lost during exsanguination. It has been reported that 33% of total blood loss was reached after 30 seconds (Anil et al. 1995a,b) and 25% was bled out after 17 seconds (Anil et al. 2006). In sheep, however, the time period is much shorter, with 50% of total blood being lost after 14 seconds and 90% after 56 seconds (Anil et al. 2004).

In conclusion, it must be remembered that blood loss, insensibility, and death will take varying lengths of time depending on the factors mentioned above. The more efficient exsanguination is, the quicker death will occur and animal welfare compromises can be minimized by ensuring rapid bleed-out at slaughter.

5.8 Legal Considerations

National and international norms such as the Office International Epizootique (OIE) standards and European regulations (1099/2009 Regulation; European Community 2009) apply to religious slaughter, with derogations. UK law concerning slaughter and killing is WASK (1995). WASK has been replaced by WATOK (Welfare of Animals at the Time of Killing in England).

References

Al-Masri, B. (1989). Animals in Islam. The Athena Trust ISBN 1–870603-01-X.

Anil, H. (2012). Effects of Slaughter Method on Carcass and Meat Characteristics in the Meat of Cattle and Sheep. English Beef and Lamb Executive (EBLEX).

Anil, M.H., Yesildere, T., Aksu, H. et al. (2004). Comparison of religious slaughter of sheep with methods that include preslaughter stunning and the lack of differences in exsanguination, packed cell volume and quality parameters. *Animal Welfare* 13 (4): 387–392.

Anil, M.H., Yesildere, T., Aksu, H. et al. (2006). Comparison of halal slaughter with captive bolt stunning and neck cutting in cattle: exsanguination and quality parameters. *Animal Welfare* 15: 325–330.

Anil, H. and Lambooij, B. (2009). Stunning and slaughter methods. In: Welfare of Production Animals: Assessment and Management of Risks, vol. 5 (eds. Food Safety Assurance and Veterinary Public Health, F. Smulders and B. Algers), 169–184. Wageningen Publishers.

Anil, M.H., McKinstry, J.L., Gregory, N.G. et al. (1995b). Welfare of calves. 2. Increase in vertebral artery blood flow following exsanguination by neck sticking and evaluation of chest sticking as an alternative slaughter method. *Meat Science* 41: 113–123.

Anil, M.H., McKinstry, J.L., Wotton, S.B., and Gregory, N.G. (1995a). Welfare of calves – 1. Investigation into some aspects of calf slaughter. *Meat Science* 41 (2): 101–112.

Brooks, J. and Tracey, I. (2005). From nociception to pain perception: imaging the spinal and supraspinal pathways. *Journal of Anatomy* 207: 19–33.

Daly, C.C., Kallweit, E., and Ellendorf, F. (1988). Cortical function in cattle during slaughter: conventional captive bolt stunning followed by exsanguination compared with shechita slaughter. *The Veterinary Record* 122: 325–329.

Dialrel (2009). Religious slaughter, improving knowledge and expertise through dialogue and debate on issues of welfare, legislation and socio-economic aspects. http://www.dialrel.eu/dialrel-results.

Dunn, C.S. (1990). Stress reactions of cattle undergoing ritual slaughter using two methods of restraint. *Veterinary Record* 126: 522–525.

EFSA (2004). Opinion of the Scientific Panel on Animal Health and Welfare. http://www.efsa.europa.eu/de/scdocs/doc/45.pdf.

European Community (1993). Directive 93/119/EC on the protection of animals at the time of slaughter or killing. *European Community Official Journal* 340: 21–34.

European Community (2009). Council Regulation (EC) No 1099/2009. http://ec.europa.eu/food/animal/welfare/slaughter/regulation_1099_2009_en.pdf.

FAWC (2003). A Report on Religious Slaughter Methods. London: Farm Animal Welfare Council.

Ferrari, S. and Bottoni, R. (2010). Legislation regarding religious slaughter in the EU member, candidate and associated countries. Dialrel Deliverable 1.4. http://www.dialrel.eu/images/report-legislation.pdf.

Gibson, T.J., Johnson, C.B., Murrel, J.C. et al. (2009a). Electroencephalographic responses of halothane anaesthetized calves to slaughter by ventral-neck incision without prior stunning. *New Zealand Veterinary Journal* 57 (2): 77–83.

Gibson, T.J., Johnson, C.B., Murrell, J.C. et al. (2009b). Components of electroencephalographic responses to slaughter in halothane-anaesthetized calves: effects of cutting neck tissues compared with major blood vessels. *New Zealand Veterinary Journal* 57 (2): 84–89.

Grandin, T. (1994). Religious slaughter and animal welfare. *Meat Focus International*: 115–123.

Grandin, T. and Regenstein, J.M. (1994). Religious slaughter and animal welfare: a discussion for meat scientists. *Meat Focus International*: 115–123.

Gregory, N.G. (2004). Physiology and Behaviour of Animal Suffering. Oxford: Blackwell Science ISBN 0–632–06468-4.

Gregory, N.G., Wenzlawowicz, M., von Holleben, K et al. (2011). Recent advances in the welfare of livestock at slaughter. HSA Centenary Symposium, Portsmouth (30 June to1 July).

Kalweit, E., Ellendorf, F., Daly, C., and Schmidt, D. (1989). Physiological reactions during slaughter of cattle and sheep with and without stunning. *Deutsche tierarztliche Wochenschrift* 96: 89–92.

Katme, A. (1986). Islam & Muslims oppose stunning! https://halalhmc.org/wp-content/uploads/2017/12/Muslims-Oppose-Stun.pdf.

Levinger, I.M. (1976). Physiological and general medical aspects of Shechita. In: (eds. M.L. Munk and E. Munk), 147–149. Jerusalem Gur Aryeh Publications.

Levinger, I.M. (1995). Shechita in the Light of the Year 2000. Maskil, L. David.

Rosen, S.D. (2004). Physiological insights into Shechita. *The Veterinary Record* 154: 759–765.

Velarde, A., Gispert, M., Diestre, A., and Manteca, X. (2003). Effect of electrical stunning on meat and carcass quality in lambs. *Meat Science* 63: 35–38.

Warriss, P.D. and Wilkins, L.J. (1987). Exsanguination of meat animals. *Seminar on Preslaughter Stunning of Food Animals, European Conference Group on the Protection of Farm Animals*, Brussels (2 June 1987).

WASK (1995). Welfare of Animals at Slaughter and Killing (UK) Welfare Quality. www.welfarequality.net.

Woolf, C.J. (2004). Pain: moving from symptom control toward mechanism-specific pharmacologic management. *Annals of Internal Medicine* 140: 441–451.

6

The Slaughter Process

With or Without Stunning

Mehmet Haluk Anil[1] and Yunes Ramadan Al-Teinaz[2]

[1] *Formerly at Bristol University, based Bristol, UK*
[2] *Independent Public Health & Environment Consultant, London, UK*

6.1 Religious Requirements and Alternative Choices

Animal welfare issues relating to handling prior to halal slaughter, acceptability or exclusion of stunning, and the effects of neck cutting for exsanguination have been referred to in Chapter 5. This chapter focuses on the slaughter process with or without pre-stunning to produce unconsciousness. Before proceeding, it is first worth reconsidering the religious requirements in this regard.

As discussed in Chapter 1 the slaughtering (*dhabh* in Arabic) rules for halal, based on Islamic teachings, ethics, and jurisprudence require the following:

- The abattoirs or processors must have involvement of an Islamic religious organization in an advisory or observatory fashion.
- The premises, equipment, and machinery must be deemed acceptable by Islamic Shariah law before any production take place.
- A trained Muslim man must slaughter the animal in a licenced slaughterhouse with implementation of all hygiene and animal welfare regulations.
- The slaughterer must be a mature and pious Muslim of sound mind who understands fully the fundamentals and conditions relating to halal slaughter and be approved by the religious authorities and licenced by the state services.
- The animals must be permissible species to eat, alive, fed on natural food, and raised under good welfare conditions.
- The animal must be free from any disease or injury at the time of slaughter, certified and checked by the official veterinary surgeon.
- The animal skin or fur and bird feathers must to be clean and the animal must be fed, not hungry or thirsty before slaughter.
- The animal must not be slaughtered in front of other animals and not in sight of blood.
- Animals must be handled gently and individually, and the knife should not be sharpened in front of any animal before slaughter.

- Stress or discomfort to any animal must be minimal.
- No stunning to kill is allowed before slaughter.
- The knife must be very sharp and clean.
- The Muslim slaughterer must recite first, 'Bismillah, Allah Akbar'
- The neck cut must be made in the correct anatomical site, severing the two carotids, the two jugulars, the windpipe, and the gullet, but without cutting the spinal cord.
- The maximum amount of blood should be exsanguinated from the carcass.
- A specific time should be allowed until the animal is dead after exsanguination.
- De-feathering, de-skinning, and evisceration must not start before movements cease.
- Any unlawful meat, such as pork, should not contaminate halal meat. Separate knives, equipment, and utensils should be used for halal meat.

6.2 Slaughter without Stunning by Neck Cutting

This method is referred to by many as the traditional method of halal slaughter. When religious slaughter without stunning is employed, exsanguination is aimed at killing animals. Exsanguination is the process of severing the blood vessels to let sufficient blood out rapidly to kill an animal. This is carried out by the traditional cut across the upper third of the neck (except camels slaughtered by a chest stick at the base of the neck) for exsanguination and the animal is killed through blood loss. Blood is also drained from the carcass for better hygiene and meat quality. In this situation, the cause of death would be loss of blood, insufficient blood supply to the brain resulting in cerebral ischaemia and loss of brain function, cardiac shock, and cessation of heart beat (cardiac arrest). The effects of slaughter methods on exsanguination have been reviewed extensively (Anil 2012; Anil and von Holleben 2014).

Severance of the neck tissues can result in potentially painful stimuli that may be perceived by the animals, if conscious. However, the issue is controversial as there are variations in the times to loss of brain function caused by differences in cutting methods and results of reported studies. In addition, not only during the cut but afterwards rubbing of wound edges or exposed tissues and large or multiple cuts are more likely to elicit pain sensation. This is discussed later in this chapter.

The Dialrel project issued the following guidelines for slaughter without stunning (Dialrel 2009):

1) The slaughterer must be ready to perform the cut before the animal is restrained.
2) The neck cut must be performed without any delay.
3) Both carotid arteries and both jugular veins must be cut without touching the bones of the spine (vertebrae) with the knife.
4) Each animal should be neck cut by a single swift or continuous back and forward movement of the knife without interruption.
5) The knife used must be sufficiently long for each type of animal to minimize the need for multiple cuts. Ideally, the length of the knife blade should be at least twice that of the width of the animal's neck.
6) The knife must be sharp for each animal. The knife should be checked by the slaughterer (or shochetim for shechita) as frequently as required for nicks and bluntness and

sharpened accordingly. An emphasis on training slaughterers to improve their knife sharpness is recommended.

7) Neck breaking must not be performed together with the cut.

6.3 Post-cut Management of Animals Slaughtered Without Stunning

If not pre-slaughter stunned, the animal becomes unconscious when brain perfusion becomes insufficient after the neck cut (Anil 2012; Anil and Holleben 2014). The time taken for unconsciousness to supervene is variable between animals.

Some studies on neck cutting in cattle have shown that delays in time to loss of consciousness can vary from a mean of 20 seconds (SD ± 33) to up to more than 120 seconds in exceptional cases.

Most sheep and goats seem to lose consciousness within 2–20 seconds after ventral neck cut, but sheep can show signs of recovery for longer times in exceptional cases. Most chickens lose consciousness within 12 and 15 seconds, but signs of recovery/consciousness are possible for up to 26 seconds after the cut.

However, as time to loss of consciousness varies between animals, clinical signs are necessary to recognize unconsciousness.

Several clinical signs have been suggested to recognize unconsciousness [7]:

- Complete loss of posture.
- No attempt to regain or to retain upright body posture.
- No reactions (e.g. retraction) to mechanical impacts on the wound (e.g. contact of the wound to parts of the head holder or pen).
- Absence of tracking by the eye of movements in the vicinity, often accompanied by spontaneous closure of the eyelid.
- Absence of response to threatening movements (e.g. rushing a hand towards the eyes does not lead to closing of the eyes or moving the head backwards).

6.4 Clinical Signs of Brain Death

The clinical signs of brain death are as follows:

- Permanent absence of cardiac activity (e.g. pulse or heart beat) when bleeding has ceased.
- Permanent absence of brain stem reflexes such as pupillary light reflex, corneal reflex, rhythmic breathing, and gagging.

6.5 Recommendations for Halal Slaughter

1) There must be no interference with the wound until the animal is unconscious, except for procedures involved with checking the adequacy of the cut. Mechanical and chemical stimuli on the wound must be minimized.

2) The cut should be inspected carefully for complete sectioning of both carotid arteries and both jugular veins, and for the efficiency of bleeding through strong flow and seeing the pulsating effect of the heartbeat on this flow. When inspecting the wound unnecessary contact with the severed edge of the skin must be avoided. Thus, visual inspection is preferable. It is understood that at times the shochet may have a religious responsibility to carry out a physical inspection on the cut and a visual inspection will not suffice. If the inspection is done by the shochet, they need to be trained to minimize or totally avoid touching the skin surfaces.

3) The animal must be assessed to be unconscious by the slaughterer (or the shochet) before it can be released from the restraint. It is suggested that signs of unconsciousness are checked at least twice, for cattle between 30 and 40 seconds post-cut, and for sheep and poultry between 15 and 25 seconds post cut. The following clinical signs should be used as a guide for monitoring (Dialrel 2009):

- No attempts to regain or retain upright body posture.
- No reaction (e.g. retraction) to mechanical impacts on the wound (e.g. contact of the wound with parts of the head-holder or pen).
- Absence of tracking by eye movements, often accompanied by spontaneous closure of the eyelid.
- Absence of response to threatening movements (e.g. the rushing of a hand towards the eyes does not lead to closing of the eyes or moving of the head backwards).
- No wing flapping in poultry.

4) In the event of inefficient bleeding or prolonged consciousness being exhibited during repeated checks after neck cutting, animals should be stunned with a suitable method as soon as possible, even if this requires the religious authorities to declare the animal to be non-kosher or haram. Optimally, this should be done within 45 seconds post cut for cattle, or within 30 seconds for small ruminants and poultry.

5) As prolonged consciousness is an indicator of poor procedure, in the event of prolonged consciousness the problem should immediately be investigated and necessary corrective action taken. Records of failure should also be documented for monitoring purposes.

6) Further dressing, scalding or electro-stimulation shall only be performed after brain death of the animal has been verified as indicated above.

7) When the cut is performed in a 180° inverted position in cattle, it may be preferable to turn the box to a position between 180° and 90° directly after the cut for better access to the head of the animal and a more relaxed position.

6.6 Exsanguination Techniques

As described in Chapter 4, during slaughter severance of tissues and blood vessels is carried out using knives of different types with varying thicknesses and lengths to cut the neck and blood vessels for bleeding the animal out. Ordinary slaughter knives of different shapes (straight and curved) and lengths (from 10 to 30 cm) can be used for halal slaughter. They are generally shorter than the shechita knives (called *chalaf*) that Jewish slaughterers use. The cutting action can be transverse, or partly stabbing and retrograde.

Two exsanguination methods are commonly used in slaughterhouses. Neck cutting is used for halal slaughter to sever both carotid arteries and jugular veins in the upper neck, often behind the mandibles (jaw bones), on cattle, sheep, rabbits, and poultry. The second method, chest or thoracic sticking, involves inserting a knife through the thoracic inlet at the base of the neck into the chest in front of the sternum, mainly to slaughter cattle and pigs after stunning whilst the animal is in a recumbent position on a cradle or conveyor or hoisted on to an overhead rail. Chest sticking is not usual for halal slaughter, with the exception of camel slaughter, instead transverse neck cutting is employed. Chest sticking lets out a large volume of blood in a short time from vessels near the aorta, leading to common carotid arteries supplying the head with oxygenated blood to the brain, which has been shown to cause the brain lose its function due to rapid profuse bleeding. However, chest sticking is not applied in halal slaughter as it is considered impractical and is probably against religious rules and tradition. Time to loss of blood and consciousness following slaughter without stunning can vary depending on species, technique, number of vessels cut, and restrictions on the rate of bleeding. In particular, cattle can have delays due to ballooning of the cut arteries.

6.7 Exsanguination and Loss of Consciousness

As discussed in Chapter 4, it is estimated that total blood volume is 8% of body weight and the brain receives 18% of cardiac output. After effective slaughter by neck cutting, 40–60% of total blood is lost until carcass dressing. During exsanguination blood pressure drops dramatically, resulting in a state of shock and leading to loss of consciousness. In cattle, the time taken for blood loss to decrease to critical levels by neck cutting is highly variable and a number of factors can influence this. Ineffective cuts to blood vessels, anatomical differences, and clots in the cut arteries can increase blood pressure. In contrast, the rate of loss is much quicker after neck cutting.

The brains of ruminants are perfused with blood from a vascular network connected to the carotid and vertebral arteries. In cattle extra branches can sometimes supply blood to the brain even after neck cutting, but this does not happen in sheep and goats. Some claim, however, that this rate of blood flow after a neck cut would not be enough to sustain brain function. It is nevertheless known that carotid arteries can develop clots after cuts and impede blood loss, prolonging the time to onset of brain isoelectric electrocorticogram (ECoG recorded waves going flat) and loss of brain function. Studies have found that if carotid artery occlusion occurs, vertebral artery blood flow to brain is maintained, delaying the onset of unconsciousness.

The sharpness of the knife and performing a good cut can also cause vasoconstriction, clotting, and ballooning, which is known also as carotid occlusion or aneurysms.

Following slaughter without stunning it is therefore important that consciousness is lost quickly. Time to loss of brain function in cattle has been studied by researchers who looked at the electrical activity of the brain using an electroencephalogram (EEG, brain waves with electrodes on head) or ECoG (EEG with implanted electrodes) to evoke responses as well as animal reactions and reflexes and found variations. In conclusion, following neck cutting, any delay in time to loss of consciousness could result in welfare problems. Slowing of blood flow after a cut could compromise animal welfare due to delayed loss of consciousness and other problems such as inhalation of blood into the trachea.

6.8 Blood Loss and Retention

Halal slaughter rules require as much blood as possible to be removed from the carcass. It is claimed that stunning reduces the rate of bleed out and total loss. However, research has shown no difference in exsanguination after comparing stunning and slaughter versus slaughter with no stunning in sheep and cattle.

Based on available studies and results, it seems that regardless of whether or not pre-slaughter stunning is used, total blood loss likely to be similar. However, there may be other differences in terms of physiological effects and carcass quality.

6.9 Carcass and Meat Quality

Stunning and slaughter can influence carcass and meat quality manifested as haemorrhages, bruising, and broken bones (Anil 2012). Blood splash (petechial haemorrhages) in the muscles can be due to bad pre-slaughter handling, electrical stunning using high voltages, and nutritional or unknown factors. These can lead to off meat. Effective neck cutting after head-only electrical stunning can minimize detrimental effects of rising blood pressure on carcass and meat quality. Neck cutting while the heart is still beating should result in up to 85% of total blood to be lost within 60 seconds, therefore prompt exsanguination following head-only stunning should lose half the circulating blood volume. Consequently, under normal circumstances blood pressure should not be responsible for haemorrhages.

6.10 Slaughter with Stunning

Effective stunning before slaughter is intended to induce unconsciousness in animals. Regarding halal slaughter, if stunning were used the method would have to be reversible.

A non-reversible stunning method such as heart stopping by an electric current (first on the head followed by application to the chest) is regarded as a killing method because animals will not regain consciousness even if bleed out is not performed. Nevertheless, death in these animals in practice occurs by exsanguination not by the stunning method itself because blood loss from sticking has a more rapid effect on the brain.

Stunning for religious slaughter requires animals to be alive at the time of slaughter. Suitable reversible stunning methods induce temporary loss of consciousness and need prompt and accurate neck cutting procedures (bleeding out) to cause death.

After effective stunning, the presence of a heartbeat can indicate the reversibility of the unconsciousness if the animal is not slaughtered.

The Dialrel project (Dialrel 2009) has issued the following recommendations:

1) The animal must be introduced in the restraining device only when the slaughterer is ready to stun the animal, and stunning must be performed without any delay.
2) Correct stunning should induce loss of consciousness without pain before, or at the same time as, the animal is slaughtered.

3) The criteria for monitoring the loss of consciousness need to be applied according to the stunning system and species to ensure that the animals do not present any signs of consciousness or sensibility in the period between the end of the stunning process and death.

Signs of successful mechanical stunning in ruminants:

- immediate collapse
- immediate onset of tonic seizure (tetanus) lasting several seconds
- prompt and persistent absence of normal rhythmic breathing
- loss of corneal reflex.

Signs of successful mechanical stunning in poultry:

- immediate collapse (this may not be applicable to poultry restrained in a cone or shackle)
- immediate onset of tonic seizure (tetanus)
- severe wing flapping due to damage to the brain
- prompt and persistent absence of normal rhythmic breathing
- loss of corneal reflex.

Signs that indicate ineffective stunning in cattle and sheep (ruminant) include flaccid muscles immediately after stunning, return of rhythmic breathing, and rotated eyeballs.
Signs of successful electrical stunning in ruminants:

- immediate collapse of free-standing animals (not applicable to animals held in a restrainer conveyor)
- immediate onset of tonic seizure (tetanus) lasting several seconds, followed by clonic seizure (kicking or uncoordinated paddling leg movements)
- apnoea (absence of breathing) lasting throughout tonic–clonic periods
- upward rotation of eyes.

Signs of successful electrical stunning in poultry:

- immediate collapse of free-standing animals (not applicable to poultry restrained in a cone or shackle)
- water bath electrical stunning leads to an immediate onset of tonic seizure (tetanus), followed by short duration clonic seizure (kicking or uncoordinated paddling leg movements)
- head-only electrical stunning leads to clonic–tonic convulsions (a reverse of the sequence seen in red meat species)
- apnoea (absence of breathing) lasting throughout tonic–clonic periods.

Indicators of ineffective stunning include escape behaviour, sometimes with vocalization, absence of the typical tonic or clonic muscle convulsions, return of rhythmic breathing, and righting attempts.

In poultry the return of eye reflexes and rhythmic breathing indicates the return of brain function after electrical stunning. During bleeding vocalization and wing flapping, head raising, spontaneous blinking and eye tracking are indicative of recovery.

An important criterion of electrical stunning is ensuring unconsciousness with potential recovery. This is only possible with the application of the correct parameters. As a general rule application of a shock of 1 A for sheep and goats, and 1.5–2 A for cattle is suitable.

For poultry, an individual bird current of around 105 mA using a frequency of 50 Hz or higher should be used. Since high frequencies do not stop the heart, it is important to ensure sufficient current is used to induce unconsciousness to protect animal welfare.

Mechanical stunning with a captive bolt is another option. However, since penetrating guns damage brain tissue there are objections to this method on grounds of no recovery, although the heart continues to beat. Non-penetrating guns with high speed have been used for halal slaughter. One problem with this method is that the guns are often ineffective, but it is still used for halal slaughter in the Far East.

6.11 Post-Cut Stun

Post-cut stunning is considered if pre-slaughter stunning is not acceptable. Post-cut stunning reduces the time to loss of consciousness and death, preventing the animal form feeling distress during neck cutting.

6.12 Recommendations

1) The post-cut stun should be performed immediately and at the latest 5 seconds after the neck cut, without further manipulation of the animal between the cut and the stunning application (except if manipulation is required to allow a relaxed bleeding position).
2) When a post-cut captive bolt stun is used, the gun must be placed in the correct position using the correct captive bolt/cartridge combination for that animal type.
3) Post-cut stunning must induce immediate loss of consciousness.

References

Anil, H. (2012). *Effects of Slaughter Method on Carcass and Meat Characteristics in the Meat of Cattle and Sheep*. English Beef and Lamb Executive, (EBLEX): UK.

Anil, M.H. and von Holleben, K. (2014). Exsanguination. In: *Encyclopedia of Meat Sciences*, 2e, vol. 1 (eds. C. Devine and M. Dikeman), 561–563. Oxford: Elsevier.

Dialrel (2009). Improving Animal Welfare during religious slaughter - recommendations for good practices. http://www.dialrel.eu/press468e.pdf?format=pdf (accessed September 2019).

7

Recent Slaughter Methods and their Impact on Authenticity and Hygiene Standards

Ibrahim H.A. Abd El-Rahim

Department of Environmental and Health Research, the Custodian of the Two Holy Mosques Institute for Hajj and Umrah Research, Umm Al-Qura University, Makkah Al-Mukaramah, Saudi Arabia
Infectious Diseases, Department of Animal Medicine, Faculty of Veterinary Medicine, Assiut University, Assiut, Egypt

7.1 Introduction

The worldwide volume and value of trade in halal meat and co-products are huge. Muslim countries alone consumed meat estimated to be worth US$57.2 billion in 2008. The commercial production of halal red meat is rapidly growing in importance and so is the controversy surrounding the slaughter without stunning that is used in producing a substantial amount of the meat (Farouk 2013).

The regulations and standards governing animals' slaughter methodology vary considerably around the world. In some communities, animal slaughter may be controlled by religious laws, most notably halal for Muslims and kosher for Jewish communities. Both these methods require that the animals being slaughtered should be conscious at the point of slaughtering without pre- or post-slaughter stunning. This method of non-stunning slaughtering may conflict with national regulations when a slaughterhouse following such religious rules is located in some Western countries. Also, the exportation of frozen large animals or chicken meat from some Western countries into one of the Islamic countries can cause conflict over the method of slaughtering used for the preparation of this meat.

Islam has imposed certain rules and instructions for slaughtering animals whose meat is permissible for Muslims to eat under Islamic Shariah law, such as cattle, camels, sheep, goats, and poultry. Islam has also developed many legal provisions required during the slaughtering process. The halal slaughter method consists of a horizontal cut by hand on the throat of a fully conscious animal, severing the oesophagus, trachea and all four vessels of the throat in order to remove all the impure blood from the animal without any kind of pre- or post-slaughter stunning. The halal slaughter should also include:

- resting the animal before slaughtering
- prevention of pain and agony for animals before slaughtering
- accessibility of animals to drinking water before slaughtering

- avoidance of slaughtering animals in front of other animals
- mentioning the name of God (Allah) during the slaughtering with sincerity and conviction
- using a sharp knife to quickly severe the four vessels of the throat
- avoidance of complete cutting of the animal's throat during slaughtering while the animal is still bleeding.

The halal slaughter of animals has an important role in preventing infectious diseases and is considered one of the main reasons for the popularity of halal products even amongst non-Muslims. It has been shown that halal slaughter protects consumers from many diseases which occur as a result of slaughter methods that involve stunning. In this chapter, the impact of recent slaughter methods on authenticity and hygiene standards are discussed.

7.2 Definition of Humane Slaughter

Several criteria define a humane slaughter method from the scientific point of view: (i) animals should not be treated cruelly, (ii) animals should not be unduly stressed, (iii) bleeding must be done as quickly and as completely as possible, (iv) carcass bruising must be minimal, and (v) the slaughter must be hygienic, economic, and safe for the operators (Swatland 2000). In addition, humane conditions must be presented during pre-slaughter handling (Roça 2002). A good animal welfare auditing system also has standards that prohibit bad practices such as dragging, dropping, throwing, puntilla, and hoisting live animals before ritual slaughter (Grandin 2010).

Humane slaughter can be defined as a set of technical procedures which guarantee animal welfare from loading at the farm up to bleeding in the slaughter plant (Roça 2002), and are intended to reduce unnecessary suffering of the slaughtered animal (Cortise 1994).

7.3 Halal Slaughter and Animal Welfare

Some people believe that there are conflicts between the halal slaughter method and animal welfare standards. Islam teaches that animals are to be slaughtered according to the mindful and attentive way (prophetic method) taught by the prophet Mohammed (peace be upon him). In the halal slaughter method, the animal should be dispatched as painlessly as possible. The halal method of slaughter is considered cruel and contrary to scientific wisdom by those who think that the animal must first be stunned in order to avoid compromising its welfare (Farouk et al. 2014). The standards which were developed by the Standardization Expert Group of the Organization of the Islamic Conference (OIC) can be summarized as follows:

a) The animal to be slaughtered has to be a halal animal, such as cattle, camel, sheep etc. and not and an unlawful (haram) animal like a pig or a dog etc.
b) The animal to be slaughtered shall be alive or deemed to be an alive at the time of slaughter. The slaughtering procedure should not cause torture to animals and should be done with animal welfare/rights consideration.

c) The slaughterer shall be a Muslim who is mentally sound and fully understands the fundamental rules and conditions related to the slaughter of animals.

d) If animals have arrived from long distance, they should first be allowed to rest before slaughtering.

e) The animal may be slaughtered after having been hung or laid preferably on its left side facing Kiblah (the direction of Makkah Al-Mukaramah). Care shall be taken to reduce the suffering of the animal while it is being hung or laid and it is not to be kept waiting long in that position.

f) At the time of slaughtering the animals, the slaughterer shall utter 'Bismillah Wallahuakbar' which means 'In the Name of Allah and Allah is the Greatest' and he should not mention any name other than Allah otherwise this makes it non-halal. Mentioning the name of Allah should be repeated on each slaughter if more than one animal is to be slaughtered or on each group being slaughtered continuously and if the continuous process is stopped for any reasons he should mention the name of Allah again.

g) Slaughtering shall be done only once to each animal. The 'sawing action' of slaughtering is permitted for as long as the slaughtering requires and the knife shall not be lifted off the animal during the slaughter.

h) The act of halal slaughter shall begin with an incision on the neck at some point just below the glottis (Adam's apple) and after the glottis for long-necked animals.

i) The slaughter act shall sever the trachea (*halqum*), oesophagus (*mari*) and both the carotid arteries and jugular veins (*wadajain*) to hasten the bleeding and death of the animals.

j) The bleeding shall be spontaneous and complete. The bleeding time must not be less than 2.5 minutes to ensure full bleeding.

k) The slaughterer should grab the head by the left hand, stretching it down tightly and cut the throat using a sharp slaughtering knife held in the right hand. The sharp edge of the knife which is used for slaughter should be not less than 12 cm in length.

7.4 Definition of Pre-slaughter Stunning

The technical process to which animals are subjected to induce unconsciousness for minimizing the pain associated with slaughter is known as pre-slaughter stunning (European Food Safety Authority (EFSA) 2006; Limon et al. 2010). This method may be helpful to allow easier handling, especially of large animals (Bergeaud-Blackler 2007).

7.5 Aims of the Stunning

The main aim of pre-slaughter stunning is to put the animal into an unconscious state which must last until bleeding (Gil and Durao 1985). According to EU law, all animals and birds must undergo pre-slaughter stunning to render them unconscious before they are slaughtered (Sante et al. 2000).

7.6 Types of Stunning

There are several means and methods of stunning used for different animals and poultry species, including:

  i) fired captive-bolt stunners
 ii) firearm/gunshot
iii) pneumatic-powered stunners
 iv) pneumatic-powered air injections stunners
  v) cash knocker
 vi) mallet
vii) cutting of the medulla
viii) electro-narcosis
 ix) electric water trough
  x) gas killing.

These are the most commonly used kinds of stunning. In Western countries, the pneumatic stunner or the captive-bolt pistol are used in the majority of cattle (beef) slaughterhouses. Application of an electric rod to the head of sheep and goats is used in most small ruminant slaughterhouses. The reversible method using an electric water trough is usually used for stunning chickens.

7.7 Stunning and Animal Welfare

Generally stunning is against animal welfare as it induces severe pain for animals and birds. In veterinary medicine, pain and other stress responses are usually measured by electroencephalography (EEG) and stress-related hormones (adrenocorticotrophic hormone, ACTH). Several research programmes and studies have confirmed that pre-slaughter stunning induces severe pain for the animal and may lead to death. On the basis of EEG data, it was found that animals subjected to penetrative mechanical stunning had the lowest alpha and beta wave intensity immediately post-stunning and at 30 seconds after throat cut, compared to both low-power non-penetrative mechanically stunned and high-power non-penetrative mechanically stunned animals. This could be explained by the animals' awareness of pain or other stressful factors attributed to the slaughtering procedure. Also, the presence of large intervals of higher frequency alpha and beta brain waves, which usually occur in conscious animals, suggest stressful conditions related to post-slaughter pain (CSIRO 2011). In another study, the animals showed a dramatic elevation in the percentage change of circulating ACTH after penetrative stunning, suggesting a physiological stress response (Zulkifli et al. 2014).

In the halal slaughter method the animal suffers loss of consciousness very quickly due to anaemia of the brain caused by simultaneous and instantaneous severance of the carotid arteries with a sharp knife. This means there is no pain sensation in halal slaughter. Experientially, a team at the University of Hannover in Germany confirmed this theory through the use of electrocardiography (ECG) and EEG records during halal and stunning slaughtering. With the halal method of slaughter there was no change in the EEG graph for the first 3 seconds after the incision was made, indicating that the animal did not feel any pain from the cut itself. The following 3 seconds were characterized by a condition of deep sleep-like unconsciousness brought about by the draining of large quantities of blood from the body. Thereafter the EEG recorded a zero reading, indicating no pain at all, yet at that time the heart was still beating and the body convulsing vigorously as a reflex reaction of the spinal cord. It is this phase which is most unpleasant to onlookers, who are falsely convinced that the animal suffers whilst in fact its brain is no longer recording any sensual messages (Mustaqim Islamic Art and Literature 2014).

7.8 General Impact of Stunning on Authenticity and Hygiene Standards

It was found that the various stunning methods have adverse effects on the carcass, meat quality, and public health, and cause downgrading as well as possible mis-stuns (Anil 2012; Farouk et al. 2014). The general impacts of stunning slaughter methods on authenticity and hygiene standards includes:

- inadequate bleeding
- spoilage of the meat
- low-quality meat
- adverse effects on public health.

7.9 Inadequate Bleeding

Inadequate bleeding means incomplete drainage of blood from the carcases. There are several factors which are responsible for bleeding efficiency after slaughtering, such as the physical state of the animal before slaughter, the stunning method, and the interval between stunning and bleeding. All diseases which debilitate the circulatory system can affect bleeding. Feverish, acute diseases promote generalized vasodilatation, impairing efficient bleeding (Petty et al. 1994). Two main factors lead to inadequate bleeding associated with the various kinds of stunning slaughter methods: stress resulting from stunning and cutting of the head (separating the head from the body) before the animal dies. Resting the animal before slaughter and avoidance of neck separation before the animal's death in halal slaughter help to ensure the complete drainage of blood (perfect bleeding). Prevention of neck separation during Islamic slaughter is very important to maintain the connection of the brain to the rest of the body via the spinal cord in order to send the nerve signals and hormonal alerts that are necessary to complete the bleeding process which removes all of the liquid blood from the carcass.

7.10 Spoilage of the Meat

As mentioned above, imperfect bleeding of the carcass which associated with pre-slaughter stunning leads to an increase of residual blood in muscles. This increase in residual blood, as well as disorder of glycolysis process, leads to an increase in meat pH and consequently water activity (Wa) of the meat is raised. These two changes may result in a proliferation of microorganisms, which cause spoilage of the meat (Lahucky et al. 1998; Hajimohammadi, et al. 2014).

7.11 Low-quality Meat

Stunning slaughter usually induces muscular haemorrhages. The subsequent rise in blood pressure following electrical stunning of lambs exacerbates the leakage of blood into tissues and the exacerbated blood becomes more apparent in the form of discrete haemorrhages or blood splash (Kirton et al. 1978). Some haemorrhages are associated with hypercontracted and disrupted muscle fibres, indicating that they are caused by severe muscular strain associated with the electrical stunning of broiler chickens. Many haemorrhages are found near the venules or veins where rupture was observed (Kranen et al. 2000).

After slaughtering, the post-mortem changes that take place when the muscle is converted into meat have a marked effect on the quality of the meat. After slaughtering, glycolysis occurs in which the glycogen in the muscle is converted into lactic acid, causing a fall in pH from an initial value of pH 6.8–7.3 to about 5.4–5.8 at rigour mortis. If animals are stressed immediately prior to slaughter, as they are when stunned, the muscle glycogen is released into the bloodstream and is rapidly broken down to lactic acid, causing drop in pH (Devine et al. 1984), while the carcass is still warm. This high level of acidity causes a partial breakdown of muscle structure, which results in pale, soft, and exudative (PSE) meat, a condition that mostly occurs in pigs. The meat loses some of its water-binding capacity, which is important in certain types of meat processing. On the other hand, long-term stress before slaughter or starvation uses up the glycogen so that less lactic acid is formed after slaughter, resulting in an abnormal muscle condition in which the muscles remains dark purplish-red on exposure to air instead of a bright red colour. This is termed dark, firm, and dry (DFD) in the case of pigs and 'dark cutting' in beef. Such meat and products spoil quickly since the low acidity favours rapid bacterial growth (Bender 1992). In the halal slaughter method, due to the resting of the animal before slaughtering and the absence of pre-slaughter stress (no pre-slaughter stunning), the glycogen content of animal muscles is maintained and the subsequent glycolysis process is normal, which keeps the meat pH within normal values. This has several advantages for meat quality, as it provides an unfavourable medium for the growth of bacteria, increases shelf-life, keeps the colour of the meat bright red, and makes the meat tasty.

From a scientific point of view, previous studies have indicated several adverse effects of stunning on meat quality, such as rapid changes in both electrolyte and amino acid metabolism (Lynch et al. 1966) and changes in some quality parameters, such as colour and water losses (Linares et al. 2007) as well as providing favourable growth conditions for various microorganisms (Dave and Ghaly 2011).

7.12 Adverse Effects on Public Health

All types of stunning lead to inadequate bleeding. This means most of the blood which should come out after halal slaughter will stay inside the muscles of the carcass. Blood is a typical enrichment media for the proliferation of different kinds of microbes, therefore its complete removal from the slaughtered animal is vital to protect consumers from infectious diseases. The ineffective bleeding associated with pre-slaughter stunning represents a source of infection with bacterial diseases.

Certain types of stunning, such as penetrating and non-penetrating captive bolts in cattle and cartridge–activated and pneumatically activated guns in sheep, lead to central nervous system (CNS) embolism in jugular blood. As the heart continues pumping for several minutes between stunning and the end of the exsanguination, some of the embolic CNS material dislodged by the penetrating captive-bolt gun might enter the venous blood vessels draining the head and consequently be disseminated to other organs/tissues. This may lead to transmission of incurable zoonotic prion diseases, such as bovine spongiform encephalopathy (BSE) from cattle and scrapie from sheep, to consumers (Anil 2012).

7.13 Specific Impact of Various Stunning Methods on Authenticity and Hygiene Standards

The use of firearms obviously induces severe pain for the animal. Bager et al. (1990) stated that the use of a cash-knocker leads to diffusive brain injury and changes in the intracerebral pressure caused by a sudden blow, resulting in a rotational deformation of the brain, with consequent lack of motor coordination and severe pain.

The use of a mallet induces both direct and indirect lesions in the animals. The direct lesions are in the form of a severe lesion of the bone tissue, with depression of the affected region. The indirect lesions are macroscopic and microscopic haemorrhages in the pons and the bulb, that is, a haemorrhage at the opposite point to the blow to the brain caused by the counter-blow of the basilar portion of the occipital bone. It produces a cranial-encephalic contusion, but not a concussion, as reported by several researchers (Roça 1999).

The head-only electrical stunning technique for sheep is commonly employed using a hand-held electrode placed between the eyes and the base of the ears on both sides. It causes an immediate and prolonged increase in the blood pressure of the stunned sheep. Electrical stunning (head to the back method), which is used for both sheep and calves, causes cessation of circulation and an immediate drop in blood pressure (Blackmore and Newhook 1982). In cattle, electrical stunning induces excessive convulsions and has adverse effects on pH and meat quality (Anil 2012).

Severe brain lacerations usually result from pneumatic-powered air injection stunners (Roça 1999). The use of captive-bolt stunners (pneumatic or cartridge-fired) causes CNS damage and spreading of CNS tissues throughout the animal's organs (Schmidt et al. 1999). In cattle, it was found that respiration ceased in all animals when they were stunned with captive-bolt stunning and did not resume (Vimini et al. 1983). Such dead animals are not lawful (haram) for Muslims.

Gas killing methods may be used on poultry and young animals. This method should guarantee that the animal is dead at the end of the exposure. If carbon monoxide is used in a confined space, the method is hazardous for operators. Inhalation of a high concentration of carbon dioxide on its own or with argon or nitrogen may be distressing to the animals (European Food Safety Authority 2004).

Post-slaughter stunning is used in some countries, such as Australia and New Zealand, through immediate thoracic sticking after the halal neck cut to avoid problems of prolonged consciousness. However, this method has the potential to cause carcass quality problems if bleeding is impaired (Pleiter 2005).

Table 7.1 Differences between halal and recent slaughter methods.

Comparison	Halal slaughter	Stunning slaughter
State of the animal	The animal should be alive	The stunned animal may die before slaughtering
Resting of the animal	The animal should be rested before slaughtering	Pre-slaughter stunning induces severe stress for the animal
Pain	Minimal pain (due to disruption of the sense centres in the brain and loss of consciousness as a result of blood shortage immediately after cutting of the common carotid arteries)	Pre-slaughter stunning induces severe pain for the animal
Bleeding	Perfect (complete bleeding process) due to severing of all throat vessels and keeping the head connected to the carcass through the spinal cord	No bleeding if the animal dies due to stunning; if still alive after stunning the bleeding is imperfect
Meat safety	Safe for human consumption	If the animal dies due to stunning, the meat becomes unsafe for human consumption
Infectious diseases	Protects consumers from infectious diseases (the complete bleeding process stops the growth and multiplication of microorganisms)	Represents a source of bacterial diseases, due to imperfect bleeding, and incurable prion diseases, due to contamination of the meat with brain tissue
Rigour mortis and glycolysis process	Normal (the breakdown of glycogen content of animal muscles into lactic acid via an anaerobic glycolytic pathway)	Abnormal
Meat pH	The meat pH is within the normal ultimate values, which provides an unfavourable medium for the growth of bacteria, increases shelf-life, keeps the colour of the meat bright red, and makes the meat tasty	Higher pH than normal results in dark, firm, and dry (DFD) meat, which has a shorter shelf life Lower pH than normal results in pale, soft, and exudative (PSE) meat, which provides a favourable medium for the growth of bacteria
Conclusions	According to the scientific basis, the halal method is the best method of slaughter because it is characterized by minimal pain sensation, complete drainage of liquid blood from the carcass, increased shelf life, and increased meat quality as well as improved meat safety and hygiene	The pre-slaughter stunning methods have disadvantages relating to animal welfare, meat safety and hygiene, and public health The use of any type of pre-slaughter stunning makes the meat unlawful (haram) for Muslims due to incomplete bleeding as well as resulting in low-quality meat

Finally, stunning chickens in an electric water trough may lead to the death of the chicken before the neck cut. The products of some poultry slaughterhouses may therefore be considered as electrically stunned.

7.14   Simple Comparison Between Halal Slaughter and Slaughter involving Stunning

Stunning slaughter methods have disadvantages relating to animal welfare, meat safety and hygiene, and public health. In addition, the stunned animal may die before slaughter. Inadequate bleeding is one of the most common disadvantages of the recent slaughter methods. The halal slaughtering method has many advantages for animal welfare, minimal pain sensation, and perfect bleeding as well as meat safety and hygiene. A comparison of these two types of slaughtering method is given in Table 7.1.

7.15   Conclusion

Methods of slaughtering such as stunning, shutting, electrical shock etc. used in Western countries hinder the bleeding process. Blood is an enrichment medium for the growth and multiplication of various microorganisms, so stunned meat may act as a source of infection of bacterial diseases for consumers. In addition, contamination of the organs and muscles with the brain tissue, due to certain types of stunning, may act as source of infection with incurable prion diseases. It has been shown that stunning can accelerate the ageing of meat and result in changes in some quality parameters, such as water loss and colour. Recent slaughter methods also impair the glycolysis process (rigor mortis) and result in low-quality meat. It has been confirmed that the halal method of slaughtering is the best method for perfect bleeding. Furthermore, the halal method is of great importance for human health as it protects consumers from infectious diseases and has a significant impact on meat safety and hygiene. It is recommended that in addition to Muslims, non-Muslims should also avoid recent slaughter methods and practice the non-stunning and hand-slaughter halal method to benefit from these advantages.

References

Anil, M.H. (2012). Effects of slaughter method on carcass and meat characteristics in the meat of cattle and sheep. http://www.eblex.org.uk/wp/wp-content/uploads/2013/04/slaughter_and_meat_quality_feb_2012-final-report.pdf (accessed on 16 September 2019).

Bager, F., Shaw, F.D., Tavener, A. et al. (1990). Comparison of EEG and ECoG for detecting cerebrocortical activity during slaughter calves. *Meat Science* 27 (3): 211–225.

Bender, A. (1992). Meat quality. In: *Meat and Meat Products in Human Nutrition in Developing Countries*. Rome: Food and Agriculture Organization (FAO).

Bergeaud-Blackler, F. (2007). New challenges for Islamic ritual slaughter: a European perspective. *Journal of Ethnic and Migration Studies* 33 (6): 965–980.

Blackmore, D.K. and Newhook, J.C. (1982). Electroencephalographic studies of stunning and slaughter of sheep and calves – Part 3: The duration of insensibility induced by electrical stunning in sheep and calves. *Meat Science* 7 (1): 19–28.

Cortise, M.L. (1994). Slaughterhouses and humane treatment. *Revue Scientifique et Tecnnique Office International des Epizooties* 13 (1): 171–193.

CSIRO (2011). Effect of slaughter method on animal welfare and meat quality. Meat Technology Update. http://www.meatupdate.csiro.au/data/MEAT_TECHNOLOGY_UPDATE_11-1.pdf (accessed on 11 March 2019).

Dave, D. and Ghaly, A.E. (2011). Meat spoilage mechanisms and preservation techniques: a critical review. *American Journal of Agricultural and Biological Sciences* 6 (4): 486–510.

Devine, C.E., Ellery, S., Wade, L., and Chrystall, B.B. (1984). Differential effects of electrical stunning on the early post-mortem glycolysis in sheep. *Meat Science* 11: 301–309.

European Food Safety Authority (2004). Welfare Aspects of Animal Stunning and Killing Methods. Scientific Report of the Scientific Panel for Animal Health and Welfare on a request from the Commission related to welfare aspects of animal stunning and killing methods.

European Food Safety Authority (2006). The welfare aspects of the main systems of stunning and killing applied to commercially farmed deer, goats, rabbits, ostriches, ducks and geese and quail. *The EFSA Journal 2006* 326: 1–18. Scientific Report. Question No. EFSA-Q-2005-005.

Farouk, M.M. (2013). Advances in the industrial production of halal and kosher red meat. *Meat Science* 95: 805–820.

Farouk, M.M., Al-Mazeedi, H.M., Sabow, A.B. et al. (2014). Halal and kosher slaughter methods and meat quality: a review. *Meat Science* 98: 505–519.

Gil, J.I. and Durao, J.C. (1985). *Manual de inspeção sanitária de carnes*. Lisboa: Fundação Caloustre Gulbenkian 563pp.

Grandin, T. (2010). Auditing animal welfare at slaughter plants. *Meat Science* 86: 56–65.

Hajimohammadi, B., Ehrampoush, M.H., and Hajimohammadi, B. (2014). Theories about effects of Islamic slaughter laws on meat hygiene. *Health Scope* 2 (4): e14376.

Kirton, A.H., Bishop, W.H., Mullord, M.M., and Frazerhurst, L.F. (1978). Relationships between the time of stunning and time of throat cutting and their effect on blood pressure and blood splash in lambs. *Meat Science* 2: 199–206.

Kranen, R.W., Lambooij, E., Veerkamp, C.H. et al. (2000). Haemorrhages in muscles of broiler chickens. *World's Poultry Science Journal* 56: 94–126.

Lahucky, R., Palanska, O., Mojto, J. et al. (1998). Effect of preslaughter handling on muscle glycogen level and selected meat quality traits in beef. *Meat Science* 50 (3): 389–393.

Limon, G., Guitian, J., and Gregory, N.G. (2010). An evaluation of the humaneness of puntilla in cattle. *Meat Science* 84 (3): 352–355.

Linares, M.B., Bornez, R., and Vergara, H. (2007). Effect of different stunning systems on meat quality of light lamb. *Meat Science* 76: 675–681.

Lynch, G.P., Oltjen, R.R., Thornton, J.W., and Hiner, R.L. (1966). Quantitative changes in the free plasma amino acids and other ninhydrin reactive substances of sheep due to preslaughter stunning. *Journal of Animal Science* 25 (4): 1133–1137. https://doi.org/10.2134/jas1966.2541133x.

Mustaqim Islamic Art and Literature (2014). The halal slaughter controversy: do animal rights activists protect the sheep or the butcher? http://www.mustaqim.co.uk/halal.htm (accessed on 16 September 2019).

Petty, D.B., Hattingh, J., Ganhao, M.F., and Bezeuidenhout, L. (1994). Factors which affect blood variables of slaughtered cattle. *Journal of the South African Veterinary Association* 65 (2): 41–45.

Pleiter, H. (2005). Electrical stunning before ritual slaughter of cattle and sheep in New Zealand. In: *Animal Welfare at Ritual Slaughter* (eds. J. Luy et al.). DVG Service GmbH.

Roça, R.O. (1999). *Abate humanitário: o ritual kasher e os métodos de insensibilização de bovinos*, 232p. Botucatu: FCA/UNESP. Tese (Livre-docência em Tecnologia dos Produtos de Origem Animal), Universidade Estadual Paulista.

Roça, R.O. (2002). Humane slaughter of bovine. *First Virtual Global Conference on Organic Beef Cattle Production* (2 September to 15 October 2002), pp. 1–14.

Sante, V., Le Pottier, G., Astruc, T. et al. (2000). Effect of stunning current frequency on carcass downgrading and meat quality of turkey. *Poultry Science* 79: 1208–1214.

Schmidt, G.R., Hossner, K.L., Yemm, R.S. et al. (1999). An enzyme-linked immunosorbent assay for glial fibrillary acidic protein as na indicator of the presence of brain or spinal cord in meat. *Journal of Food Protection*, Desmonines 62 (4): 394–397.

Swatland, H.J. (2000). Slaughtering. 10 pp. http://www.bert.aps.uoguelph.ca swatland/ch1.9.htm (accessed on 16 September 2019).

Vimini, R.J., Field, R.A., Riley, M.L., and Varnell, T.R. (1983). Effect of delayed bleeding after captive bolt stunning on heart activity and blood removal in beef cattle. *Journal of Animal Science* 57 (3): 628–631. https://doi.org/10.2134/jas1983.573628x.

Zulkifli, I., Goh, Y.M., Norbaiyah, B. et al. (2014). Changes in blood parameters and electroencephalogram of cattle as affected by different stunning and slaughter methods in cattle. *Animal Production Science* 54: 187–193.

8

The Religious Slaughter of Animals

Joe M. Regenstein

Department of Food Science, College of Agriculture and Life Sciences, Cornell University, Ithaca, NY, USA

8.1 Introduction

Muslim and Jewish religions are both religions of law. For people of these faith communities the consumption of food is regulated by a detailed set of laws that are part of a larger set of laws that influence all aspects of their daily life. To be able to understand the impact and importance of these laws on the slaughter of meats for these communities, it would be helpful to understand the broad outlines of the details of these laws and see how these particular laws affect the choice of method used to slaughter animals. This chapter will focus on introducing the Muslim dietary laws and then on the challenges and needs for the slaughter of animals to meet Muslim requirements. However, appropriate practices in the Jewish religion will be considered when they can be helpful in understanding and/or improving Muslim religious slaughter from an animal welfare perspective. The major challenges to the religious slaughter of animals come from the secular world, which tries to lump both Muslim and Jewish religious slaughter together, generally to attack them. So the two communities need to work together to understand what they are each doing, to determine what are the religious limitations to any changes, and to improve the religious slaughter of animals so it can meet both the highest religious standards and the highest animal welfare standards. This author strongly believes that this is eminently possible and that the two communities would benefit from addressing this issue collectively.

A comprehensive review of both the kosher and halal food laws, targeting those in the food industry, was written by Regenstein et al. (2003). The current chapter will summarize some of the material in that paper, but will be much more focused and detailed with respect to the religious slaughter of animals and the current issues surrounding religious slaughter.

For Muslims, one of the major components of obtaining lawful, halal food is the concern that the meat of an animal is considered acceptable. This starts with determining which animals are lawful. The most commercially relevant mammals are beef, sheep, and goat, although water buffalo, bison, deer, elk, camel, and rabbit are raised commercially and,

The Halal Food Handbook, First Edition. Edited by Yunes Ramadan Al-Teinaz, Stuart Spear, and Ibrahim H. A. Abd El-Rahim.
© 2020 John Wiley & Sons Ltd. Published 2020 by John Wiley & Sons Ltd.

therefore, may also be slaughtered and available for food. These are all halal animals. The pig and most animals people would consider carnivorous animals or those which are a 'pest' are prohibited, i.e. haram. There is some controversy within the Muslim community as to the acceptability of horse meat.

Once the animal is determined to be a halal animal, the slaughter of such an animal must then be done in accordance with the halal requirements. Where do these laws come from? The Muslim laws are found in the Holy Quran, the recitation of the angel Gabriel to the Prophet Mohammed (peace be upon him, pbuh) from 610 to 632. In addition the Hadiths and the Sunna are the record of the sayings and doings of the Prophet Mohammed (pbuh) during his time as the leader of the initial Muslim community. Those Hadiths and Sunna that have been authenticated by Muslim scholars serve as the basis for Muslim law, or Shariah, of which the laws of halal are a subset. A short time after the death of the Prophet (pbuh), the Muslim community split into two sects based on a difference of opinion on how the future leadership of the community would be chosen. This led to the Shia/Sunni split, with about 15% of Muslims being Shia. Within the Sunni community a number of 'schools' arose reflecting the thoughts of a leader and his teachings, generally in geographically dispersed parts of the Muslim world. The four major Sunni schools have some differences with respect to halal food laws, although the big picture is fairly consistent. In the Shia community there are also subgroupings, which approach some of the food laws differently.

Given the spread of Islam from West Africa to China and Southeast Asia, this meant that these communities were often isolated from each other, and the customs and interpretations of the local Muslim leaders led to some additional differences in the Muslim community. The most dramatic differences, as far as this author can determine, deal with what are acceptable fish and seafood (see below).

Arriving at a consensus is a desirable goal within the Muslim community. Thus, there have been and continue to be many efforts to come up with a single global standard for halal foods. Although that is a wonderful goal, it is unrealistic after so many years of separation and the intensity of support for practices that are so much a part of each local (national) community's identity. So, these possible standards are then subject to ongoing discussions by Muslim scholars.

However, the author of this chapter believes that the Muslim community would be better served if they were to recognize some of these differences as real and have a sufficient depth of feeling that they need to be incorporated into any global halal standard. Thus, the global halal standard would need to identify these differences, accept them as authentic, and most importantly apply a trade standard to then assure that a consistent system of marking food products exists globally so that Muslim consumers around the world can make an informed choice that respects and permits these differences to be honoured.

How does kosher and halal meat differ from meat in the secular Western world? Because the kosher and halal requirements require that slaughter be different from the normative activities of the meat industry, at least in the Western world, the marketplace provides meat specifically marked as kosher and/or halal for these two religious groups and this often serves as a value-added niche market for the meat industry. Obviously, in many countries with large Muslim populations the religious slaughter of animals may be the normative standard. Because these countries are assumed to operate with a halal standard, the average

Muslim consumer may not be aware of the details of how his/her meat is being prepared with respect to any differences of belief within the Muslim community. In fact, research by the Muslim Council of Britain clearly showed that the industrial practices with respect to the slaughter of animals for halal meat and the religious desires of Muslim consumers are not necessarily in sync (see below).

What is the broader framework for halal? It is important to understand the broader set of rules governing halal practices so the reader can understand how the slaughter of animals and processing of meat fit into the overall market for these products.

The halal dietary laws deal predominantly with three broad issues. Two of these laws are focused on the animal kingdom:

A) Allowed animals
B) Prohibition of blood
C) Prohibition of ethyl alcohol

We'll have a brief look at the last item before focusing on the first two items. The prohibition of ethyl alcohol (ethanol) is mainly associated with the plant kingdom, although there have been and continue to be attempts to produce alcohol from whey that suggests that this generalization will not hold in the future. The use of non-beverage, industrial alcohol in food production remains controversial, especially when it remains in food products at very low levels.

8.2 Allowed Animals

For Muslims, as indicated above, the pig is uniquely prohibited. Thus, many Muslims will wish to avoid all contact with the pig and its by-products regardless of whether the application is food related or not. Thus, using pig by-products for leather, medicines, and paint brushes, for example, can negatively impact the ability to serve the Muslim market.

With respect to poultry, which has not yet been covered, the traditional domestic birds, i.e. chicken, turkey, squab, duck, and goose, are halal along with the new commercially available category of birds in the ratite category: ostrich, emu, and rhea. Again birds of prey are not acceptable.

Seafood, as previously mentioned, is the area that seems to have the greatest diversity of views. Permitted seafood according to the Holy Quran is those sea animals that have spent their entire life in the water. The interpretation of theses Quranic verses varies widely in the Muslim community and reflects both the major divisions of Islam, i.e. Sunni and Shia, along with the schools within both of these major divisions, and also is influenced by the customs of people in different parts of the Muslim world. Thus, determining what seafood is acceptable in a given community is vital for the food industry – a list of the key commercial fish/seafood annotated to indicate their acceptability in different Muslim communities around the world would be a useful document to have.

Most insects are not halal. The Muslim community considers most insects as pests but allows insects in the locust/grasshopper family for food based on the tradition that the Prophet Mohammed (pbuh) consumed them. Assuring the integrity of halal food products with respect to the absence of insects, especially those that are visible, is something that has not received any commercial attention. This might deserve some consideration and offer some commercial opportunities. The Jewish community considered whole insects unacceptable and has been working with the food industry to create products that are religiously free of bugs. Is there any market for such products or technology in the Muslim community?

The whole area of the food use of insect products is somewhat controversial and includes issues around products such as lac resin (shellac), which is obtained as an exudate from an insect, and carmine/cochineal, which is the pigment obtained from the shell of an insect. Again the halal status of these materials seems, according to most authorities, to be permitted. Honey and other products obtained from bees also seem to be accepted as they do not require the pest/vermin to be consumed.

With respect to domestic animals under the control of humans, there is a concern that the animals should not be fed any filth, particularly prior to slaughter. The time of concern varies depending on the animal and the different traditions, but in some cases for up to 40 days prior to slaughter it seems that the animal should not be intentionally fed filth, which is often defined as animal by-products including rendered meals, i.e. even those that have had proper heat treatment to make them safe for animals to consume, and any manure-based products, which have been proposed as animal feeds although the use of manure is currently quite limited. However, most lists of 'filth' (*najis*) are not consistent, so again some clarification of this category in terms of what is and is not acceptable in different countries would be helpful for the food industry.

8.3 Prohibition of Blood

Blood has traditionally been considered as the life fluid for humans and animals. Thus, it is prohibited for consumption and use in Islam as it is in Judaism. In addition to animal welfare considerations, the goal of the religious slaughter of animals is to ensure that the slaughter leads to the removal of as much blood as possible. This has led to an emphasis in the religious community on being sure that the animal is alive at the time of slaughter so that its heart is still pumping, which is consistent with the Muslim definition of life as requiring a functioning heart. Whether more blood is actually removed or not is difficult to prove, but the limited data available suggest that the amount of blood removed is fairly similar regardless of the method of slaughter used, whether religious or secular variations. In most experiments there have been no statistical differences, and although a few do show a small but statistically significant difference, from a practical point of view the differences are quite small and are probably not meaningful.

Before proceeding, a few key words need to be discussed/defined.

- *Consciousness/unconsciousness.* When an animal is conscious it is able to feel pain. Once it becomes unconscious it cannot feel pain. The normal practical test for unconsciousness is the loss of righting ability, i.e. the ability of an animal to stand up on its own. At this point it is appropriate to hang an animal by one leg, i.e. shackle it to the normal commercial system used to process meat. If the animal is upside down, determining unconsciousness becomes more difficult and research on determining unconsciousness *in situ* with various religious slaughter of animal systems is recommended.

- *Sensible/insensible.* When an animal is slaughtered, there are voluntary and involuntary movements. Often one sees involuntary movements that appear to suggest that the animal is still alive. In fact many of these movements are not relevant to a discussion on sensibility. The voluntary movements in the head are the key to whether the animal is still sensible. Generally, the last sense to go is the eye reflex. So, if an animal is poked on the eyelid and does not blink, then it can be defined as insensible. This is normally when further cutting of the animal can be started. Insensibility is probably equivalent to brain death.

- *Death.* Defining this term can be controversial. Traditionally death has been defined as the loss of heart function, but in recent years the definition has focused on brain death. Defining death in terms of brain death, which usually comes before heart death, allows for the removal of a heart from a person so that it can be used for heart transplants. For Muslims, heart death remains the normative standard for humans, but how this is defined in conjunction with the religious slaughter of animals remains for Muslim scholars to decide. It might be noted that in the 20th century, a large part of the Orthodox Jewish community has gone to brain death because of the importance of saving a human life.

- *Stunned/unstunned.* In modern Western society, an animal is generally made unconscious prior to cutting its blood vessels. In many cases this may also lead to insensibility, usually with a method that is irreversible, i.e. stun to kill. On the other hand, some of these techniques are in principle reversible, i.e. stun to stun. However, determining that every animal has only been stunned can be difficult. For Muslims this is a critical issue because the animal must be alive at the time of slaughter. The demonstration of reversibility under ideal conditions is not sufficient to ensure that all animals are alive at the time of slaughter in a slaughter house with many different sizes and conditions of animals, and the wear and tear on equipment and people in the course of ongoing slaughter. In addition to religious slaughter, many other forms of killing of animals are done without prior intervention, e.g. hunting, many on-farm forms of animal slaughter, and bullfighting in countries where the bull is permitted to be killed. Like religious slaughter, the goal of these other forms of slaughter, except possibly hunting, is for the kill to specifically provide for a quick induction of unconsciousness. This also raises the issue of how to evaluate the scientific literature that is focused on the slaughter of an animal with a cut across the neck without prior intervention. In many cases, the procedures used for the unstunned slaughter of an animal may not be relevant to religious slaughter, which has specific restrictions, and thus cannot be logically extrapolated to the religious slaughter of animals, although this may be done by researchers. An example of the misuse of work on the slaughter of animals without prior intervention will be discussed later in this chapter.

The actual vocabulary used to describe issues related to the religious slaughter of animals can have a large impact on how people perceive the actions behind the words. In the appendix to this chapter, preliminary data taken from classes at Cornell is used to show how important these considerations can be and how the choice of words influences people's perception of what is taking place. Thus, it is important that the words used in these discussions are carefully selected. It is also clear that many people do not understand these words and that opinions can be changed by providing factual information describing the various aspects of the slaughter of animals. Hopefully, this chapter will help work towards the greater understanding and tolerance of different approaches to the slaughter of animals. Thus, this author is intentionally avoiding as far as possible using the words 'unstunned slaughter'.

The removal of blood is, of course, the essence of the religious slaughter of animals. Some of the extremely detailed requirements that have been worked out for kosher slaughter are shared here because they may serve as a model for potential improvements in halal slaughter.

Ruminants and fowl must be slaughtered according to Jewish law by a specially trained religious slaughterer (the shochet) using a special knife designed specifically for the purpose of slaughter (the *chalef*). The knife must be extremely sharp and have a very straight blade that is at least twice the diameter of the neck of the animal to be slaughtered. This requirement for the size and straight shape of the knife is now being recognized as an important consideration with respect to the religious slaughter of animals and has slowly been moving into Western regulatory language, e.g. the recent English and Welsh rules to implement EU regulations for improving animal welfare in slaughter houses contains language specifically defining the length of the knife although the language was removed following the public comment period. Dr Temple Grandin of Colorado State University strongly believes that relating the knife length to the neck length helps prevent gouging and other potential problems during slaughter.

Dr Temple Grandin is a Professor of Animal Science at Colorado State University. She has designed much of the special equipment used for religious slaughter as well as designing most of the slaughter plants in the USA (over 50% of animal slaughter in the USA occurs on systems she designed). Much of the rest of the world also uses slaughter and other animal handling facilities that either she has designed or that follow her principles of animal welfare.

A great deal of the training of a Jewish slaughterer goes into making sure the knife is razor sharp and being absolutely certain that the knife is free of nicks. The knife is checked by running its entire working blade along a finger nail both before and after each slaughter of ruminants, which, with practice, allows any nicks to be detected. For poultry, the knife may not be checked before and after each slaughter, but as a result all animals slaughtered between the checking of the knives may be declared unacceptable for kosher. A trained shochet will find 'nicks' in the knife that most secularly trained knife sharpeners will miss. It is the author's hope that in the future this extensive and intensive training on knife sharpening can be extended to the Muslim community.

The need for a sharp knife is pretty universally accepted when it comes to the slaughter of animals, but the importance of having no nicks on the knife has not been fully appreciated nor has much research been done on measuring the impact of this factor. The assumption is that the presence of nicks during cutting irritates the cut vessels and the irritation might be responsible for any 'pain' signals associated with the cutting which would be absent if the nicks were not present. A recent series of papers by Gibson et al. (2009a–d) on unstunned slaughter is hard to evaluate as to its relevance, if any, to religious slaughter of animals because the knife was too short and machine sharpened (without any details being given beyond that). Thus, the work may not represent good conditions for the religious slaughter of animals, which makes it difficult to understand the relevance of these papers to any discussion of religious slaughter although they have been widely misused for this purpose (Gibson et al., 2009a–d).

The shochet will rapidly cut the jugular veins and the carotid arteries along with the oesophagus and trachea without burrowing, tearing, or ripping the animal. 'This process when done properly leads to a rapid death of the animal. A sharp cut is also known to be less painful.' (Grandin and Regenstein 1994).

Recently a company in New Zealand developed an instrument capable of determining knife sharpness (Figure 8.1). We have had the opportunity to test this equipment and not only does it quantitatively determine sharpness, it also seems to be able to show the presence or absence of nicks. Figure 8.2a shows a typical knife and illustrates the challenges of properly sharpening a knife. Figure 8.2b is data collected by the author. The top trace is for a Jewish slaughter knife that was ready for use. The bottom trace is for a special Muslim slaughter knife that we have developed (shown in Figure 8.3) that meets the traits to give an improved halal knife. It had been 'sharpened' but not specifically for slaughtering an animal. The shochet who sharpened the knife in the top trace then spent 15 minutes sharpening the Muslim slaughter knife and declared it better but not yet ready to be used to slaughter an animal. Our effort is now focused on how to confirm this information and then begin to use the instrument in working with the Muslim community to introduce the knife and the instrument to improve halal slaughter.

Before the slaughter occurs, the shochet will quickly check the neck of the animal to be sure it is clean and there is nothing on the neck that could harm the knife. If there is a problem, the neck of the animal needs to be washed, which slows down the slaughter process. It may also serve to stress the animal but does ensure a clean cut. It may be desirable to arrange for animals destined for religious slaughter to be washed just prior to entering the slaughter area so that time is not lost washing the neck. This will presumably also be less stressful for the animal as they will have time to recover from the stress of either washing and/or the removal of wool prior to slaughter. This is another area where the Muslim community may wish to consider changes in their practical approach to handling animals. Again, there is research work being done both academically and industrially in New

(a) (b)

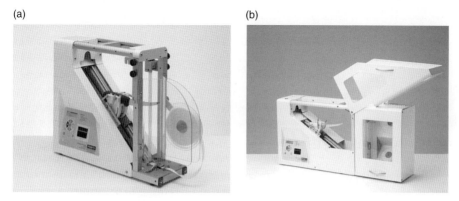

Figure 8.1   The Anago knife sharpness tester: (a) an overview and (b) a close-up of the knife interacting with the test mesh.

(a)

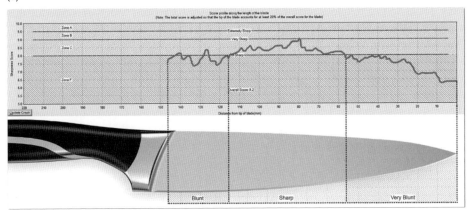

(b)

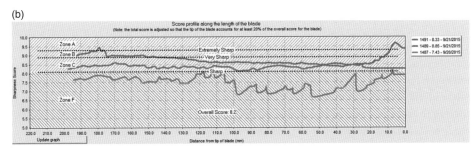

Figure 8.2 A comparison of knives using the Anago knife sharpness tester: (a) a regular knife showing how the shape of the knife affects the sharpness and (b) the results of three tests. The top knife trace is that of a very sharp knife ready to be used for the religious slaughter of sheep and goats. The bottom knife trace is that of the knife shown in Figure 8.3. After 15 minutes of sharpening the bottom knife was much better but still not ready for use in the slaughtering of a sheep or a goat.

Figure 8.3 A Muslim knife designed to meet the standards discussed in this chapter.

Zealand (and possibly elsewhere) to design equipment specifically to wash animals prior to slaughter. Its relevance to the religious slaughter of animals needs to be evaluated.

During the actual slaughter the following five kosher laws must be observed by the shochet. These also represent considerations that might become part of Muslim slaughter:

- *No pausing (shehiyyah)*. The cut can be multiple continuous strokes (Muslim requirements are similar), but generally speaking Dr Grandin has learned that fewer more aggressive strokes are better. The author does note that in general the Muslim slaughterers seem to be more aggressive than the shochets. Hopefully, this positive trait can be more actively encouraged in the Jewish community.
- *No pressure (derasah)*. There is the concern that the head may fall back on the knife. This, of course, makes slaughter more difficult and possibly painful. The sharpness of the knife should be what is doing the cutting. The use of a proper head-holder is critical here, particularly with upright slaughter. However, a head-holder is also needed for upside-down slaughter.
- *No burrowing (haladah)*. The knife has to do its job by cutting.
- *No deviating (hagrama)*. There is a correct area for cutting. Work by Dr Temple Grandin suggests that the upper limit of this allowed range is the best for overall animal welfare. The area for cutting for Muslims is similar. The problem is that some researchers suggest an even higher cut (the C1 area), outside the area allowed religiously. This may then lead to cutting of the larynx and the bones that are found in the larynx. This would not be good for the knife and the slaughter would then not meet religious requirements.
- *No tearing (ikkur)*. If the neck is stretched too tight, tearing may occur before cutting. If it is too loose then pressure on the knife may occur. It is for this reason that Dr Grandin has developed a special head-holder for religious slaughter that is designed to sit away from the cutting area on the neck, giving the slaughterer good access to the critical area (Figure 8.4). The head-holder also gives the right amount of tension, so one gets neither tearing nor covering of the cut, and allows the eyes to be observed as the eyes are the last organ in the head to become insensible. The question of whether the current placement should be changed to be slightly less taut is being considered by Dr Grandin (personal communication).

If the initial cut is not done fully, certain additional cuts can be made by the non-Jewish helper to facilitate the rapid removal of blood, e.g. the blood vessels may need to be more

Figure 8.4 A head-holder designed by Dr Temple Grandin.

completely cut. In some slaughter plants (e.g. those using light stunning) additional cuts are made after the throat cut during Muslim slaughter while in the case of camels, where a special form of slaughter (*nahr*) is done, the additional cuts are made before the horizontal cut across the throat.

Who can slaughter halal? All adult Muslims with normal mental health are permitted to slaughter. They are taught to use a sharp knife. The traits of this knife and its sharpness are not specified in as great detail as those for the chalef. Thus, further emphasis and training on getting knives for Muslim slaughterers to the same high sharpness and nick-free status as kosher knives is needed. This may be an area where the two communities can work together for everyone's benefit.

All the preliminary steps to slaughter must be optimized to ensure that animals will be 'unstressed' at the time of slaughter. This is important for all slaughter systems, as ideally calm animals are wanted so that higher quality meat is obtained. Again, Dr Grandin is the recognized expert on designing and managing systems to get animals to and into the critical restraining systems with minimal stress.

There is then a need to consider the various systems and equipment that are available to help improve the religious slaughter of animals. Some restrainers are actually quite simple and low cost; others can be designed to meet the most demanding high-speed production requirements. A few of the halal slaughter systems available are discussed below.

The American Meat Institute (AMI, now the North American Meat Association, NAMI) has for many years made available slaughter guidelines. These have been widely accepted for a number of years in the USA and are even accepted by almost all animal activist groups. These guidelines have called for an upright religious slaughter of animals, using one of the many restraining devices available for this purpose. For some groups within the Jewish and Muslim communities, however, upright slaughter may be unacceptable. Upside-down slaughter is either preferred as better because it reflect traditional slaughter and/or is felt to be better in ensuring that the rules of religious slaughter of animals are not violated. Thus, the author, working with Dr Grandin, who has written the AMI standards, arranged to have these standards modified to account for this need, although the long-term goal of NAMI, Dr Grandin, and the author is to move more, if not all, of the religious slaughter of animals to the upright position.

In both religions the normative practice is that the animal is not stunned prior to slaughter, which is controversial in the Western world where the requirement for stunning has become a matter of secular faith. Many people simply cannot believe that a well done religious slaughter may actually be better (currently a hypothesis) than the current stunning systems. One critical component of this discussion is the failure rate of current stunning systems. The current NAMI standard permits 5% of the animals to be mis-stunned on the first try and the actual figures in the best slaughterhouses in the USA now tend to be below 2% (T. Grandin, personal communication). Having an animal mis-stunned will lead to an animal that is under very high stress, which is not good animal welfare and will have a negative effect on meat quality.

It is important to recognize that religious slaughter of animals takes more effort to do right and is slower. The use of upside-down slaughter systems makes it even more difficult to do right, although it can be done with proper training of slaughterers and good support from management. For upside-down slaughter the slaughterhouse needs more expensive equipment that is more difficult to maintain, and the animal to be slaughtered, even if handled perfectly, is likely to become perturbed by the upside-down system. However, if done quickly after turning, Dr Grandin has postulated that there is a short time period (probably about 10 seconds) where the animal does not realize what has happened. So if at all possible the upside-down slaughter should be done quickly, ideally as the upside-down box is just getting into position. Handling systems that turn the animal upside down exist but some of the older versions are not very acceptable. The most modern versions, properly operated, may be acceptable if the rest of the animal handling is well done. These need to be designed to properly support the animal when upside down and the mechanicals need to be fast and quiet.

In all cases, most of the animal welfare and quality requirements for high-quality slaughter outcomes are not directly related to meeting the religious requirements. Therefore, these requirements can and should be met and animals slaughtered to the highest modern animal welfare and quality requirements. For both humane and safety reasons, plants which do religious slaughter of animals need to install the most modern restraining equipment to hold the animal in place with a proper head-holder. The practice of hanging live cattle, calves, or sheep upside down (i.e. shackle and hoisting and any variants of that practice) needs to be eliminated, first for cattle but also for sheep, goat, and veal in the near future. There are many different types of humane restraint devices available at all different price ranges.

Note that poultry is a different story and upside-down hanging is not nearly as traumatic for birds. Remember that many birds can fly almost straight down! The standard shackling line is also permitted to be used for religious slaughter of poultry, i.e. the animals can be shackled prior to slaughter although most kosher and halal slaughter is done with the animal being held by the slaughterer or his helper to do the cut. Subsequently the animals are then put on the shackles or into a bleeding cone.

Examples of acceptable upright restraint systems for mammals include the American Society for the Prevention of Cruelty to Animals (ASPCA) pen. This device is highly desirable for reasonable commercial rates of the religious slaughter of animals and is probably the most common equipment used for this purpose. It consists of a narrow stall with an opening in the front for the animal's head. The instructions for operating this pen are based on the work of Dr Grandin. After the animal enters the box, it is nudged forward with a pusher gate and a belly lift comes up under the brisket but does not lift the animal off the floor. The head is restrained and lifted to the right tension level, as determined by the religious authorities, using a chin lift so that the head is properly positioned and the neck area readily available for the slaughterer. Vertical travel of the belly lift should be restricted to 71 cm (28 in.) so that it does not lift the animal off the floor.

If lifting the animal off the ground is required for religious reasons, the belly lift could be modified to support the animal, but the belly lift is not designed to be comfortable or strong enough for this at this time. One possibility would be to put a small 'double rail' (see below) in place, so that the animal is comfortable off the ground with its body properly and comfortably supported. An alternative is to lift the whole pen or to tilt the entire pen forward so it is off the ground, although the animal's legs in this case would still be on the floor of the pen. The rear pusher gate should be equipped with either a separate pressure regulator or special pilot-operated check values to allow the operator to control the amount of pressure exerted on the animal. The pen should be operated from the rear towards the front. Restraining of the head is the last step. The operator should avoid sudden jerking of the controls. All hydraulic equipment needs to be properly baffled as the noise itself may disturb the animal. Many cattle will stand still if the box is slowly closed up around them and less pressure will be required to hold them. Religious slaughter of the animals should be done immediately after the head is in the head-holder.

If the animals are too large for the pen, the pen size may need to be adjusted. At the very least the rear pusher gate should probably not be used. This pen has a maximum capacity of about 100 cattle per hour and it works best at about 75 head per hour. Dr Grandin suggests that a smaller version of this pen could be easily built for calf plants.

For higher speed operations some type of conveyor restrainer system is needed and this includes both a V restrainer (Figure 8.5) and a centre track conveyor. The V restrainer is very common for sheep and smaller animals but not as common for cattle because of the difficulty of maintaining the system, i.e. for cattle you generally need to have two sets of 'shingles' that allow the system to hold an animal in between the two belts, the two must be synchronized and must be kept free of any rough surfaces. This device is similar to some baggage carousels in major airports. The centre track conveyor moves the animal supported by two metal supports that allow it to circle around, so these must fit well and again not pinch the animals. These two sets of supports ('double rail') can be used for holding cattle, sheep, or calves in an upright position during religious slaughter (Figure 8.6b–e). The conveyor systems must completely support the animal's body in a comfortable upright position. The restrainer is stopped for each animal and a head-holder similar to that for the ASPCA pen holds the head for the religious slaughterer. Research in Holland indicates that the centre track design provides the advantage of reducing

Figure 8.5 A V-restrainer system courtesy of Shanghai One-Stop Engineering Co., Ltd.

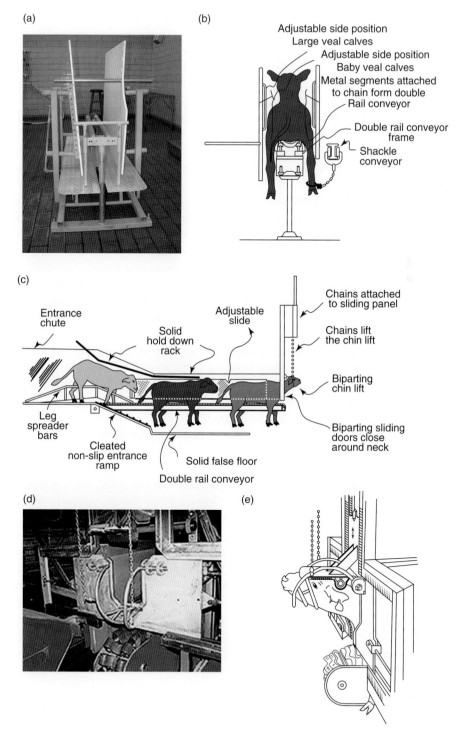

Figure 8.6 The double-rail system: (a) a small-scale double-rail restrainer for sheep and goats, (b) view of the restrainer from the back, (c) view of the restrainer from the side, (d) the head-holder, and (e) the head-holder and 'cleats'. *Source:* Pictures (b) to (e) are courtesy of Dr Temple Grandin.

bloodspots in the meat. In both of these systems the animal's feet are off the ground. Having the feet off the ground when the animal is comfortably supported seems to have a calming effect on the animal and therefore is highly recommended. A small-scale pen for sheep, goats, and small cattle that has the double rail in a static mode has been built (www.spiritofhumane.com) (Figure 8.6a). It uses pipe rails to create the double-rail effect. When the floor is lowered and the animal is supported by the double rail, the animal visibly calms down, which presumably has a very positive effect on meat quality and allows the religious slaughter to be done with less stress.

Regardless of the system used for the religious slaughter, the animals must be allowed to bleed out and become completely insensible before any other slaughter procedure is performed.

According to Dr Grandin (personal communication), in medium-sized plants where religious slaughter is not routinely done, but which want to do religious slaughter for a limited period of time, a modified SPCA pen could be assembled and installed in a plant on a weekend and could be disassembled and removed on a weekend so that the plant is ready to go on Monday morning without any lost time for the transition to or from animal welfare appropriate religious slaughter systems. Thus, if it were desired, a high-quality religious slaughter system could be set up in many plants around the world that are much more practical and lower in cost than the systems used by the larger plants, but better than the practices used to date in such plants.

From an animal welfare point of view, it is important to recognize that during the slaughter there are two rules that are practiced by Muslims. These need to be taken into account when designing slaughter equipment. The first is that the knife is not to be sharpened in the presence of (i.e. visible to) an animal that will be slaughtered and second that one animal should not see another animal being slaughtered.

Dr Grandin has summarized her attitude to the religious slaughter of animals in the following quote:

> Recently, I participated in a ritual kosher slaughter – in this ritual, the way it was meant to be done, I must say. This was at a plant where the management really understood the importance and significance of what they were doing, and communicated this to their employees – and to the animals as well, I believe. As each steer entered the kosher restraining box, I manipulated the controls to gently position the animal. After some practice, I learned that the animals would stand quietly and not resist being restrained if I eased the chin-lift up under the animal's chin. Jerking the controls or causing the apparatus to make sudden movements made the cattle jump ... Some cattle were held so loosely by the head-holder and the rear pusher gate that they could easily have pulled away from the rabbi's knife. I was relieved and surprised to discover that the animals don't even feel the super-sharp blade as it touches their skin. They made no attempt to pull away. I felt peaceful and calm. (Regenstein and Grandin 1992)

Notice how positive Dr Grandin is about religious slaughter when it is done properly. The behaviour of the animal suggests that death occurs without pain. In fact, there are various forms of anecdotal evidence that support the idea that opiate-type compounds called endorphins are released when an unstressed animal is cut with a very sharp knife with no nicks. These compounds then allow the animal to die on a 'drug high'. That the animal dies comfortably is exactly the goal of the religious slaughter and also ensures the highest quality of meat. In such a case, the time to death, if reasonable, is not the critical element as the quality of the death is what should be of greatest concern.

The animal welfare issues associated with religious slaughter are controversial and it is often difficult to separate the impact of pre-slaughter handling from those aspects directly related to the religious slaughter of the animal itself and to the work of the religious slaughterers. In addition, much of the current research on religious slaughter has been done on slaughter systems that are not fully described in the literature and which in many cases may represent systems that need to be upgraded. In many cases generalizations about all religious slaughter are made from very specifically selected and poorly operated facilities or laboratory research that did not meet some or all of the critical religious requirements, e.g. the previously machine-sharpened knife that is much shorter than a chalef described above.

A consensus paper describing some of the key components of reporting about kosher and halal slaughter needs to be prepared so that the quality of the research in this area can be upgraded and scientists can more critically evaluate the work of their peers. Just because a paper is peer-reviewed does not mean that some critical problems with the research do not still exist. If all the scientists in a specialty field have the same bias, then a paper may be accepted that was actually not well done but supports the prevailing dogma.

In contrast, a paper by Barnett et al. (2007) reviews a poultry slaughter operation and is unfortunately unique in that it does so with a very complete description of what happens during slaughter at the particular plant being studied. An example of doing the science right!

There are three basic issues that need to be considered when evaluating the religious slaughter of animals. They are the stressfulness of the restraint methods, which is the responsibility of the processing plant and the plant's management, the pain perception during the incision and what happens with respect to stress and pain until the animal is unconsciousness, which is the actual religious part of the slaughter and the point when an animal may be shackled, and, finally, the point when the animal is completely insensible so that further processing can occur, which again is a plant issue but is impacted by the quality of the religious slaughter.

Assuming that the animal arrives at the restraint system unstressed (which may not be true), which is the point at which religious slaughter departs from secular slaughter, then a key scientific consideration that can be difficult to study in practice is the separation of the variable of restraint stress from the animal's reaction to the slaughter procedure. Stressful or painful methods of restraint mask the animal's reactions to the throat cut. In some parts of the world kosher and halal slaughter plants use very stressful methods of restraint such as the previously mentioned shackling and hoisting of fully conscious animals by one rear leg. (Its use with cattle is no longer tolerated in the USA and most Western countries, but is acceptable in some countries that ship meat into Western countries.) In some countries practices even less acceptable than shackle and hoisting are used for religious slaughter and this remains totally unacceptable.

In somewhat more detail the following needs to be considered: Observations of [Dr Grandin] indicate that cattle restrained in this manner [shackle and hoisting] often struggle and bellow, and the rear leg is often bruised. ... In Europe, the use of casting pens which invert cattle onto their backs completely masks reactions to the throat cut. Cattle resist inversion and twist their necks in an attempt to right their heads. Earlier versions of the Weinberg casting pen are more stressful than an upright restraint device (Dunn 1990). An improved casting pen, called the Facomia pen, is probably less stressful than the older Weinberg pens but a well-designed upright restraint system would be more comfortable for cattle. An even newer casting pen has been built in Ireland and is recommended where upside-down slaughter is required. Another problem with all types of casting pens is that both cattle and calves will aspirate blood after the incision. This does not occur when the animal is held in an upright position.

Unfortunately, some poorly designed upright ASPCA restraint boxes apply excessive pressure to the thoracic and neck areas of cattle. In the interest of animal welfare the use of any stressful method of restraint should be eliminated. A properly designed and operated upright restraint system will cause minimum stress. Poorly designed systems can cause great stress.

Many stress problems are caused by rough handling and excessive use of electric prods. The very best mechanical systems will cause distress if operated by abusive, uncaring people.

(Note: The following text is taken from the AVMA Humans Slaughter Panel guidelines, with permission.)

> In Europe there has been much concern about the stressfulness of restraint devices used for both conventional slaughter (where the bovid is stunned) and religious slaughter. Ewbank et al. (1992) found that cattle restrained in a poorly designed head holder, for example one for which over 30 seconds was required to drive the animal into the holder, had higher cortisol (a common measure of stress in animals) levels in their blood than cattle stunned with their heads free. Cattle will voluntarily place their heads in a well-designed head-restraint device that is properly operated by a trained operator (Grandin 1992). Tume and Shaw (1992) reported that very low cortisol levels, e.g. about $15\,\mathrm{ng\,ml^{-1}}$, were found in cattle during stunning and slaughter. Their measurements were made in cattle held in a head restraint (Shaw, personal communication). Cortisol levels during on-farm restraint of extensively reared cattle ranged from 25 to $63\,\mathrm{ng\,ml^{-1}}$ (Mitchell et al. 1988; Zavy et al. 1992). Thus, some of the treatments given to animals on the farm were more stressful than the slaughter!

> ... For ritual slaughter [or captive bolt stunning with a non-penetrating stunner] devices to restrain the body are strongly recommended. Animals remain calmer in head restraint devices when the body is also restrained, which may not be the case for animals held for non-penetrating stunning. Stunning or slaughter must occur within 10 seconds after the head is restrained. *(Grandin and Regenstein 1994)*

> The variable of reactions to the incision must be separated from the variable of the time required for the animal to become completely insensible. Recordings of EEG or evoked potentials measure the time required for the animal to lose consciousness.

They are not measurements of pain. Careful observations of the animal's behavioural reactions to the cut are one of the best ways to determine if cutting the throat without prior stunning is perceived as painful by the animal. The time required for the animals to become unconscious will be discussed later.

Observations of over 3000 cattle and formula-fed veal calves were made [by Dr. Grandin] in three different U.S. kosher slaughter plants. The plants had state of the art upright restraint systems. The systems have been described in detail by Dr Grandin (1988, 1991, 1992, 1993, 1994). The cattle were held in either a modified SPCA pen or a double rail (center track) conveyor restrainer.

This equipment was operated by [Dr Grandin] or a person under her direct supervision. Very little pressure was applied to the animals by the rear pusher gate in the SPCA pen. Head holders were equipped with pressure limiting devices. The animals were handled gently and calmly. It is impossible to observe reactions to the incision in an agitated or excited animal. Blood on the equipment did not appear to upset the cattle. They voluntarily entered the box when the rear gate was opened. Some cattle licked the blood.

In all three restraint systems, the animals had little or no reaction to the throat cut. There was a slight flinch when the blade first touched the throat. This flinch was much less vigorous than an animal's reaction to an ear-tag punch. There was no further reaction as the cut proceeded. Both carotids were severed in all animals. Some animals in the modified SPCA pen were held so loosely by the head holder and the rear pusher gate that they could have easily pulled away from the knife.

These animals made no attempt to pull away. In all three slaughter plants there was almost no visible reaction of the animal's body or legs during the throat cut. Body and leg movements can be easily observed in the double rail restrainer because it lacks a pusher gate and very little pressure is applied to the body. Body reactions during the throat cut were much fewer than the body reactions and squirming that occurred during testing of various chin lifts and forehead hold-down brackets. Testing of a new chin lift required deep, prolonged invasion of the animal's flight zone by a person. Penetration of the flight zone of an extensively raised animal by people will cause the animal to attempt to move away (Grandin 1993). The throat cut caused a much smaller reaction than penetration of the flight zone. [Dr Grandin has shown that animals (and humans) have a distance around their head where they feel violated if anyone approaches closer than that. Dr Grandin has designed extensive handling of animal procedures by taking advantage of their flight zone and their point of balance, i.e. when penetration of the flight zone will get an animal to move forward versus backward].

It appears that the animal is not aware that its throat has been cut. Bager et al. (1992) reported a similar observation with calves. Further observations of 20 Holstein, Angus, and Charolais bulls indicated that they did not react to the cut. The bulls were held in a comfortable head restraint with all body restraints released. They stood still during the cut and did not resist the head restraint. After the cut the chin lift was lowered, the animal either immediately collapsed or it looked around like a normal alert animal. Within 5–60 seconds, the animals went into a hypoxic spasm and sensibility appeared to be lost. Calm animals had almost no spasms and excited cattle had very vigorous spasms, which may contribute to blood splash (rupture of capillaries in the flesh). Calm cattle collapsed more quickly and appeared to have a more rapid onset of insensibility. Munk et al. (1976) reported similar observations with respect to the onset of spasms. The spasms were similar to the hypoxic spasms [that] occur when cattle become unconscious in a V-shaped stanchion due to pressure on the lower neck. Observations in feed-yards by [Dr Grandin] during handling for routine husbandry procedures indicated that pressure on the carotid arteries and surrounding areas of the neck can kill cattle within 30 seconds' (Grandin and Regenstein, 1994).

The details spelled out in Jewish law concerning the design of the knife and the cutting method appeared to be important in preventing the animal from reacting to the cut. The fact that the knife is razor sharp and free of nicks may be the critical factors in getting the reactions detailed above. As previously mentioned, the cut must be made continuously without hesitation or delay. It is also prohibited for the incision to close back over the knife during the cut. This is called 'covering' (Epstein 1948). The prohibition against covering appears to be important in reducing the animal's reaction to the cut.

Further observations of kosher slaughter conducted in a poorly designed head holder, i.e. one which allowed the incision to close back over the knife during the cut, resulted in vigorous reactions from the cattle during the cut. The animals kicked violently, twisted sideways, and shook the restraining device. Cattle that entered the poorly designed head holder in an already excited, agitated state had a more vigorous reaction to the throat cut than calm animals. These observations indicated that head holding devices must be designed so that the incision is held open during and immediately after the cut. Occasionally, a very wild, agitated animal went into a spasm [that] resembled an epileptic seizure immediately after the cut. This almost never occurred in calm cattle.

The issue of time to unconsciousness is also an important issue. 'Scientific researchers agree that sheep lose consciousness within 2 to 15 seconds after both carotid arteries are cut (Nangeroni and Kennett 1963; Blackmore 1984; Gregory and Wotton 1984). However, studies with cattle and calves indicate that most animals lose consciousness rapidly. However, some animals may have a period of prolonged sensibility (Blackmore 1984; Daly et al. 1988) that lasted for over a minute. Other studies with bovids also indicate that the time required for them to become unconscious is more variable than for sheep and goats (Munk et al. 1976; Gregory and Wotten 1984). The differences between cattle and sheep can be explained by differences in the anatomy of their blood vessels, i.e. bovids have additional small blood vessels in the back of their head that are not cut during the slaughter process.

Observations by [Dr. Grandin], of both calf and cattle slaughter, indicate that problems with prolonged consciousness can be corrected. When a shochet uses a rapid cutting stroke, 95% of the calves collapse almost immediately (Grandin 1987). When a slower, less decisive stroke was used, there was an increased incidence of prolonged sensibility. Approximately 30% of the calves cut with a slow knife stroke had a righting reflex and retained the ability to walk for up to 30 seconds.

Gregory (1988) provided a possible explanation for the delayed onset of unconsciousness. A slow knife stroke may be more likely to stretch the arteries and induce an occlusion. Rapid loss of consciousness will occur more readily if the cut is made as close to the jawbone as religious law will permit, and the head holder is partially loosened immediately after the cut. The chin lift should, however, remain up so the cut is still open. Excessive pressure applied to the chest by the rear pusher gate will slowly bleed out. Gentle operation of the restrainer is essential. Observations indicate that calm cattle lose consciousness more rapidly and they are less likely to have contracted occluded blood vessels, i.e. blood splash. Calm cattle will usually collapse within 10–15 seconds. Dr. Grandin recently scored time to unconsciousness (drop to the ground) in a glatt (a higher standard for kosher meat based on lung inspection) kosher plant in North America and found that 34/36 cattle were insensible in less than 10 seconds!

Captive-bolt and electric stunning will induce instantaneous insensibility when they are properly applied. However, improper application can result in significant stress. All stunning methods trigger a massive secretion of epinephrine (Warrington 1974; Van der Wal 1978). This outpouring of epinephrine is greater than the secretion that would be triggered by an environmental stressor or a restraint method. Since the animal is expected to be unconscious, it does not feel the stress. One can definitely conclude that improperly applied stunning methods would be much more stressful than religious slaughter with a long, straight, razor-sharp knife. Kilgour (1978), one of the pioneers in animal welfare research, came to a similar conclusion on stunning and slaughter.

In some religious slaughter plants animal welfare is compromised when animals are pulled out of the restraint box before they have lost consciousness. Observations clearly indicated that disturbance of the incision or allowing the cut edges to be touched caused the animal to react strongly. Dragging the cut incision of a conscious animal against the bottom of the head-opening device is likely to cause pain. Animals must remain in the restraint device with the head-holder and body restraint loosened until they collapse. The belly lift must remain up during bleed-out to prevent bumping of the incision against the head opening when the animal collapses.

Since animals cannot communicate, it is impossible to completely rule out the possibility that a correctly made incision may cause some unpleasant sensation. However, it is possible to definitely conclude that poor cutting methods and stressful restraint methods are not acceptable, especially for the religious slaughter of animals. Poor cutting technique often causes vigorous struggling. When the cut is done correctly, behavioural reactions to the cut are much less than reactions to air hissing, metal clanging noises, inversion, or excessive pressure applied to the body. Discomfort during a properly made religious slaughter cut is probably minimal because cattle will stand still and do not resist a comfortable head-restraint device. Religious slaughter is a procedure that can be greatly improved by the use of a total quality management approach of continual incremental improvements in the process. In plants with existing upright restraint equipment significant improvements in animal welfare and reductions in petechial haemorrhages can be made by making the following changes:

- training employees in gentle calm cattle handling
- modifying the restrainer to calmly hold the animal
- eliminating distractions which make animals baulk
- giving careful attention to the exact cutting method.

There needs to be continual monitoring and improvements in technique to achieve rapid onset of insensibility. Poor cutting technique, rough handling, excessive pressure on the animal from the restraint device, and/or agitated or excited animals cause a high incidence of prolonged sensibility and probably a lower quality meat.

Once the animal is slaughtered, the Muslim community returns the animal to the secular system. There is an extensive set of further processing issues for kosher slaughtered animals that will not be reviewed here.

In modern times, to decrease the amount of blood splash (capillary rupture in the meat, particularly in the hindquarters) it has been determined that a post-slaughter non-penetrating captive-bolt stun will greatly reduce blood splash. This is not accepted by the normative mainstream of the Jewish community, although some more liberal Orthodox rabbis do accept it. The Muslim community is more divided on this issue, although most seem to oppose it as it leads to a question of whether the cut or the stun caused death. This procedure allows the animal to be hung on the shackle more rapidly, i.e. it speeds up the time to unconsciousness and probably also the time to insensibility. The question of what is actually happening scientifically to prevent blood splash with this post-slaughter stun has not been answered but it could be postulated that the animal's involuntary reflex kicking that occurs while the animal has limited movement because it is in the restrainer, i.e. being tightly held inside the box, might lead to higher blood splash. By being able to remove the animal from the restrainer sooner, that activity is not changed but the animal is not doing the kicking in the air.

Any ingredients or materials, like tallow, that might be derived from animal sources are generally prohibited for Muslims because of the difficulty of obtaining them exclusively from halal animals, although with the expansion of the halal market there may be opportunities to develop halal versions of many of these currently unacceptable products.

8.4   The Prohibition of Alcohol

The prohibition on the consumption of alcohol is not only for alcoholic beverages but extends to the use of industrial alcohol in food processing. The normative standard as espoused by the largest Muslim certification organization (the Islamic Food and Nutrition Council of America, US based) is that ingredients need to have less than 0.5% industrial alcohol while finished products need to be less than 0.1% industrial alcohol. The use of beverage alcohol is prohibited. The natural presence of greater than 0.1% alcohol is also acceptable. Other agencies, and in some cases other countries, are examining their standards for alcohol in finished products and coming up with different numbers. The advantage of having a guideline would be that finished ingredients and products could be tested for alcohol content in the absence of inspection of production, although the latter is to be highly preferred. However, even with periodic inspection, it is helpful to have a test that can serve to deter producers from trying to do things improperly when inspectors are not present.

This then reflects a rather limited review of the kosher and halal laws with a more complete emphasis on meats, especially halal meat. Many other laws and special interpretations have been skipped. The only remaining topic is how to deal with equipment.

8.5 Equipment Preparation

There are generally three ways to make equipment kosher and these are more extensive than is required for halal. Generally if the equipment is kosher, it is suitable for halal production, but again the final decision on this rests with the halal certifying agency. For more details about equipment kosherization, please see Regenstein et al. (2003).

For Muslims the procedure is much simpler. Normally a good cleaning is sufficient but there are some occasions, particularly if the equipment was used with pork, where a clean-up requires the use of an abrasive agent on contaminated surfaces.

8.6 Meat of Animals Killed by the Ahl-al-Kitab

May a person other than a Muslim slaughter an animal? From a religious point of view the issue of prayer at the time of slaughter is one that at least ought to be understood as one looks to possible synergisms between the Jewish and Muslim communities. Prior to undertaking a period of slaughter the shochet says a prayer. The Muslim slaughterer, on the other hand, according to most schools, is required to say a prayer over each animal, although a few schools are more lenient on this point.

There has been much discussion and controversy amongst Muslim consumers as well as Islamic scholars over the issue of the permissibility of consuming the meat of animals killed by the Ahl-al-Kitab or people of the book, meaning, amongst certain other faith communities, particularly Jews and Christians as these are religions that are based on scripture. The issue focuses on whether meat prepared in the manner practiced by either faith would be permitted for Muslims.

In the Holy Quran, this issue is presented only once in Surah V, verse 5, in the following words:

> 'This day all good things are made lawful for you. The food of those who have received the Scripture is lawful for you, and your food is lawful for them.'

This verse addresses Muslims and seems to establish a social context where Muslims, Jews, and Christians could interact with each other. The majority of Islamic scholars are of the opinion that the food of the Ahl-al-Kitab must meet the criteria established for halal and wholesome halal food including proper slaughter of animals. They believe that the following verse establishes a strict requirement for Muslims:

> 'And eat not of that whereupon Allah's name hath not been mentioned, for lo! It is abomination.'
>
> *(Holy Quran VI:121)*

However, some Islamic scholars are of the opinion that this verse does not apply to the food of Ahl-al-Kitab and there is no need to mention the name of God at the time of slaughter (Al-Qaradawi 1984). However, in recent years some members of the Orthodox Jewish rabbinate have ruled that the saying of the Muslim takbir, i.e. the blessing 'Allah is great (Bismillah Allau Aqaba)' in Arabic by the Jewish slaughterer, i.e. the shochet, is permitted. This would permit meat to be shared between the two communities, which has some positive economic implications for both communities as the post-slaughter inspection and processing of animals to meet Jewish religious requirements leads to a high rejection rate. Such meat after passing secular government inspection would be available as halal.

8.7 Gelatin

Gelatin is an important ingredient that is used in many food products, but is probably the most controversial of all modern religious food ingredients. Gelatin can be derived from pork skin, beef bones, or beef skin. These were originally the only sources permitted in the Food Chemical Codex. However, in a process the author was involved in, the definition has been extended in more recent years to include other animal products including sheep, goat, poultry, and fish. In addition, some gelatins from fish skins have entered the commercial market. Fish gelatins, depending on the species selected, can be produced as kosher and/or halal with proper supervision, and would be acceptable to almost all of the mainstream religious supervision organizations in both religions.

Most currently available gelatins – even if called 'kosher' – are not acceptable to the mainstream US kosher supervision organizations and to Islamic scholars. Many are, in fact, totally unacceptable to halal consumers because they may be pork-based.

8.A Appendix

How one choses to express critical ideas in words with respect to the religious slaughter of animals can have a bearing on the response one receives about the religious slaughter of animals. In classes at Cornell on kosher and halal food regulations and on animal welfare, students were asked to respond to questions using a polling device, i.e. they could answer anonymously and the results could easily be captured for a whole class. They were given similar questions before the topic of slaughter (or religious slaughter) of animals was brought up and then at the end of the talk the same questions were asked. These questions were asked in a series without discussion between questions. What is interesting is that students answered the different questions differently even though they covered the same topic and they presumably were aware of this.

(Note: The following text is taken from Regenstein (2012).)

'... the vocabulary used in discussing these issues can have a significant impact as to how the consumer understands the issues and how the scientists frame the research.

'In a subject as sensitive as religious slaughter of animals, this vocabulary can be a source of tension. Thus, calling the process 'ritual' slaughter versus 'religious' slaughter of animals gives it a different tone. Other members of the religious communities recommend the term 'traditional' slaughter, which encompasses other traditions that cause unconsciousness by a neck/throat cut such as many on-farm slaughters, but with a much wider range of acceptable practices. But this also distracts from the idea that in the case of kosher and halal slaughter, the slaughter of animals is tied to a higher religious purpose. The use of the term 'shechita' in the scientific literature is clearly designed to give it a foreign, 'other' context. Thus we prefer the terms traditional religious slaughter, either traditional kosher slaughter or traditional halal slaughter when writing about these processes in the scientific literature. For general purposes it might be suggested to use 'the prophetic method of slaughter' for halal and 'the Jewish religion's humane slaughter of animals' for kosher.

'And as may be clear from the text to this point, the author has mostly avoided the terms 'stunned' and 'unstunned', which are particularly problematic. The framing of these two words is a polarization of terms. The goal in all cases is to humanely make the animal unconscious. And the religious slaughter does so using a trained religious person, respectfully slaughtering the animal. The methods of stunning could be described as 'cracking the skull', 'electrocuting' the animal or 'putting it into a gas chamber'! Those do not sound anywhere as nice as 'respectfully hand slaughtered with respect for the animals'; so words do matter.

'We actually tested wording with students before and immediately after an extensive discussion of religious slaughter. Using polling devices provides a rapid, relatively anonymous solicitation of opinions. The first question used a balance vocabulary but framing all the methods as ways to make the animal unconscious. From the following choices, which form of slaughter do you consider most humane? Use of a penetrating stunner going through the skull to cause unconsciousness, 29 votes; by using a non-penetrating stunner to crack the skull to cause unconsciousness, 12 votes; use of gases to cause unconsciousness, 42 votes; use of an electrical current to the head to cause unconsciousness, 22 votes and use of a sharp knife to cut the neck to cause unconsciousness, 47 votes. It was interesting that gassing received the highest number of votes from amongst the secular slaughter methods.

'In the second poll taken immediately after the first polling without showing the results and with no comments, a number of students switched their votes to religious slaughter when a less balanced wording was used. From the following choices, which form of slaughter do you consider most humane? By smashing the animal over the head to crack its skull, 3 votes; by smashing through the skull, 14 votes; by electrocuting the animal, 12 votes; by using a gas chamber, 17 votes; and tradition-ally hand slaughtered with respect for the animal, 109 votes.

'Following the lecture, which was admittedly favourable to religious slaughter and covered some of the same material that has and will be covered in this paper, the students when presented with the neutral words still voted even more strongly than before the talk for the traditional religious slaughter. From the following choices, which form of slaughter do you consider most humane? Use of a penetrating stun-ner going through the skull to cause unconsciousness, 10 votes; by using a non-pen-etrating stunner to crack the skull to cause unconsciousness, 7 votes; use of gases to cause unconsciousness, 8 votes; use of an electrical current to the head to cause unconsciousness, 4 votes and use of a sharp knife to cut the neck to cause uncon-sciousness, 124 votes.

As always, care in defining and then using all of these words consistently would prevent some of the unnecessary confusion. (Regenstein 2012)

Acknowledgements

We specifically thank Dr Muhammad Chaudry of the Islamic Food and Nutrition Council of America (IFANCA) for his extensive input over the years. This chapter incorporates significant material from two of the authors' previous papers: Regenstein et al. (2003) and Grandin and Regenstein (1994).

References

Al-Qaradawi, Y. (1984). The Lawful and the Prohibited in Islam. (trans. K. El-Helbawy, M.M. Siddiqui, and S. Shukry). Indianapolis, IN: American Trust Publications.

Bager, F., Braggins, T.J., Devine, C.E. et al. (1992). Onset of insensibility in calves: effects of electropletic seizure and exsanguinations on the spontaneous electrocortical activity and indices of cerebral metabolism. *Research in Veterinary Science* 52: 162–173.

Barnett, J.L., Cronin, G.M., and Scott, P.C. (2007). Behavioural responses of poultry during kosher slaughter and their implications for the birds' welfare. *Veterinary Record* 160: 45–49.

Blackmore, D.K. (1984). Differences in the behaviour of sheep and calves during slaughter. *Research in Veterinary Science* 37: 223–226.

Daly, C.C., Kallweit, E., and Ellendorf, F. (1988). Cortical function in cattle during slaughter: conventional captive bolt stunning followed by exsanguinations compared to shechita slaughter. *The Veterinary Record* 122: 325–329.

Dunn, C.S. (1990). Stress reactions of cattle undergoing ritual slaughter using two methods of restraint. *The Veterinary Record* 126: 522–525.

Epstein, I. (ed.) (1948). The Babylonian Talmud. London: Soncino Press.

Ewbank, R., Parker, M.J., and Mason, C.W. (1992). Reactions of cattle to head restraint at stunning: a practical dilemma. *Animal Welfare* 1: 55–63.

Gibson, T.J., Johnson, C.B., Murrell, J.C. et al. (2009a). Components of electroencephalographic responses to slaughter in halothane-anaesthetised calves: effects of cutting neck tissues compared with major blood vessels. *New Zealand Veterinary Journal* 57: 84–89.

Gibson, T.J., Johnson, C.B., Murrell, J.C. et al. (2009b). Electroencephalographic responses of halothane-anaesthetised calves to slaughter by ventral-neck incision without prior stunning. *New Zealand Veterinary Journal* 57: 77–83.

Gibson, T.J., Johnson, C.B., Murrell, J.C. et al. (2009c). Amelioration of electroencephalographic responses to slaughter by non-penetrative captive-bolt stunning after ventral-neck incision in halothane-anaesthetised calves. *New Zealand Veterinary Journal* 57: 96–101.

Gibson, T.J., Johnson, C.B., Murrell, J.C. et al. (2009d). Electroencephalographic responses to concussive non-penetrative captive-bolt stunning in halothane anaesthetized calves. *New Zealand Veterinary Journal* 57: 90–95.

Grandin, T. (1987). High speed double rail restrainer for stunning or ritual slaughter. *International Congress of Meat Scientists and Technology*: 102–104.

Grandin, T. (1988). Double rail restrainer for livestock handling. *Journal of Agricultural Engineering Research* 41: 327–338.

Grandin, T. (1991). Double rail restrainer for handling beef cattle. Technical paper 915004. American Society for Agricultural Engineering, Joseph, MI.

Grandin, T. (1992). Observations of cattle restraint devices for stunning and slaughtering. *Animal Welfare* 1: 85–91.

Grandin, T. (1993). Management commitment to incremental improvements greatly improves livestock handling. *Meat Focus* (Oct): 450–453.

Grandin, T. (1994). Euthanasia and slaughter of livestock. *Journal of the American Veterinary Medical Association* 204: 1354–1360. [Kosher: 1358–1359].

Grandin, T. and Regenstein, J.M. (1994). Religious slaughter and animal welfare: a discussion for meat scientists. *Meat Focus International* 3: 115–123.

Gregory, N. (1988). Published Discussion, 34th International Congress of Meat Science and Technology, Workshop on Stunning of Livestock. Brisbane, Australia: CSIRO Meat Research Laboratory, p. 27.

Gregory, G. and Wotton, S.D. (1984). Time of loss of brain responsiveness following exsanguinations in calves. *Research in Veterinary Science* 37: 141–143.

Kilgour, R. (1978). The application of animal behavior and the humane care of farm animals. *Journal of Animal Science* 46: 1479–1486.

Mitchell, G., Hahingh, J., and Ganhao, M. (1988). Stress in cattle assessed after handling, transport and slaughter. *The Veterinary Record* 123: 201–205.

Munk, M.L., Munk, E., and Levinger, I.M. (1976). Shechita: Religious and Historical Research on the Jewish Method of Slaughter and Medical Aspects of Shechita. Jerusalem: Feldheim Distributors.

Nangeroni, L.L. and Kennett, P.D. (1963). An electroencephalographic study of the effect of shechita slaughter on cortical function of ruminants. Unpublished report. Ithaca, NY: Department of Physiology, NY State Veterinary College, Cornell University.

Regenstein, J.M. (2012). The politics of religious slaughter – how science can be misused. *Proceedings 65th Annual Reciprocal Meat Conference*, 7 pp.

Regenstein, J.M. and Grandin, T. (1992). Religious slaughter and animal welfare – an introduction for animal scientists. *Proceedings 45th Annual Reciprocal Meat Conference*, pp. 155–159.

Regenstein, J.M., Chaudry, M.M., and Regenstein, C.E. (2003). The kosher and halal food laws. *Comprehensive Reviews in Food Science and Food Safety* 2 (3): 111–127.

Tume, R.K. and Shaw, F.D. (1992). Beta endorphin and cortisol concentration in plasma of blood samples collected during exsanguination of cattle. *Meat Science* 31: 211–217.

Van der Wal, P.G. (1978). Chemical and physiological aspects of pig stunning in relation to meat quality. A review. *Meat Science* 2: 19–30.

Warrington, R. (1974). Electrical stunning: a review of the literature. *The Veterinary Bulletin* 44: 617–633.

Zavy, M.T., Juniewicz, P.E., Phillips, W.A., and Von Tungeln, D.L. (1992). Effect of initial restraint, weaning and transport stress on baseline ACTH stimulated cortisol response in beef calves of different genotypes. *American Journal of Veterinary Research* 53: 551–557.

Further Reading

Blech, Z. (2004). Royal jelly. In: Kosher Food Production. Ames, Iowa: Blackwell.

Chaudry, M.M. (1992). Islamic food laws: philosophical basis and practical implications. *Food Technology* 46 (10): 92.

Chaudry, M.M. and Regenstein, J.M. (1994). Implications of biotechnology and genetic engineering for kosher and halal foods. *Trends in Food Science and Technology* 5: 165–168.

Chaudry, M.M. and Regenstein, J.M. (2000). Muslim dietary laws: food processing and marketing. *Encyclopedia of Food Science*: 1682–1684.

Egan, M. (2002). Overview of halal from Agri-Canada perspective. Presented at the Fourth International Halal Food Conference, Sheraton Gateway Hotel, Toronto, Canada (April 21–23).

Giger, W., Prince, R.P., Westervelt, R.G., and Kinsman, D.M. (1977). Equipment for low stress animal slaughter. *Transactions of the American Society of Agricultural Engineers* 20: 571–578.

Govoni, J.J., West, M.A., Zivotofsky, D. et al. (2004). Ontogeny of squamation in swordfish, *Xiphias gladius. Copeia* 2004 (2): 390–395.

Grandin, T. (1991). Recommended Animal Handling Guidelines for Meat Packers. Washington, DC: American Meat Institute.

Grandin, T. (1996). Factors that impede animal movement at slaughter plants. *Journal of the American Veterinary Medical Association* 209: 757–759.

Grandin, T. (2000). Livestock Handling and Transport, 2e. Wallingford: CAB International.

Grandin, T. (2001). Welfare of cattle during slaughter and the prevention of nonambulatory (downer) cattle. *Journal of the American Veterinary Medical Association* 219: 1377–1382. [Kosher: 1379–1380].

Grandin, T. (2002). Good Management Practices for Animal Handling and Stunning, 2e. Washington, DC: American Meat Institute.

Grandin, T. (2003). Getting religious about slaughter. *Meat and Poultry* 8: 76.

Grunfeld, I. (1972). The Jewish Dietary Laws, 11–12. London: The Soncino Press.

Jackson, M.A. (2000). Getting religion – for your product that is. *Food Technology* 54 (7): 60–66.

Khan, G.M. (1991). Al-Dhabah, Slaying Animals for Food the Islamic Way, 19–20. Jeddah: Abul Qasim Bookstore.

Larsen, J. (1995). Ask the Dietitian. Hopkins, MN: Hopkins Technology, LLC http://www.dietitian.com/alcohol.html (accessed 24 April 2003).

Masri, A.-H.B.A. (2007). Animal Welfare in Islam. Leichestershire: The Islamic Foundation.

Ratzersdorfer, M., Regenstein, J.M., and Letson, L.M. (1988). Appendix 5: Poultry plant visits. In: A Shopping Guide for the Kosher Consumer (eds. J.M. Regenstein, C.E. Regenstein and L.M. Letson) for Governor Cuomo, 16–24. Governor, State of New York.

Regenstein, J.M. (1994). Health aspects of kosher foods. *Activities Report and Minutes of Work Groups & Sub-Work Groups of the R & D Associates* 46 (1): 77–83.

Regenstein, J.M. and Grandin, T. (2002). Kosher and halal animal welfare standards. *Institute of Food Technologists Religious and Ethnic Foods Division Newsletter* 5 (1): 3–16.

Regenstein, J.M. and Regenstein, C.E. (1979). An introduction to the kosher (dietary) laws for food scientists and food processors. *Food Technology* 33 (1): 89–99.

Regenstein, J.M. and Regenstein, C.E. (1988). The kosher dietary laws and their implementation in the food industry. *Food Technology* 42 (6): 86+88–86+94.

Regenstein, J.M. and Regenstein, C.E. (2000). Kosher foods and food processing. *Encyclopedia of Food Science*: 1449–1453.

Regenstein, J.M. and Regenstein, C.E. (2002a). The story behind kosher dairy products such as kosher cheese and whey cream. *Cheese Reporter* 127 (4): 8, 16, 20.

Regenstein, J.M. and Regenstein, C.E. (2002b). What kosher cheese entails. *Cheese Marketing News* 22 (31): 4–10.

Regenstein, J.M. and Regenstein, C.E. (2002c). Kosher byproducts requirements. *Cheese Marketing News* 22 (32): 4–12.

Usmani, M.M.T. (2006). The Islamic Laws of Animal Slaughter. (trans. A. Toft). Santa Barbara, CA: White Thread Press.

Weiner, M. (2008). The Divine Code. (trans. M. Schulman), 291–367. Pittsburgh, PA: Ask Noah International.

Westervelt, R.G., Kinsman, D., Prince, R.P., and Giger, W. (1976). Physiological stress measurement during slaughter of calves and lambs. *Journal of Animal Science* 42: 831–834.

Part III

Halal Ingredients and Food Production

9

Factory Farming and Halal Ethics

Faqir Muhammad Anjum[1], Muhammad Sajid Arshad[2] and Shahzad Hussain[3]

[1] University of the Gambia, Banjul, The Gambia
[2] Institute of Home and Food Sciences, Government College University Faisalabad, Pakistan
[3] College of Food and Agricultural Sciences, King Saud University, Riyadh, Saudi Arabia

9.1   Introduction

The fundamental principle for a Muslim's diet is that food not only has to be halal (permissible, Shariah compliant), but also *toyyiban*, which means wholesome (healthy, safe, nutritious, good quality). This principle arises from one of the scriptures in the Holy Quran, which says 'O ye people! Eat of what is on earth, Halal and Toyyib; and do not follow the footsteps of the Evil One, for he is to you an avowed enemy' (Al-Baqarah). The notion of halal in Islam has very precise motives. They are to preserve the purity of religion, to safeguard the Islamic approach, to preserve life, to defend property, to protect forthcoming generations, and to preserve self-respect and integrity (Riaz and Chaudry 2004). Most consumers who choose to use halal meat products are unaware of the halal supply chain principles. Currently, consumers of halal meat habitually purchase products with a halal symbol posted on the packaging without doubting its authenticity and trusting the suppliers. Most of them are unaware of the halal food requirements involved in the supply chain and its logistics. Nevertheless, it is suggested that most of consumers do not have a choice when they purchase halal meat products (Hawkes 2008). For example, in many countries halal meat products are sold in display racks beside non-halal meat products (such as pork) which invalidate one of the halal principles. In Singapore and Malaysia, one of the halal issues that concerns consumers is whether there is segregation and appropriate handling of halal meat products and the non-halal meat products, and whether companies conducting halal meat production are fully knowledgeable. Since there is no way consumers can be completely sure that the meat that they are purchasing is halal, they have no choice but to take it at face value. Halal ethics includes everything, even conditions for and the treatment of animals before slaughtering.

9.2   Good Animal Husbandry Practices and Animal Welfare

Good animal husbandry practices are mandatory at farm level. This is to guarantee that animals are raised to achieve a certain standard that certifies animal welfare all along the production line. They are also intended to monitor the security and superiority of the animal product to ensure they are fit for human consumption. Good animal husbandry practice (GAHP) emphasizes the following requirements:

1) Satisfactory facilities to shelter and housing animals from weather extremes while maintaining the air and water quality in the environment.
2) Well-kept facilities to allow safe, civilized, and effective movement of animals.
3) Qualified and well-trained employees to care for the animals and handle each stage of manufacture with no tolerance of ill-treatment.
4) Access to a high-quality water supply and a nutritionally well-adjusted diet for each class of animals.
5) Ensuring that basic food and water needs are being met and identifying illness or injury.
6) Herd health should be maintained by a veterinary practices.
7) Quick veterinary medical attention provided whenever requisite.
8) A humane method is used to treat sick or injured animals not responding to medication.

9.3   Good Governance in Halal Slaughtering

Halal slaughtering of an animal involves the following pre-requisites:

1) Restraint then stunning (if used) and cutting of the trachea (*halqum*), oesophagus (*mari'*) and both the carotid arteries and jugular veins (*wadajain*).
2) An essential obligation for halal slaughtering is that the slaughterer should be a committed Muslim who is of sound mind and full age, and has a complete understanding of the basics of halal slaughtering and situations related to the slaughter of animals, is registered, skilled, and overseen by the Halal Certification Body, and is knowledgeable in Shariah compliance.
3) The animal to be slaughtered should be alive or believed to be alive, healthy and have been certified by the capable authority.
4) Before slaughtering unhealthy and unfit animals should be removed from the halal slaughter line.
5) The slaughtering knife should be sharp and clean to ensure that smooth slaughtering is achieved by the sharpness of the blade not by the animal's weight. The cutting action of the slaughtering is legitimate as long as the slaughtering knife is not lifted off the animal during the slaughtering.
6) It must be confirmed that the animal is absolutely dead prior to dressing. Carcass dressing operations can only be conducted once the animal is completely dead after the slaughtering.

9.4 Good Governance for Slaughtering of Livestock for *Qurban*

The Department of Veterinary Services (DVS) in Malaysia has published animal slaughter (*qurban*) guidelines. These guidelines were formulated and introduced with the intention of harmonizing, regulating, and blending the rules and requirements of slaughtering (*qurban*) animals. The guidelines were developed based on the existing regulations (Animal Act 1953 (Revised 2006), state enactments, and Government Gazette, and local authority bylaws that focus on animal welfare, the slaughtering process, the slaughtering method, animal waste management, transportation, and storage, personal hygiene, and slaughter premises/houses. The guidelines emphasize the importance of protecting the rights and welfare of animals from prior to slaughter to the point of distribution. Among other things they emphasize the ethics when dealing with animals:

1) Water should be offered to the animal before slaughter and it should not be slaughtered when hungry.
2) The knife should be hidden from the animal and slaughtering should be done out of sight of other animals waiting to be slaughtered.
3) The animal must be slaughtered using a sharp knife.
4) The slaughtering must be done in one stroke without lifting the knife. The knife should not be placed and lifted when slaughtering the animal.
5) The knife should not be sharpened in front of the animal.
6) Skinning or cutting any part of the animal is not allowed before the animal is completely dead.

9.5 Animal Housing and Management

The animals' housing must be precisely designed, built, equipped, and preserved to ensure a good standard of animal care and should follow satisfactory standards of animal welfare for the species concerned and should fulfil scientific requirements. In identifying the standards of animal care, the criterion should be animal well-being rather than the mere capability of surviving under poor conditions such as environmental extremes or higher population.

9.6 Veterinary Care

Institutions should provide adequate veterinary care and equip the attending veterinarian appropriately. This should include provision of appropriate facilities, equipment, personnel, and services to execute the guiding principles, use of appropriate procedures to control (e.g. vaccination and other prophylaxis, isolation and quarantine), diagnose, and treat diseases and injuries, and daily observation of all animals to evaluate their health and well-being. Certain manipulations or other tasks related to the handling and care of animals must be performed only by a qualified veterinarian.

9.7 Cruelty to Animals Under Malaysian Law

Malaysia has in place vibrant legal powers concerning animal welfare. Section 43 of the Animal Act 1953 (Revised 2006) (Malaysia) defines an animal as any living creature other than a human being and includes any beast, bird, fish, reptile or insect, whether wild or tame. Section 44 of the Act provides that a person is guilty of brutality to an animal and shall be subject to fine or imprisonment if that person:

a) 'cruelly beats, kicks, ill-treats, overrides, overdrives, overloads, tortures, infuriates, or terrifies any animal;
b) causes or procures or, being the owner, permits any animal to be so used;
c) being in charge of any animal in confinement or in course of transport from one place to another neglects to supply such animal with sufficient food or water;
d) by want only or unreasonably doing or omitting to do any act, causes any unnecessary pain or suffering, or, being the owner, permits any unnecessary pain or suffering to any animal;
e) causes or procures or, being the owner, permits to be confined, conveyed, lifted, or carried any animal in such manner or position as to subject it to unnecessary pain or suffering;
f) employs or causes or procures or, being the owner, permits to be employed in any work or labour, any animal which in consequence of any disease, infirmity, wound or sore, or otherwise is unfit to be so employed; or
g) causes, procures or assists at the fighting or baiting of any animal, or keeps, uses, manages, or acts or assists in the management of any premises or place for the purpose, or partly for the purpose, of fighting or baiting any animal, or permits any premises or place to be so kept, managed or used, or receives or causes or procures any person to receive, money for the admission of any person to such premises or place.'

9.8 Islamic Law in Modern Animal Slaughtering Practices

Slaughter denotes the practice of killing animals for human consumption (Merriam Webster.com). According to Ibn Manzur (n.d.), slaughtering is also known as *al-tazkiyyah* or *al-zakah*, which means completeness. There are three categories of halal animal slaughter. The first is *al-nahr*, i.e. cutting the throat of long-necked animals like camels and giraffes, and one part of the body for other animals like horses (al-Khalili 1997). The second category is *al-labbah*, i.e. the cutting of the lowermost part of the neck between the neck and chest (Ibrahim Fadhil 1997). The third category is *al-zabh*, i.e. the cutting of the trachea (*halqum*), the oesophagus (*mari*) and both the carotid arteries and jugular veins (*wadajain*) of an animal (Jafri et al. 2011).

There are five Shariah requirements that need to be satisfied and practically observed to make the meat of a slaughtered animal halal to Muslims. The first is the intention (*niyyah* for Allah), the second is that the slaughterer should be a Muslim, the third is that the animal must be legitimate and alive, fourth is use of permitted piercing tools, except for nails and teeth (Al-Khan and Al-Bugha 2008), and last concerns the place for the slaughter.

The practice of animal slaughtering has seen rapid growth from the customary to the more cultured method. According to Fitzgerald (2010), methods of animal slaughtering have evolved from small scale in the public area, such as a backyard, to a bigger scale in a centralized modern slaughterhouse. In addition, farmers no longer use traditional methods but have adopted a new and more sophisticated method reflecting the shift from traditional to mechanical slaughtering, the adoption of stunning, and the practice of thoracic sticking. This has been driven by the continuous demand from consumers for meat, which has led to the need to increase meat production. Apart from market demands (Khadijah et al. 2012), the development of animal slaughtering practices is also believed to minimize animal handling, reduce costs and time constraints, and safeguard animal welfare (Jafri et al. 2011).

9.9 Modern Methods of Animal Slaughtering

There are three methods of animal slaughtering that have been approved in modern slaughtering practice: stunning, mechanical slaughtering and thoracic sticking (Jafri et al. 2011). For the purpose of this discussion, we shall focus only on two procedures, stunning and mechanical slaughtering, because these are the methods that are adopted in most slaughterhouses. In order to certify that these slaughtering methods are in line with Shariah requirements, several organizations, such as the Food and Agriculture Organization of the United Nations (FAO), Malaysian Halal Standard (Jabatan Kemajuan Islam Malaysia (JAKIM) 1500:2009), and religious bodies, have produced halal standards.

i) Stunning prior to slaughtering

The practice of stunning, which renders the animal unconscious prior to slaughter, has been adopted by most developed countries. The adoption of pre-slaughter stunning started in the West in the 20th century (Bergeaud-Blacker 2007). There are three types of stunning: chemical, mechanical, and electrical (Flectcher 1999). The most popular methods practiced are mechanical and electrical stunning (Abdul Salam Babji et al. 2006). While the practice of stunning is legally required in most European countries, conventional slaughter is also permissible for Muslim and Jewish cultures (Bergeaud-Blacker 2007). In Malaysia, the practice of stunning has also been adopted and has been set out in the Malaysian halal standard (Jafri et al. 2011).

As a new method of slaughtering, stunning invited controversy amongst Muslims. In 1988 the Fatwa Committee of the National Council of Malaysia (1988) decided that electrical and water stunning are allowable for cattle and chickens, respectively. The Council has further declared that pneumatic-percussive stunning is also allowed on the condition that the skull of the animal is not split and the animal must die from the slaughtering process and not because of the stunning (Fatwa Committee of the National Council of Malaysia 2006). In 1997, the *Majma' al-Fiqh al-Islami* in its tenth meeting at Makkah declared that electric shock stunning is allowed as long as the animal is still alive after the stunning process, even though this process is painful for the animal. In fact, the practice of stunning is acceptable if all the Shariah requirements are satisfied.

ii) Mechanical Slaughtering

Mechanical slaughtering uses rotary blades to cut the throat of small animals, mostly birds and chickens. Before the slaughtering process takes place, the birds or chickens are stunned using the water-bath stunning technique. The role of the slaughterer is to control the on/off button to start/stop the slaughtering process. This has raised a number of questions with regard to the halalness of the slaughtered birds or chickens since the slaughtering process is carried out by a machine rather than a man.

The Muslim jurists have a variety of opinions on this. One group argues that the practice should be stopped because the slaughtered animals or chickens are not rendered halal because the throat cutting is not done by a man but by a machine and so no appropriate intention is given (Mufti Kerajaan Brunei 2000). Another group, however, argues that the practice is permissible and the slaughtered birds or chickens are halal because the throats are correctly cut (International Fiqh Academy of Jeddah 1997; National Fatwa Committee Council of Malaysia 2000).

9.10 The Halal Meat Chain

The meat chain that meets all approved religious criteria is very complex and, as well as there being disagreement on some processing issues, there is the threat of cross-contamination at all stages, for example halal meat becomes haram if it comes into contact with pork meat. In addition, halal meat safety and wholesomeness in terms of its halal status is difficult to verify by consumers before purchase, during consumption, and even after, resulting in potentially uncertainty for them. In this situation, implementation of a quality assurance scheme is a prerequisite so that stakeholders involved in the meat chain can be sure that halal meat fulfils the defined quality requirements.

Hazard Analysis Critical Control Points (HACCP) (Codex Alimentarius Commission 2003) is a worldwide documented and applied quality assurance system used by companies that is intigated at different levels of the agro-food chain. It consists of seven principles constituting a stepwise approach to identifying probable hazards and critical control points (CCPs) where operational failures might create or fail to eliminate ultimate hazards. It has become the internationally renowned standard for achieving the maximum possible levels of food security throughout the food chain (EurLex 93/43/EEG). The use of quality assurance systems has changed over time from management tools to assure food safety into more wide-ranging approaches allowing the guarantee and safeguarding of process standards relating to, for example, animal welfare and certified production methods such as organic or halal (Wood et al. 1998; Fearne et al. 2001; Juska et al. 2003; Ten Eyck et al. 2006). Moreover, the identification of CCPs, particularly halal control points (HCPs), is important for ensuring food safety, and this is a way of managing self-regulation, requiring assessment and auditing by a third party with self-governing inspectors (Bolton et al., 2001).

Increasing demand for formally slaughtered meat and the necessity for good practice from an animal welfare viewpoint were recently examined by Cenci-Goga et al. (2004), who established that proper handling necessitates continuous measurement, monitoring, and management. Riaz and Chaudry (2004) introduced a HACCP method with several HCPs at the slaughterhouse level to ensure the halal status of meat.

However, a total or integrated halal quality approach requires that the entire halal meat chain is controlled in accordance with HACCP principles. In this respect, Zadernowski et al. (2001) and Snijders and van Knapen (2002) have stressed that intervention should not focus on the abattoir or food processing stages only; it should also target risks associated with upstream preharvest production and downstream retailing in the farm to fork chain. According to these authors, properly structured HACCP-like methods applied from the farm to the kitchen offer the best available approach to optimize meat inspection. This viewpoint was adopted by Wyss and Brandt (2005) for organic foods, resulting in the organic HACCP concept. This organic HACCP differs from standard HACCP in three aspects:

- It encompasses the complete chain, not just one part.
- It is responsible for safeguarding a series of qualities, including taste, credibility, and legitimacy, not just protection in the sense of avoiding a potential health hazard.

9.11 Halal Breeding: HCP1

The animal must be of an acceptable species. Prohibited species such as pork cannot be made halal through halal slaughter. Although most authors stress the slaughter conditions in Islam, the breeding of animals should be halal as well. In the hadith, according to Abu-Dawud, it is recorded: 'The messenger of Allah forbade eating the animal and drinking its milk, which nourishes on filth' (Book 27, 3776). Religious scholars agree that it is prohibited for Muslims to ingest animals that have eaten filth, which is considered an unnatural, non-vegetarian diet. For such animals, the term 'jallalah' is used. An animal can be consumed only if it is fed a natural, vegetarian diet for at least three months. Consequently, animals that are fed the meat of other animals or animal protein are prohibited.

In addition, when cross-contamination between allowed and forbidden meat occurs, the acceptable meat becomes prohibited. Therefore, halal and haram meat should be segregated at all stages of the halal meat chain. Genetically modified (GM) foods could also be a source of cross-contamination. It has not yet been decided whether these foods are permissible or forbidden for Muslims. God requires Muslims to eat of the good things: 'Eat of the good things wherewith We have provided you, and transgress not in respect thereof lest My wrath come upon you; and he on whom My wrath cometh, he is lost indeed' (Holy Quran 20:81). All unlawful foods are prescribed in the Holy Quran or in the hadith. Since biotechnology did not exist in prehistoric times, only religious researchers can tell whether GM foods are haram or halal. However, the religious rules that would permit or prohibit these foods have not amended (Regenstein et al. 2003), leaving individuals free to impose the Holy Quran and hadith on biotechnology. The most likely interpretation is that GM foods containing only derivatives from halal foods are halal and GM foods containing any derivatives from haram foods are haram. This interpretation is supported by the Islamic Food and Nutrition Council of America and the Islamic Jurisprudence Council in the USA. It is, however, not always clear whether a GM food contains a halal or a haram derivative and therefore GM foods could be inferred as being 'mashbooh', meaning their provenance is uncertain and thus they should be avoided. Most of the religious people interviewed by

Maarabouni (2002) support this view. In addition to these two conceivable interpretations, Maarabouni (2002) adds that Muslims must take care of nature and let nothing harm the environment since earth and nature are given to them by God. If GM foods were to have a negative influence on nature, then they would be immoral for mankind and therefore forbidden. Moses (1999) agrees that consumer doubts about GM products may arise from concerns about possible hazards as well as ethical contemplations. Based on faith and belief, people may regard biotechnology as being disrespectful to nature, something that is beyond the right of man and should be left to God.

9.12 Animal Welfare: HCP2

The second point of control in the halal meat chain relates to animal welfare. Islam teaches humanitarian treatment to animals prior, during and after slaughter. Animals should be treated so that they are not stressed or excited prior to slaughter; they should be fed and well rested. Drinking water must be available in slaughterhouses. In addition, the knife should not be sharpened in front of the animal and no animal should be able to watch the slaughter of another animal. Numerous verses in the hadith support these stipulations:

- Shaddid Bin Aus said: 'Two are the things which I remember Allah's Messenger (may peace be upon him) having said: Verily Allah has commanded goodness to everything; so when you kill, kill in a good way and when you slaughter, slaughter in a good way. So, every one of you should sharpen his knife, and let the slaughtered animal die comfortably.' (Sahih Muslim, Book 021, Number 4810)
- Narrated Hisham bin Zaid: 'Jabir Bin Abdullah testified that Allah's Messenger (may peace be upon him) prohibited that any beast should be killed after it has been tied.' (Sahih Muslim, Book 021, Number 4817)

9.13 Stunning: HCP3

The question of whether or not stunning is allowed before halal slaughter is still unresolved both inside and outside the Muslim community. According to Riaz and Chaudry (2004), stunning should not be used since the animal must be alive at the time of slaughter and must die of bleeding rather than as a result of stunning. Aldeeb (2001), however, notes that Islamic dietary laws do not forbid stunning; they ban the consumption of blood and dead animals, and require kind handling before and during slaughter. Stunning, as long as it does not slay the animal, could thus be said to reduce suffering and therefore meet the religious requirement of humane handling. Additionally, several fatwas have been written by religious scholars who confirm Aldeeb's position towards stunning. A fatwa from 1987, for example, issued by the Egyptian fatwa commission, states that stunning is allowed when it is used to reduce pain during slaughter without causing the death of the animal.

In many European countries, such as Belgium, the UK, France, Germany, and the Netherlands, guidelines on animal welfare necessitate that all animals must be rendered unresponsive before being slaughtered, excluding for religious or ritual slaughter. Other

European countries do not grant exceptions from stunning for halal and kosher slaughter. In response to the international market development and prospects for halal meat, New Zealand, the largest exporter of halal slaughtered sheep meat and an important exporter of halal slaughtered beef, supported research in the late 1970s to develop slaughter and stunning techniques that would meet both Muslim requirements and animal welfare concerns. As a result, head-only electrical stunning was considered acceptable, in which the animal is only momentarily unconscious. If the animal were not slaughtered, it would thus recover consciousness. Most Muslims, however, are opposed to stunning since they believe it is strictly banned by Islamic rulings.

9.14   Knife: HCP4

The fourth HCP is the tool used for slaughtering. This must be so sharp that the animal does not sense the pain of the cut, particularly when no stunning is used. The size of the knife should be balanced to the size of the neck. As specified before, the knife should not be sharpened in the presence of the animal for animal welfare reasons.

9.15   Slaughterer: HCP5

The slaughterer must be a stable, healthy, adult Muslim male or female, or someone from 'the people of the book', namely a Jew or a Christian. Two schools of thought of Islamic jurisprudence claim that although Jews and Christians are people of the book, the meat which is slaughtered by them is unlawful unless the name of God is spoken while slaughtering. Another school of thought considers the meat slaughtered by Jews or Christians halal without restriction since it is allowed by the Holy Quran and they claim that the prophet Mohammed (pbuh) used to eat meat slaughtered by Jews or Christians (Sakr 1971; Hussaini 1993b). The Muslim must invoke the name of God before eating this meat since his name was not invoked during slaughter: 'This day are (all) good things made lawful for you. The food (ta'am) of those who have received the Scriptures is legitimate for you, and your food is lawful for them ...' (Holy Quran, chapter 5, verse 5). The application of the right type of cut is fundamental to the humane handling of the animal during slaughter. This requires that Muslim slaughterers are skilled and qualified for their job. Although the competent authorities licence slaughterers, monitoring of slaughter methods is often lacking (Cenci-Coga et al. 2004).

9.16   Slaughter Method: HCP6

The animal should be slaughtered by cutting the anterior part of the neck, and cutting the carotids, jugulars, trachea, and oesophagus without reaching the bone in the neck, all in one cut. It is preferable to turn the animal or bird to face towards Makkah before slaughtering, but this is not obligatory. Generally, the slaughtering of ruminants and poultry should be done by hand. Slaughtering by hand is chosen by all Muslims and extensively followed in countries where Muslims manage slaughterhouses. However, in Western countries mechanical or machine slaughtering of birds is attaining approval amongst Muslims.

9.17   Invocation: HCP7

The name of Allah should be invoked while severing. The usual formula is 'In the name of Allah; Allah is the greatest' (Bismillah, Allahu Akbar). There are two main reasons for saying this blessing during slaughter. The first reason is to remind the slaughterer of his accountability in noticing the agreed requirements and to remove any uncertainty regarding to whom the animals are dedicated (Abdussalam 1981). Second, it strengthens the notion that the animal is being slaughtered in the name of Allah for food and not for fun:

- 'And do not eat of that on which Allah's name has not been cited, and that is most surely a disobedience; and most surely the Shaitans recommend to their friends that they should compete with you; and if you obey them, you shall most surely be polytheists.' (Holy Quran, chapter 6, verse 121).
- 'Therefore, eat of that on which Allah's name has been cited if you are devotees in His communications.' (Holy Quran, chapter 6, verse 118).

The blessing must be invoked when passing the knife over the neck of the animal. If the slaughterer is someone from 'the people of the book', he should not invoke another name than God. If he invokes the name of Jesus or Abraham, the meat is haram. All schools of thought agree that if the name of a person instead of God is stated, the meat is entirely forbidden (Sakr 1971). Benkheira (2002:77), however, mentions that invocation is only a secondary condition and that if the slaughterer should forget, the meat does not become haram. In the case of mechanical Islamic slaughter, the following actions should be performed. A Muslim switches on the machine while pronouncing the name of God. One Muslim slaughterer is located behind the machine to make an incision on the neck if the machine misses a bird or if the cut is not satisfactory for proper bleeding. This person invokes continuously the name of God during slaughter.

9.18   Packaging and Labelling: HCP8

For meat to be categorized correctly as halal, all the HCPs in the halal meat chain should be assessed by a reputable regulatory organization, which acts as a third and independent control official recognition body. Each slaughter should be halal certified individually unless the slaughterhouse is solely a halal-slaughtering facility.

9.19   Retailing: HCP9

Although most authors focus on the slaughtering process itself, dissemination and selling of halal meat must also be considered in order to avoid cross-contamination. In practice, three distribution outlets are available for halal meat: Islamic butchers, supermarkets, and farms or slaughterhouses. The first and important distribution channel is Islamic butcher shops. It is estimated that 80% of halal meat is bought at Islamic butchers in France (Haut Conseil a'l_Inte'gration, 2000) and 75% in the Netherlands (Foquz 1998). A survey of Moroccan families in Belgium revealed that 94% of families always purchase meat from an Islamic butcher. In contrast, 10% of halal meat in France is bought at the supermarket. In the Netherlands and Belgium, supermarkets only account for 3–4% of the halal meat market. Belgian supermarkets do not sell fresh halal meat, although some sell frozen halal processed meats. Finally, some Muslims go straight to the slaughterhouse to buy their halal meat or purchase animals from the farmer to slaughter themselves at home (which is an unlawful practice in most European countries) or at the farm. In the Netherlands, 10–13% of the halal meat is bought straight from the farmer and slaughtered on the farm or at home. In France, this distribution channel is thought to account for 10% of the halal meat market. About 68% of Moroccan families in Ghent occasionally buy an animal from a farm. However, almost every Muslim family annually buys a sheep for Eid-el-Adha at a farm.

9.20 A Simplified EU Legislative Outline for Animal Welfare

A simplified European Union (EU) legislative framework for animal welfare standards for all animals kept in the context of an economic activity, including where appropriate pet animals, with emphasis on interpretation, reduction of administrative burden, and the validation of welfare standards as a means to improve the competitiveness of the EU food industry with special reference to animal welfare. It considers the following:

a) The use of science-based animal prosperity indicators as a possible resource to simplify the legal framework and allow tractability to improve the competitiveness of livestock producers.

b) A new EU outline to increase the clarity and sufficiency of information to consumers on animal welfare for their purchase choice.

c) The foundation of a European network of reference centres.

d) The creation of common rules for the competence of workers who handle the animals.

a) The use of outcome-based animal welfare indicators
   The possibility of using scientifically authenticated result-based indicators to supplement inflexible requirements in EU legislation will be deliberated when necessary, with a specific attention to the contribution of such an approach to the simplification of the acquis. Animal-based indicators have been introduced in two recent pieces of EU animal welfare legislation (Directive 2007/43/EC laying down minimum rules for the protection of chickens kept for meat production and Regulation [EC] No. 1099/2009 on the protection of animals at the time of killing). Principles developed by the Welfare Quality® project22 associated with a risk valuation system as applied in the food safety area (see Food Law 23) will be inspected. The European Food Safety Authority (EFSA) Scientific Opinions on the expansion of welfare pointers should be taken into account

together with socio-economic factors. The use of outcome-based animal welfare pointers is also recognized at international level by organizations such as the World Organization for Animal Health (OIE).

b) A new EU framework to increase the transparency and adequacy of information to consumers on animal welfare for their purchase choice

The revised EU legislative framework for animal welfare could provide a method for guaranteeing to consumers that animal welfare entitlements are crystal clear and scientifically applicable. Convergent and synergistic actions with similar advantages at EU level in other relevant policy areas will be determined to enhance consumer authorization.

c) A European network of reference centres

The notion of a network of reference centres for animal welfare has already been discussed by the EFSA. The intention is to ensure that the competent authorities follow comprehensive and uniform procedural information on the way the EU legislation should be executed, especially in the context of outcome-based animal welfare indicators.

The network could be established through the co-financing of existing scientific and technical national assets on animal welfare. The role of the network would be to supplement and not duplicate the role of the European Food Safety Authority and the activity of the Joint Research Centre of the EU. It would be organized to reflect the current structure of EU legislation in order to ensure the following at EU level:

- Support for the EFSA and state members with technical proficiency, especially in the context of the use of outcome-based animal welfare pointers.
- Organization of training courses for the benefit of staff from relevant authorities and experts from third countries where necessary.
- Contribution to the distribution of research findings and technical inventions amongst EU investors and the international scientific community.
- Coordination of research in collaboration, when required, with existing EU-funded research structures

d) Common necessities for competence of workers handling animals

The basic EU legislative framework for animal welfare could amalgamate in a single text the requirements for competence that already exist in certain pieces of EU legislation. General principles to prove aptitude would be developed on the basis of an impact assessment.

Common EU requirements for competence for staff handling animals would aim to ensure that handlers possess the ability to identify, prevent or restrict animals' pain and suffering as well as knowing their legal responsibilities related to the protection and welfare of animals.

In addition, an adequate level of competence could be measured for people accountable for the design of processes, conveniences or equipment that apply to animals.

Animal welfare education should be developed to identify animal welfare topics and should be included in the curriculum of professions involved with animals to identify what actions need to be taken to improve responsiveness amongst these professions.

9.21   An Overview of Animal Welfare in the World

Within recent decades, several wide-ranging pieces of animal welfare regulation have been introduced in Europe, many of them limiting the concentrated confinement of farm animals. Switzerland has some of the oldest and most inflexible animal welfare laws. The Swiss Federal Act on Animal Protection of 9 March 1978 (as per 1 July 1995) and Swiss Animal Protection Ordinance of 27 May 1981 (as per 1 November 1998) include basic ethics that animals should be treated in the manner that best conforms to their needs (Article 1), and that 'mass produced housing systems and systems for the keeping of farm animals may not be publicized and sold without aforementioned authorization from authority nominated by the Federal Council. The authorization shall only be granted if such systems provide suitable living conditions for animals.'

The costs of these testing procedures are paid by the applicant (Article 5). EU Member States have increasingly controlled husbandry practices, such as eliminating conventional battery cages for laying hens and incubation crates for propagates, and demanding group housing for veal calves over eight weeks of age (Wilkins 1997). A significant development for animal safety in Europe was the Treaty of Amsterdam, which came into effect in 1999, and which states:

'Eager to ensure improved safety and respect for the welfare of animals as conscious beings; [have agreed upon] the following provision, which shall be annexed to the Treaty inaugurating the European Community, in articulating and employing the Community's agricultural, transportation, internal market and exploration policies, the Community and state members shall pay full honour to the welfare necessities of animals, while regarding the legislative or administrative provisions and customs of the Member States relating in particular to religious rites, cultural ethnicities and regional legacy.'

Hence, the standard creates clear legal compulsions to address animal welfare issues arising in policies and also includes a requirement to support animal welfare research. An EU action strategy for animal welfare was adopted in January 2006 for 2006–2010, with the objectives of:

1) promotion of the standards relating to animal welfare
2) encouraging research and alternative tactics to animal testing
3) introducing standardized animal welfare pointers
4) better informing animal handlers and public on animal welfare concerns
5) supporting international enterprises for the protection of animals.

However, in most other countries, animal safety laws have usually been based on the value of animals as assets, and hence the focus of most laws has been to protect animal owners against losses, rather than to defend animals from suffering (Wiskerke 2003). Similarly, animal cruelty laws were traditionally adopted to protect the public from violent individuals, rather than to prevent animal suffering (perhaps echoing Thomas Aquinas' 12th-century sentiments about human cruelty to animals ultimately leading to cruelty to other humans [Aquinas 2004]). For example, animal cruelty is debated in the property section of the Canadian Criminal Code, which states 'it is a federal crime to consciously cause or permit to be caused, by being the owner, unnecessary pain, suffering or injury to an animal or by willful negligence, cause damage or harm to animals while they are being transported or conveyed'.

Another significant factor that has affected the European response to animal protection is the use of the protective principle for political decisions involving risk administration. The precautionary principle has its foundation in Germany and is used explicitly in EU risk management decisions about health, welfare, and the environment (Commission of the European Communities 2000). The precautionary principle is used 'where scientific information is inadequate, indecisive, or uncertain and where there are suggestions that the possible effects on the environment, or human, animal or plant health may be theoretically dangerous and unpredictable with the chosen level of protection' (Commission of the European Communities 2000, p. 8). The burden of proof about whether particular farming practices cause suffering is placed on those wishing to use them. Hence, this approach is risky/hostile, there must be evidence that harm has not occurred, rather than simply a lack of evidence that harm is caused just because the necessary studies have not been conducted. A country with a particularly advanced approach to animal protection that compliments the creativity and distinctiveness of producers while preserving quality standards is Sweden. The Swedish Animal Welfare Law of 1988 includes some precise husbandry criteria as follows:

1) Animals are placed according to specific behavioural requirements, for instance cattle are grazing animals.
2) The animal-based health and enactment audit rewards stockperson skills and devotion to management.
3) An on-farm animal welfare standard is used to gather bird-related consequences on broiler farms, such as mortality records, culls because of leg malformations, and footpad dermatitis (Algers and Berg 2001).
4) Justifications to improve housing and management are provided by correlating the maximum stocking density permitted at the time of slaughter in each broiler house to the total animal welfare score received.

With an increasingly global economy, it is impossible to anticipate the future of farm animal welfare without taking into account the trade in animal products. For countries with categorized animal welfare standards, and thus greater costs of production, domestic industries are susceptible to imports from countries with fewer guidelines. Consequently, how concerns about animal welfare can be lodged within the framework of the World Trade Organization (WTO) has been an active topic during discussions (Bowles and Fisher 2000). However, it is important to note that animal protection is also receiving attention in developing countries (Favre and Hall 2004), although understanding and implementation of regulations varies amongst countries. For example, the Taiwan Animal Protection Law was proposed by the president in 1998, specifying that persons owning or caring for animals must be 15 years of age or older, and must 'provide adequate food, water, and sufficient space for activities for the animal, and other proper care to prevent the animal from superfluous harassment, ill-treatment or hurt' (Article 5). Furthermore, Article 9 specifies that 'while carrying [transporting] an animal, it shall be prevented from being terrified or hurt'.

Similarly, the Philippines Animal Welfare Act (1998) offers some regulation of the treatment of conscious animal species and also permits religious and tribal ethnicities of animal sacrifice that continue to be practiced on some of the islands (Favre and Hall 2004). More recently, animal welfare has been addressed by the World Organization for Animal Health (OIE 2004), with member states drafting standards for the protection of animals during transport and slaughter.

9.22 Farm Animals Welfare

In contrast with the EU nations, the USA currently has comparatively little legislation intended to regulate farm animal welfare. The perceived inconsistencies in the level of animal protection have fuelled scrutiny and criticism of US animal production industry standards, primarily by animal protection organizations. Many of these criticisms have focused on quality of life issues arising from housing animals in close confinement, traditionally referred to as factory farming, the term popularized by Ruth Harrison (1964). However, recognizing and ranking the needs of farm animals has been tremendously problematic for the scientific community. For example, the literature evaluating the extent to which incubation stalls meet the needs of propagates compared with other housing systems is debatable, with European scientists recommending elimination of incubation stalls on the grounds of behavioural deficiency and affective states, whereas Australian scientists reviewing the same literature deemed these stalls to improve sow welfare on the grounds of biological function (Fraser 2003). In addition to the underlying value judgments, the replication of studies is often difficult (McGlone et al. 2004b) because there can be considerable variation in the welfare measurements used and in their interpretation. Nonetheless, the explanation for using gestation crates for sows, battery cages for laying hens, and crates for veal calves is particularly questioned because the ability of such preventive rearing systems to meet the behavioural requirements of animals housed within them is uncertain and because the animals are subjected to high levels of confinement for virtually their entire lives. It is estimated that 95% of sows in the USA are housed in confinement, mainly in gestation stalls (Bowman et al. 1996). This type of housing has been favoured because it facilitates caretaker safety and efficiency, maximizes the efficiency of space utilization, requires reduced capital investment relative to other sow housing systems, and has notable benefits for sow welfare. For example, pigs housed individually have shown better growth rates than those that are group housed (Petherick et al. 1989). Other studies have reported similar production levels in sows kept in gestation stalls versus sows housed in large group pens (Morrison 2002) and small pens (Pajor 2002) as well as outdoors and in hoop barns (Honeyman 2002). Moreover, gestation stalls prevent potentially harmful agonistic behaviours because the restricted feeding practices that are necessary to prevent obesity can also result in increased competition and fighting when sows are group-housed (Deen 2005). However, economic constraints have resulted in gestation crates that are too small to permit sows to turn around. Although they may stand up and lie down, sows are unable to fully adjust their posture (Arey 1999) and, depending on their size, many are unable to even lie down comfortably.

9.23   Voluntary Guideline of Farm Animal Welfare

Islam is a religion of ethics. When we talk about halal meat, halal ethics should be included as animal farming demands not only that animal production industries provide healthy and affordable products to their consumers, but also that animal interests are given proper consideration. Recent and ongoing attempts to legalize farm animal husbandry policies appear to be intended to achieve this goal. Because legitimate issues and doubts about actual information can delay legislation and make laws difficult to adapt to new knowledge, it is inevitable that the animal production industries will move towards self-regulating their policies and practices in keeping with the new social ethic of sympathetic animal farming. The United Egg Producers is one commodity group that has taken a practical stand by commissioning an independent scientific review of their production practices in terms of existing knowledge about poultry welfare and then developing an action plan to identify areas for which action was warranted and where research funds should be directed (Mench 2003; Fraser 2006).

From the consumer viewpoint, progress regarding animal welfare is approaching a appoint where stakeholders themselves will take action to address welfare issues (Bracke et al. 2004). In the USA, Certified Humane was developed as an animal welfare certification program, including third-party audits and a suggested scientific community. Several of the most encouraging initiatives have begun with efforts to develop authenticated means of monitoring farm animal welfare. According to Thompson (2005), development of animal production industry standards that are based on sound scientific and ethical principles may provide the best alternative to compulsory regulation of modern farm animal husbandry, but only if some key conditions are met:

1) It must be clear that the ethical objectives and values place appropriate weight on the welfare and benefits of the animals themselves at the same time as they identify the role of animal agriculture in satisfying dynamic human needs.
2) Consumers must have trust that the standards are taken seriously and that livestock producers faithfully follow the suggested practices.
3) Producers must trust that the standards are documented and administered fairly. Although some mix of market incentives, government regulations, and self-administered industry standards may eventually emerge to address the new challenges of ethical animal husbandry, only a system that meets all three of these criteria can truly be said to be ethically justified. Steps have already been taken in this direction.

In fact, the recent food animal welfare initiatives implemented by the fast-food and supermarket industries may represent a shift in animal agriculture similar to that predicted by many scientists. They have provided unprecedented motivation for farm animal welfare reform in the USA in the absence of legislation. This is referred to as 'the arrival of politics by other means', suggesting that once companies believe that their consumers value a particular aspect of a product (e.g. welfare-friendly animal husbandry), it becomes possible for policy promoters to influence enough of the market to form regulations that may provide a competitive edge (Schweikhardt and Browne 2001). Animal welfare may offer that differentiating characteristic for progressive producers and retailers in the USA. For example, in the late 1990s to early 2000s, under the guidance of animal welfare scientists, the McDonald's corporation presented animal welfare guidelines for its producers (Fraser 2006). This action was subsequently matched by its competitor, Burger King, and later by other fast-food chains. By 2003, the National Council of Chain Restaurants (NCCR) and the Food Marketing Institute (FMI), which represent over 26 000 food retail stores and over 120 000 restaurants, franchises, and cooperators, followed suit, drafting a program to develop and support industry efforts at improving farm animal welfare (NCCR 2003). These minimum welfare standards, arising mainly because of pressure applied to multinational corporations by animal protestors within the USA and abroad, have effected change in US farm animal welfare more quickly and substantially than government actions would probably have done (Estevez 2002). It is unlikely that these efforts borne fruit followed without a community ethic similar to that described by Rollin (1995) that extends some level of moral consideration to farm animals. Nonetheless, Bracke et al. (2004) point out that the monitoring of farm animal welfare may involve important con-flicts of interest for the stakeholders, who may have contrasting goals. This may clash with the producer's desire to maintain independence and to validate success in encouraging animal welfare. Further conflict arises if welfare groups involved in the monitoring proce-dure decide that intensive animal production systems should be significantly modified or discontinued altogether (Bracke et al. 2004). How monitoring of welfare is done also should be seriously considered because there are several ways to doing this, including intermittent monitoring of indiscriminately selected farms and voluntary monitoring for animal welfare certification.

9.24 Factory Farming

Factory farming is a way of farming where livestock is raised in confinement in large num-bers (and at high density) in a factory-like environment.

The purpose of factory farming is to produce as much meat, eggs, or dairy at the lowest possible cost. For the system to work, it needs high volumes of inexpensive animal feed as well as antibiotics and pesticides to alleviate the spread of disease intensified by the crowded living conditions. Animals are often confined to small areas and physically restrained to control or limit movement. Food is supplied inside and is characterized by high protein concentration levels. A wide variety of methods are used to maintain animal health, including growth hormones and antimicrobial agents. Often these organizations employ breeding programs to produce more productive animals suited to the confined con-ditions. The cheap meat produced has substantial hidden costs.

9.24.1 Fish Farming

Intensive fish farming, whereby large numbers of fish are limited to a small area, causes severe welfare problems. Practices in the industry that can cause suffering include keep-ing density, water quality and path of flow, food shortage before slaughter, movement and transportation of fish, and treatment with detrimental chemicals, biotechnology, and genetic engineering techniques involving chromosome manipulation. Cruel and intoler-able slaughter methods are still permitted for fish (such as choking, bleeding without stunning, and stunning using carbon dioxide gas). As with other farmed animals, only slaughter methods that cause immediate death or render the animal instantly unresponsive to pain until death should be permitted.

9.24.2 Veal Farming

Most veal originates from calves that are compelled to live in tiny barren crates, without any contact with other calves. These crates permit little movement, only a few steps forward or backward. The calves are compelled to either lie down or stand up. White veal is produced by keeping the calves anaemic by feeding them on a liquid-only diet.

9.24.3 Cattle Farming

Many countries (including the USA) use feedlots for 'finishing' prior to slaughter. These cluster animals together to live with no grass because they are fed on concentrates. Welfare problems arise from mixing animals and the change of diet, as well as the congested and unhygienic conditions.

9.24.4 Turkey Farming

Numerous turkeys are factory farmed. They are selectively bred for rapid weight gain, which causes health problems and ailments, and incapability to mate naturally so artificial insemination is routinely used. Also, the heavy live turkeys are shackled upside down at slaughter, causing intolerable pain and trauma.

9.24.5 Dairy Farming

Dairy cows are submitted to the restrictions to produce ever-increasing quantities of milk. Cows are enforced to breed at an early age, with calves detached as soon as one day after birth. The dairy cow suffers the shock of having her calf taken away, and often bellows for days. After this, the cow is drained for milk to high volume for about 10 months, after which she is impregnated again. The cow is pregnant and lactating for seven months. A few weeks before she gives birth, the milking is stopped. The cycle is usually repeated two or three times before the cow becomes unhealthy or 'uneconomic' and is slaughtered. The cows' natural life span is over 20 years. Other causes for poor welfare in dairy cows include:

1) discriminatory breeding for exceptionally large udders (associated with hind foot lameness)
2) overfeeding of starchy, high-protein foods, which can cause digestive problems and lead to lameness
3) poor housing in cubicle sheds
4) intolerable levels of mastitis
5) high rates of lameness.

Bovine somatotropin (BST), a genetically engineered variety of the cow's growth hormone, is injected into some cows to increase milk yield in some countries, although it has been banned in the EU.

9.25 Impacts on Economy

A decrease in the labour force results in a decrease in unemployment hence the economy of a country is affected. A contributory factor in reducing the labour force is a reduction in the number of people needed to produce the same amount of food, if not more, and amplified competition by producing a large amount of meat and dairy at low cost. Smaller farms frequently have little option but to either strengthen their own production or go out of business. The introduction of factory farming into a specific livestock sector tends to drive down prices throughout the sector, making it difficult for farmers to resist it. One of the principle methods for reducing costs within factory farming is to reduce the amount spent on salaries. This is achieved by mechanization, increasing the size and scale of farms and production procedures, and reducing the income, wages, and conditions of farmers and workers.

9.26 Impact on Environment

The principle influence of factory farming on both local inhabitants and the natural environment in the UK derives from the mounds of animal waste that result from having large numbers of livestock congested in relatively small spaces. Animal waste from factory farms is usually put into a pit to be disposed of later. These pits can be gigantic – the proposed 8100-head Nocton dairy would produce around 187 000 cubic metres of manure per year, enough to fill about 75 Olympic swimming pools. These lakes of animal waste can damage both water and soil through run-off and leakage. Nitrogen from the waste causes both eutrophication and oxygen exhaustion in water, damaging biodiversity and killing fish. Approximately 30% of the nitrogen that pollutes water in the EU is due to livestock. In addition to nitrogen, animal waste can also pollute water with pathogens (such as Salmonella and *Escherichia coli*), antibiotics and hormones, heavy metals and sediments (through soil erosion). In addition to water-based pollution, aerial pollutants can also pose a threat to both workers and nearby residents.

9.27 Antibiotics

The high stocking density, the stress of factory farming on animals, and the low level of genetic diversity all increase the potential for the spread of disease amongst livestock. To stop the spread of disease, factory farms usually use high levels of antibiotics, often to prevent disease rather than cure existing conditions. The large amount of antibiotics used in factory farming is a significant factor in the resistance of many common pathogens to the antibiotics used to treat infections in humans. An example is the statement by UK government scientists that a new, almost fatal, type of antibiotic resistance in *E. coli*, known as extended-spectrum beta-lactamase (ESBL) resistance, has spread from the handful of farms on which it had been identified to more than one in three of all dairy farms in England and Wales. A recent study into the increase in ESBL *E. coli* on farms has related it to the increasing farm use of modern antibiotics categorized by the WHO as it is important in human medicine. Antibiotic usage is most predominant amongst the two most intensively farmed animals (cows and sheep).

9.28 Water

Livestock production uses 15% of all irrigated water globally. Around two billion people currently suffer from water shortage, with this figure set to increase to between four and seven billion by 2050, more than half of the estimated world population. The water used by livestock production is projected to increase by 50% by 2025.

9.29 Climate Change

Factory farming is sometimes advanced as a way of reducing greenhouse gas discharges because compared to grazing animals less methane is released, but this ignores important sources of emissions related to concentrated farming. The livestock sector produces a significant proportion of greenhouse gas emissions throughout the manufacturing process, 18% of the global total. The conversion of forest and grassland to cropland discharges stored carbon and lessens the global volume for absorbing carbon dioxide. Globally, the land-use alteration set in motion by livestock farming led to the release of 2.4 billion tons of carbon dioxide a year, equivalent to around 6% of global greenhouse discharges and a third of all of the livestock sector emissions.

9.30 Impact of Confinement on Animal Welfare

Today's intensive animal production systems are devoted to producing meat as economically as possible while attaining certain standards of taste, consistency, and efficacy. Confinement systems are intended to produce animals of sought-after weight in less time and with a lower occurrence of some diseases. When the animals are kept indoors, problems due to weather are reduced. The downside is that animals are kept in more congested conditions, are subject to a number of prolonged and production-related ailments, and are unable to show natural behaviours. In addition, the animals are often physically adapted or restrained to prevent injury to themselves or workers. Confined animals are generally raised indoors and, in some cases (e.g. poultry, laying hens, hogs), the group size, when raised indoors, is larger than outdoors. In other cases (e.g. veal crates or gestation crates for sows), animals are segregated and confined to spaces that allow only minimal movement. The major welfare concern is the ability of the animal to express natural behaviours, for instance having natural materials to walk or lie on, having sufficient floor space to travel around with some freedom, and rooting (for hogs). Crates, battery cages, and other such systems do not allow even these minimal normal behaviours.

Animal administration practices that have been introduced include feeding and nutrition. For example, beef cattle are characteristically fed grains rather than forage (grass, hay, and other roughage), even though their digestive systems are intended to metabolize forage diets. The result is that beef cattle put on weight quicker, but they also often experience internal boils. Some laying hens have their feed restricted at regular intervals in order to induce moulting or encourage egg laying (although this practice is mostly phased out, according to United Egg Producers standards). Most animals are physically altered without pain relief when raised in intense, confined production systems (as well as in some more open systems), even though it is widely accepted that such alteration causes pain. For example, hogs have their tails cut to avoid tail biting by other hogs in close proximity. Laying hens and broilers have their toenails, spurs, and beaks clipped. Dairy cows may have their horns detached or their tails docked. The purpose of such alteration is to avoid injury to the animal or to make the animal easier to handle or to meet market demands on alteration, such as castration of bulls raised for beef, and so these practices are common throughout animal agriculture.

References

Abdussalam, M. (1981). Muslim attitudes to the slaughter of food animals. *Animal Reg. Studies* 3: 217–222.

Aldeeb Abu-Sahlieh, S.A. (2001). *AVIS sur le'tourdissement des animaux avant leur abattage. Avis 01–162*. Lausanne, Switzerland: Institut Suisse de droit compare', ISDC.

Algers, B. and Berg, C. (2001). Monitoring animal welfare on commercial broiler farms in Sweden. *Acta Agric. Scand., Sect. Anim. Sci.* 51 (Suppl. 30): 88–92.

Al-Khalili, A.B.H. (1997). Al-Zabaaih wa al-Tarqu Syar'iyyah fi al-Najazi al-Zakati, *Majallah Majma' al-Fiqh al-Islamih (Daurah al-'Asyirah, Vol.1)*. Jeddah.

Al-Khan, M. and Al-Bugha, M. (2008). *Al-Fiqhu al-Manhaji*, 9e. Damascus: Dar Al-Qalam.

Alqudsi, S.G. (2014). Awareness and demand for 100% halal supply chain meat products. *Procedings of the Society for Behavioural Science* 130: 167–178.

al-Rubuu, I.F. (1997). Al-Zabaih wa al-Tarqu al-Syar'iyyah fi Injazi al-Zakah, *Majallah Majma' al-Fiqh al-Islamiy (Daurah al-'Asyirah, Vol.1)*. Jeddah.

Aquinas, T. (2004). Animals are not rational creatures. In: *Animal Rights: A Historical Anthology* (eds. A. Linzey and P.B. Clarke), 7–12. New York: Columbia Univeristy Press.

Arey, D.S. (1999). Time course for the formation and disruption of social organization in group-housed sows. *Appl. Anim. Behav. Sci.* 62: 199–207.

Babji, A.S., Barnett, J.B., and Rahim, A.A. (2006). Halal meat and the islamic slaughter. *Jurnal Halal* 1: 9–18.

Bakri, N. (2010). *Dimensi Budaya Teknologi dalam Era Globalisasi*. Skudai: Penerbit UTM Press.

Bergeaud-Blacker, F. (2007). New challenges for islamic ritual slaughter: a European perspective. *J. Ethn. Migr. Stud.* 33 (6): 965–980.

Bolton, D.J., Doherty, A.M., and Sheridan, J.J. (2001). Beef HACCP: Intervention and non-intervention systems. *Int. J. Food Microbiol.* 66: 119–129.

Bowles, D. and Fisher, C. (2000). Trade liberalization in agriculture: the likely implications for European farm animal welfare. In: *Negotiating the Future of Agricultural Policies: Agricultural Trade and the Millennium WTO Round* (eds. S. Bilal and P. Pezaros), 199–210. Alphen aan den Rijn: Kluwer Law International.

Bowman, G.L., Ott, S.L., and Bush, E.J. (1996). Management effects of pre-weaning mortality: a report of the NAHMS National swine survey. *Swine Health Prod.* 4: 25–32.

Bracke, M.B.M., De Greef, K.H., and Hopster, H. (2004). Qualitative stakeholder analysis for the development of sustainable monitoring systems for farm animal welfare. *J. Agric. Environ. Ethics* 18: 27–56.

Brunei, J.M.K. (2000). Penyembelihan Ayam dengan Mesin (siri 27/1983). In: *Penyembelihan Binatang dan Pengendalian Daging: Himpunan Fatwa Mufti Kerajaan Negara Brunei Darussalam Mengenai Pengendalian Daging 1962–1999*. Brunei: Jabatan Mufti Kerajaan, & Jabatan Perdana Menteri.

Cenci-Goga, B.T., Ortenzi, R., Bartocci, E. et al. (2004). Ritual slaughter: Where the paradox lays. In: *Science, Ethics and Society*. Preprints of the 5th Congress of the European Society for Agricultural and Food Ethics, September 2–4, 2004 (eds. J. De Tavernier and S. Aerts), 55–58. Leuven, Belgium: CABME, Centre for Agricultural Bio- and Environmental Ethics.

Codex Alimentarius (2003). *Codex principles and guidelines on foods derived from biotechnology*. Rome: Codex Alimentarius Commission, Joint FAO/WHO Food Standards Programme, Food and Agriculture Organisation.

Commission of the European Communities (2000). Communication from the commission on the precautionary principle. Brussels, 02.02.2000, COM (2000) 1.

Deen, J. (2005). Sow housing: opportunities, constraints, and unknowns. *J. Am. Vet. Med. Assoc.* 226: 1331–1334.

Estevez, I. (2002). Poultry welfare issues. Poultry Digest. http://www.wattnet.com/library/DownLoad/PD2aw.pdf 3 (2)1–12 (accessed 29 June 2006).

European Commission (2009). Feasibility study on animal welfare labelling and establishing a community reference centre for animal protection and welfare. Part 1: Animal welfare labelling. Health & Consumer Protection Directorate-General. Brussels. http://ec.europa.eu/food/animal/welfare/farm/aw_labelling_report_part1.pdf (accessed January 2010).

European Parliament and Council (2002). Regulation (EC) No 178/2002 of the European Parliament and of the Council of 28 January 2002 laying down the general principles and requirements of food law, establishing the European Food Safety Authority and laying down procedures in matters of food safety. *Official J. Eur. Commun.* 31: 1–24.

Fatwa Committee of the National Council (1988). The use of electrical stunning in the slaughtering cattle. Retrieved on 2nd January 2013 from http://www.e-fatwa.gov.my/fatwa-kebangsaan/penggunaan-electrical-stunning-dalam-penyembelihan-lembu.

Fatwa Committee of the National Council (2000). Guidelines on the production, preparation, handling and storage of halal food. Retrieved on 3rd January 2013 from http://www.e-fatwa.gov.my/fatwa-kebangsaan/garis-panduan-mengenai-pengeluaran-penyediaan-pengendalian-dan-penyimpanan-makanan.

Fatwa Committee National Council of Islamic Religious Affairs Malaysia (2006). Animals died due to slaughtering not because of stunning.

Favre, D. and Hall, C.F. (2004). *Comparative National Animal Welfare Laws*. East Lansing, MI: Michigan State University College of Law, Animal Legal and Historical Center.

Fearne, A., Hornibrook, S., and Dedman, S. (2001). The management of perceived risk in the food supply chain: a comparative study of retailer-led beef quality assurance schemes in Germany and Italy. *Int. Food Agribus. Man. Rev.* 4 (1): 19–36.

Foquz (1998). *Vleesaankopen allochtonen*. Foquz: Utrecht, the Netherlands.

Fitzgerald, A.J. (2010). A social history of the slaughterhouse: from inception to contemporary implications. *Human Ecol. Rev.* 17 (1): 58–69.

Fletcher, D.L. (1999). Slaughter technology. *Poult. Sci.* 78: 277–281.

Fox, J.G., Anderson, L.C., Loew, F.M., and Quimby, F.W. (2002). Laws, regulation and policies affecting the use of laboratory animals. In: *Laboratory Animal Medicine* (ed. L.C. Anderson), 19–32. Maryland: Academic Press.

Fraser, D. (2003). Assessing animal welfare at the farm and group level: the interplay of science and values. *Anim. Welf.* 12: 433–443.

Fraser, D. (2006). Animal welfare assurance programs in food production: a framework for assessing the options. *Anim. Welf.* 15: 93–104.

Harrison, R. (1964). *Animal Machines: The New Factory Farming*. London: Stuart.

Hawkes, D.T., Warnick, D.R., and Ensign, M.D. (2008). US Patent 7,445,627. Washington, DC: US Patent and Trademark Office.

Hodges, J. (2003). Why livestock, ethics and quality of life? In: *Livestock, Ethics and Quality of Life* (eds. J. Hodges and I.K. Han), 1–26. Wallingford: CABI Publishing.

Honeyman, M.S. (2002). Sow well-being in extensive gestating sow housing: Outdoor and hoop barn system. *Proceedings of the Symposium on Swine Housing and Well-being*, Des Moines, IA, pp. 45–51.

Hussaini, M.M. (1993b). *Islamic Dietary Concepts & Practices*. Bedford Park, Illinois: The Islamic Food & Nutrition Council of America.

International Fiqh Academy of Jeddah (1997). Slaughtering, No.95 (10/3). http://www.fiqhacademy.org.sa/qrarat/10-3.htm.

Jafri, A., Suhaimi Ab, R., and Zaidah Mohd, N. (2011). Sembelihan Halal dalam Industri Makanan Halal. In: *Pengurusan Produk Halal di Malaysia* (eds. R. Suhaimi Ab and A. Jafri), 26–50. Serdang: Penerbit Universiti Putra Malaysia.

Juska, A., Gouveia, L., Gabriel, J., and Stanley, K.P. (2003). Manufacturing bacteriological contamination outbreaks in industrialized meat production systems: The case of *E. coli* O157:H7. *Agric. Human Values* 20 (1): 3–19.

Maarabouni, A. (2002). *Pour une gestion e'thique des OGM. LIslam et les OGM*. Montre'al, Canada: La Commission de l'Ethique de la Science et de la Technologie.

Malaysia Standard MS 1500:2009. Halal food-production, preparation, handling, and storage-general guidelines.

McGlone, J.J., von Borell, E.H., Deen, J. et al. (2004b). Review: compilation of the scientific literature comparing housing systems for gestating sows and gilts using measures of physiology, behavior, performance, and health. *Prof. Anim. Sci.* 20: 105–117.

Mench, J.A. (2003). Assessing animal welfare at the farm and group level: a United States perspective. *Anim. Welf.* 12: 493–503.

Morrison, R. (2002). Large group systems for gestating sows. *Proceedings of the Symposium on Swine Housing and Well-being*, Des Moines, IA, pp. 53–54.

Moses, V. (1999). Biotechnology products and European consumers. *Biotechnol. Adv.* 17: 647–678.

Nakyinsige, K., Man, Y.C., and Sazili, A.Q. (2012). Halal authenticity issues in meat and meat products. *Meat Sci.* 91: 207–214.

NCCR (2003). FMI January 2003 report: FMI-NCCR Animal Welfare Program. http://www. nccr.net/newsite/download/june2002reportfinalletterheadv2.doc (accessed 18 April 2003).

Office International des Epizooties (OIE) (2004). *Global Conference on Animal Welfare: An OIE Perspective*. Office for Official Publications of the European Communities, Luxembourg.

Olfert, E.D. and A. A, M.W. (1993). Control of animal pain in research, teaching and Testing. In: *Guide to the Care and Use of Experimental Animals*, vol. 1 (ed. H.L. Amyx), 188–189. Ottawa: Canadian Council on Animal Care.

Pajor, E.A. (2002). Group housing of sows in small pens: Advantages, disadvantages and recent research. *Proceedings of the Symposium on Swine Housing and Well-being*, Des Moines, IA, pp. 37–44.

Petherick, J.C., Beattie, A.W., and Bodero, D.A.V. (1989). The effect of group size on the performance of growing pigs. *Anim. Prod.* 49: 497–502.

Regenstein, J.M., Chaudry, M.M., and Regenstein, C.E. (2003). The kosher and halal food laws. *Comp. Rev. Food Sci. Food Safety* 2 (3): 111–127.

Riaz, M.N. and Chaudry, M.M. (2004). *Halal Food Production*. Boca Raton, Louisiana: CRC Press.

Rollin, B. (1995). *Farm Animal Welfare: Social, Bioethical and Research Issues*. Ames: Iowa State University Press.

Ruys, T. (1991). Laboratory animal facilities. In: *Handbook of Facilities Planning* (ed. T. Ruys), 193–197. New York: Van Nostrand Reinhold.

Sakr, A.H. (1971). Dietary regulations and food habits of Muslims. *J. Am. Dietetic Assoc.* 58 (2): 123–126.

Schweikhardt, D.B. and Browne, W.P. (2001). Politics by other means: the emergence of a new politics of food in the United States. *Rev. Agric. Econ.* 23: 302–318.

Shariff, S.M. and Lah, N.A.A. (2014). Halal certification on chocolate products: A case study. *Procedia – Social Behav. Sci.* 121: 104–112.

Department of Veterinary Services Malaysia (2012). Slaughtering Guidelines for Qurban. Retrieved on 18 February 2006 from www.gov.islam.

Snijders, J.M.A. and Van Knapen, F. (2002). Prevention of human diseases by an integrated quality control system. *Livestock Prod. Sci.* 76 (3): 203–206.

Standing Committee on Agricultural Research (SCAR) collaborative working group on animal health and welfare research and the Animal Health and Welfare ERA-Net (ANIHWA). Nugent, N., & Rhinard, M. (2015). The european commission. Macmillan International Higher Education.

Ten Eyck, T.A., Thede, D., Bode, G., and Bourquin, L. (2006). Is HACCP nothing? A disjoint constitution between inspectors, processors, and consumers and the cider industry in Michigan. *Agric. Human Values* 23 (2): 205–214.

Thompson, P.B. (2005). Animal agriculture and the welfare of animals. *J. Am. Vet. Med. Assoc.* 226: 1325–1327.

van den Belt, H. and Gremmen, B. (2002). Between precautionary principle and "sound science": distributing the burdens of proof. *J. Agric. Environ. Ethics* 15: 103–122.

Wilkins, D.B. (1997). *Animal Welfare Legislation in Europe: European Legislation and Concerns*. London: Kluwer Law International.

Wiskerke, J.S.C. (2003). On promising niches and constraining socio-technical regimes: The case of Dutch wheat and bread. *Environment and Planning* 35: 429–448.

Wood, J.D., Holder, J.S., and Main, D.C.J. (1998). Quality assurance schemes. *Meat Sci.* 49 (1): 191–203.

Wyss, G. and Brandt, K. (2005). Assessment of current procedures for animal food production chains and critical control points regarding their safety and quality: preliminary results from the Organic HACCP-project. Systems Development: Quality and Safety, pp. 127–132. *ISBN* 07049 (9851): 3.

Zadernowski, M.R., Verbeke, W., Verhe, R., and Babuchowski, A. (2001). Toward meat traceability critical control point analysis in the Polish pork chain. *J. Int. Food Agribus. Market.* 12 (4): 5–22.

10

Halal Ingredients in Food Processing and Food Additives

Yunes Ramadan Al-Teinaz

Independent Public Health & Environment Consultant, London, UK

10.1   Introduction

Consuming halal products produced by halal means is an important obligation of Muslims wherever they live. As per the Islamic tenets, it is the responsibility of every Muslim to follow the norms and values provided by the Holy Quran and hadiths as best as they can in their daily life, even though it is challenging to assimilate Islamic norms into a non-Muslim majority surrounding. The industry of halal is not only about slaughtering animals in accordance with Islamic law, it also includes halal food, pharmaceuticals, cosmetics, lifestyle, and even halal services (Elasrag 2016). Halal in Islam means 'legal' or 'permitted'. It can be defined as anything that is permitted and upon which no restriction exists and the doing of which is allowed by Allah, God Almighty (see Chapter 1).

Food additives are not a recent discovery and have been used by mankind for centuries. Our ancestors used salt to preserve meats and fish, added herbs and spices to improve the flavour of foods, preserved fruit with sugar, and pickled olives and cucumbers in a vinegar solution.

Today, with the advent of processed foods, there has been a massive explosion in the chemical adulteration of foods with additives. Considerable controversy has been associated with the potential threats and possible benefits of food additives.

Food processing is a set of methods and techniques used to transform raw ingredients into food or to transform food into other forms for consumption by humans or animals either in the home or in the food industry.

Today, with the advent of processed foods, there has been a massive explosion in the chemical adulteration of foods with additives. Considerable controversy has been associated with the potential threats and possible benefits of food additives. Islam lays down clear guidelines and legal principles concerning the legality of materials from haram sources, whether from animal or unclean sources.

The Halal Food Handbook, First Edition. Edited by Yunes Ramadan Al-Teinaz, Stuart Spear, and Ibrahim H. A. Abd El-Rahim.
© 2020 John Wiley & Sons Ltd. Published 2020 by John Wiley & Sons Ltd.

10.2 Why Use Additives?

Food made at home is usually at its best when eaten straightaway. Food produced on the large scale that is needed to supply supermarkets and other food shops has to be transported and stored before it is consumed. It has to stay in good condition over a much longer period of time than home-cooked food.

Food additives are substances added intentionally to foodstuffs to perform certain technological functions, for example to colour, sweeten or preserve. They are so essential that additives are used even in certain organic foods.

In many countries, lots of food is lost because it 'goes off' due to microbial growth before it can be eaten. Food poisoning also shows the dangers of contaminated food and without the use of preservatives, it would quite likely be more common. A food additive is defined as any natural or artificial material, other than the basic raw ingredients, used in the production of a food item to improve the final product or any substance that may affect the characteristics of any food, including those used in the production, processing, treatment, packaging, transportation, or storage of food.

In the European Union (EU) food additives are often referred to by E numbers because in EU countries additives are numbered with the prefix E. The E thus refers to an approved additive. Additives are not used to cover problems (such as spoiling) in the food, but are often used to prevent spoilage or other loss of quality. All additives are tested for toxicity and safety, but side effects can never completely be excluded. An E number means that a food additive has passed safety tests and is approved for use throughout the EU.

All the foods we eat consist of chemicals in one form or another. Many food additives are chemicals which exist in nature, such as antioxidants, ascorbic acid (vitamin C), and citric acid, found in citrus fruits. With technological advancements, many other additives are now made synthetically to perform certain technological functions. Whether or not the chemicals used in additives exist in nature, they are subject to the same safety evaluations by the European Food Safety Authority (EFSA).

Some consumers think of food additives (E numbers) as a modern invention used to make cheap foods. In reality, food additives have a long history of consumption and are used in many traditional foods (Armanios and Ergene 2018). For example, wines, including Champagne, contain sulphites, and bacon contains the preservatives nitrates and nitrites to prevent the growth of bacteria *Clostridium botulinum* (botulism).

10.2.1 Aims of Food Processing

Generally, food processing aims to:

- ensure food products are safe
- improve the safety and freshness of products
- improve and control the functionality of food constituents
- increase the storage life of food products
- improve or maintain nutritional values
- maintain the quality of food products
- develop traditional and innovative food products
- isolate specific compounds and exploit the biological potential of constituents.

10.2.2 Food Ingredients Sources

The common sources of food ingredients are:

- animals (milk, eggs, meat, seafood)
- plants (fruits, vegetables, spices, seafood)
- synthetic (flavours, colours, additives)
- fermented (organic acids, cultures, enzymes).

10.2.3 Groups of Food Ingredients

Food ingredients include:

- preservatives
- sweeteners
- colour additives
- flavours, spices
- flavour enhancers
- fat replacers
- nutrients
- emulsifiers
- stabilizers, thickeners, binders, and texturizers
- leavening agents
- anti-caking agents
- humectants
- yeast nutrients
- dough strengtheners and conditioners
- firming agents
- enzyme preparations
- gases.

10.2.4 Sources of Halal Ingredients

The sources of halal ingredients can be summarized as follow:

- vegetables: all vegetables are halal except those that are intoxicating (e.g. fermented apples, fermented grapes etc.)
- animals from halal species.
- halal animals slaughtered by a sane Muslim: ensure complete removal of blood from carcass, humane handling to be practiced
- synthetic ingredients.

10.2.5 Haram Ingredients

Haram foods are foods or drinks strictly prohibited by the Holy Quran and Sunnah. Haram foods include those containing:

- pork
- alcohol

- blood
- dead animals
- animals slaughtered without invoking the name of Allah.

10.2.6 Questionable/Mashbooh Ingredients

Mashbooh is an Arabic term that means 'suspected' or 'doubtful'. An ingredient is mashbooh if we are not sure of its source. Such ingredients include the following:

- gelatin: from pork, beef, fish
- glycerine/glycerol: from the saponification of animal fats
- emulsifiers: from animals
- enzymes: animal, microbial, biotechnological
- dairy ingredients: whey, cheese
- alcoholic drinks
- animal protein/fat
- flavourings and compound mixtures
- taurine: often used in energy drinks, mostly derived from pig gall
- pepsin, clarifiers, and stabilizers: used to make drinks look clear
- cloudifiers: used to make juice look cloudy
- active carbon and flavours: for fruity aromas

10.3 GMOs and Biotechnology

Islamic dietary law contains no specific mention of genetically modified organisms (GMOs) and genetic engineering in food and ingredients because these scientific developments are very recent, so the products are categorized according to conventional halal guidelines. Genetically modified or engineered products containing genes from haram animals would have great difficulty being accepted by Muslims. Since pork is prohibited, by extension any products made with pig genes are considered haram by many scholars and may not be accepted by Muslim consumers (Fadzlillah et al. 2011). A scientist can explain a new scientific development; likewise, a religious scholar can interpret whether or not the development violates any of the tenets of Islam. The following points are important when dealing with GMO food (Kurient 2002):

1) Haram is usually associated with what is harmful and unhealthy, and the safety of GMO foods is a key issue for halal. If it is determined beyond doubt that any of the foods or ingredients developed through genetic modification are harmful or unhealthy, the regulatory bodies, including Islamic organizations, will not approve them.
2) Good intentions do not make haram foods become halal. For example, if scientists try to make pigs cleaner and disease-free or grow pig organs for food in the laboratory, such organs would still be haram.
3) There is always a better replacement for something that is haram, especially through biotechnology. For example, until the mid-1980s porcine pepsin was used to extend the supply of calf rennet in cheese manufacture. Since the introduction of GMO-derived

chymosin, the use of pepsin as a replacement for calf rennet has practically vanished. This is a big issue for biotechnology in the area of halal foods (Chaudry and Regenstein 1994).

4) Muslim consumers are required to avoid doubtful things. If a Muslim consumer feels that GMO foods are doubtful, then he or she must avoid them. Currently, doubtful GMO foods are those modified with the use of the genes from prohibited animals. Biotechnology is an extension of plant and animal breeding and genetics, which have been practiced for decades and in some cases for centuries (Khattak et al. 2011). For example, animal breeding dates back to prehistoric times, when a donkey and a mare were crossbred to produce a mule. The meat of a donkey is not accepted as halal food and therefore neither is the meat of a mule.

5) Since genes were first identified, scientists have learned how to move a gene from one species to a more distant species. Currently, genes from fish, insects, or pigs can be introduced into plant species without affecting appearance or taste but giving the plants better disease resistance or nutritional advantages. Muslim scholars are striving to come to an acceptable decision about this kind of product.

6) Currently, food safety is the responsibility of government agencies and organizations like the United Nations Food and Agriculture Organization (FAO) and the World Health Organization (WHO).

10.4 E Codes

E codes indicate an ingredient which is some type of food additive. These codes are sometimes found on food labels in the EU. The 'E' indicates that it is an EU approved food additive. Other countries have different food labelling laws.

10.4.1 E Code Groups

Table 10.1 shows a list of E numbers of different food ingredients.

Questionable and haram E numbers are shown in Table 10.2. The following ingredients are considered as haram:

- E120: cochineal (red colour from insects), according to Hanoi Muzhab
- E124: if cochineal red A is used
- E441: gelatin, if it is from pork
- E542: edible bone phosphate if it is from pork bones.

10.5 Requirements for Halal Food Processing

The requirements for halal food processing are (Riaz 1998):

- Ingredients used in the processing of halal products must be from halal sources.
- Processing must be carried out according to Islamic rules and regulations.
- Products must be checked to see if there has been any involvement of an alcoholic product during processing.
- In the final product alcoholic ingredients must not exceed the permissible limit.

Table 10.1 E numbers of different food ingredients

E number	Food ingredient
E100	Colouring agents
E200	Preservatives
E300	Antioxidants
E400	Thickeners, stabilizers, gelling agents, emulsifiers
E500	Agents for physical characteristics
E600	Flavour enhancers
E900	Glazing agents, improving agents
E1100	Stabilizers, preservatives
E1200	Stabilizers
E1400	Thickening agents
E1500	Humectants

Table 10.2 Mashbooh and haram E numbers

E number	Food ingredient	Mashbooh/haram
E106	Riboflavin 5-sodium phosphate	Mashbooh
E120	Cochineal, carmines (animal)	Haram
E140	Chlorophyll, chlorophy	Haram
E161b	Lutein	Haram
E252	Potassium nitrate, saltpetre	Haram
E304	Fatty acid esters of ascorbic acid	Mashbooh
E322	Lecithin from animal fat	Mashbooh
E431	Polyoxyethylene	Haram

- Packaging material should not contain any haram ingredient.
- Cross-contamination with any haram ingredient must be avoided.
- Equipment must be washed with permissible detergents.

10.6 Hygiene and Cross-contamination

To maintain good hygiene practices and to avoid cross-contamination with non-halal ingredients, the following points must be taken into consideration during the processing of halal food (Riaz and Chaudhry, 2004):

- All equipment must be clean according to a visual/laboratory inspection.
- All food processing areas and equipment must be cleaned thoroughly after non-halal ingredient production (if both halal and haram foods are processed with the same equipment).
- All halal products must be segregated during storage to avoid cross-contamination.

10.7 Halal Markets

Halal markets include halal, ethnic, and specialty stores, supermarket chains, food services, universities, schools, offices, prisons, airlines, and the armed forces. Halal labelling on the processed product is important for clarification of doubt regarding ingredients and saving time (reading the labels) as well as peace of mind and satisfaction (Fischer 2011).

10.8 Some Food Ingredients

Gelatin, coagulating enzymes, emulsifiers, and stabilizers are commonly used food ingredients. Gelatin is a protein obtained from the bones, cartilage, tendons, and skin of animals. It has the following properties:

- gelling ability
- elasticity
- water-holding capacity
- emulsifying ability
- adhesive ability.

Rennet is a coagulating enzyme obtained from a young animal's stomach (usually a calf). It is used to form curds in cheese making. Whey is a watery liquid that separates from the solids (curds) of milk in cheese-making.

Emulsifiers are food additives that encourage the suspension of one liquid in another, as in the mixture of oil and water in margarine, shortening, ice cream, and salad dressings. The molecules in emulsifiers have one end that likes to be in an oily environment and one that likes an aqueous environment. The most frequently used raw materials for emulsifiers are:

- palm oil
- rapeseed oil
- soy bean oil
- sunflower oil
- lard/tallow
- egg.

Stabilizers are substances which make it possible to maintain the physicochemical state of a foodstuff. They include substances that can maintain a homogenous dispersion of two or more immiscible substances and also substances which stabilize, retain or intensify the existing colour of a foodstuff. The following hydrocolloids are commonly used as stabilizers:

- alginate
- agar
- carrageen
- gelatin
- xanthan gum
- guar gum

- gum Arabic
- pectin
- starch
- cellulose and cellulose derivatives.

10.9 Food Processing Aids

Food processing aids help to improve flavour, taste, nutrition value, appearance, freshness and safety. Commonly used food processing aids are:

- colours
- preservatives
- antioxidants
- sweeteners
- emulsifiers, stabilizers, thickeners, and gelling agents.

10.9.1 Food Colours

The primary reasons for adding colours to foods are:

- to offset colour loss due to exposure to light, air, extremes of temperature, moisture, and storage conditions
- to compensate for natural or seasonal variations in food raw materials or the effects of processing and storage to meet consumer expectations (masking or disguising inferior quality, however, are unacceptable uses of colours)
- to enhance colours that occur naturally but at levels weaker than those usually associated with a given food.

Commonly used colours include caramel (E150a), which is used in products such as gravy and soft drinks, and curcumin (E100), a yellow colour extracted from turmeric roots.

Some people think that adding colour makes food look more attractive, while other people think added colours are unnecessary and misleading.

10.9.2 Preservatives

Preservatives help to stop food from going off and ensure that food can be kept safe for a longer period. Most food that has a long shelf-life is likely to include preservatives, unless another method of preserving has been used, such as freezing, canning or drying. For example, to stop mould or bacteria growing, dried fruit is often treated with sulphur dioxide (E220), and bacon, ham, corned beef, and other cured meats are often treated with nitrite and nitrate (E249–E252) during the curing process. More traditional preservatives such as sugar, salt, and vinegar are also still used to preserve some foods.

10.9.3 Antioxidants

Any food made using fats or oils – from meat pies to mayonnaise – is likely to contain antioxidants. These make foods last longer by helping to stop the fats, oils, and certain vitamins from combining with oxygen in the air – this is what makes food taste 'off', when it becomes rancid and loses colour.

Vitamin C, also called ascorbic acid or E300, is one of the most widely used antioxidants.

10.9.4   Sweeteners

The desire for the pleasure of sweetness has a strong influence on what people choose to eat and drink. Since early times, people have sought out foods with a sweet taste, for example drawings on the walls of Egyptian tombs show bee-keepers collecting honey and sugar cane was grown in India some 2000 years ago.

Today, sucrose, or table sugar, is the taste standard by which all other sweeteners are measured. An 'ideal' sweetener tastes like sucrose and is colourless, odourless, readily soluble, stable, and economical. Some sweeteners, like sugar, contain calories, but some are low-calorie or calorie-free Sweeteners are lower in calories and safer for teeth, and are often used instead of sugar in products such as fizzy drinks, yoghurt, and chewing gum. However, consumption of some sweeteners, even in very small amounts, for a long time can cause cancer and or multiple sclerosis (MS).

Intense sweeteners, such as aspartame (E951), saccharin (E954), and acesulfame-K (E950), are many times sweeter than sugar and so only very small amounts are used.

Bulk sweeteners, such as sorbitol (E420), have about the same sweetness as sugar and so they are used in similar quantities to sugar. If you give concentrated soft drinks that contain sweeteners to children aged under four, it is important to dilute them more than you would for an adult. This is to avoid children having large amounts of sweetener.

10.9.5   Emulsifiers, Stabilizers, Thickeners, and Gelling Agents

Add oil to water and the two liquids will never mix, at least not until an emulsifier is added. Emulsifiers are molecules with one water-loving (hydrophilic) and one oil-loving (hydrophobic) end. They make it possible for water and oil to become finely dispersed in each other, creating a stable, homogenous, smooth emulsion.

Emulsifiers such as lecithins (E322) help to mix ingredients together that would normally separate, such as oil and water. Stabilizers, such as locust bean gum (E410) made from carob beans, help stop these ingredients from separating again.

Emulsifiers and stabilizers also give foods a consistent texture. They are used in foods such as low-fat spreads and other sweet and savoury foods. The most common gelling agent is pectin (E440), which is used to make jam. Gelling agents are used to change the consistency of food. Thickeners help to give body to food in the same way as adding flour thickens a sauce.

10.9.6   Flavour Enhancers and Flavourings

Flavour enhancers are used to bring out the flavour in a wide range of savoury and sweet foods without adding a flavour of their own. For example, monosodium glutamate (E621), known as MSG, is added to processed foods, especially soups, sauces, and sausages. Flavour enhancers are also used in a wide range of other foods, including savoury snacks, ready meals, and condiments.

Flavourings, in contrast, are added to a wide range of foods, usually in very small amounts, to give a particular taste or smell. Flavourings do not have E numbers because they are controlled by different laws to other food additives. Ingredient lists will say if flavourings have been used, but individual flavourings might not be named. Salt, although not classed as a food additive, is the most widely used flavour enhancer.

10.10 Food Conservation and Additives

10.10.1 Food Conservation

Food processing and conservation can be improved by modified packaging (sterile, impenetrable to water, etc.). Methods used for food conservation are:

- refrigeration
- freezing
- pasteurization
- sterilization
- UHT processing
- concentration (for liquid food, such as milk)
- drying
- filtration
- high pressure
- ultrasound
- pulsed electrical fields.

10.10.2 E Numbers and Additives of Animal Origin

Table 10.3 lists food additives by E number and gives their origins, which may be an animal or a non-animal source. Additives that contain fatty acids are of plant origin, but animal origin cannot be excluded. As the products are chemically identical, the producers can give information on their exact origin.

Religious (e.g. Muslim, Jew, Hindu) or other (e.g. vegetarians, vegans) groups can use the list In Table 10.3 to determine whether or not to accept an additive. In addition, the following three questions should be asked:

Which E numbers are halal/haram?
Which E numbers are allowed by Islam?
Which products contain ethanol and are haram?

10.10.3 Forbidden Additives

The list of prohibited E numbers is very short. E120 and E904 are prohibited because they are made of or contain insects. E901 is made by insects, like honey, but does not contain insects and thus is generally considered halal.

All other E numbers are basically permitted and are widely used in Islamic countries. However, this does not mean that all additives are *always* halal. Fatty acids are used in the

Table 10.3 E numbers of additives and their origin

E number	Name	Origin
E120	Carmine, cochineal	Colour isolated from the insect *Coccus cacti*
E322	Lecithin	Soy beans and for some purposes chicken eggs
E430	Polyoxyethylene (8) stearate	
E431	Polyoxyethylene (40) stearate	
E432	Polyoxyethylene-20-sorbitan monolaurate	
E433	Polyoxyethylene-20-sorbitan mono-oleate	
E434	Polyoxyethylene-20-sorbitan monopalmitate	
E435	Polyoxyethylene-20-sorbitan monostearate	
E436	Polyoxyethylene-20-sorbitan tristearate	
E441 (invalid)	Gelatin	From animal bones. Since the BSE crisis mainly from pork, but other animal bones are used. Halal gelatin is available in specialized shops.
E470	Fatty acid salts	
E471	Mono- and di-glycerides of fatty acids	
E472	Esters of mono- and di-glycerides	
E473	Sugar esters of fatty acids	
E474	Sugar glycerides	Combination of sugar and fatty acids[a]
E475	Polyglycerol esters of fatty acids	
E477	Propylene-glycol esters of fatty acids	
478	Mixture of glycerol- and propylene-glycol esters of lactic acid and fatty acids	
E479 and 479b	Esterified soy oil	[a]
E481/2	Natrium(sodium)/calcium-stearyl lactate	Mixture of lactic acid and stearic acid, a fatty acid[a]
E483	Stearyl tartrate	Mixture of tartaric acid and stearic acid, a fatty acid[a]
484	Stearyl citrate	Mixture of citric acid and stearic acid, a fatty acid[a]
E485 (invalid number)	Gelatin	From animal bones. Since the BSE crisis mainly from pork, but other animal bones are used. Halal gelatin is available in specialized shops.

(Continued)

Table 10.3 (Continued)

E number	Name	Origin
E491–495	Combinations of sorbitol and fatty acids	
542	Edible bone phosphate	From animal bones. Since the BSE crisis mainly from pork, but other animal bones are used.
E570–573	Stearic acid and stearate	Stearic acid is a fatty acid[a]
E626–629	Guanylic acid and guanyllinates	Mainly from yeast, also from sardines and meat
E630–635	Inosinic acid and inosinates	Mainly from meat and fish, also made with bacteria
636 and 637	Maltose and isomaltose	From malt (barley), sometimes also from heating milk sugar
E640	Glycine	Mainly from gelatine (see 441 above), also made synthetically
E901	Bees wax	Made by bees, but does not contain insects
E904	Shellac	Natural polymer derived from certain species of lice from India. Insects get trapped in the resin.
913	Lanolin	A wax from sheep. It is excreted by the skin of the sheep and extracted from the wool.
920–21	L-Cysteine/L-cystine	Derived from proteins, including animal protein and hair
E966	Lactitol	Made from milk sugar
1000	Cholic acid	From beef (bile)
E1105	Lysozyme	From chicken eggs

[a] *Source:* taken from http://www.food-info.net.

production of many additives and it is a matter of concern for many Muslims where these come from. If the fatty acids are of plant origin, they are halal, but if they are of animal origin they may be halal or haram, depending on the animal species. Chemically, fatty acids are identical, but from their chemical composition it cannot be determined whether animal or vegetable fat has been used. Only the producer and/or ingredient supplier is able to provide this information (Indrasti et al. 2010).

Another complication is that additives can be listed by their chemical name or by their E number. In the EU the producer has to provide the name or the number or both. In other countries, the producers sometimes use only the chemical name or the number without the letter E and in some countries only the chemical names are used. It is better to use the E numbers and/or the chemical name as in EU countries.

For example, in the UK chicken sold in Indian and Chinese restaurants and some cafes may be bulked up with a water and chemical mix containing pork or beef (Poulter 2009). The chicken breasts are injected with the mix or placed with the mix in a device similar to a washing machine. This tactic fools customers into thinking they are getting more for their money. However, consumers are eating beef and pork – generally waste skin harvested from carcasses – without being told. This is not good for the country's millions of Muslims who eat halal meat from approved butchers, but may have unwittingly eaten food containing pork, which is haram (Lever and Miele 2012). The problem of chicken secretly injected with beef and pork products was highlighted in 2009 by the Food Standards Agency (FSA), which carried out DNA testing of the chemical powders used to bulk up chicken in three production plants – two in the UK and one in continental Europe (Poulter 2009).

10.10.4   Fat Additives

Fats, whether from plant or animal origin, consist of glycerol and generally three fatty acids. Fats can be split into fatty acids and glycerol (the same reaction also takes place in the intestine when fats are digested). The fatty acids can be purified and reconnected to glycerol as mono- di- or tri-glycerides (glycerol with one, two or three fatty acids, respectively). Many additives consist of these semi-natural fats, which act as emulsifiers. These fats are degraded and metabolized in the body, just like normal fat.

Chemically the fatty acids of animal or plant origin are identical, therefore the origin is of no importance for the function of the fatty acid in the food. Producers thus normally choose the cheapest oils to make these fats. This is generally some vegetable oil, which makes the additives halal. However, sometimes animal fats may not be completely excluded and thus the same additive may sometimes be haram.

In the Islamic world there are several discussions on additives containing fatty acids. Here we will mention the two points of issue, but unless there is agreement amongst Islamic scholars, we do not further elaborate on the arguments of all parties concerned.

The main issue is the presence of animal fatty acids. If this fatty acid originates from pork, it is generally considered haram. If it is from other animals, it is generally considered halal. However, there is a discussion concerning whether animal fat from animals other than pigs is halal if the animal has not been slaughtered in the prescribed Islamic way.

10.10.5   Alcohol and Ethanol

According to Islam the use of alcohol is forbidden, as it may influence the mind of the person and thus his behaviour. Unfortunately, when the Holy Quran was written, the word alcohol meant only ethanol. Nowadays, chemically, alcohol means all chemical components with an –OH (alcohol) group. In common daily language, therefore, alcohol means either ethanol or any drink containing a certain percentage of ethanol. For religious purposes and in daily life ethanol and alcohol thus are identical, but for chemists ethanol is just one of many alcohols.

Vinegar is a product that is traditionally prepared from wine or other fermented liquids. During the fermentation, the ethanol is converted by acetic acid bacteria into acetic acid. The final vinegar thus contains only traces of ethanol. It is not possible to get drunk from vinegar and so vinegar is considered halal. Modern vinegar can also be made chemically and never contained ethanol. Traditional vinegar is therefore often referred to as wine vinegar or similar, which may be confusing. Wine vinegar does not contain ethanol, the name simply indicates that is made from wine and not made chemically.

Sugar alcohols, also known as polyols, are sweeteners. They include sorbitol, xylitol, and a number of other products, all with names ending in -itol. These products are made of sugars and chemically the aldehyde or ketone group in the sugar is converted into an alcohol group (=O is converted into –OH). Here 'alcohol' means the chemical group –OH and not ethanol, the forbidden form of alcohol. Sugar alcohols thus are halal.

10.10.6 Ice Cream

Ice cream is a food that would not exist were it not for emulsifiers. The added fruits or chocolate sometimes contain alcohol and/or emulsifiers. Some of the flavours used may also be of concern.

Ice cream is both a foam and an emulsion, and its texture results from the ice crystals and unfrozen water it contains.

The emulsifiers that are used commercially in ice cream come from both natural and synthetic sources. They include:

Lecithin (E322) are mixtures of phospholipids such as phosphatidyl choline and phosphatidylethanolamine and are usually extracted from sources such as egg yolk and soybeans. The precise composition of the phospholipids depends on the source. Uses include salad dressings, baked goods, and chocolate.

Esters of monoglycerides of fatty acids (E472a–f) are made from natural fatty acids, glycerol, and an organic acid such as acetic, citric, lactic, or tartaric. The fatty acids are usually from a vegetable source, though animal fats can be used. Products that use them include ice cream, cakes, and crisps.

Mono- and diglycerides of fatty acids (E471) are semi-synthetic emulsifiers made from glycerol and natural fatty acids, which can be from either plant or animal sources. They are used in products like breads, cakes, and margarines.

10.10.7 Drinks

In drinks, the combination of sodium benzoate or potassium benzoate (E212) and ascorbic acid (vitamin C, both naturally occurring and the additive E300) can result in the formation of *benzene*, a known carcinogen. Exposing the bottle to heat or light during transport or storage can boost the amount of benzene formed.

Food Standards Australia New Zealand (FSANZ) tested 68 flavoured drinks, including flavoured mineral waters, cordial, fruit juice, and fruit drinks, and found 38 of the samples contained trace levels of benzene. While the majority had benzene levels below WHO guidelines for drinking water – 10 parts per billion (ppb) – some contained levels up to 40 ppb.

10.10.8 Chewing Gum

Some chewing gums contain glycerine.

10.10.9   Fruits and Vegetables

Oranges, lemons, apples and other fruits are often sprayed. Exotic fruits, e.g. bananas, that are transported long distances are often treated with chemical substances to stop them ripening and make them last longer.

10.10.10   Cube Sugar

Bone char, often referred to as natural carbon, is made from the bones of cattle. It is widely used by the sugar industry as a decolorizing filter, which allows the sugar cane to achieve its desirable white colour. Other types of filters involve granular carbon or an ion-exchange system (Fairclough 1998).

   Bone char is also used for processing other types of sugar. Brown sugar is created by adding molasses to refined sugar, so companies that use bone char in the production of their regular sugar also use it in the production of their brown sugar. Confectioner's sugar – refined sugar mixed with cornstarch – also involves the use of bone char. Fructose processing may, but does not typically, involve a bone-char filtration. Supermarket brands of sugar are obtained from several different refineries, making it impossible to know whether or not they have been filtered with bone char.

10.10.11   Medication

Animals have been used as medicinal resources for the treatment and relief of a myriad of illnesses and diseases in practically every human culture. Although considered by many as superstition, the pertinence of traditional medicines based on animals cannot be denied since they have been methodically tested by pharmaceutical companies as sources of drugs in the modern medical science. The phenomenon of zootherapy represents strong evidence of the medicinal use of animal resources. Indeed, drug companies and agribusiness firms have been evaluating animals for decades without paying anything to the countries where these genetic resources are found. The use of animal body parts as folk medicines is relevant because it places additional pressure on critical wild populations. It is argued that many animal species have been overexploited as sources of traditional medicines.

   Some Muslims scholars believe that if the medicine is necessary for treatment then the ingredients are immaterial and all can be used.

10.10.12   Antibiotics in Animal Feeds

The food-producing animal and poultry industries have undergone a dramatic change that began around 1950. What was an extensive industry became extremely intensive: units increased in animal concentration, both physically and numerically. Utilization of the beneficial responses of feed-additive antibiotics in improved growth and feed efficiency developed concurrently with the intensification of the animal and poultry industry. It has been proposed that feed-additive antibiotic usage was an integral part of this revolution in animal-production technology (Elasrag 2016).

10.10.13 Toothpaste

Toothpaste may contain substances from pigs (especially fat), and some may contain phosphates from animal bones.

10.10.14 Soap, Shampoo, and Cosmetics

Many toiletries and cosmetics may contain proteins. Proteins generally are made of slaughter remnants, blood or carcases.

Detergents may contain enzymes (for removing stains). Today's enzymes are produced synthetically but in the past came from pig stomach tissue. Some detergent ingredients are:

- alcohol ethoxylate (AE): a non-ionic surfactant
- alkyl (or alcohol) ethoxy sulphate (AES) and alkyl sulphate (AS): anionic surfactants
- amine oxide
- carboxymethyl cellulose (CMC)
- citric acid
- cyclodextrin
- diethyl ester dimethyl ammonium chloride (DEEDMAC)
- ethanol.

Halal/haram ingredients

Some of the emulsifiers found in canned and packaged foods are sometimes made from an unlawful (haram) source(s). Table 10.4 lists the E numbers and gives a brief description of haram emulsifiers that should be avoided.

Table 10.4 E number of haram emulsifiers

E number	Description of the emulsifier
E120	Cochineal: fatty acids and egg yolks of dried female *Dactilopius coccus* and 10% carminic acid
E140	Fatty acids and phosphates
E141	Fatty acids
E252	Waste animal and vegetable material
E422	Fatty acids, by-products in the manufacturing of soaps
E430	Fatty acid molecules
E431	Fatty acids
E470	Emulsifiers and stabilizers - salts or esters of fatty acids from pork fat
E471	Glycerine and fatty acids
E472(a)	Acetic acid esters of mono- and diglycerides of fatty acids

Table 10.4 (Continued)

E number	Description of the emulsifier
E472(b)	Lactic acid ester of mono- and diglycerides of fatty acids
E472(c.)	Citric acid ester of mono and di-glycerides of fatty acids
E472(d)	Tartaric acid ester of mono- and diglycerides of fatty acids
E472(e)	Mono-diaceryltartaric acid ester of mono- and diglycerides of fatty acids
E473	Sucrose esters of fatty acids
E474	Lard (pig fat), tallow (hard animals fat), palm-oil etc.
E475	Polyglycerol esters of fatty acids
E477	Proplylene glycol esters of fatty acids
E478	Lactylated acid esters of glycerol and propene-1,2-diol
E481	Sodium stearoyl-2 lactylate
E482	Calcium stearoyl-2 lactylate
E483	Stearyl tartarate
E491	Sorbitan monostearate
E492	Sorbitan tristearate (span 65)
E494	Sorbitan mono-oleate (span 80)
The following without 'E' prefixes are also haram:	
476	Polyglycerol esters of polycondensed acids of castor oil
542	Edible bone phosphates
570	Stearic acid
572	Magnesium stearate
631	Meat extracts, dried sardines, and sodium 5-insosinate
Other haram emulsifers:	
120, 141, 160(A), 161, 252, 300, 301, 433, 435, 422, 430, 431, 471, 472(a.e), 436, 441, 470, 476, 477, 473, 474, 475, 491, 492, 481, 482, 483, 570, 572, 494, 542	

Source: https://www.halalcertifiering.se/halal_haram_e_nummer.pdf and the London Central Mosque Trust and the Islamic Cultural Centre.

10.11   Conclusions

Halal food processing ingredients must come from a halal source and the processing must be carried out in accordance with Islamic rules and regulations. Final compositions must be checked if there is involvement of any alcoholic product during processing. In the final composition, alcoholic ingredients must not exceed the permissible limits. Packaging materials should not contain any haram ingredients. Cross-contamination with non-halal materials must be avoided. Equipment must be thoroughly washed with permissible detergents.

In food processing, all equipment must be cleaned after non-halal ingredient production (if both halal and haram foods are processed with the same equipment). All halal products must be segregated during storage to avoid cross-contamination.

The adulteration of pig sources in food production is prohibited in Islam. From the Islamic point of view, this prohibition includes all parts of the pig, i.e. flesh, skin, and their derivatives (lard, enzymes, etc.) (Sakr 1991). Products that contain lard must have this clearly stated on the label. Recently, several biotechnological techniques have been developed for detection of food adulteration and these will help Muslim consumers when choosing halal food products.

From the halal perspective, the compliance of some ingredients and products in the food and pharmaceutical industries remains unresolved. A challenge to the Islamic community is to set up a food laboratory where ingredients can be analysed and then the results made available to the Muslim community. It is important to ensure that food additives are from acceptable sources and processed according to halal requirements without the use of alcohol-based carriers (Ruževičius 2012).

Packaging materials should not be prepared, processed or manufactured using equipment that is contaminated with things that are not permitted by Islamic dietary law (Eliasi 2002). During the preparation, processing, storage or transportation of the packaging material, it should be physically separated from any packaging material that does not meet halal requirements. In addition, packaging materials should not contain any raw materials that are considered hazardous to human health. The packing process shall be carried out in a clean and hygienic manner and in sound sanitary conditions.

References

Armanios, F. and Ergene, B. (2018). Halal Food: A History. Oxford University Press.

Chaudry, M.M. and Regenstein, J.M. (1994). Implication of biotechnology and genetic engineering for kosher and halal foods. *Trends Food Sci. Technol.* 5: 165.

Elasrag, H. (2016). Halal Industry: Key Challenges and Opportunities. ISBN: ISBN-13: 978-1530029976.

Eliasi, J.R. (2002). Kosher and halal: religious observances affecting dietary intakes. *J. Am. Diet. Assoc.* 101 (7): 911–913.

Fadzlillah, N.A., Man, Y.B.C., Jamaludin, M.A., et al. (2011). Halal food issues from Islamic and modern science perspectives. 2nd International Conference on Humanities, Historical and Social Sciences. IPEDR vol.17 (2011). IACSIT Press, Singapore. https://www.academia. edu/4089569/Halal_Food_Issues_from_Islamic_and_Modern_Science_Perspectives (accessed September 2019).

Fairclough, G. (5 March 1998). A special background report on trends in industry and finance. *The Wall Street Journal, Section A*: 1.

Fischer, J. (2011). The Halal Frontier Muslim Consumers in a Globalized Market. Pulau Pinang: Persatuan Pengguna Pulau Pinang, CAP.

Indrasti, D., Che Man, Y.B., Mustafa, S., and Hashim, D.M. (2010). Lard detection based on fatty acids profile using comprehensive gas chromatography hyphenated with time-of-flight mass spectrometry. *Food Chem.* 122: 1273–1277.

Khattak, J.Z.K., Mir, A., Anwar, Z. et al. (2011). Concept of halal food and biotechnology. *Adv. J. Food Sci. Technol.* 3 (5): 385.

Kurient, D. (2002). Malaysia: studying GM foods acceptability of Islam. Dow Jones, Online News via News Edge Corporation, Kuala Lumpur, Malaysia (9 August 2002).

Lever, J. and Miele, M. (2012). The growth of halal meat markets in Europe: an exploration of the supply side theory of religion. *J. Rural. Stud.* 28 (4): 528–537.

Poulter, S. (2009). Chicken secretly injected with beef and pork products served in UK restaurants. *Daily Mail* (5 June 2009).

Riaz, M.N. (1998). Halal food. An insight into a growing food industry segment. *Int. Food Mark. Technol.* 12 (6): 6.

Riaz, M.N. and Chaudhry, M.M. (2004). Halal Food Production, 251. CRC Press.

Ruževičius, J. (2012). Products quality religious–ethnical requirements and certification. *Econo. Manage.* 17 (2): 761–767.

Sakr, A.H. (1991). Pork: Possible Reasons for Its Prohibition. Lombard: Foundation for Islamic Knowledge.

Further Reading

Consumers Association of Penang (2006). Halal Haram: An Important Book for Muslim Consumers: A Guide, 200 p. Consumers Association of Penang https://books.google.com/books/about/Halal_Haram.html?id=IDTYAAAAMAAJ.

11

Halal and Genetically Modified Ingredients
Majed Alhariri

Independent Halal Researcher in Food Science and Technology, Cairo University, Cairo, Egypt

11.1   What is a Genetically Modified Organism?

A genetically modified organism (GMO) is one in which an alteration has been made to the material that is inherited by the offspring of the organism. Such alterations can be in the form of additions to, subtractions from, or substitutions to the existing material passed on. This material, known as deoxyribonucleic acid (DNA), is common to all known living things and is the means by which the specific characteristics of a parent organism are transmitted to its offspring. Changes to the DNA will typically result in changes to the inherited characteristics of the organism.

A 'gene' is often described as the string of DNA that contains all the necessary information for the making of a specific protein, the machinery in cells. DNA is a long string of four repeating molecular building blocks that are brought together in different combinations to be the code of all genes. The length of DNA and the code for a gene are specific to the function of the protein encoded for.

Genetically modified (GM) foods are foods derived from organisms whose genetic material (DNA) has been modified in a way that does not occur naturally, e.g. through the introduction of a gene from a different organism using genetic modification techniques.

These techniques, generally known as recombinant DNA technology, use DNA molecules from different sources, which are combined into one molecule to create a new set of genes. This DNA is then transferred into an organism, giving it modified or novel genes.

11.2   How Does Genetic Modification Work?

Using genetic modification, scientists take genes from bacteria, viruses, or other sources and force them into the DNA of a plant. The process of genetic modification requires the successful completion of a series of five steps.

The Halal Food Handbook, First Edition. Edited by Yunes Ramadan Al-Teinaz, Stuart Spear, and Ibrahim H. A. Abd El-Rahim.
© 2020 John Wiley & Sons Ltd. Published 2020 by John Wiley & Sons Ltd.

Step 1: DNA extraction: DNA extraction is the first step in the genetic modification process. In order to work with DNA, scientists must extract it from the desired organism. A sample of an organism containing the gene of interest is taken through a series of steps to remove the DNA.

Step 2: Gene cloning: The second step of the genetic modification process is gene cloning. During DNA extraction, all of the DNA from the organism is extracted at once. Scientists use gene cloning to separate the single gene of interest from the rest of the genes extracted and make thousands of copies of it.

Step 3: Gene design: Once a gene has been cloned, genetic engineers begin the third step, designing the gene to work once inside a different organism. This is done in a test-tube by cutting the gene apart with enzymes and replacing gene regions.

Step 4: Transformation: The new gene is inserted into some of the cells using various techniques. Some of the more common methods include the gene gun, agrobacterium, microfibers, and electroporation. The main goal of each of these methods is to transport the new gene(s) and deliver them into the nucleus of a cell without killing it. Transformed plant cells are then regenerated into transgenic plants. The transgenic plants are grown to maturity in greenhouses and the seed they produce, which has inherited the transgene, is collected.

Step 5: Backcross breeding: The fifth and final part of producing a genetically modified crop is backcross breeding. Transgenic plants are crossed with elite breeding lines using traditional plant breeding methods to combine the desired traits of elite parents and the transgene into a single line. The offspring are repeatedly crossed back to the elite line to obtain a high yielding transgenic line. The result will be a plant with a yield potential close to current hybrids that expresses the trait encoded by the new transgene.

11.3 Currently Commercialized GM Crops

There are over 40 plant varieties which are at commercialization stage and a large amount of foodstuff on a supermarket shelf already contains GM food or its derivatives. Microbial rennet (used in many dairy products, including cheese), grains such as maize, corn or rice, and even everyday vegetables like spinach and tomatoes (or derivatives such as tomato sauce or other types of cooking sauces containing tomatoes) can be from a GM source. Let us not forget livestock, which can often be fed on GM grains.

The four major GM crops are soy beans, corn, cotton, and canola (Table 11.1). Other GM crops include sugar beets, zucchini, rice, tomato, potato, alfafa, peas, melon etc.

Table 11.1 Planted area of major GM crops

GM crop	Planted area (million hectares)	% of global planting
Soy beans	84.5	79
Cotton	23.9	70
Corn	57.4	32
Canola	8.2	24

Biotech crop hectares increased by more than 100-fold from 1.7 million hectares in 1996 to over 175 million hectares in 2013. In 2013, 1.5 billion hectares of all crops were planted worldwide.

Governments around the world are working hard to establish regulatory processes to monitor the effects of GM, approving new varieties of GM plants and labelling regulations for foods containing GM, depending on the political, social, and economic climate within each country or region. In the European Union (EU), for example, if a food contains GMOs, or contains ingredients produced from GMOs, this must be indicated on the label. This means products such as flour, oils, and glucose syrups have to be labelled as GM if they are from a GM source. However, products produced with GM technology (cheese produced with GM enzymes, for example) do not have to be labelled. Products such as meat, milk, and eggs from animals fed on GM animal feed also do not need to be labelled. Due to major global inconsistencies on the regulation of GM food, it is impossible to ascertain which everyday foodstuffs are GM-derived; only local food regulatory authorities within each country can provide clarification on this matter.

11.4   GM Crop Benefits

Most existing GM crops have been developed to improve yield and quality. There are many potential applications of GM crops, such as pest resistance, herbicide tolerance, disease resistance, cold tolerance, drought tolerance, salinity tolerance, faster maturation, vitamin enrichment, and altered fatty acid composition.

The 1970s saw the start of genetic modification technology, which in essence takes genes from one species and forces it into the DNA of another species to produce GM food or organisms. These 'modified' genes produce proteins which generate specific characteristics or traits, such as highly productive and resilient crops, vegetables growing in the desert, and vitamin-fortified grains. Thus, genetic modification promises significant advantages, such as feeding millions of starving people in the world, as well as environmental benefits by reducing the amount of land needed to grow crops, thus allowing more land to be used for natural habitats.

Biochemical pesticides or biopesticides is another application of GM food. Biopesticides are naturally occurring substances produced by GM plants to control pests by non-toxic mechanisms. GM plants genetically modified with natural materials such as bacteria genes will allow GM crops to survive by producing substances such as insect sex pheromones that interfere with mating, as well as various scented plant extracts that attract insect pests to traps. GM plants that produce biopesticides will increase their resistance to many different kinds of pests. Conventional pesticides, by contrast, are generally synthetic materials that directly kill or inactivate the pest and may affect different existing organisms, from birds to insects to mammals. In addition, biopesticides are often effective in very small quantities and often decompose quickly, thereby resulting in lower exposures and largely avoiding environmental problems (pollution) that can be caused by conventional pesticides.

11.5   Concerns about Food Safety and Human Health

Questions have been raised about the long-term safety of GM food for human consumption. Scientists agree that there is a possibility of 'accidental changes' in genetically modified plants, which may produce unexpectedly high doses of plant toxins. GM crops might have 'increased levels of known naturally occurring toxins', and the 'appearance of new, not previously identified' toxins (Mathews 1991).

The same mechanism can also produce allergens, carcinogens, or substances that inhibit assimilation of nutrients. Laboratory tests as well as anecdotal human studies show genetic modification to be linked to toxic reactions in the digestive tract and liver damage. Animals fed on GM feed were shown to have higher death rates and organ damage. Signs of reproductive failure and infant mortality, sterility in livestock as well as GM crops triggering immune reactions and increasing the incidence of allergies in humans have all been reported.

Without adequate testing of the long-term effects of ingesting GM products, there is no way of guaranteeing that there is no long-term health issues associated with a particular type of genetic modification. Just because one type of GM does not cause problems does not infer that another type of modification will be free of health risks. Since GM foods are not properly tested before they enter the market, we consumers are the guinea pigs. With this knowledge in the public domain, it comes as no surprise that the sale of organic-certified foodstuffs (that do not allow any GM ingredients) is going through the roof around the world, as cautious consumers are deliberately avoiding foodstuffs that may contain GM crops.

A number of publications, including books and articles, with strong arguments against GM technology and GM foods have been published in recent years (Institute for Responsible Technology, 2015; Malatesta et al. 2008; Hammond et al. 2006; Mahgoub Salah 2016). Scientists believe that what is known about GMOs is very little compared with what is unknown and that governments use consumers as experimental animals for testing GM foods.

The authors of the report *GMO myths and truths* (Fagan et al. 2014) believe that GM crops and foods are neither safe nor necessary to feed the world. The three authors, who are genetic engineers, discuss the different issues related to GM crops and GM foods and try to differentiate between the myths and truths surrounding GM foods. They state that the GMO debate is far from over and that the evidence of risk and actual harm from GM foods and crops to health and the environment has grown in the last few years. This report represents very strong opposition to, and criticism and condemnation of GM crops and GM foods.

Engdahl (2009) reports that the American Academy of Environmental Medicine (AAEM) states that 'GM foods pose a serious health risk in the areas of toxicology, allergy, and immune function, reproductive health, and metabolic, physiologic, and genetic health.'

GM foods might be riskier than traditional foods. The health effects of consuming GM foods are not yet fully understood. The most pertinent general question asked by many consumers, which needs a satisfying answer, is: 'Is it safe to consume GM foods?'. The potential risks of GM foods related to health have not yet been adequately investigated. There is a need for improved and long-term food safety tests, in addition to reliable technologies and protocols to better identify and manage potential risks.

Critics and opponents of GM technology in general, and its application to produce GM foods in particular, have presented arguments that support their stance against this technology. The following is some of the points which reflect the point of view that GM technology is hazardous and has many risks.

1. *It is unnatural, has unpredictable results, and can lead to unexpected and unintended side effects*

Friends of the Earth (Friends of the Earth 2014) are worried that GE creates whole new life forms. The organisms resulting from this technology are alive and can mutate, multiply, breed with other living things, and continue breeding for generations to come. This trend has been observed and recorded all around the world. FoE cites the following words of Dr Michael Antoniou (Senior Lecturer in Molecular Biology, London) to support their argument: 'This is an imperfect technology with inherent dangers .... It is the unpredictability of the outcomes that is most worrying'. The food produced using such unnatural techniques is expected to be unnatural too. People fear eating foods that are not natural.

2. *Genetic uncertainties and the disturbance of nature's boundaries*

Natural boundaries are violated: crossing animals with plants, strawberries with fish, grains, nuts, seeds, and legumes with bacteria, viruses, and fungi, or human genes with swine (Batalion 2000). This leads to a 'control' of living nature and imposing a 'non-living' model on it. Critics believe that the term 'bioengineering' itself is a contradiction. 'Bio' refers to 'life' – something that is not mechanically predictable or controllable – and 'engineering' implies developing the basics for machines that are predictable, not alive but 'dead', so the contradiction.

3. *Unpredictability and the unknown*

Insertion of genetic material in a host could lead to uncontrolled and erratic behavior in the host. This is because DNA is complex and there is the potential for complex interactions. These interactions can cause gene suppression or over-expression, causing unpredictable and uncommon changes. The potential hazards are difficult to predict with any certainty. The technique of using a 'gunshot' to blast DNA fragments through cell membranes is cited as an example, leading to unpredictable consequences. The foreign genetic material is shot in a random, unpredictable way, possibly resulting in unknown products.

Séralini et al. (2012) studied the long-term toxicity of GM maize on rats. Results showed that all treated female groups died one to three times more often than controls, and more rapidly. Females developed large mammary tumours almost always more often than and before controls, the pituitary was the second most disabled organ; the sex hormonal balance was modified by consumption of GMO and Roundup treatments. In treated males, liver congestion and necrosis were 2.5–5.5 times higher. Marked and severe kidney nephropathies were also generally 1.3–2.3 times greater. Males presented four times more large palpable tumours than controls, in which only one tumour was noted, and they occurred up to 600 days earlier. Additionally, biochemistry data confirmed very significant kidney chronic deficiencies; for all treatments and both sexes, 76% of the altered parameters were kidney related. Results also showed that sex hormonal balance was modified by consumption of GM maize.

Results also showed that by the beginning of the 24th month, 50–80% of the female animals had developed tumours in all treated groups, with up to three tumours per animal (Figure 11.1).

A study conducted on female rats fed on GM soy showed that more than 50% of the offspring died within 3 weeks. Feeding rats on GM soy caused a dramatic reduction in average weight. In addition, a histopathology study on liver and testicles sections illustrated that rats fed GM soy showed colour change in their liver and testicles, as well as changes in the cell structures.

Another example of a GM product is recombinant bovine growth hormone (rBGH). rBGH is a genetically modified variant of the natural growth hormone produced by cows. Manufactured by Monsanto, it is sold to dairy farmers under the trade name Posilac. The consumption of milk from cows' injected with rBGH leads to an increase in Insulin-like growth factor-1 (IGF-1) (a hormone that normally helps some types of cells to grow) in humans. Several epidemiological studies have indicated a relationship between dairy consumption and colon, breast, and prostate cancer risk (Figure 11.2; Dona and Arvanitoyannis 2009; Dohoo 2003).

The critics of GM technology argue that GM products, including foods, are hazardous, and their long-term use can produce ill-effects on human and animal health. The USA has signed the Monsanto Protection Act prevents government restrictions on GMO foods found to cause health risks (Lanphier 2014).

Genetic modification and its application to produce GM foods have the potential to create unpredictable dangers to human health that may be short term or long term, direct or indirect. Specific potential food safety and human health problems that have been expressed are centered on the following points:

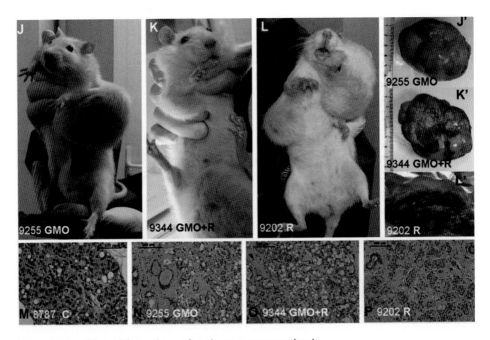

Figure 11.1 Effect of GM maize on female rat mammary glands.

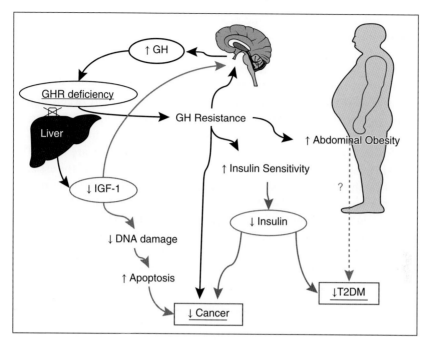

Figure 11.2 Effect of IGF on human health.

1) *GM food may cause allergies*

A concern that consumers might be inadvertently exposed to allergens has been expressed. The potential of a GM food to produce an allergic reaction is one of the safety considerations that needs careful attention. If a person is allergic to a particular food and a gene from it is transferred to another food, that person could also become allergic to the second food. Transgenic products may adversely affect people suffering from allergies. Testing for allergens in GM foods is part of the research and development of GMOs intended for food. It is essential that tests for allergenicity are improved and that such potential allergens present in food are labelled.

Perhaps the number one health concern over GM technology is its capacity to create new allergens in our food supply. Proteins are the main components that bring about the observed allergic reactions. It is very common that transfer of genetic material from one living organism to another leads to the creation of a novel protein. Genetic modification affects the presence of allergens in different ways. It can increase the levels of a naturally occurring allergen already available in a food or create new allergenic properties in a food that did not originally contain them. Genetic modification also has the potential to introduce brand new allergens humans have not known before. Some plants naturally produce toxins as a defence mechanism to drive away pests (Malarkey 2003). These toxins either are present in very small amounts that would not harm humans or animals when consumed or are inactivated by processing. Genetic modification can lower or remove these toxins or allergens, but at the same time it may increase their levels. GMOs make it possible for allergens in one food type to emerge in a completely different species (Union of Concerned Scientists 2012). Each new food item produced contains many new potentially allergenic proteins. Novel proteins produced through genetic modification could result in new allergy outbreaks in humans. One such example is lectin, which is a protein used in beans to prevent aphids from attacking potato crops. Some people are allergic to lectin.

2) *GMOs are toxic*

A serious concern has been expressed by some consumers about toxicity of the protein products of transgenes themselves or the potential that these proteins might have to induce unintended effects on plant metabolism, leading to the production of toxins.

The AAEM has come to the conclusion that GMOs and GM foods are toxic. The academy cited results of some scientific and clinical studies that indicate the association of GM foods and specific disease processes. According to the AAEM, animal studies showed immune dysregulation, altered structure and function of the liver, kidney, pancreas, and spleen, and cellular changes that could lead to accelerated ageing. Studies also showed intestinal damage in animals fed GM foods, including proliferative cell growth and disruption of the intestinal immune system. The AAEM concluded that GM foods pose a serious health risk in the areas of toxicology, allergy, immune function, reproductive health, and metabolic, physiological, and genetic health.

3) *Interior toxins in GM foods*

The terms 'interior toxins' and 'pesticidal foods' have been used by Batalion (2000) to describe the toxic pesticides engineered inside plants and the foods produced from such plants, respectively. The food is engineered to produce its own built-in pesticide in every cell to kill pests. The potential long-term health impacts of consuming such foods are little known, and people continue to ingest the interior plant toxin from GM foods (Snell et al. 2011). Pesticide-related illnesses continue to rise in farm workers due to the increased use of pesticides because of pesticide-resistant transgenic crops and the contamination of food and drinking water.

4) *Vectors can infect mammalian cells and resist breakdown in the gut*

There is a concern about the use of marker genes in GMOs. Marker genes are sequences of DNA used in genetic modification to help researchers find out which organisms have taken up the introduced genes. The concern is that the marker genes may allow their recipient organism to produce new proteins. Would these new proteins be present in the GM food? Could the marker gene be transferred to the human gut? Does it have the potential to cause harm to consumers?

Two extremely important factors in the safety of transgenic organisms used as food are the extent to which DNA, in particular vector DNA, can resist breakdown in the gut and the extent to which it can infect the cells of higher organisms. Some studies have shown that bacterial plasmids carrying a mammalian virus are able to infect cultured mammalian cells, which then continue to synthesize the virus. Bacterial viruses can also be taken up by mammalian cells. Greenpeace advises that people should be extremely cautious about ingesting transgenic foods, as foreign DNA can resist digestion and may be taken up by gut bacteria or absorbed through the gut wall into the bloodstream. DNA uptake into cells could then lead to the regeneration of viruses, or if the DNA integrates into the cell's genome, a range of harmful effects may result, including cancer. Moreover, one cannot assume, without adequate data, that DNA is automatically degraded in processed transgenic foods. GMOs are found to transfer genetically altered DNA into the DNA of bacteria living in the human stomach, reproducing indefinitely.

5) *Direct and indirect cancer and degenerative disease links*

The possible carcinogenic effect of GM foods is another concern that has been voiced. Is there the potential for consumption of GM foods to cause cancer in individuals who eat those foods?

Rees (2006) and Batalion (2000) warned about possible direct cancers and other degenerative diseases as a result of consuming GM foods. The example of the GM growth hormone rBGH, used to stimulate more milk production in cows, and the possibility that potent chemical substances present in such milk cause cancer in humans is cited by Batalion (2000). The conclusions drawn by a number of scientists who support this claim have also been reported. The possibility that the recent increase in cancer rates might be attributed to the consumption of GM foods has also been raised by Batalion (2000).

6) *Resurgence of infectious diseases*

A number of studies reported by Batalion (2000) indicate that gene technology may be implicated in the resurgence of infectious diseases. The growing resistance to antibiotics used in bioengineering, the formation of new and unknown viral strains, and the lowering of immunity due to consumption of processed and GM foods have been cited as ways in which the infectious disease rate increases.

7) *Antibiotic threat via milk*

According to the Center of Food Safety, as reported by Batalion, cows injected with rBGH have a 25% increase in the frequency of udder infections. Farmers would consequently use more antibiotics, which may eventually end up in the different dairy products people consume. High levels of antibiotic residues in milk may lead to some allergic reactions and could reduce the effects of other antibiotics.

8) *Antibiotic resistance through plants*

Many concerns have been expressed about the use of antibiotic-resistant marker genes, which could reduce the efficacy of antibiotics in treating human and animal diseases. Current transgenic crops may contain antibiotic-resistant marker genes. Such GM crops might contribute to increasing dissemination of antibiotic resistance as an issue with environmental and health implications. If animals are fed antibiotic-resistant plants, the antibiotic-resistant gene might be transferred to bacteria in the animal's gut and subsequently transmit antibiotic-resistant organisms to humans through the food chain.

When applied to crops, GM technology uses markers to help track where the gene goes into the plant cell. The most common marker used is a gene for antibiotic resistance. This gene can be transferred to the food chain. A large number of GM food products contain this gene. The resistant qualities of GM bacteria in food can be transferred to other bacteria in the environment and throughout the human body.

9) *The safety of GM products is uncertain and has not yet been verified*

The consumption of GM foods started in the mid-1990s. There are no long-term data on how GM foods affect human health. Clinical trials carried out on animals fed GM crops were of short duration, and the results and conclusions showed wide variations. Short-term studies on the effects of GM foods, for example for 4 weeks, are not expected to give conclusive results on the safety of these foods for human consumption. Feeding rats GM corn caused negative effects in the function of the liver, kidney, heart, adrenal gland, and spleen (de Vendômois et al. 2009). There are no reports of research conducted on the health and safety of people who consumed GM foods. The critics of GM foods argue that it is not logical to say that GM foods are safe simply because people have been eating them for several years with no ill-effects. This is because GM foods are not labelled, which makes it difficult to differentiate between GM and non-GM foods. They state that in the USA food-derived illness has doubled since the introduction of GM foods. Another argument is that the safety testing carried out on GM foods has been inadequate so far, and most of the testing is conducted by companies producing the food or benefiting from its production. An urgent call for independent, regulated, and long-term studies on the safety of GM foods is made by opponents of GM foods. The current situation of GM food safety as viewed by the geneticist Richard Steinbrecher and reported by Batalion (2000) is summarized as 'to use genetic modification to manipulate plants, release them into the environment and introduce them into our food chains is scientifically premature, unsafe and irresponsible'.

10) *Chemical contamination of GM foods*

GM crops used to produce GM foods are treated with different chemicals, for example herbicides and pesticides, to promote their growth and production. These chemicals have the potential to cause harm to humans and lead to negative health consequences. One example cited by GM food critics is the soybean, which is treated with chemical herbicides genetically developed to kill all weeds and other plants in the farm field without harming the soy crop. This chemical treatment might result in the food produced from such crops containing traces of these toxic chemicals. Some pesticides have been associated with some health problems such as nervous system disturbances, visual and skin irritations, endocrine problems, and an increase in some types of cancer.

11.6   GMOs from the Halal and Tayyib Point of View

The Islamic perspective on GM foods is complex and goes deeper than simply a determination of whether a certain food is halal or not (although that is certainly part of it). There are three main objections to genetic modification: it interfers with divine work, it may cause harm to humans, animals, and the environment, and it uses genes from haram sources.

11.6.1   Interfering with Divine Work

Genetic modification is considered a controversial issue by scholars and jurisprudents. Genetic modification manipulates the Creation, which means it interferes in Divine work and, thus, is illegitimate. This view is based on the verse 'And I (devil) will mislead them, and I (devil) will arouse in them [sinful] desires, and I (devil) will command them so they will slit the ears of cattle, and I (devil) will command them so they will change the creation of Allah'.

One overriding and forceful normative trope in Muslim ethics is the preservation of naturalness. To preserve the uncontaminated nature coupled with the celebration of a healthy or sound human nature is viewed as the highest of Islamic ideals. These multiple meanings are captured by the notion of 'fitra', a concept central to Muslim ethics and spirituality. Fitra is the inborn, intuitive ability to discern between right and wrong. 'So direct your face steadfastly to faith', instructs the Holy Quran, 'as God-given nature [fitra], according to which God created humanity: there is no altering the creation of God' (30:30). God in the Holy Quran is also described as the Original Creator (fatir), one who creates the heavens and the earth without any model. The upshot is this: even nature has an ingrained disposition that determines its order. Nevertheless, fitra is susceptible to distortion and corruption through sin and disobedience. It goes back to the standoff between God and Satan in prelapsarian time. Satan, on refusing to bow to Adam, earned divine wrath. Then Satan asked for respite to tempt the offspring of Adam, saying in the words of the Holy Quran: 'And I will order them to alter the creation of God' (Holy Quran, 4: 119). Even though altering the creation of God here means that humans will corrupt their inner natures through disobedience and rebellion, this passage lends itself to multiple readings. While most classical exegetes have held to a metaphorical meaning, more contemporary lay and specialist readings of these verses claim that any nonremedial physical alteration to the human body or nature amounts to one acting on a Satanic inspiration (Brunk and Coward 2009).

11.6.2 Causing Harm and Corruption

Islam accepts and allows the use of all science and innovations for the benefit of mankind as long as they achieved the benefits and do not lead to harm and damage. There are many concerns about the safety of GM crops and their side effects on human health.

Another related issue that we should note is that while focus on the issue of permissibility of food in Islam has always been on the halal criterion, many people forget that in the Holy Quran the concept of 'halal' always come together with the concept of 'tayyib'. In the Holy Quran, it is stated clearly that 'O mankind, eat from whatever is on earth [that is] lawful and good and do not follow the footsteps of Satan. Indeed, he is to you a clear enemy'.

The Holy Quran uses the phrase 'halalan tayyiban' which is translated into 'lawful' and 'good'. In other words, when we discuss the Islamic perspective on food and other products, it must be remembered that there are two criteria being given emphasis: (i) halal (must be lawful or permissible according to Islamic law and (ii) tayyib (must be of good quality and safe).

Much scientific research provides a basis for prohibiting biotechnology utilization, which some GMO products can cause harmful effects to health and the environment.

An example of a disadvantage of GM plant is plants that are genetically modified to allow a high survival rate under very high doses of herbicides (giving farmers more flexibility in weeding), which leads to using even stronger and potentially more dangerous chemicals as the weeds become more resistant to one weed killer, thus needing an even stronger one. This may cause damage to wildlife and the wider ecosystem. Furthermore, with cross-pollination it is possible that these strains may 'leak' out, causing permanent and unpredictable damage to our environment.

In the Holy Quran, Allah Ta'ala says, 'And when he goes away, he strives throughout the land to cause corruption therein and destroy crops and animals. And Allah does not like corruption.

11.6.3 Using Genes from Haram Sources

The possible introduction of animal genes into food plants also presents considerable ethical difficulties for Muslims and members of other religions which forbid the eating of certain animals or their by-products. At present, due to lack of labelling regulations, it is impossible to tell whether the GM foodstuff on our supermarket shelves already contains animal genes. However, present commercial technology appears to be more focused on splicing bacterial genes into plant genes, rather than animal genes.

Experimentally, pig genes have already been planted into plants and plant genes have been planted in pigs (Niemann 2004). It is hypothesised that human genes may also be put into plants as part of the 'functional food' fad of biotech companies. Such foods are designed for specific health problems. Scientists are concerned about putting human genes into plants, which they argue would go beyond crossing the species barrier as it would cross the kingdom barrier (animal to plant). This barrier has already been broken, with transfers of firefly genes to corn, hamster genes to tobacco, and flounder genes to tomatoes (Fitz 1999).

When considering whether GM food as a product of genetic modification technology is halal or not, we should also bear in mind that genetic modification as a process applies not only to foodstuffs, but also promises to improve the quality of human life. For instance, it has been used medicinally to manufacture cheaper vaccines and to provide an alternative to natural animal insulin (usually derived from the pancreases of pigs or cattle) to treat diabetes, thus providing a potential halal source of medicine for the treatment of diabetes (one of the fastest growing diseases in the Muslim world).

Theoretically, GM products can be derived from transferring genes of haram origin, as well as from halal sources. If the transferred gene source is from unlawful source, the genetically modified food product will be haram.

11.7 Conclusion

The general themes around which opposition to GM technology and GM foods revolves from halal point of view can be summarized as follows:

1) Changing the nature (*fitre*) of the organism or the crop can be considered to be interfering with Divine work.
2) Using genes from haram (forbidden) sources.
3) GM technology can cause harm and corruption and it has many drawbacks, hazards, risks, and threats to humans, animals, agriculture, and the environment (International Workshop for Islamic Scholars on Agribiotechnology 2010). More laboratory and clinical research on genetically modified foods should be conducted to determine the extent of harm to human health.

References

Batalion, N.B. (2000). *50 Harmful Effects of Genetically Modified Foods*. Oneonta, NY: Americans for Safe Food.

Brunk, C. and Coward, H. (2009). *Acceptable Genes? Religious Traditions and Genetically Modified Foods*. Albany, NY: State University of New York Press.

de Vendômois, J.S., Roullier, F., Cellier, D., and Séralini, G.E. (2009). A comparison of the effects of three GM corn varieties on mammalian health. *International Journal of Biological Sciences* 5: 706–726.

Dohoo, R., DesCôteaux, L., Leslie, K. et al. (2003). A meta-analysis review of the effects of recombinant bovine somatotropin 2. Effects on animal health, reproductive performance, and culling. *Canadian Journal of Veterinary Research* 67 (4): 252–264.

Dona, A. and Arvanitoyannis, S.I. (2009). Health risks of genetically modified foods. *Critical Reviews in Food Science and Nutrition* 49:2, 164–175.

Mathews, E.J. (1991). Memorandum to the Toxicology Section of the Biotechnology Working Group. Subject: Analysis of the Major Plant Toxicants, 28 October 1991.

Engdahl, F.W. (2009). Moratorium on Genetically Manipulated (GMo) foods. Global Research. http://globalresearch.ca/a-moratorium-on-genetically-manipulated-gmo-foods/13701 (accessed 12 September 2015).

Fagan, J., Antoniou, M., and Robinson, C. (2014). *GMO Myths and Truths: An Evidence-Based Examination of the Claims Made for the Safety and Efficacy of Genetically Modified Crops and Foods*, 2e, Version 1.0. Earth Open Source.

Fitz, G. (1999) Will human genes be spliced into food for people? Synthesis/Regeneration 19. http://www.greens.org/s-r/19/19-08.html (accessed 16 September 2019).

Friends of the Earth (2014). Genetic engineering. http://www.foe.org/projects/food-and-technology/genetic-engineering (accessed 26 August 2014).

Séralini, G.-E., Clair, E., Mesnage, R. et al. (2012). Long term toxicity of a roundup herbicide and a roundup-tolerant genetically modified maize. *Food and Chemical Toxicology* 50: 4221–4231.

Hammond, B.G., Dudek, R., Lemen, J.K., and Nemeth, M.A. (2006). Results of a 90 day safety assurance study with rats fed grain from corn borer-protected corn. *Food and Chemical Toxicology* 44: 1092–1099.

Malaysia Biotechnology Information Center (2010). *Proceedings of the International Workshop for Islamic Scholars on Agribiotechnology: Shariah Compliance*, Georgetown, Penang, Malaysia (1–2 December 2010). International Service for the Acquisition of Agri-biotech Applications: Los Baños, Laguna, Philippines.

Lanphier, L. (2014). More GMO info: What you must know. Exhibit health: a better way to health and wellness. http://www.exhibithealth.com/health-education/more-gmo-info-what-you-must-know-1518 (accessed 17 September 2015).

Mahgoub Salah, E.O. (2016). *Genetically Modified Foods; Basics, Applications, and Controversy*. Boca Raton: CRC Press.

Malatesta, M., Boraldi, F., Annovi, G. et al. (2008). A long-term study on female mice fed on a genetically modified soybean: effects on liver ageing. *Histochemistry and Cell Biology* 130: 967–977.

Niemann, H. (2004). Transgenic pigs expressing plant genes. *Proceedings of the National Academy of Sciences of the United States of America* 101 (19): 7211–7212.

Rees, A. (2006). *Genetically Modified Food: A Short Guide for the Confused*. Organic Consumer Association.

Snell, C., Bernheim, A., Bergé, J.-B. et al. (2011). Assessment of the health impact of GM plant diets in long-term and multigenerational animal feeding trials: a literature review. *Food and Chemical Toxicology* 50: 1134–1148.

Institute for Responsible Technology (2015). State of the science on the health risks of GM foods. http://www.saynotogmos.org/paper.pdf (accessed 8 August 2015).

Malarkey, T. (2003). Human health concerns with GM crops. *Mutation Research* 544: 217–221.

Union of Concerned Scientists (2012) Genetic engineering in agriculture. http://www.ucsusa.org/food_and_agriculture/our-failing-food-system/genetic-engineering/#monarch (accessed 1 September 2014).

12

Halal Personal Hygiene and Cosmetics

Mah Hussain-Gambles

Consultant to Pharmaceutical Personal care and Halal Industry, based Bristol, UK

12.1   Introduction

Halal food and lifestyle products are a serious matter for Muslims and there is now a world-wide demand for certified halal products. It is obligatory for every Muslim to consume only halal food and avoid food that is non-halal or haram, or food that contains *mushboo* or questionable ingredients.

With an increasing global demand for halal personal care products and the complexity of ingredients used in these products, it would be naive for a modern Muslim consumer to just look for 'pig-free' and 'alcohol-free' ingredients, which frankly are now few and far between, thanks mainly to vegetarian/vegan consumer lobbying around the globe.

Personal care products include skincare and body care products, shampoos, hair conditioners, hair dyes, colour cosmetics, and other personal hygiene products such as antiperspirants, toothpaste, and mouthwash. The industry is highly regulated in terms of product safety and manufacturing, but many of these products contain synthetic or man-made ingredients, which can have questionable (*mushboo*) origins and this has implications in terms of religious compliance. This is what is discussed next.

Increased consumer awareness of harmful ingredients found in personal care products has, in part, necessitated a rise in stricter regulation of this market. REACH Regulations (https://echa.europa.eu/regulations/reach/understanding-reach) (which aim to ensure a high level of protection of human health and the environment from the risks that can be posed by chemicals), the EU Cosmetics Directive (76/768/EEC) (http://www.legislation.gov.uk/uksi/2009/796/made), and the UK Cosmetic Products (Safety) Regulations stipulate that cosmetics now also have to be manufactured under good manufacturing practice (GMP) conditions (http://www.ctpa.org.uk/content.aspx?pageid=428). These regulations set very high requirements to ensure consumer safety. The regulations concern all participants in the cosmetics products chain, including European and non-European producers. Ingredients producers, final product assemblers, distributors, and importers/exporters are all involved and have new and defined responsibilities. Increasing globalization means that these regulations have to be adhered to by manufacturers around the world to survive the fiercely competitive and fast-growing personal care market, as any imported products also have to be compliant with these regulations. These relatively recent regulations are in place to improve product quality and traceability to protect human life and to ensure products are safe, pure, and healthy (*tayyib*).

In the pharmaceutical industry, however, quality assurance/control requirements for actives (drug) and excipients (ingredients used to bind the drug in a tablet/capsule or syrup form) are the foundation of GMP and have been closely regulated for aeons. Unlike the cosmetic industry, where 'natural' or organic ingredients are often used due to consumer demand for natural products, the pharmaceutical industry has to play it differently. To control for purity and accurate 'active' levels, most pharmaceutical ingredients tend to be synthetically/laboratory produced so that each batch is consistent in terms of appearance, quality, and 'activity'. These are known as 'pharmaceutical grade' ingredients, or United States Pharmacopeia (USP) or National Formulary (NF) grade.

Synthetically produced pharmaceutical drugs by their very nature tend to be chemical entities, unless they are deliberately derived from animal sources, e.g. adjuvants (substances that are formulated as part of a vaccine to enhance its ability to induce protection against infection) in some vaccinations, gelatine capsules, and animal-derived insulin. These can be from any animal species, including porcine sources.

The personal care industry is a slightly different story. Due to consumer demand for pure ingredients and animal rights activists in the West, animal testing and animal-derived ingredients are increasingly being replaced by natural ingredients. Natural ingredients are therefore increasingly of commercial importance to manufacturers of cosmetic products in this fast-growing sector. Companies are formulating and reformulating with more natural materials and according to Organic Monitor around 70–80% of all new cosmetic active ingredients are based on natural ingredients. Natural ingredients tend to be more expensive to produce compared to their synthetic counterparts, and with increasing industry focus on quality control and consumer safety, it will be interesting to see how long this trend for natural ingredients will to last (https://www.cosmeticsdesign.com/Article/2011/04/20/Sourcing-natural-ingredients-could-backfire-Organic-Monitor).

Unlike synthetically manufactured pharmaceutical ingredients, natural ingredients are difficult to control in terms of reproducibility/quality and activity, and for this reason natural essential oils are seldom used in commercially produced perfumes. Naturally derived essential oils are either fractionated into their individual components or synthetically created. This ensures consistency in terms of fragrance and colour each time a batch of perfume is manufactured on a large scale. The trend towards synthetically produced 'natural' ingredients by the larger cosmetics manufacturers is partly due to the increasing cost of natural raw materials, but also due to the regulatory focus on the safety and quality control of the final cosmetic product (http://www.mdpi.com/journal/cosmetics/special_issues/quality-control-cosmetics).

The consumer demand for animal-free ingredients stems not just from growing numbers of traditional vegetarians and vegans, but from people who are concerned about diseases in certain animal species. Scares such as mad cow disease (bovine spongiform encephalopathy, BSE) in cattle, foot and mouth disease, and avian flu in poultry have also prompted consumers to buy animal-free products. Ingredient manufacturers have had to develop synthetic or vegetarian alternatives and have to provide a transmissible spongiform encephalopathy (TSE)/BSE-free statement to their suppliers (https://ec.europa.eu/food/safety/biosafety/food_borne_diseases/tse_bse_en).

Although the manufacturing of cosmetics and pharmaceuticals is highly regulated, thus giving the consumer confidence in the safety of the product, this does not necessarily mean that the final product will be halal compliant. The use of alcohol in cleaning equipment can contaminate the final product. Halal auditors not trained in GMP may miss potential points of contamination with non-halal products during manufacturing. Some halal certifiers allow genetically modified organisms (GMOs), others don't. Irradiation of raw materials or final products by lethal doses of gamma radiation as well as the recent use of nanoparticles in skincare formulations are also not currently aligned on as far as halal certifiers are concerned.

My inspiration to be the first to write halal standards for personal care products arose from a visit to a maternity ward in the UK back in 2003. I was faced with the dilemma of whether or not, as a Muslim, I should be using the alcohol-based hand sanitizer provided on the ward to protect my friend and her newborn baby against the methicillin-resistant *Staphylococcus aureus* (MRSA) virus. Whilst driving home, I could still smell the alcohol on my hands. How much of the alcohol crossed my skin barrier is open to scientific questioning, but my lungs certainly got a large whiff of alcohol.

As a scientist, I have no doubt about the importance of sanitizing your hands to protect vulnerable patients, and if alcohol is the only thing on offer, then, as a Muslim, my personal beliefs are that it is permissible since there is no alternative. However, I know for a fact that there was an alternative available on the market, a halal-compliant hand sanitizer, manufactured in the UK, that was not only better for the environment and more potent than alcohol, but also cheaper than the industry standard alcohol-based hand sanitizer.

Despite there being a halal alternative to an alcoholic hand sanitizer on the market, many Muslims were not given the choice, so what is the solution? One cannot expect the NHS to automatically provide a halal sanitizer for the Muslim population unless it is made aware of the need for it. Another example of this is halal hospital food. Just 10 years ago, halal food was unheard of in the NHS and it was through the hard work of lobbyists and increasing awareness of the growing demand for halal compliance that it has now become the norm on hospital menus.

As a pharmaceutical industry insider, I do not yet see any obvious signs of the awareness amongst Western pharmaceutical companies for the need for halal-certified pharmaceutical products, although market research shows the demand for halal products is a growing global phenomena. Large pharmaceutical companies may therefore be losing out if they do not engage in this growing global market. The first step is to open up a dialogue with Muslim specialists in the field. Awareness and knowledge of what constitutes a halal product can open doors for the development of halal-friendly personal care and pharmaceutical products.

In all fairness, pharmaceutical companies cannot be lobbied to manufacture halal drugs without an authentic halal certification body and a global unified check procedure in place. I have come across halal certification bodies who have willingly given certificates and charged large sums of money without any due diligence. For instance, in the case of supplements, only the ingredients were checked and not the manufacturing plant. In another case, no annual audits were carried out and certificates were reissued without checking any changes made to the products. However, I am pleased to see a recent global impetus in the policing of such rogue certifiers.

Due to my lack of confidence in the halal certification system at the present time, I have set up a not-for-profit halal certification body for personal care products/ingredients and health supplements. This organization consists of a board made up of a pharmaceutical industry specialist, a PhD in organic chemistry, and a medical doctor. The board used to meet regularly with the imam and scrutinized the whole process from manufacturing of raw materials to the final product. Once satisfied, the factory was audited to check for segregation and manufacturing. An annual certificate was only granted if the applicant satisfied the strict criteria laid down by the specialist board. Since 2004, the halal certification world has come a long way, I am pleased to say. I will cover the certification industry in more depth later, but first it is important to review two major aspects of what constitutes a halal personal care product: ingredients and the use of alcohol.

There are far too many cosmetic and pharmaceutical ingredients to list in this chapter but fundamentally, replacing animal derived ingredients with vegan or vegetarian substitutes is a good starting point. There are now readily available vegetable-derived glycerin and porcine gelatin alternatives on the market, and replacing these in standard formulations is not technically difficult to achieve, although it may increase the cost of the final product and requires comparative trials. As far as manufacturing is concerned, there are alcohol-free cleaning solutions available that can easily replace alcohol sanitizers and some of these alternatives have been shown to be more effective at killing spores (http:// alcoholfreesanitiser.co.uk), (http://www.phs.co.uk/our-products/washrooms/handcare-products/no-germs-alcohol-free-hand-sanitiser). Vigilance regarding ingredients and their origins as well as the manufacturing process to check for potential points of contamination with haram products is a basic concept shared by most halal certifiers of personal care and pharmaceutical products.

12.2 Personal Care Ingredients

Reading the ingredients list on food labels is the norm for most of us, but how many of us take the time to scrutinize the ingredients list on the back of everyday personal care products? Having said that, even with a PhD in science, the complexity of ingredients lists can be daunting and at times I have had to contact the manufacturer directly to get clarification on the origins of an ingredient and the process of manufacture to rule out any contamination from porcine sources. Identifying dubious or potentially haram ingredients in a long list of cosmetic ingredients is not an easy task. The solution lies in either directly asking the manufacturer for the origins of the questionable ingredient or buying halal-certified personal care products.

There are thousands of technical and patented names for cosmetic ingredients and additives (some also known as E numbers) (http://www.food.gov.uk/business-guidance/eu-approved-additives-and-e-numbers). Many ingredients (known by one name) can be of animal, vegetable, or synthetic origin. The prime example of this is glycerin, which is used in personal care products. Glycerin can be from a vegetable or animal source. The term 'natural source' can mean an animal or vegetable source, and often in the industry it means an animal source, such as animal elastin, glands, fat, protein, or oil (animal by-products tend to be cheaper than plant-based ingredients). Adding to the confusion over whether or not a natural ingredient is of animal origin is the fact that many companies have removed the word 'animal' from their ingredient labels to avoid putting off conscientious consumers. For example, rather than use the term 'hydrolyzed animal protein', companies may use another term such as 'hydrolyzed collagen'.

Table 12.1 provides a short 'watch out' list of some commonly used E numbers in food, cosmetics, and the pharmaceuticals industry. Many additives contain fatty acids (glycerol/glycerides or emulsifier) and the origins of these fatty acids may not be a halal animal. In such cases, only the manufacturer of the E number can provide assurance about its halal integrity. A large number of websites provide more detailed information on E numbers that are potentially haram.

It is beyond the scope of this chapter to enter into a debate about the halal integrity of certain E numbers as the current scholarly ruling is unclear. If the fatty acid is of porcine origin, then it is considered haram; if it is from a halal animal that has not been slaughtered in the prescribed Islamic way, then that may cause an issue for some Muslims. It is also impossible to ascertain from just reading the International Nomenclature of Cosmetic Ingredients (INCI) list on the packaging whether or not an ingredient is of plant or animal origin. The chemical names of many ingredients are completely meaningless to most of us. However, with some fundamental awareness it should be easier to avoid products with animal ingredients.

There are many other animal ingredients used in a variety of cosmetic products. Lipsticks may contain animal-derived fats and oils. Shellac (coating or glaze derived from the hardened, resinous material secreted by the lac insect) may be used in nail varnish. Civet (from the anal pouch of the civet cat), musk (from male musk deer), castoreum (the anal sex gland of the beaver), and ambergris (whale excretion) may be used in perfumes. Chitin (from insects and crustacea) and keratin (protein from hair, horn, hoof, and feathers) may be used in shampoos, conditioners, and skincare products. Tallow is commonly used in soaps. Gelatin, elastin, and squalene (from the liver of sharks or olive oil) may be used in anti-ageing cosmetics. The commonly used cosmetic ingredient glycerin or glycerol can be of animal origin. In addition, animal by-products may be used in the processing of raw materials, such as the use of dried bones as filters or deodorisers, or dried bones as a source of phosphate in toothpastes.

It is also important to remember that defining halal is far more complex than just pork-free ingredients and this can often be overlooked due to the fact that the focus has been largely on halal foods. There are a large number of religious questions associated with deciding what is or is not halal. Perhaps the most important factor in deciding whether or not a product or ingredient is halal should be the premise of preserving human health.

One of the most basic premises in Islam is to preserve human well-being. Based on this, is it halal to use skin lightening ingredients such as hydroquinone or kojic acid, when hydroquinone is banned from cosmetics in the EU because it causes cancer and kojic acid has been shown to cause cancer in laboratory animals? Technically speaking, both these ingredients can be classed as halal due to their non-animal source origin, but is it halal to use ingredients that clearly can cause harm to our bodies?

Table 12.1 Some commonly used E numbers in the food, cosmetics, and pharmaceuticals industries

Allantoin

This can be derived from uric acid from cows and other mammals. Allantoin is used in treatment of wounds and ulcers, and in cosmetics (especially creams and lotions).

Amino acids

The building blocks of protein in all animals and plants. They are used in some cosmetics, vitamins, supplements, shampoos, etc.

Arachidonic acid

A liquid unsaturated fatty acid that is found in the liver, brain, glands, and fat of animals and humans. Generally isolated from animal liver. Used in some skin creams and lotions to soothe eczema and rashes.

Collagen

Usually derived from animal tissue.

Elastin

Protein found in the neck ligaments and aortas of cows. Similar to collagen.

Gelatine

Protein obtained by boiling skin, tendons, ligaments, and/or bones with water. From cows and pigs. Used in shampoos, face masks, and other cosmetics.

Glycerine

A by-product of soap manufacture (normally uses animal fat). Used in cosmetics, foods, mouthwashes, chewing gum, toothpastes, soaps, ointments, and medicines.

Hydrolyzed animal protein

Used in cosmetics, especially shampoo and hair treatments.

Keratin

Protein from the ground-up horns, hooves, feathers, quills, and hair of various animals. Used in hair rinses, shampoos, and permanent wave solutions.

Lard

Fat from hog abdomens. Used in shaving creams, soaps, and cosmetics.

Lipids

Fats and fat-like substances that are found in animals and plants.

Myristic acid

A type of acid found in most animal and vegetable fats. Used in shampoos, creams, cosmetics etc.

Oleic acid

Obtained from various animal and vegetable fats and oils. Usually obtained commercially from inedible tallow. Found in some soft soaps, bar soap, permanent wave solutions, creams, nail polish, lipsticks, many other skin preparations.

Stearic acid

Fat obtained from cows and sheep. Most often refers to a fatty substance taken from the stomachs of pigs. Used in cosmetics, soaps, lubricants, candles, hairspray, conditioners, deodorants, creams, chewing gum, and food flavouring.

Tallow

Rendered beef fat, used in candles, soaps, lipsticks, shaving creams, and other cosmetics.

Looking good on the outside has never been so hazardous for our health and well-being. Thousands of personal care products and ingredients have come under the microscope and some have now been banned in Europe. These include phthalates, commonly found in many personal care products, including deodorants and perfumes, and linked to reproductive abnormalities and 'feminising' boys. They may appear on an ingredients list as dimethyl phthalate (DMP), diethyl phthalate (DEP), butyl cyclohexyl phthalate (BCP), di-*n*-pentyl phthalate (DNPP) and so on (http://www.safecosmetics.org/get-the-facts/chemicals-of-concern/phthalates). Phthalates are not of animal origin.

Similar to phthalates, many other petrochemical-based ingredients can be classed as halal due to their non-animal origin, but they also tend to be fat soluble and can go through the skin barrier into your bloodstream and end up being stored in your body's fat cells. These chemicals can also be stored in fatty breast tissue and be passed onto a baby through breast feeding. Many of these are now banned in the EU (https://www.cosmopolitan.com/style-beauty/beauty/g7597249/banned-cosmetic-ingredients). Identifying questionable or potentially haram ingredients in the long list of personal care ingredients is not very easy. Different halal certifiers have different religious rulings on these questionable ingredients. At present there does not seem to be alignment or a central database that halal certifiers can adhere to.

The following sections provide information on some of the 'actives' used in personal care formulations and explore those which may be of dubious nature and potentially haram.

The anti-ageing products segment is the fastest growing in the personal care industry. One of the more commonly used anti-ageing ingredients in cosmetic preparations is collagen or hydrolysed collagen. Apart from skin application, it is also used to strengthen nails and hair. Collagen can be derived from bone and cartilage, and can also be produced from fish bones, which is potentially acceptable by Muslims. Marine-based collagen is a safer alternative for Muslims, albeit a lot more expensive.

Typically, in more advanced formulations, it is usually the excipients that are of doubtful nature, as many can be derived from animals. The common muscle relaxant treatment Botox (which was originally used for medical reasons), which on the surface appears to be halal due to its origin being a bacterial toxin, is usually combined with pig-derived ingredients for delivery as an injectable. 'Fillers' or injections to fill out deep wrinkles can also be cross-linked with animal-based proteins, usually porcine-based.

Other commonly used anti-ageing ingredients include hyaluronic acid, which is derived from sodium hyaluronate and is available in a form that is suitable for vegetarians. However, sometimes hyaluronic acid is derived from chicken or other animal sources. Only the manufacturer can advise about the origin of any of these ingredients.

For halal consumers, the focus should also be on the non-active ingredients, such as emulsifiers (for mixing oil with water), preservatives (many of which are banned in the EU for health reasons), colouring, foaming agents, and even packaging.

Glycerine is perhaps one of the most commonly used ingredients in all personal care products, after water. It is used as a humectant (attracts water to the skin) in anything from skin creams to shampoos and toothpaste. Polyethylene glycol (recognizable as PEG in the ingredients list) and its derivatives butylene glycol, ethoxydiglycol, and dipropylene glycol etc. are commonly used to help stop water and oil from separating. They are manufactured chemically and therefore there should not be any concerns about their halal status, but they are associated with skin sensitivities, toxicity, and even potential carcinogenicity. When reacted with stearic acid (from either an animal or a vegetable source), they can form other ingredients used in personal care products such as sorbitan stearates, sorbitan oleate, sorbitan palmitate, and sorbitan tristearate. As the origin of the stearic acid may be animal or vegetable, these ingredients warrant further investigation.

Personal care products also contain silicones to impart a soft feel/touch to the formulation and subsequently the skin or hair. Examples of silicones include dimethicone and methicone. They are considered halal as they are mineral-based and chemically manufactured.

Other ingredients used in the personal care industry include paraffin and other microcrystalline waxes. These are synthetically manufactured 'butters' and not likely to be of concern from a halal point of view. They are cheaper alternatives to natural cocoa butter or shea butter to give hardness and are used as emollients in creams and lipsticks. Lanolin also falls into this category and is natural as it is obtained from the wool of sheep and is generally considered halal. Other ingredients worth mentioning are surfactants or foaming agents used in soaps, shampoos, detergents, and toothpastes. Sodium laureth sulfate and sodium lauryl ethyl sulfate (SLES) and its derivatives are found in many personal care products and are made by mixing sulfuric acid, monododecyl ester, and sodium salt. Monododecyl ester may be from coconut or palm oil, or from an animal source. Because of this, these ingredients should be considered questionable. Glyceryl stearate is another common emulsifier and imparts a smooth feel to the skin. It is made by reacting glycerine with stearic acid (both can be of either animal or vegetable origin) and unless the source of both the glycerine and the stearic acid is known, glyceryl stearate can also be of doubtful origin.

Traditional soaps can be made from tallow, which is fat from animal source. Vegetable alternatives originating from palm or coconut oil are readily available; if in doubt it is advisable to use vegetarian or vegan approved products.

Lastly, we should look at ingredients that are used as preservatives in many personal care systems. All personal care formulations (unless they do not contain any water) need to be adequately preserved for consumer safety, quality, and extended shelf life. Water-based formulations that do not contain preservatives can be dangerous as the product can grow fungi and bacteria, and there have been cases of people going blind with a particular type of bacteria that grows in unpreserved face and eye creams. Commonly used preservatives are listed in Table 12.2 and, to the best of my knowledge, these are all from synthetic origin and suitable for vegetarians and vegans. However, I am doubtful whether some of them should be considered halal due to their documented harmful nature.

Some personal care products and ingredients are also irradiated with high levels of gamma radiation, basically to kill off any bugs. There is some consumer concern about irradiated products having long-term health implications, similar to GMOs (http://www.todaysdietitian.com/newarchives/011209p32.shtml). However, certain raw materials, such as titanium oxide, commonly used in colour cosmetics and suncare products, can use gamma radiation as currently there is no viable antimicrobial process available on the market. It can be argued that in some rare cases product safety overrides potential environmental or religious implications.

Table 12.2 Some common cosmetic preservatives

Butylated hydroxyanisole (BHA) and related chemicals are also nicknamed gender benders because they mimic the female hormone oestrogen.

Methylchloroisothiazolinone and methylisothiazolinone can trigger an immune system response that includes itching, burning, scaling, hives, and blistering of skin. They can also trigger an immune system response that includes asthma attacks or other problems with the lungs and airways.

Parabens are a family of chemicals that includes methyl paraben, ethyl paraben, propyl paraben, butyl paraben, and isobutyl paraben. They are known to mimic female hormones, but there are some scientifically credible studies to show that they are amongst the safest and gentlest preservatives around and do not mimic female hormones. Parabens are of chemical origin and are naturally found abundantly in fruits and vegetables. Some types of parabens are banned in certain European countries.

Imidazolidinyl urea and diazolidinyl urea are cheap preservatives that work by slowly releasing formaldehyde in the formulation to kill any bacterial or fungal growth. Note that formaldehyde is a known carcinogen and a banned substance for use in personal care preparations in Europe.

12.3   Alcohol-free

Many personal care products claim to be 'alcohol-free'. This is important for people who wish to avoid the drying effects alcohol has on the skin but also for Muslims seeking halal-compliant products as Islam prohibits the consumption of intoxicating alcohol (*khamr*) in any quantity. As recorded in a well-known Hadith, 'whatever intoxicates in large quantity, and then a small quantity of it is also forbidden'. Some halal certifiers allow small amounts of alcohol, but the majority follow stricter guidelines. Use of alcohol in any concentration in an industrial process may be acceptable for technical reasons, but only if viable alternatives are not available. If alcohol has to be used in the production process, then it must be allowed sufficient time to evaporate or the equipment rinsed off with potable water to remove any residues of alcohol. Alternatively, alcohol-free or halal-certified cleaners can be used.

It is more prudent to err on the side of caution where alcohol is concerned and avoid it totally as this also follows the majority of scholarly views and upcoming global halal standards, i.e. no amount of 'alcohol' is allowed in halal-certified products if avoidable.

Alcohol is a generic term for ethanol, which is a particular type of alcohol produced by the fermentation of foodstuffs such as barley, hops, and grapes. Other types of intoxicating alcohols commonly available include methanol (common in glass cleaners and highly toxic), isopropyl alcohol (rubbing alcohol), denatured alcohol (around 90% ethanol, with the rest made up of methanol and a small amount of other chemicals to make it unpalatable and to induce vomiting), methylated spirits etc. These alcohols produce intoxication because of their effect on the brain since they contain mostly ethanol. Typically, short-chained *aliphatic alcohols* tend to cause intoxication and therefore are not permissible in Islam. Most other alcohols used in the personal care industry are more complex in structure than ethanol and do not conform to the same general formula. They also have very different properties.

Some 'alcohol-free' products may appear to actually contain alcohols, such as cetyl alcohol or benzyl alcohol, but they are permissible as they do not cause intoxication.

It takes a chemist to really know which of the ingredients in personal care products containing 'alcohol' are actually intoxicating and therefore haram. If we go through some of the common ingredients with alcohol in the name in more detail we can see the complexity of the situation.

Benzyl alcohol is produced naturally by many plants and is commonly found in fruits and teas. It is also found in a variety of essential oils, including jasmine, hyacinth, and ylang-ylang. It is used as a bactericide (kills bacteria) in personal care formulations and it does not cause intoxication, therefore it is classed as halal, even though its chemical name contains the word 'alcohol'.

Stearyl alcohol, cetearyl alcohol, myristyl alcohol, behenyl alcohol and cetyl alcohol are emulsifiers, which mean they help the oily ingredients in the formulation dissolve in the water phase to give creams a smooth consistency. They tend to be white, waxy solids, and are sometimes known as fatty alcohols, and occur naturally in some plants and animals. These could be halal, even though the name suggests otherwise, because if ingested, they do not cause intoxication. However, there is no way of knowing by the names only if these 'alcohols' are originally derived from animal or plant sources.

Phenoxy ethanol is a commonly used bacteriocide in many preparations, including baby products, as it is classed as fairly safe and gentle. Phenoxy ethanol is a thick syrup in appearance and is made synthetically from petrochemicals, therefore it can be argued that it is halal. Personal care products containing these types of alcohols are alcohol-free.

The reason for the apparent confusion is the difference between the terms used by scientists and those used by the general public. To the layman, beer, wines, spirits etc. contain alcohol; to the scientist, they contain ethanol. To the layman, alcohol is a single substance, but scientifically speaking the term describes a whole group of chemical substances or ingredients with differing properties.

Simple alcohols, like ethanol, are defined as having a general chemical formula of $C_nH_{2n+1}OH$, where n equals any number from 1 upwards. If $n = 1$, the compound is CH_3OH, known as methanol (often used as antifreeze) or methyl alcohol. In the case of 'alcohol' (ethanol), $n = 2$ and the formula is C_2H_5OH. This group of chemicals is also known as aliphatic alcohols and they cause intoxication when ingested. There is a religious ruling of prohibition of aliphatic alcohols if ingested and therefore they are classed as haram. Having said that, some halal certifiers argue that synthetically manufactured ethanol is permissible and only ethanol obtained made from grain and grape should be considered haram in personal care products.

In summary, the layman's 'alcohol' means ethanol and products that are 'alcohol-free' are actually ethanol-free.

Currently there is no single or harmonised global standard for halal personal care manufacturing, although standards are being written by independent Muslim organizations and governments all over the world. These standards appear to work within the frameworks of existing European standards such as International Organisation for Standards (ISO) Guidelines, European Union Cosmetic Products Regulations, and Organic Cosmetics Standards, thus providing a good platform on which to base halal cosmetic manufacturing standards.

The challenge is to align on a single internationally recognized standard that not only conforms to the safety of cosmetic products but also to halal principles, which extend beyond the obvious references to food and incorporate not harming the body, protecting the environment, humane treatment of animals, and ethical business.

Alignment is also imperative on the use of alcohol in halal personal care products, in terms of both ingredients and manufacturing processes. Some standards take a zero-tolerance policy on alcohol, whilst other organizations allow varying percentages of alcohol. Some allow a restricted percentage of alcohol made from dates and grains, others argue with religious back-up that if a product is fermented for three days or less, then the alcohol generated from such a process is permissible in Islam. Such varying thoughts on permissible levels of alcohol in the final product not only confuse the consumer but complicate an already religiously/politically charged situation.

Aliphatic alcohols are traditionally used during the cleaning of equipment used in the manufacture of personal care products, as well as being used as hand sanitizers. Use of gloves by operators during the manufacture of halal cosmetics is an alternative solution, but total avoidance of alcoholic hand sanitizers is the preferred option, especially as there are now products available on the market which are alcohol-free and certified as halal.

12.4 Halal Certification of Personal Care Products

Halal manufacture should be about ensuring that the manufacturing processes, equipment and/or machinery have been cleansed according to Islamic law and final products (as well as ingredients) do not come into contact with objectionable ingredients. Halal raw materials and finished products should be stored in designated marked bays to ensure against cross-contamination with non-halal products. Product packaging that is not contaminated with non-halal products, such as animal-based glue, should also be used. The principles discussed next cover skincare products, toiletries, colour cosmetics, personal hygiene products, perfumes etc. However, to monitor and ensure compliance with halal standards, policing in the form of annual halal certification is crucial, including spot checks on finished goods in retail outlets. There is technology now available (laboratory on a stick technology) to check for porcine/animal DNA in situ.

12.5 Certification Processing

It is highly recommended that for the cosmetic product to display the halal symbol on its packaging or any marketing literature, it must undergo a certification process by a recognized halal certification body. Certification is a two-part system. The first part involves the submission of all raw materials and formulations to the certification body, along with raw material supplier signed documents for each ingredient. Upon approval of ingredients and formulations, the next phase is a site inspection. Following a satisfactory site inspection by a trained auditor, the manufacturer would be issued with an annual certificate of registration, a trading schedule specifying the range of products certified as halal, and an annual licence to use the halal symbol.

12.6 Inspection

The auditor should check the operation to make sure that it meets halal standards. In the case of non-compliance, sanctions should be imposed depending on the severity of the non-compliance and recommendations made to remedy any minor non-compliance within an agreed time period. Only when a satisfactory audit report is obtained should the licence to use the halal symbol be granted. Additional inspections would need to be undertaken if a new brand is to be added or there are any additions to the product line. In these cases raw material data and formulations would need to be submitted prior to approval. Inspections would also need to be carried out following a move to new manufacturing premises or investigation of a quality-related complaint regarding the business.

The manufacturer must have procedures in place to maintain the halal integrity of products from buying ingredients to finished goods out. They must work to good practice guidelines, such as GMP guidelines, and operate high standards of hygiene on the premises. The staff need to be properly trained in halal criteria and follow high standards of personal hygiene. The manufacturer needs to have quality management systems in place to ensure that halal ingredients and finished cosmetic products are not contaminated, for example with non-halal ingredients and finished products, cleaning and pest control products, packaging materials and foreign bodies, pests, pathogens or spoilage microorganisms.

12.7 Staff Training

The manufacturer must ensure that those involved in processing halal personal care products are fully trained for the tasks they are carrying out and are aware of halal standards. They should also have an understanding of the importance of maintaining halal integrity throughout the manufacturing cycle.

12.8 Segregation

The manufacturer must keep halal and non-halal products separate at all stages of manufacturing and processing. It is recommended that halal products are processed in separate areas using separate equipment that handles only halal products. If this is not possible for practical reasons, then it is the manufacturer's responsibility to minimize risk of contamination by assessing the risks and putting in place adequate controls to avoid them.

12.9 Storage and Warehousing

Manufacturing areas and racking should be labelled with the word 'HALAL' to show that it is for storing halal products only. All halal ingredients need to be clearly labelled to avoid accidental contamination and there should be sufficient space or ideally a barrier around halal storage to avoid accidental contamination. The staff need to use containers and storage bins dedicated to halal ingredients only. The stores need to be cleaned regularly (and cleaning records kept) so that there is no residue that could contaminate halal products or encourage pests.

12.10   Transporting

To prevent contamination, mixing or substitution of halal and non-halal products, the manufacturer must transport halal goods in closed packaging or containers in suitable vehicles. They need to ensure the loading equipment and the vehicles are clean and have been cleaned in ways allowed by the halal standard, and record details of all collection runs and the results of all checks made. Where it is not possible to transport halal products in a separate vehicle, there needs to be a space of at least 10 cm between the halal and non-halal pallets.

In this chapter, I have attempted to give a flavour of the complexity of the halal certification procedure and I cannot emphasize enough how important it is to have not only trained and knowledgeable auditors, but also a robust standard with clear-cut and aligned religious rulings on ingredients and processes, including alcohol, GMOs, irradiation, and nanotechnology. Last but not least, is my personal agenda, which is to embrace an eco-ethical philosophy. I believe that the principles of halal no longer mean just alcohol and pork-free, but encompass a wholesome concept that embraces a pure and eco-ethical lifestyle.

The halal personal care sector has opened a new frontier in the halal market. Muslim consumers today prefer halal-endorsed products and are choosing to spend money on lifestyle products that meet their religious and cultural requirements. Educated and conscientious Muslim consumers are reaching out for environmentally friendly and organic lifestyle products in line with their non-Muslim peers in the West. Due to the growing vegetarian and vegan movement in the West, consumers in general are becoming more discerning and want to know what goes into the products they purchase.

The term eco-ethical means free from animal cruelty, caring for the environment, not harming one's body by using naturally derived personal care products, and eating natural and organically grown products considered free from pesticides or ingredients deemed harmful to the body. Fulfilling corporate social responsibility, including Fairtrade and no exploitation of workers, is an added bonus and a feelgood factor. All these principles are also in line with the teachings of Islam.

Caring for the body, the environment, and the animal kingdom is the modern take on halal, and, in my personal opinion, ought to form the foundation of halal-certified products. This ethos was reflected when I launched my own organic and halal-certified eco-ethical skincare range, Saaf Pure Skincare in 2004. Back then I dreamt of the day when the halal personal care industry would become more mainstream, like the organic market. I received many requests from consumers about halal hair dyes, nail varnishes, colour cosmetics, perfume, bleaching creams, hair-removing creams, and intimate personal hygiene products as well as deodorants. Many of these products are now on the market as halal approved. However, it deeply saddens me to see so many stalls promoting 'halal' perfumes that are direct copies of Chanel or other luxury brands and made with a base of petroleum-derived propylene glycol. This is wrong on two accounts. Theft of someone else's brand is not only illegal but also morally wrong, and substituting alcohol with cheap antifreeze is not exactly tayyib or wholesome!

To conclude, gone are the days when pig placenta was used in anti-ageing cosmetics, thanks to increasing consumer demand for cruelty-free, eco-ethical, and animal-free products. Increasing use of synthetic ingredients (regulation led) to replace many animal-derived ingredients means more choice for the Muslim consumer seeking halal-compliant pharmaceutical or personal care products.

There are many thousands of personal care ingredients on the market today and the use of many of these ingredients can be hidden and not transparent. The use of alcohol in cleaning equipment, which potentially contaminates the final product, is another contentious issue amongst certifiers. Halal auditors not trained in GMP may miss potential points of contamination by non-halal products during manufacturing. Ensuring that halal auditors and technical experts have the necessary knowledge and training is essential for the growth of halal-certified cosmetics and pharmaceuticals.

12.11   Conclusion

Some halal certifiers allow the use of GMOs, others don't. There does not appear to be alignment on the use or level of 'permissible' alcohol in halal-certified products. Irradiation of raw materials or final products and the use of nanoparticles in formulations are also not globally aligned as far as halal certifiers are concerned. Although generally certifiers around the globe follow the same basic process of certification I have outline in this chapter, many don't seem to recognize each other's standards. This becomes particularly financially problematic for companies who may have to pay different certifiers to approve their products so that they can export to different countries. If global harmonziation is not achieved soon, there is a danger that Muslim consumers and non-Muslim companies will lose confidence in the halal certification process.

There is also a need for further research into the development of halal-compliant ingredients/products. The onus should not just be on non-Muslim multinationals to carry out private research, the drive and funding opportunity for research into this crucial area also needs to come from within Muslim organizations and governments.

Part IV

Halal Standards, Procedures, and Certification

13

Halal and HACCP

Guidelines for the Halal Food Industry[1]

Hani Mansour M. Al-Mazeedi[1], Yunes Ramadan Al-Teinaz[2] and John Pointing[3]

[1] Kuwait Institute for Scientific Research, Kuwait City, Kuwait
[2] Independent Public Health & Environment Consultant, London, UK
[3] Barrister, London, UK

13.1   Introduction

The hazard analysis and critical control point (HACCP) system was developed by the Pillsbury Corporation for the National Aeronautics and Space Administration (NASA) with the aim of ensuring food safety for the first manned space missions back in the 1960s. Its efficiency was proven in the missions so the World Health Organization (WHO) issued the HACCP principles in Codex Alimentarius in 1963. With implementation into the Codex the aim was changed to identifying and eliminating any potential hazard in processing food to ensure food safety and quality. Before 1996, HACCP was advisable but compliance was not mandatory for the food industry (HACCP for excellence 2009). However, following an outbreak of *Escherichia coli 0157* in Scotland in 1996, the Pennington Report recommended that HACCP be adopted by all food businesses to ensure food safety (Pennington 1997).

The halal food industry has seen tremendous growth over the past few years, with increasing demand for halal products coming from both Muslim and non-Muslim consumers (Pointing et al. 2008, p. 208). Halal has a wider application than food: it refers generally to things or actions that are permissible or lawful under Shariah law. Shariah law forms the body of Islamic law, meaning 'way' or 'path'. It provides the legal framework within which public and some private aspects of life are regulated for those living in accordance with Islam.

Halal dietary laws define food products as halal (permitted) or as haram (prohibited) and deal with the following issues:

- prohibited animals
- prohibition of carrion
- prohibition of blood

---

1 This chapter is based on an article written by the authors and published in *The Meat Hygienist* in April 2015.

- methods of slaughtering/blessing
- prohibition of intoxicants.

Halal food must also comply with the requirements of domestic state regulation. For Member States of the EU, food law is governed by the General Food Law of the EU. The HACCP system is an international standard based on a cost-effective and preventative approach to food safety assurance (Motarjemi et al. 1996). It is established in Member States by Article 5 of the Food Hygiene Regulation (2004).

HACCP provides a systematic approach to identifying and controlling hazards (i.e. microbiological, chemical or physical) that could pose a threat to the preparation of safe food. HACCP involves identifying what can go wrong and planning to prevent hazards from materializing by ensuring good practice. In simple terms, it involves controlling ingredients and supplies coming into a food business and tracking what is done with them thereafter. Before implementing HACCP, basic food hygiene conditions and practices (called prerequisites) must to be in place in a food business.

This chapter will consider the requirements and essentials of HACCP in relation to halal. Potential hazards may arise from taking non-halal processing steps or from the use, or suspected use, of haram or non-tayyib (unwholesome) ingredients in the preparation of halal foods.

13.2 Why HACCP?

HACCP was first developed as a quality assurance system in the 1960s to enhance food safety (Pal et al. 2016). It is concerned with all stages involved in the preparation and sale of safe food. It has become internationally accepted as the system of choice for food safety management (WHO 1995) and has evolved to manage risks associated with the international trade of food (Al-Kandari and Jukes 2011).

The food hygiene law of the EU – based on the Food Hygiene Regulation (2004) – applies to all Member States of the EU. It is likely that it will continue to apply to the UK after it leaves the EU if transposed into domestic law. Article 5 of the Food Hygiene Regulation (2004) requires food business operators, including meat plant operators, to implement and maintain hygiene procedures based on HACCP principles.

The elements and characteristics comprising HACCP are wide-ranging and include the following:

1) It identifies what should be done to make food safe.
2) It ensures that what has been planned for is being applied properly.
3) It implements improvements to enhance food safety.
4) It is a proactive system that deals with problems early on, as an alternative to testing final products (i.e. it minimizes the number of rejected food items).
5) It is a system that requires a number of controls to prevent or minimize the incidence of hazards.
6) It is a system that provides a practical means of educating employees in food establishments.

7) It requires examination of the business to ensure all members of the HACCP team are familiar with daily food preparation procedures, plant operations, systems, and the food premises.

The following matters need to be considered in establishing compliance with HACCP requirements:

- cleaning and sanitation
- maintenance
- personal hygiene
- pest control
- plant and equipment
- premises and structure
- services (compressed air, ice, steam, ventilation, water etc.)
- storage, distribution, and transport
- waste management
- zoning (physical separation of activities to prevent potential food contamination).

13.3   Halal and HACCP

Muslims are under a duty only to consume food, pharmaceuticals, nutrients, and cosmetic products that are halal and tayyib (wholesome). Halal characteristics complement the HACCP system, so providing an additional layer of protection for consumers. Together they provide a complete system for analysing any food operation to identify potential hazards within the three traditional categories: microbial, chemical, and physical. This system applies to all manufacturing and production steps, and to inspection and control measures, so ensuring that safe and hygienic conditions prevail at all stages of production.

In addition to the three traditional hazard categories, a fourth category (itself broken down into four concepts) has drawn the attention of Muslim scientists and is of a religious nature. This hazard category arises with the use of non-halal processing systems and it involves the presence of any of the following four characteristics.

Haram
Some foods, such as pork and alcohol, are forbidden. Some objects, foods or actions which are normally halal can, under some conditions, become haram, for example halal food and drinks consumed at noon-time during the holy month of Ramadan, and cattle or other halal animals that are not slaughtered in the Islamic way and in the name of Allah (God).

Najis
This means unclean, so food should not contain any ingredients that are najis according to Shariah law.

Mashbooh
This is a food designation in Islam, literally meaning 'doubtful' or 'suspect'. Foods are labelled as 'mashbooh' when it is unclear whether they are halal or haram.

Makruh

In Islam the word 'makruh' is defined as anything that is undesirable or inappropriate. Muslims believe that makruh food is determined by their own innate guidance. It is makruh to eat the meat of a horse, donkey or mule.

HACCP emphasizes the importance of evaluating prerequisite programs (PRPs). PRPs for food safety management systems are the tayyib aspect of halal. These are programs and practices put in place to ensure that the environment is safe, clean, sanitary, and appropriate for manufacturing safe products. PRPs include procedures for cleaning and sanitizing, management of cross-contamination, waste disposal, deciding equipment suitability (cleaning and maintenance), pest control, management of purchased materials, personal hygiene, and warehousing. The philosophy of HACCP is to ensure that no operation affecting the safety of food at any stage is skipped. This philosophy applies throughout the processing, including raw materials, product combination, production circumstances, storage, distribution, preparation, marketing, and personnel.

13.4 Application of HACCP to the Halal Food Industry

HACCP consists of five requirements and seven essentials. The requirements are:

1) Assembling the HACCP team.
2) Description of product.
3) Identification of intended use.
4) Construction of a flow diagram.
5) On-site confirmation of the flow diagram.

The seven essentials are:

1) Conducting a hazard analysis.
2) Determining critical control points (CCPs).
3) Establishing critical limits for each CCP.
4) Establishing a monitoring system for each CCP.
5) Establishing corrective actions.
6) Establishing documentation and record-keeping systems.
7) Establishing verification procedures.

How can we apply HACCP in the halal food industry? Potential hazards of a religious nature must be identified so that appropriate procedures can be designed to monitor, control, and remove any hazard. Potential religious hazards are divided into two categories: those that can be and those that cannot be tolerated. Potential hazards that cannot be tolerated include stages or points in slaughtering or processing that clearly contradict the requirements of Shariah law. These are based on the texts of the Holy Quran or Sunnah (the way of life prescribed as normative for Muslims on the basis of the teachings and practices of the Prophet Mohammed (pbuh) and interpretations of the Holy Quran). Matters that cannot be tolerated include those so decided by a consensus of Muslim scholars. Examples of hazards that cannot be tolerated include the presence of meat, fat, or gelatin or any other parts of a pig, the presence of gushed blood in a manufactured product, the absence of acceptable religious cleansing (taharah) of production lines, and the use of stunning that led to death or of methods not in accordance with the prescribed conditions for halal slaughter.

Potential religious hazards that can be tolerated include forgetting (non-deliberately) to utter the name of Allah at the time of slaughter (in some schools) and forgetting to direct the bird or animal at the time of slaughter towards Qibla (Makkah).

13.5 Critical Control Points

CCPs are steps in processing where loss of control or lack of monitoring may lead to unacceptable hazards (or unacceptable decontamination). Examples include:

- the presence of a non-halal ingredient of animal origin in raw materials
- possible contamination with non-halal materials during processing
- presence of non-halal carcasses or meat near to halal carcasses
- use of very low quality meat or low quality raw materials
- use of non-halal slaughtering methods.

By using the HACCP decision tree to identify CCPs we can determine whether a step is a CCP for an identified hazard (Wallace 2014). In small establishments, the use of the HACCP decision tree may be substituted by asking this question: If we lost control, can this cause a hazard to the product? If the answer is 'yes', it is a CCP. It is important to remember that a CCP hazard will not be removed by subsequent production steps.

Critical limits provide a qualitative or quantitative value to a criterion that distinguishes between what is acceptable and what is not acceptable. These values are used to ensure that when deviation occurs corrective actions are taken. The presence of najis materials is an example where critical limits can be used to ensure that they do not become incorporated.

13.6 Conclusion

This chapter demonstrates that HACCP is compatible with the religious requirements of Shariah law. By introducing a fourth, religious element to the microbial, chemical, or physical elements found in secular HACCP, an additional layer of regulation and protection is afforded for those wishing to consume halal food and products.

We end this chapter with the following supplication which has come to us from the early Muslims:

> O ALLAH, make us independent of Thy Haram with Thy Halal, of disobedience to Thee with obedience to Thee, and of any other than Thee with Thy bounty.

> All praise is for ALMIGHTY ALLAH Subhanahu wa Ta'ala, Who guided us to this; had He not given us guidance, we would not have been guided.

References

Al-Kandari, D. and Jukes, D.J. (2011). Incorporating HACCP into national food control systems – analyzing progress in the United Arab Emirates. *Food Control* 22: 851–861.

HACCP for Excellence (2009). History of HACCP. http://www.haccpforexcellence.com/home/history_of_haccp (accessed June 2018).

Motarjemi, Y., Kaferstein, F., Moy, G. et al. (1996). Importance of HACCP for public health and development: the role of the World Health Organization. *Food Control* 7: 77–85.

Pal, M., Gebregabiher, W., and Singh, R. (2016). The role of hazard analysis critical control point in food safety. *Beverage & Food World* 43 (4): 34.

Pennington, T.H. (1997). The Pennington Group: Report on the Circumstances Leading to the 1996 Outbreak of Infection with E. coli O157 in Central Scotland, the Implications for Food Safety and the Lessons to be Learned. Edinburgh: Stationery Office.

Pointing, J., Teinaz, Y., and Shafi, S. (2008). Illegal labelling and sales of halal meat and food products. *Journal of Criminal Law* 72 (3): 206–213.

Wallace, C.A. (2014). Food safety assurance systems: Hazard Analysis and Critical Control Point System (HACCP): principles and practice. *Encyclopedia of Food Safety* 4: 226–239.

WHO (1995). Hazard analysis critical control point system: concept and application. Geneva: World Health Organization.

Further Reading

Alieva, A. (2015). Halal boosts fresh meat performance. In: Doing Business in the Halal Market, 7–9. London: Euromonitor International. https://www.foodreview.co.za/wp-content/uploads/2015/09/WP_Halal-Market_1%204-0715.pdf

Al-Mazeedi, H.M., Regenstein, J.M., and Riaz, M.N. (2013). The issue of undeclared ingredients in halal and kosher food production: a focus on processing aids. *Comprehensive Reviews in Food Science and Food Safety* 12 (2): 228–233.

Department of Statistics Malaysia (2010). *Taburan Penduduk dan Ciri-cir Asas Demografi*, Government of Malaysia. http://www.statistics.gov.my/portal/download_Population/files/census2010/Taburan_Penduduk_dan_Ciri-ciri_Asas_Demografi.pdf (accessed June 2018).

Fazira, E. (2015a). Certification and infrastructure are essential to cater to Muslim targeting businesses. In: Doing Business in the Halal Market, 2–4. London: Euromonitor International. https://www.foodreview.co.za/wp-content/uploads/2015/09/WP_Halal-Market_1%204-0715.pdf

Fazira, E. (2015b). Demand for halal food rising in tandem with growth of Southeast Asia. In: Doing Business in the Halal Market, 11. London: Euromonitor International.

Huang, J. (2015). Halal foods in China: trends and opportunities. In: Doing Business in the Halal Market, 13–15. London: Euromonitor International. https://www.foodreview.co.za/wp-content/uploads/2015/09/WP_Halal-Market_1%204-0715.pdf

Jabatan Standard Malaysia (2007). Food Safety According to Hazard Analysis and Critical Control Point (HACCP) System, MS 1480:2007. Putrajaya: Jabatan Standard Malaysia.

Jabatan Standard Malaysia (2009). Halal Food – Production, Preparation, Handling and Storage – General Guidelines, MS 1500:2009. Cyberjaya: Department of Standards Malaysia.

14

Halal International Standards and Certification

Mariam Abdul Latif

Faculty of Food Science and Nutrition University, Sabah, Malaysia

14.1   Introduction

In this era of globalization, food travels across the globe faster than people and enters any country where there is a demand. Advances in science and technology have meant being able to create niche foods to meet the needs of all types of consumers, including religious and health requirements. The obligation for Muslims to consume halal food was established 1400 years ago as laid out in the Holy Quran. Chapter 1 of this book looks at where those dietary obligations come from and what halal means to the Muslim community.

Of course, halal goes beyond food. It has expanded from products to services, responsibility, integrity, and a lifestyle, which is where the opportunity for business to meet these needs begins. The halal industry has now expanded to include non-food products such as cosmetics, pharmaceuticals, health products, toiletries, and medical devices as well as service sectors such as logistics, marketing, media, packaging, and banking. The global halal industry today is estimated to be worth around US$2.3 trillion (excluding Islamic finance).

To date, there are only two international halal food standards in existence. One has been produced by the Codex Alimentarius Commission (CAC), the other by the Organization of Islamic Cooperation (OIC, http://oic.org). The CAC's idea of developing an international halal standard began in Malaysia in 1979 and was continued by the Codex Coordinating Committee for Asia in Manila, who went on to focus on the labelling of processed meat according to religious requirements. All this effort finally bore fruit when the draft guideline was adopted by the CAC in Geneva in 1997, making it the first international halal food standard. The standard produced by the OIC is more recent, having been adopted in 2011.

14.2   Harmonization of Halal Standards

The World Trade Organisation (WTO) Sanitary and Phytosanitary (SPS) Agreement defines 'harmonisation' as 'the establishment, recognition and application of common SPS measures by different members' (WTO 1995). The International Organisation for Standardisation

The Halal Food Handbook, First Edition. Edited by Yunes Ramadan Al-Teinaz, Stuart Spear, and Ibrahim H. A. Abd El-Rahim.

(ISO) defines 'standard' as 'a document that provides requirements, specifications, guidelines or characteristics that can be used consistently to ensure that materials, products, processes and services are fit for their purpose'.

In this context, harmonizing halal standards means the acceptance or adoption of a unified or common standard for all stakeholders of the halal industry, be it at national or international level. When it comes to halal standards, the key stakeholders are the government as a policy maker and the competent authority, the industry that is responsible for producing halal products and services, and the consumer who is at the end of the process.

Studies assessing the impacts of standards on trade by Swann (2010) show that there is often, but not always, a positive relationship between the establishment of international standards and growth in international trade. The research found that international players find it relatively easy to come to a consensus over international halal standards. However, when it comes to setting up country-specific national standards, Swann (2010) found that while they may also provide a positive boost to trade they can also produce a negative effect by creating trade barriers. We are witnessing trade barriers within the halal industry today as there are many national standards, but at the international level there is still a gap when it comes to creating a consensus. It is important to note that a standard does not need to be 'international' to be recognized and accepted worldwide. An example is the American Society for Testing and Materials (ASTM) standards.

To date, halal regulatory bodies worldwide have found the process of developing and harmonizing international halal standards for various industry products and services a frustrating task. Halal certification bodies (HCBs) around the world often make common mistakes that end up exasperating the problem by actually undermining the integrity of the halal supply chain. The fact remains that minor differences do exist between different Islamic schools of thought on halal food production and halal food ingredients. These differences have created a divergence amongst halal producers and have affected consumer confidence to such a degree that the competencies of the certification bodies themselves are being questioned. On the other hand, we are witnessing a better understanding of the concepts of 'halal' and 'tayyib' amongst Muslim consumers, and there is now an increasing need to address halal industry concerns regarding halal standards based on Shariah principles.

It is accepted worldwide that each Islamic country has the sovereign right to protect its population by creating specific laws or standards on the safe production of halal food. The challenge when it comes to harmonizing international standards is due to the diversity of people in their approach to religion, culture, and health as well as their different economic status and levels of education. Despite this, the harmonization of halal standards at the international level is achievable as long as communities focus on what they have in common and put aside the minor differences that exist between the scholars of different schools of thought.

14.3 Halal Standards

According to Orriss and Whitehead (2000), the globalization of the food trade has in the past focused on strengthening measures that ensure food quality and safety. When it comes to imported foods, quality and safety have been the main areas of concern. Governments have first to set standards, then create legislation, and finally put in place enforcement regimes to ensure effective controls to achieve food quality and safety. The food industry also plays a key role when it comes to implementing quality assurance systems. In the case of international trade, companies are also required to meet the standards and legislation of their company's country of origin.

Sine 2000 Malaysia has been developing relevant standards for the halal industry which have had the effect of increasing consumer confidence. The standards also helped to provide guidelines to both industry and regulatory agencies at the national and international level (SIRIM, 2004). The Malaysian halal standards have been used as the basis of halal standards in many other countries, including the OIC, and they have also been used as the foundation of halal certification processes.

The following national halal food standards have been produced to help stakeholders:

1) Malaysian Standard MS1500:2009, Halal Food – Production, Preparation, Handling and Storage – General Guidelines (Second Revision), (Department of Standards Malaysia, 2009).
2) Brunei Standard PBD 24:2007 Brunei Darussalam Standard Halal Food
3) Singapore MUIS Halal Certification Standard, General Guidelines for the Handling and Processing of Halal Food, 2005
4) Saudi Standard SASO 2172:2003, General Requirements for Halal Food
5) Bosnia and Herzegovina Standard BAS 1049:2010 Halal Foods: Requirements and Measures
6) Austria Standard ONR 142000:2009 Halal food – Requirements for the food chain
7) Pakistan Standard PS 3733:2010 Halal Food Management Systems: Requirements for any Organization in the Food Chain (First Revision)

14.4 Halal International Standards

While these standards are voluntary, they still have a significant impact on national technical regulations as well as on global trade. Since its establishment in 1995, the WTO, SPS, and Technical Barriers to Trade (TBT) Agreements have recognized the CAC as the international reference for food standards when it comes to protecting human health and the resolution of trade disputes between member countries. Annex 3 of the WTO/TBT Agreement and the growing globalization of trade have been the main drivers when it comes to increasing the importance and significance of international standards.

The Codex Alimentarius was established by the Food and Agriculture Organization (FAO) and World Health Organization (WHO) in 1963, and today it plays a vital role in consumer protection and facilitating international trade through the development of international food standards, guidelines and codes of practices. 'Codex Alimentarius' is Latin for 'food code' or 'food law'. Codex is a collection of internationally adopted food standards presented uniformly.

The SPS Agreement covers food safety and animal and plant health protection, and gives a government the right to give priority to health protection over trade. The need for trade restrictions to ensure health protection, however, must be scientifically justified or based on risk assessment. The TBT Agreement covers mandatory technical regulations, voluntary standards and conformity assessment procedures, with the right of governments to apply the regulations needed to achieve legitimate objectives, including protection from deceptive practices.

As mentioned earlier, the current international halal standards that address food labelling and halal food production are:

1) CAC, CAC/CL 24-1997 General Guidelines for Use of the Term 'Halal'
2) Organization of Islamic Conference, Organization of Islamic Cooperation/the Standards and Metrology Institute for the Islamic Countries (OIC/SMIIC) 1:2011 General Guidelines on Halal Food.

14.5   Codex General Guideline for Use of the Term Halal

In 1979, Malaysia took the lead in developing the Codex General Guideline for Use of the Term Halal, which was eventually adopted in 1997 to complement the Codex General Standard on Packaged Food (CAC, 1979).

This guideline covering packaged food has provided the basis for how food can be produced and then labelled as halal. It was the availability of this first international halal guideline, along with rising consumer demand, which prompted the food industry to address how they sourced their ingredients and how they processed their products to ensure they fully complied with the new requirements in the halal standard. The driver was their desire to capture a piece of the growing global halal market.

Table 14.1 shows a brief chronology of some of the highlights of the development of Codex Halal guideline between 1979 and 1997.

Table 14.1   Chronology of the Codex Halal Guideline 1979–1997.

Year
Development
20–26 March 1979
Malaysia presented a Conference Room Document entitled 'Specific Labelling Provision for Processed Meat products about Islamic religious requirements' in Malaysia, at the Second Session of Codex Coordinating Committee for Asia, Manila, Philippines.
The document highlighted that the concept of consumer protection should not be confined to health and technical factors only but must also include the protection of the cultural values, traditions and attitudes of the consumer. The need to protect Muslim consumers who have to abide by Islamic religious requirements is an example. The acceptance and inclusion of the cultural factor into the concept of consumer protection should be worldwide and not limited to national level only since it involves a large portion of the world community. A working group was formed, led by Saudi Arabia, to discuss the format and specific labelling provisions for processed meat products about Islamic religious requirements.
2–8 February 1982
The 3rd Session of Codex Coordinating Committee for Asia held in Colombo, Sri Lanka, was informed that Working Group 1 would not convene pending completion of a study by the Islamic Centre in Brazil into Islamic rules and requirements governing the slaughter of animals.

Table 14.1 (Continued)

5–8 December 1985

The WHO Regional Office for the Eastern Mediterranean and the Muslim World League sponsored a meeting on Islamic Requirements of Food of Animal Origin, held in Jeddah, Saudi Arabia. The meeting was attended by an international group of widely recognised Muslim scholars, as well as by observers from the FAO/WHO Collaborating Centre, Berlin (West) and Muslim associations.

8–14 April 1986

The 6th Session of Codex Coordinating Committee for Asia held in Yogyakarta, Indonesia was informed about the meeting held in Saudi Arabia on 5–8 December 1985. It was also reported that the meeting had prepared a list of food animals, the meat of which is permitted for consumption by Muslims. The essentials of Islamic methods of slaughter were defined, and methods for stunning before slaughter, especially by the use of a captive bolt, or by electric or carbon dioxide methods, had been discussed. The meeting had appointed a Committee of four members to study the aspects of electric stunning and to report in six months to the organizers of the meeting.

25–26 June 1987

The 34th Session of the Executive Committee (CCEXEC) of the CAC held in Rome, Italy pointed out that in view of the fact that the WHO Eastern Mediterranean Regional Office jointly with the Muslim World League had issued the report of a meeting on Islamic rules governing food of animal origin in 1986, the Secretariat felt that no further action was required by the CAC. However, the committee was informed that while Australia followed a Code of Practice for slaughtering animals according to Islamic rules the requirements they used were not fully identical with those contained in the Jeddah meeting report. The Committee took note of this information but agreed with the Secretariat proposal of not taking further action in this matter.

26 January–1 February 1988

The 6th Session of the Codex Coordinating Committee for Asia held in Denpasar, Indonesia reported that no further action was required by the Codex Alimentarius Commission, given the fact that the report of the meeting held in Jeddah had since been published. The Committee also noted that the following documents were made available:

 i) Report of Joint Meeting of Muslim World League and WHO on Islamic Rules Governing Foods of Animal Origin (copies can be obtained from WHO)
 ii) A draft Arab Standard 'Requirements of Animal Slaughter', Islamic method elaborated by Arab Organization for Standardization and Metrology
 iii) Instructions and Prerequisites of Slaughter According to Shariya of Islam issued by the Brazil Islamic Center
 iv) Overview of the labelling provisions for processed meat products about Islamic religious requirements presented at the 2nd Session of CCCAsia by Malaysia.

The Committee commended Malaysia for initiating the study on labelling provisions for processed meat products about Islamic religious requirements (CAC, 1989).

11–15 March 1991

The 21st Session of the Codex Committee on Food Labelling (CCFL) held in Ottawa, Canada noted a proposal forwarded by the 7th Session of the CCCAsia concerning the implementation of labelling guidelines specific to the Asian region, taking into account the Codex General Standard and other Codex labelling requirements. The Secretariat noted that the elaboration of labelling guidelines specific to the Asian region could create a variety of problems, especially for the creation of trade barriers. However, if the object of the CCASIA was to create guidelines for their regional use in addition to Codex Standards, this could be acceptable to the CCFL and Commission. The Committee agreed to indicate to the CAC that if the CCCAsia wished to elaborate labelling guidelines addressing regional needs, the CCFL should be entrusted to examine any proposals for review and endorsement.

(Continued)

Table 14.1 (Continued)

27–31 January 1992

The 8th Session of the Codex Coordinating Committee for Asia held in Kuala Lumpur, Malaysia agreed to continue the elaboration of requirements to ensure proper labelling of food produced according to halal conditions. This decision was taken with the understanding that the Executive Committee would approve of this procedure, and the Codex Committee on Food Labelling would be informed. It was also agreed that Malaysia would prepare the document for circulation and government comment at Step 3 well before the next session of the Committee (CAC, 1993).

24–27 May 1994

The 9th Session of the Codex Coordinating Committee for Asia held in Beijing, China considered the proposed draft guidelines presented by Malaysia. The draft emphasised the size of the international trade of halal food with a huge market and business opportunities. With the enormous market potential throughout the world, it was imperative that the food industry understood the requirements of the consumers in this area. It was most timely that guidelines for labelling of halal food be developed by the CAC to facilitate international trade.

The Committee recommended to the Executive Committee the elaboration of the Guidelines for Use of the Term 'Halal', with the understanding that the Guidelines would be elaborated by the CCFL based on the Malaysian draft.

24–28 October 1994

The 23rd Session of the CCFL held in Ottawa, Canada considered the proposed draft Guidelines for Use of the Term 'Halal', as presented in CL 1994/19-FL, by a proposal arising from the Codex Coordinating Committee for Asia, and as confirmed by the 41st Session of the Executive Committee. Government comments submitted at Step 3 were received from Canada, the Czech Republic, and Sweden. In presenting the proposed draft Guidelines, the Malaysian delegation informed the Committee that rapidly increasing Muslim and non-Muslim consumer demand for products produced under halal religious requirements had greatly increased the economic benefits for exporters. It was noted that expanding marketing efforts on the part of exporters had helped to provide a highly profitable product that helped to fulfil religious requirements for the Muslim population. The Malaysian delegation felt that such requirements would greatly enhance and complement Section 5.1(iv) of the Codex General Guidelines on Claims.

The Committee, while discussing the Guidelines point by point, noted that an introductory paragraph was included to acknowledge minor differences of opinion in the interpretation of Islamic schools of thought. As various aspects of the Guidelines were subject to the interpretation of individual importing countries, the Committee decided to rename the Guidelines the Codex General Guidelines on Use of the Term 'Halal'. In response to concerns expressed on the conditions outlined in Section 2 (Definitions) of the Guidelines, the delegation from Malaysia explained that halal products could be produced in non-halal facilities as long as proper cleaning procedures were followed. Given this discussion, the Committee decided to add new Sections 2.2(i) and 2.2(ii) to the Guidelines to clarify parameters for the production of halal foods in facilities where non-halal foods were produced.

While noting that halal products could be sold without indicating such on food product labels, the Committee agreed to revise and clarify Section 4.1 of the Guidelines to indicate that when a claim was made that food was halal, the word 'halal' or equivalent terms should appear on the label. About the approval, inspection, certification, and transport of halal products, the Malaysian delegation noted that these requirements were normally addressed in bilateral agreements between national authorities. The Committee also noted that the Guidelines applied to foods produced through biotechnology.

The Committee agreed to forward the proposed draft General Guidelines on Use of the Term 'Halal' to the 21st Session of the Codex Alimentarius Commission for adoption at Step 5 (CAC, 1995).

Table 14.1 (Continued)

14–17 May 1996

The 24th Session of the CCFL held in Ottawa, Canada revised the Draft Guidelines in the light of comments received. It noted that the prohibition of the use of hazardous and intoxicating plants in the preparation of food additives, such plant products, could be used when the toxin or hazard was eliminated by further processing, and the text was amended accordingly. The Committee agreed to advance the General Guidelines for Use of the Term 'Halal' to Step 8 of the Procedure, subject to the advice of the Executive Committee.

4–7 June 1996

The 43rd Session of the Executive Committee of the CAC held in Geneva, Switzerland noted that the Draft General Guidelines for Use of the Term 'Halal' had been developed in the interest of promoting fair practices in the food trade. The Committee agreed that the Draft Guidelines would be considered at Step 8 by the 22nd Session of the CAC, as had been proposed by the CCFL.

23–28 June 1997

The 22nd Session of the CAC held in Geneva, Switzerland adopted the Draft General Guidelines for Use of the Term 'Halal' (CAC, 1997).

As the international halal trade has increased over the years, and so have the HCBs which issue halal certificates to those qualified applicants to allow them to use the halal logo on their food packaging. In light of this, at the 7th Codex Coordinating Committee for the Near East (CCCNEA) (held in Beirut from 21 to 25 January 2013), Egypt presented a project document regarding halal products. The proposals, concentrating on the gaps found in Codex General Guidelines for Use of the Term 'Halal', (CAC/GL 24 1997) (REP13/NEA, paras 111–117), were reported as a proposal for the elaboration of a new standard at the 36th Session of the CAC held in Rome, Italy on 1–5 July 2013 (CAC, 2014).

The purpose of this new work is to develop regional/international directives for halal foods concerning the following:

- General principal/directives on halal foods to be followed during any process in the food chain, i.e. receiving materials, preparation, manufacturing, sorting, determining, packaging, fixing marks, monitoring, processing, transportation, distribution, storing, and finally handling halal foods and products according to Islamic Shariah law.
- Directives to be followed by HCBs for foods as well as determining the requirements for the procedures of issuing a Halal certificate.
- Directives for accredited bodies that approve HCBs for foods which include the general directives and procedures of the halal accreditation body. In the light of principles and directives issued by the OIC in this regard, the aim was to reduce the obstacles facing intra-regional trade for these products as well as assuring the practices in food trade and protecting the health and safety of consumers.

The 36th Session of the CAC recalled that the Codex Executive Committee had recognized the importance of the consumption and trade of halal products worldwide. The CAC agreed that, in the context of this new work proposal, the project document should be re-scoped to identify gaps with existing relevant Codex texts. The Member proposing new work should seek the advice of the Codex Committee on Food Labelling (CCFL) and the

Codex Committee on Food Import and Export Inspection and Certification System (CCFICS) to assist the CCEXEC and Commission to take a decision.

The recent 38th Session of the CAC, held in Geneva, Switzerland on 6–11 July 2015, noted the report from the 8th Session of the CCCNEA that they will continue to work by electronic means to revise the project document for consideration by the CCFL. The revised project document will address gaps about the labelling of halal food so that the CCFL can make a decision on the possible revision of the Guidelines for Use of the Term 'Halal'. The CCCNEA stressed the need for comprehensive work on halal food within Codex to address all relevant aspects related to the production of halal food (CAC, 2015). Several delegations highlighted the following:

1) the importance of halal food for Muslim countries
2) the global nature of halal food
3) work on halal food should be limited to labelling for the time being
4) any work in Codex on halal food should take into account the standards and related texts developed at the level of Islamic countries, in particular the OIC/SMIIC as reference documents
5) work on halal food should have full participation of all interested member countries.

Based on the above discussion, the Committee agreed to recommend that work on halal food should be limited to the revision of the Guidelines for Use of the Term 'Halal' in the CCFL with full participation from member countries, taking into account the standards and related texts developed by the OIC/SMIIC as reference documents. In May 2016, the Codex Committee on Food Labelling discussed the proposal to revise the General Guidelines for Use of the Term 'Halal' and agreed that a revision of the Guidelines is not necessary as the current Guidelines are sufficient in that they provide common principles for labelling foods as "halal". The Committee agreed not to proceed with the revision as proposed, however, the proposal raised a question on how to deal with consumer preference claims in a broader way (CAC, 2016).

14.6   OIC/SMIIC 1:2011 General Guidelines on Halal Food

The other international halal standard that addresses halal food production is the OIC/SMIIC 1:2011 General Guidelines on Halal Food. The Standards and Metrology Institute for the Islamic Countries (SMIIC) is a regional standardization organization that aims to realize harmonized standards, achieving uniformity in metrology, laboratory testing, and establishing accreditation schemes for the purpose of expediting the exchange of materials, manufactured goods, and products in the Muslim world under the umbrella of the OIC. The OIC has 57 member countries spread over four continents and is the second-largest inter-governmental organization after the United Nations. It forms the platform for the Muslim world in ensuring the safeguarding and protection of the interests of the Muslim population and promoting international peace and harmony amongst various populations of the world.

The idea to establish unified halal standards amongst Islamic countries can be traced back to the 1st Meeting of the Economic and Commercial Cooperation Standing Committee (COMCEC) of the OIC in 1984. The Standardization Experts Group for Islamic Countries (SEG), which was established in 1985 for this purpose, worked to this end and its work led to the approval of the SMIIC Statute at the 14th COMCEC Meeting in 1998. The Statute of the SMIIC was first submitted to the member countries for signature during the 15th

COMCEC Meeting held in Istanbul, Turkey, on 4–7 November 1999. The SEG was mandated to develop halal food standards and procedures at the 23rd Meeting of COMCEC in 2007 and prepared three draft standards by getting consensus through mutual work and contribution amongst its stakeholders, namely OIC member countries, the International Islamic Fıqh Academy (IIFA), and other interested parties. The Statute entered into force after fulfilling the ratification requirement by 10 OIC member countries in May 2010. Eventually, the SMIIC was officially established in August 2010 after its first inaugural General Assembly meeting. As of December 2013, it has 28 member countries and one member country as an observer.

The OIC/SMIIC 1:2011 General Guidelines on Halal Food was adopted on 17 May 2011. This standard defines the basic requirements that should be followed at any stage of the food chain, including receiving, preparation, processing, sorting, determination, packaging, labelling, marking, controlling, handling, transportation, distribution, storage, and service of halal food and its products based on Islamic rules. The Guidelines are not for those countries which have already provision for a halal food standard.

A brief comparison between Codex and OIC/SMIIC Halal Food Standards is summarized in Table 14.2.

Table 14.2 Comparison between Codex and OIC/SMIIC Halal Food Standards.

CODEX
OIC: SMIIC 1

1 Scope

- These guidelines recommend measures to be taken on the use of halal claims in food labelling.
- These guidelines apply to the use of the term 'halal' and equivalent terms in claims as defined in the General Standard for the Labelling of Prepackaged Foods and include its use in trademarks, brand names, and business names.
- These guidelines are intended to supplement the Codex General Guidelines on Claims and do not supersede any prohibition contained therein.

2 Scope

- This standard defines the basic requirements that shall be followed at any stage of the food chain, including receiving, preparation, processing, sorting, determination, packaging, labelling, marking, controlling, handling, transportation, distribution, storage, and service of halal food and its products based on Islamic rules.
- All requirements of this standard are generic and are intended to be applicable to all organizations in the food chain regardless of size and complexity. This includes organizations directly involved in one or more steps of the food chain.
- Guidelines on the application of this standard in all organizations are contained in the standard for halal food certification rules.

3 Definition

Halal food means food permitted under Islamic law and should fulfil the following conditions:

3.1.1 does not consist of or contain anything which is considered to be unlawful according to Islamic law;

3.1.2 has not been prepared, processed, transported or stored using any appliance or facility that was not free from anything unlawful according to Islamic law; and

(Continued)

Table 14.2 (Continued)

3.1.3 has not in the course of preparation, processing, transportation or storage been in direct contact with any food that fails to satisfy 3.1.1 and 3.1.2 above.

3.2 Notwithstanding Section 3.1 above:

3.2.1 halal food can be prepared, processed or stored in different sections or lines within the same premises where non-halal foods are produced, provided that necessary measures are taken to prevent any contact between halal and non-halal foods;

3.2.2 halal food can be prepared, processed, transported or stored using facilities which have been previously used for non-halal foods provided that proper cleaning procedures, according to Islamic requirements, have been observed.

4 Terms and Definition

4.1 Lawful Food

The term 'halal' may be used for foods which are considered lawful. Under Islamic law, all sources of food are lawful except the following sources, including their products and derivatives which are considered unlawful:

4.1.1 Food of Animal Origin

a) Pigs and boars.
b) Dogs, snakes and monkeys.
c) Carnivorous animals with claws and fangs such as lions, tigers, bears, and other similar animals.
d) Birds of prey with claws such as eagles, vultures, and other similar birds.
e) Pests such as rats, centipedes, scorpions, and other similar animals.
f) Animals forbidden to be killed in Islam, i.e. ants, bees, and woodpecker birds.
g) Animals which are considered repulsive generally like lice, flies, maggots, and other similar animals.
h) Animals that live both on land and in water such as frogs, crocodiles, and other similar animals.
i) Mules and domestic donkeys.
j) All poisonous and hazardous aquatic animals.
k) Any other animals not slaughtered according to Islamic law.
l) Blood.

4.1.2 Food of Plant Origin

Intoxicating and hazardous plants except where the toxin or hazard can be eliminated during processing.

4.1.3 Drink

a) Alcoholic drinks.
b) All forms of intoxicating and hazardous drinks.

4.1.4 Food Additives

All food additives derived from Items 4.1.1, 4.1.2, and 4.1.3.

4.2 Halal Food

Halal food is the food, including drinks, which is allowed to be consumed according to Islamic rules and that complies with the requirements mentioned in this standard.

4.3 Slaughtering

All lawful land animals should be slaughtered in compliance with the rules laid down in the Codex Recommended Code of Hygienic Practice for Fresh Meat and the following requirements:

Table 14.2 (Continued)

4.3.1 The person should be a Muslim who is mentally sound and knowledgeable of the Islamic slaughtering procedures.

4.3.2 The animal to be slaughtered should be lawful according to Islamic law.

4.3.3 The animal to be slaughtered should be alive or deemed to be alive at the time of slaughtering.

4.3.4 The phrase 'Bismillah' (in the Name of Allah) should be invoked immediately before the slaughter of each animal.

4.3.5 The slaughtering device should be sharp and should not be lifted off the animal during the slaughter act.

4.3.6 The slaughter act should sever the trachea, oesophagus, and main arteries and veins of the neck region.

4.4 Preparation, Processing, Packaging, Transportation and Storage

All food should be prepared, processed, packaged, transported, and stored in such a manner that it complies with Section 3.2.1 and Section 3.2.2 above and the Codex General Principles of Food Hygiene and other relevant Codex Standards.

5 Requirements

5.1 Sources of Food

5.1.1 Food of Animal Origin

5.1.1.1 Halal Animals

The following are considered as halal animals:

a) Domesticated animals such as cattle, buffalo, sheep, goats, camels, chickens, geese, ducks, and turkeys.

b) Non-predatory wild animals such as deer, antelope, chamois, and wild cattle.

c) Non-predatory birds such as pigeons, sparrows, quails, starlings, and ostriches.

5.1.1.2 Non-halal Animals

The following are considered as non-halal animals:

a) Pigs, dogs, and their descendants.

b) Animals not slaughtered in the name of Allah.

c) Animals not slaughtered according to Islamic rules.

d) Animals that died by themselves.

e) Animals with long pointed teeth or tusks which are used to kill prey or defend themselves such as bears, elephants, monkeys, wolves, lions, tigers, panthers, cats, jackals, foxes, squirrels, martens, weasels, and moles, crocodiles, and alligators etc.

f) Predatory birds with sharp claws such as hawks, falcons, eagles, vultures, ravens, crows, kites, and owls.

g) Pests and venomous animals such as rats, centipedes, scorpions, snake, wasps, mice, and other similar animals.

h) Animals which are considered repulsive such as lizards, snails, insects, and their larva stages and other similar animals.

i) Animals that are forbidden to be killed in Islam such as honeybees and hoopoe.

j) Donkeys and mules.

k) Any ingredient derived from a non-halal animal is not halal.

l) Farmed halal animals which are intentionally and continually fed with non-halal food.

(Continued)

Table 14.2 (Continued)

5.1.2 Aquatic Animals

a) All kinds of fish with scales, shrimp, and fish egg of fish with scales including their byproducts are halal. All other aquatic animals including their byproducts are halal.

b) All poisonous water animals that are harmful to health are non-halal, unless the poisonous and harmful materials are removed.

5.1.3 Amphibious Animals

All amphibious animals are non-halal.

5.1.4 Food of Plant Origin

Plants and their products are halal except poisonous and harmful plants, unless the poisonous and harmful materials are removed.

5.2 Rules of Slaughtering

5.2.1 Requirements of the Animals to be Slaughtered:

a) The animal to be slaughtered has to be an animal that is halal.

b) A certificate must be issued by a veterinary authority which attests that the animals to be slaughtered are healthy.

c) The animal to be slaughtered shall be alive at the time of slaughter. The slaughtering procedure should not cause torture to animals.

d) Only animals fed on halal feed are permitted for slaughtering following the standard veterinary procedure.

e) If animals have arrived from long distance, they should first be allowed to rest before slaughtering.

5.2.2 Slaughterer

a) The slaughterer shall be an adult Muslim who is mentally sound and fully understands the fundamental rules and conditions related to the slaughter of animals.

b) The slaughterer shall have a certificate of halal slaughtering issued by a competent authority supervising matters relating to health, hygiene, sanitation, and rules of halal slaughtering.

5.2.3 Slaughtering Tools and Utensils

a) Slaughtering lines, food-grade tools, and utensils shall be clean and used for the purpose of halal slaughter only.

b) Slaughtering tools used for beheading shall be sharp and made of steel (stainless steel).

c) Slaughtering tools shall cut by their edge, not by weight.

d) Bones, nails, and teeth shall not be used as slaughtering tools.

5.2.4 Stunning

a) All forms of stunning and concussion (loss of consciousness) shall be prohibited. However when the use of the electric shock becomes necessary and expedient (such as calming down or resisting violence by the animal), the allowed period and the electric current value for stunning shall be in accordance with Annex A of this standard.

b) Stunning (loss of consciousness) of poultry is prohibited, however if it is necessary and expedient the following conditions shall be met:

   i) Poultry shall be alive and in a stable condition during and after stunning (loss of consciousness) and upon slaughtering.

   ii) The current and duration of the electric shock, if it is used, shall be as specified in Annex A.

   iii) Any poultry that die before the act of slaughtering shall be considered as dead and unlawful.

   iv) Shall be proven to be humane.

   v) Shall not reduce the amount of blood after slaughtering.

Table 14.2 (Continued)

5.2.5 Slaughtering Procedure

5.2.5.1 Slaughtering Procedure of Animals

In addition to Clause 5.2.1, the following requirements are applied.

5.2.5.1.1 Health Checks of Animals before Slaughtering

In addition to antemortem control, the following requirements are also applied.

a) Animals to be slaughtered should be checked by a qualified veterinarian following the standard inspection methodologies.

b) Animals which have completed a third of their pregnancy shall not be slaughtered.

5.2.5.1.2 Procedure

a) The animal shall be slaughtered after having been raised or laid on its left side facing Kiblah (the direction of Makkah). Care shall be given to reduce suffering of the animal while it is being raised or laid and it should not be kept waiting long in that position.

b) At the time of slaughtering the animals, the slaughterer shall utter tasmiyah 'Bismillah' which means 'In the Name of Allah' and he shall not mention any name other than Allah otherwise this makes it non-halal. Mentioning the name of Allah shall be on each and every carcass 'zabiha' (slaughtered animal).

c) Slaughtering shall be done only once to each animal. The 'sawing action' of the slaughtering is permitted as long as the slaughtering knife is not lifted off the animal during the slaughter.

d) The act of halal slaughter shall begin with an incision on the neck at some point just below the glottis (Adam's apple) and after the glottis for long-necked animals.

e) The slaughter act shall sever the trachea (halqum), oesophagus (mari) and both the carotid arteries and jugular veins (wadajain) to enhance the bleeding and death of the animals. The bleeding shall be spontaneous and complete. The bleeding time shall be sufficient to ensure full bleeding and complete death of animal.

5.2.5.1.21 Mechanical Slaughter

Mechanical slaughter could be used with existence of a validation system. Proper labelling shall be applied on the product showing that it has been mechanically slaughtered.

a) The operator of the mechanical knife shall be an adult Muslim.

b) The slaughterer shall recite the tasmiyah 'Bismillah' prior to switching on the mechanical knife and shall not leave the slaughter area.

c) Should the slaughterer leave the slaughter area, he shall stop the machine line and switch off the mechanical knife. To restart the operation he or another Muslim slaughterer shall recite the tasmiyah 'Bismillah' before switching on the line and mechanical knife.

d) The slaughterer shall repeat the tasmiyah 'Bismillah' during each slaughtering operation as long as it is possible and not only at the time of operating the machine. It is not also allowed to use a recording device.

e) The knife used shall be of single blade type and shall be sharp, and be made of steel (stainless steel).

f) The slaughter act shall sever the trachea (halqum), oesophagus (mari) and both the carotid arteries and jugular veins (wadajain) to hasten the bleeding and death of the animal.

g) The slaughterer is required to check that each poultry is properly slaughtered and any poultry that missed the mechanical knife shall be slaughtered manually.

h) If the heads are removed completely by the mechanical blade, the poultry and their heads shall be considered non-halal.

i) Bleeding period shall be a minimum of 180 seconds.

(Continued)

Table 14.2 (Continued)

5.2.5.1.2.2 Hand Slaughtering on Automated Poultry Processing Plants

Hand slaughtering could be used with existence of a validation system. Proper labelling shall be applied on the product showing that it is hand slaughtered.

a) The slaughterer shall be an adult Muslim.

b) The slaughterer shall recite the tasmiyah 'Bismillah' for each bird.

c) The knife used shall be of single blade type and shall be sharp and be made of steel (stainless steel). The knife shall be moved horizontally and shall cut by its edge not by weight.

d) The slaughter act shall sever the trachea (halqum), oesophagus (mari) and both the carotid arteries and jugular veins (wadajain) to enhance the bleeding and death of the poultry.

e) The slaughterer is required to check that each poultry is properly slaughtered.

f) Bleeding period shall be a minimum of 180 seconds.

5.3 Beverages

a) All kinds of water and non-alcoholic beverages are halal except those that are poisonous, intoxicating or hazardous to health.

b) All products or beverages containing alcohol are prohibited according to Islamic rules even for cooking purposes or filling in candies.

c) Food additives such as colorants, preservatives, etc. used in beverages shall not have been produced from non-food-grade and non-halal ingredients

5.4 Food Additives

Food additives are regarded as food. Food additives which are derived from non-halal ingredients are not halal.

5. 5 Other Products

Products which are not included in the above sub-articles (Clause 5.1.1–5.1.4) shall not have been produced from non-halal ingredients and shall not be processed with alcohol and alcohol products.

6 Food Processing

All processed food is halal if it meets the following requirements:

a) The products or ingredients do not contain any sources that are non-halal by Islamic rules.

b) The products do not contain anything in any quantity that is decreed as non-halal by Islamic rules.

c) The product and its ingredients shall be safe.

d) The product is prepared, processed or manufactured using equipment and facilities that are free from contamination with non-halal materials.

e) During its preparation, processing, packaging, storage or transportation the product shall be physically separated from any other food that does not meet the requirements specified in items (a), (b), (c) and (d) or any other materials that are described as non-halal by Islamic rules.

7 Storage, Display, Service and Transport

a) All halal food that is stored, displayed, sold or served, and during transport shall be categorized and labelled as halal and segregated at every stage so as to prevent it from being mixed or contaminated with materials that are not halal.

b) Transport should be compatible with the nature of the product. Transport vehicles should satisfy hygiene and sanitation rules.

Table 14.2 (Continued)

7.1 Packaging and Labelling

7.1.1 Packaging

a) Halal food shall be suitably packed using packaging materials that fulfil Clause 5.14.

b) Packaging process shall be carried out in a clean and hygienic manner and in sound sanitary conditions and the temperature shall ensure the safety and quality of the product.

c) Carcasses shall be appropriately packed in clean, new, sound, odourless packages that shall in no way adversely affect the quality and safety of the meat.

7.1.2 Labelling

7.1.2.1 In addition to requirements specified in ISO 22000 or Codex CAC/RCP 1 and CODEX STAN 1 each package shall be marked legibly and indelibly or a label shall be attached to the package with the following information:

a) name of the product

b) list of ingredients

c) date of expiry

d) net content expressed in metric system (SI) units

e) name and address of the manufacturer, importer, and/or distributor and trademark

f) code number identifying date and/or batch number of manufacture for traceability

g) country of origin

h) instruction of use, where applicable

i) if any food product contains fats, meat derivatives or extracts such as gelatine and rennet, this animal originated content should be declared on the product label

j) if a food product contains GMO, this fact shall be explicitly stated

k) when a halal mark is used, the authority and certificate number should be placed on the product

l) the nature of product (dried, fresh, frozen, smoked etc.)

m) all kinds of fish with scales, shrimp, and fish egg of fish with scales including their by-products shall be properly labelled as 'scaled fish' and all other aquatic animals, including their by-products, shall be properly labelled as 'non-scaled fish and others'.

7.1.2.2 For primary meat products, in addition to requirements specified in ISO 22000 or Codex CAC/RCP 1 the label or mark shall also include the following information:

a) date of slaughter

b) date of processing

c) the number of the veterinary health report/certificate carrying the corresponding information on the carcass

d) the stamp shall be tamper proof and the branding ink shall be stable and of food grade

e) each carcass (chilled or frozen) final packages of special meat cuts shall be branded by the authorized organization's official stamp and by the authorized person to indicate that slaughter has been carried out under the supervision of that competent authority.

f) when the halal mark is used, the authorization and certificate number should be placed on the product.

14.7   Halal Certification

Halal certification has been an effective tool for Muslim consumers, giving them an informed choice, for industries, giving them a marketing tool as well as facilitating the supply and sale of more halal products and encouraging more halal businesses, and for regulatory agencies for legal enforcement.

The other relevant standard which should be adopted by all HCBs is be the OIC/SMIIC 2:2011 Guidelines for Bodies Providing Halal Certification, which specifies the rules that the HCB should fulfil and the requirements for the execution of halal certification activities. The guidelines are the most complicated standard because, at a practical level, the HCB needs to create the halal certification system based on this document. Competent halal auditors and effective certification processes are few and require knowledge, training and experience to ensure the integrity of the certification (Dug, 2012).

14.8   Halal Accreditation Standard

Finally, the OIC/SMIIC 3:2011 Guidelines for the Halal Accreditation Body prescribe the general guidance and procedures for the halal accreditation body assessing and accrediting the HCB.

Given the benefits of harmonization of standards in expanding the international halal market, the OIC/SMIIC halal standards are highly relevant and timely to be adopted and practised by all OIC countries, including the Gulf countries, and also by all related HCBs in non-OIC countries. At the national level, halal standards can be harmonized in all countries by incorporating OIC/SMIIC standards into the local standards, but common/regional halal logos will be problematic as companies may seek to take advantage of the lack of traceability and accountability. Hence national logos underpinned by the same standard would allow for better accountability.

14.9   International Halal Certification Model

Halal certification is a process where a competent HCB certifies that the products and/or services offered by a company meet the specified halal standard. In the case of halal food certification, every stage of the food processing is examined and verified on site by the Shariah and technical auditors. Halal certificates are awarded to companies that meet the requirements in the set standard, and they are allowed to use the halal logo on their products. The benefit of halal certification is to assure Muslims that they can lawfully consume a company's products based on Shariah principles. Halal certification provides consumer confidence and clarifies any confusion about the halal status of a product (Abdul Latif, 2004).

The halal product certification system also involves testing and inspection activities. The HCB offering halal certification may subcontract the testing or inspection activity to another party. However, it is crucial for the HCB to ensure the subcontracted body or person is competent and complies with stipulated requirements relevant to testing, inspection or other technicalities.

Besides the development and harmonization of halal standards for product certification, halal international certification models need to be studied to produce a comprehensive framework, which includes the conformity assessment and capability building in developing a competent halal certification system (Abdul Latif, 2011). This system can be referred to OIC/SMIIC 2:2011 Guidelines for Bodies Providing Halal Certification, which is then to be accredited by an independent body to verify such HCB is competent enough to validate the halal integrity of the halal supply chain.

The flow chart of the halal international certification model based on the existing OIC/SMIIC standards is shown in Figure 14.1.

At a practical level, there are three stages in a halal certification process: the application process, the site audit, and issuing of halal certificates (Figure 14.2).

There are three types of process flow in any halal certification model (Figure 14.3). The first model, which is commonly practised, starts with the application, the site audit, and the approval/issuing of halal certificates. The second model starts with self-training (pre-audit) for compliance with the set standards followed by the formal application, the site audit, and the approval/issuing of halal certificates. The third model is where applications are

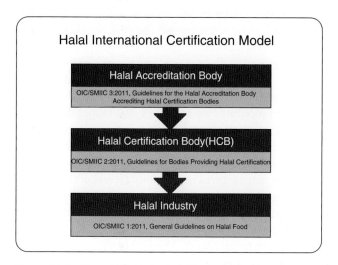

Figure 14.1 Flow chart for the halal international certification model based on the existing OIC/SMIIC standards.

Figure 14.2 The halal certification process.

Halal Certification Process

Application Process

Site Audit

Issuance of Certificate

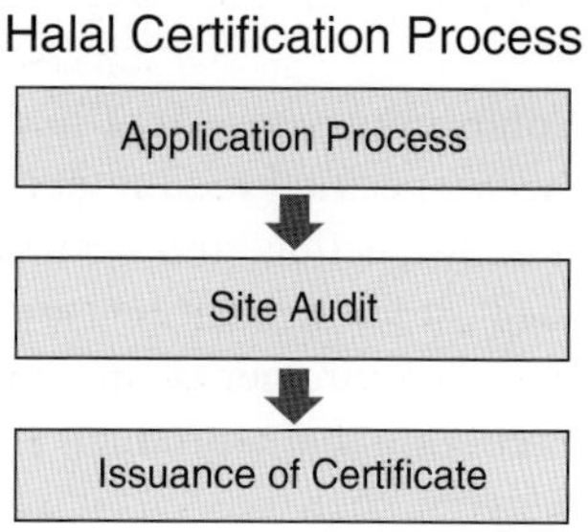

3 Types Certification Processes

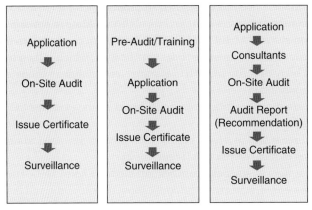

Figure 14.3 Three types of halal certification processes.

handed to consultants who will train the company and prepare the audit reports of applicants for recommendation to be approved as halal by the national authority. The differences are in the mechanism of the certification processes and the roles the HCB plays in the country, as a certifier, approval entity, and enforcer.

With regards to control of the halal certification, there should be a government-supported HCB (e.g. JAKIM, Malaysia, http://www.halal.gov.my) that mandates all halal certified companies to establish their own internal halal committees and appoint a halal executive to ensure the halal policy, standard, and procedures are adhered to at all times. Many food companies incorporate this requirement into their existing HACCP committee and appoint a senior Muslim manager to chair the committee. This requirement is even more stringent in halal certified abattoirs, where trained full-time Muslim auditors are stationed in the plant, providing a continuous audit. This management responsibility is one of the requirements in the standard, which forces companies to be self-compliant and ready for any spot audits by the HCB. The HCB will also conduct unannounced spot audits, or a raid if there is a serious complaint from the public on any halal-certified product or premise. The guilty offenders will be prosecuted in court, and they served with a heavy penalty under the halal law. Thus it is crucial to target the frequency of audits (continues versus spot) depending on the level of 'risk'.

14.10 Conformity Assessment

Conformity assessment is the process used to show that a product, service or system meets specified requirements. The main forms of conformity assessment are certification, inspection, and testing. Although testing is the most widely used, certification is the best known. In the case of halal certification for food products and food premises, these requirements are contained in the OIC/SMIIC 1:2011 halal food standard. Certifying that a product, service or system meets certain halal requirements has some benefits, such as:

- it provides Muslim consumers with added confidence
- it gives the halal-certified company a competitive edge by extending its markets
- it helps regulators to ensure that the integrity of the halal supply chain or the status of halal and tayyib conditions is met.

A study of HCBs by Abdul Latif in 2012 ('Halal Certification Bodies: An Observation', unpublished) showed that management commitment, competent Shariah advisors, competent auditors (Shariah and technical), adequate knowledge and training (e.g. halal ingredients), competent laboratories, and a comprehensive halal certification system are lacking in many HCBs. There is a need for more effective coordination and enforcement in halal conformity assessment amongst HCBs throughout the world. Many HCBs conduct halal certification at both international and country level, and the practices in implementing the certification scheme seem to differ from country to country, although in some cases they vary only marginally. All HCBs should be encouraged to harmonize their procedures in their halal certification system by aligning to the OIC/SMIIC 2:2011 Halal Standard. In supporting this system, highly capable testing facilities should be built by the government or research centres. The benefit of this development is that it will provide more facilities and choice for the industry. However, if resources are limited, and these efforts are not coordinated, there is bound to be overlapping and duplication of analytical roles amongst these agencies, which eventually may not be cost effective to maintain.

14.11 The Lack of Credibility of Halal Certification Bodies

i) Impartiality and Conflict of Interest

The lack of credibility of a certification body is an issue that has been discussed extensively internationally, and many initiatives have been taken to address this issue. It started with HCBs operating the quality management system certification to ISO 9000, where these HCBs also carried out consultancy and training activities. These multiple roles gave rise to conflicts of interest and compromised their impartiality. The relevant international guides and standards specifying the requirements for the operation of a certification body have stated that there must be a delineation of these activities. These are prerequisites that must be met by an HCB if it wants to be deemed credible and competent.

The HCB should also be transparent about making public its ownership structure to pre-empt rumours of vested commercial interests.

ii) Competency

The credibility of a certification is also dependent on the competency of the certification or inspection personnel, which is within the responsibility of the HCB. Competency extends to the ability to technically understand the halal system, to obtain detailed Shariah rulings, and to deploy them during the audit. Accreditation of HCBs is a formal declaration that the body is competent to certify halal products or services. It is therefore critical that all HCBs that are offering halal certification be accredited or assessed to the OIC/SMIIC 3:2011 accreditation standards and guidelines. The above requirements apply not only to the HCB but also to all its subcontractors and recognized partners to maintain the integrity of the halal logo and to instil confidence within the consumer community. It should also be a requirement to be transparent about the credentials and qualifications of the HCB's Shariah advisors by openly declaring its Shariah Board on its website or other suitable platform.

14.12 Capability Building: Consultants and Training Providers

Issues that have been raised relating to competency include the competency of consultants and training providers. There is a proliferation of halal consultants and trainers without the appropriate credentials and qualifications. At the moment there is no specific organization with the responsibility to ensure that these people are competent to provide these services from both the technical and Shariah aspects. A body is needed to develop the criteria and registration mechanism for these consultants and trainers.

14.13 The Way Forward

The adoption of the ISO/SMIIC as the certification criteria by all OIC countries is new. Manufacturers have yet to embrace this standard fully in their environment. There is a need to develop guidelines to interpret and implement the standard to ensure uniform and consistent adoption. The industries, especially small and medium-sized companies, will require assistance in the form of training and extension services to be able to comply with the requirements of the standard in a more systematic and effective manner.

The following recommendations are important as an immediate follow up to the above scenario:

I) Development of a comprehensive halal standards education and training programme such as that offered by the Halal Industry Development Corporation (HDC, http://hdcglobal.com) Malaysia, expanded to cater for various target groups:

   i) consultants and training providers, who in turn will provide training and extension services to the industries
   ii) auditors, assessors, and inspectors involved in halal conformity assessment (certification bodies, Islamic religious bodies, veterinary officers, testing personnel).

II) Development of interpretation guidelines and procedures to explain the OIC/SMIIC standards on halal to ensure uniformity in implementation, certification, testing, and accreditation.

III) Establishment of a registration scheme for qualified consultants and auditors/assessors/inspectors involved in halal implementation and conformity assessment. This qualification program will enhance the credibility of the consultants and auditors for halal certification work.

14.14   Conclusion

The focus now should be on harmonizing international halal standards for various subjects alongside establishing a halal model framework encompassing the development of
a model of competent halal certification body as well as the accreditation of the halal
certification system/body. Ultimately, every importing country should develop its own
halal import control system to ensure that all imported halal products are verified to
protect the interests of its Muslim consumers. In a nutshell, all stakeholders of the halal
industry need to cooperate, adopt the international OIC/SMIIC halal standards, and
develop a multilateral network for communication, integrity, impartiality, transparency,
and equivalency with their counterparts in the exporting countries for the greater good
of all humanity.

References

Abdul Latif, M. (2004). Requirements for the development of Halal Food Management System.
   MSc thesis. Universiti Putra Malaysia.
Abdul Latif, M. (2011). Malaysian Halal Certification. HDC Halal Compendium.
Codex Alimentarius Commission (1979). Codex Coordinating Committee for Asia. Alinorm 79/15.
Codex Alimentarius Commission (1989). Codex Coordinating Committee for Asia. Alinorm 89/15.
Codex Alimentarius Commission (1993). Codex Coordinating Committee for Asia. Alinorm 93/15.
Codex Alimentarius Commission (1995a). Codex Coordinating Committee for Asia. Alinorm
   95/15.
Codex Alimentarius Commission (1995b). Codex Committee on Food Labelling. Alinorm 95/22.
Codex Alimentarius Commission (1997a). General Guidelines for Use of the Term 'Halal'.
   CAC/GL 24-1997.
Codex Alimentarius Commission (1997b). Executive Committee of Codex Alimentarius
   Commission. Alinorm 97/3.
Codex Alimentarius Commission (1997c). Codex Alimentarius Commission. CX/CAC 13/36/9.
Codex Alimentarius Commission (1997d). Codex Committee on Food Labelling. Alinorm 97/22.
Codex Alimentarius Commission (2014). Codex Committee on Food Labelling. CX/FL 14/42/2.
Codex Alimentarius Commission (2015). Codex Alimentarius Commission. REP 15/NEA.
Codex Alimentarius Commission (2016). Report of the 43rd Session of the Codex Committee
   of Food Labelling, REP16/FL.
Department of Standards Malaysia (2009). Halal Food – Production, Preparation, Handling
   and Storage – General Guidelines, MS 1500: 2009. Cyberjaya: Jabatan Standard Malaysia.
Dug, H. (2012). SMIIC and Halal Food Standards. Paper presented at the First International
   Conference and Exhibition on Halal Food Control, Saudi Food and Drug Authority, Saudi
   Arabia (12–15 February 2012).
OIC (2011). OIC Standards – General Guidelines on Halal Food.
Orriss, D.G. and Whitehead, J.A. (2000). Hazard Analysis and Critical Control Point
   (HACCP) as a part of an overall quality assurance system in international food trade.
   Food Control 11: 345–351.

SIRIM (2004). Standards & Quality News: Standardisation for Halal Food, July–August 2004, Vol. 11, No. 4. SIRIM Berhad, Malaysia.

Swann, G.P. (2010). International Standards and Trade: A Review of the Empirical Literature, OECD Trade Policy Working Papers, vol. 97. OECD Publishing https://doi.org/10.1787/5kmdbg9xktwg-en.

WTO (1995). Text of the Agreement: The WTO Agreement on the Application of Sanitary and Phytosanitary Measures (SPS Agreement), Geneva, 1995.

15

Halal Certification and International Halal Standards

Yunes Ramadan Al-Teinaz[1] and Hani Mansour M. Al-Mazeedi[2]

[1] *Independent Public Health & Environment Consultant, London, UK*
[2] *Kuwait Institute for Scientific Research, Kuwait City, Kuwait*

15.1   Introduction

In addition to Islam, many other religions also forbid certain foods or have specific require-ments related to food. There are Jewish, Islamic (see Chapters 1 and 21), Buddist and Hindu dietary laws. Food has always been the subject of taboos and obligations. Which food we prefer and what we consider fit for (human) consumption differs depending on the place and time we live and the faith we adhere to.

Religious dietary laws are important to observant Muslim and Jewish populations, although not all the faithful comply with them.

Islamic dietary laws determine which foods are permitted for Muslims. Halal means permit-ted, whereas haram means prohibited. Several foods are considered harmful for humans to consume and are forbidden. This is expressed by the prohibition of the consumption of pork, blood, alcohol, carrion, and meat that has not been slaughtered according to Islamic prescrip-tions. Meat is the most strictly regulated food. The concept of 'halal' is not limited to food and food products, but also covers non-food products. The halal brand incorporates a variety of products and services in hospitality, banking and finance, insurance, beverages, and cosmetics.

The demand for halal produce is related to a belief that halal food is healthier, safer, and tastier, which appeals to both Muslims and non-Muslim consumers alike, assured through production to consumption, or from farm to table (see Chapter 1). For care products and cosmetics, due to increasing market internalization, more market product choices consist-ing of new ingredient formulation have become available to consumers. Some of these ingredients are rendered unsuitable for Muslim consumers based on Islamic law due to their source and method of processing (Ahmad et al. 2015) (see Chapter 11).

Halal food which must meet Islamic legal standards of manufacture is big business, and international food companies are recognizing its market potential. Demand for halal food products from Muslims in Europe and around the world has grown, increasing the halal food market. Halal is an Arabic word meaning 'allowable, permitted, and lawful', and

The Halal Food Handbook, First Edition. Edited by Yunes Ramadan Al-Teinaz, Stuart Spear, and Ibrahim H. A. Abd El-Rahim.
© 2020 John Wiley & Sons Ltd. Published 2020 by John Wiley & Sons Ltd.

comes from the Holy Quran, in which the term 'halal', in terms of food, means food which is permitted according to Islamic dietary law. Such foods are considered to be halal, except for foods clearly identified in the Holy Quran or Hadith as non-halal or haram (meaning 'forbidden' for consumption by Muslims).

The Holy Quran is the Islamic Holy Book. It contains the words of Almighty God (Allah), as dictated to the Prophet Mohammed (pbuh), and is written in Arabic. Hadiths are the collections of reports of the teachings, deeds, and sayings of the Islamic Prophet Mohammed (pbuh). This term comes from the Arabic language. According to the Holy Quran, all food products in principle are considered halal, except for some special products with special conditions, such as pigs and pork products, blood and blood products, alcohol, meat from cadavers, and meat from animals not slaughtered according to Islamic law (see Chapter 1). Halal standards and requirements determine which foods are permitted or prohibited for Muslims. 'Excepted products', which are not halal, are as stated in the Holy Quran 1400 years ago, when they were clear for Muslims of that time. Today, excepted products are still considered non-halal, and are explicitly prohibited in Islamic law. Food industry development makes the production of designated food products more complicated, while food technology is nowadays more sophisticated. Therefore, in respect of halal standards and requirements, it is very important and necessary to keep pace with developments in the food industry. Halal standards and requirements will help the food manufacturing industry to produce halal food products and to ensure that halal food products meet all halal standards and requirements to protect consumers of halal foods. Producers are increasingly aware of the need for halal standards and certification, and bring that knowledge to the fore when promoting their exports. Despite increasing worldwide demand for halal foods, halal food businesses in general cannot grow any faster because no agreement on worldwide halal food standards exists, each country currently adopting its own interpretation and having its own standards (Ambali and Bakar 2014). The big challenge for halal standards is the wide disparity between different Islamic schools of thought, leading to differences of opinion in the interpretation of what is lawful and what is unlawful, thus encouraging some countries to issue their own national standards, e.g. Malaysia, Indonesia and the Gulf Cooperation Council (GCC), which is composed of six member states: Saudi Arabia, Kuwait, Bahrain, Qatar, Oman, and the United Arab Emirates. At the international level, the Codex Alimentarius Commission has proposed guidelines related to the use of the term 'halal' in food labelling, such as Guidelines for Use of the Term 'Halal' (CAC/GL 24-1997), which supports and facilitates food trade between Muslim countries.

The General Guidelines for Use of the Term 'Halal' were adopted by the Codex Alimentarius Commission at its 22nd Session in 1997 (Codex 1997). The GCC consists of the states of Bahrain, Kuwait, Oman, Qatar, Saudi Arabia, and the United Arab Emirates. Halal food can be found in all places around the world where there are Muslim consumers. The basic concept of halal food is generally known to Muslims because it is explicitly mentioned in the Holy Quran. This chapter shows that differences in halal policies, cultures, and maslahah (public interest) are the main factors in the different halal standards in those six GCC countries. The implications of these differences are an imbalance in the acknowledgement of halal certification, less stringent control of the industry, and negative perceptions and double standards in halal certification. Differences in halal policies also led to confusion for Muslim consumers themselves as there are various halal logos from different countries (Fisher 2009).

15.2 The Halal Certificate

A halal certificate is an official document requiring food manufacturers to have observed and implemented Islamic requirements in producing halal foods in accordance with Islamic law. A halal certificate is issued by a halal certification organization after an inspection has been conducted to verify that the required standards have been met. When a halal certificate is issued by a halal certification organization accredited by an international halal organization (such as Majelis Ulama Indonesia (MUI5), the Malaysian Department of Islamic Development (JAKIM), or the Gulf Cooperation Council (GCC)), it provides more trust and acceptability as a halal certification organization, which could lead to requests for their services and supervision from halal food producers. MUI is the best-known Indonesian halal certifier. It has verified and accepted that many halal certification organizations around the world have created much confusion about how quality assurance for manufacturers/exporters should be carried out (Ratanamaneichat and Rakkarn 2013). JAKIM is responsible for issuing halal certificates and the execution of halal policy related to food and non-food products in Malaysia. The JAKIM halal certification approval panel consists of experts in Shariah (Islamic) law and scientists on technical matters (JAKIM 2010).

The halal certificate is valid for one to five years from its date of issue, and is conditional on individual contracts, with audit schedules during the contract period, confirming that halal standards and criteria have been met.

Halal certification benefits food manufacturers and food premises wishing to increase the demand and acceptance of their products to consumers of halal products. Moreover, the production of halal food helps the global halal market to provide assurances to halal consumers. Furthermore, it provides a great opportunity to food manufacturers to increase their customers by exporting their products to over 1.6 billion halal consumers worldwide (Pew Research Center 2009). Also, halal certification of food products provides opportunities for halal consumers, particularly Muslims, to increase their trust of halal food and know that the ingredients used and the production process are according to Islamic law, thus ensuring that they are always able to eat halal food without contravening Islamic dietary law by eating non-halal food.

Halal food is big business and international food companies are recognizing its market potential. Demand for halal food products from Muslims in Europe and around the world has grown, increasing the halal food market (Lever and Miele 2012). The halal food market has exploded in the past decade and is now valued at an estimated US$667 billion annually, currently 16% of the entire global food industry (see Chapter 16).

15.3 What is Halal Meat?

Halal is an Arabic word that literally means anything that is permissible or lawful when used in relation to meat. Such meats must be derived from specific animals slaughtered in accordance with requirements specified in the Holy Quran and Hadith. It is generally agreed within the Muslim community that for meat to be acceptable for consumption by Muslims, the animal must be a species that is accepted for halal, more importantly it must be fit and well at the time of slaughter, and sufficient time must be allowed for the loss of blood, which leads to irreversible loss of brain function (Farouk et al. 2014). The Holy

Quran expressly forbids Muslims from consuming blood. This may be due to the role residual blood (in the carcass) plays in the spoilage (and palatability) of meat, particularly against the background that at the time the Holy Quran was revealed, there were no advanced technologies such as refrigeration systems for the preservation of meat (see Chapter 1).

Despite the fact that Islamic authorities around the world unanimously agree on some of the requirements of halal slaughter, there are other aspects that have divided opinion amongst Islamic jurists, leading to confusion amongst halal consumers, food business operators, and other stakeholders in the halal industry as to what is authentic. For instance, European Council regulation EC 1099/2009 requires the pre-slaughter stunning of all animals to induce immediate loss of consciousness, but paragraph 18 EC1099/2009 gives derogation for slaughter without stunning, which is enshrined within this legislation as well as in some member states for faith groups (Council Regulation (EC) No 1099/2009). Some Islamic jurists have vehemently argued against the use of any form of stunning for halal slaughter, whilst others are of the opinion that pre-slaughter stunning is halal compliant, on condition that the stunning itself is fully recoverable and does not lead to injury (to the animal). Opponents of pre-slaughter stunning for halal slaughter have often cited the possibility of animals dying following stunning and before exsanguination as the main reason for believing that pre-slaughter stunning contradicts Islamic dietary rules. Others are of the opinion that the stunning of animals prior to slaughter results in the retention of more blood in the carcass in comparison with those slaughtered without stunning. However, repeated research has demonstrated that there is no difference in animals that are slaughtered either with or without pre-slaughter stunning in terms of the total blood lost at exsanguination (Anil et al. 2006; Khalid et al. 2015). In addition, some methods of stunning, e.g. head-only electrical stunning, support the full recovery of animals if they are not slaughtered post-stunning (Wotton et al. 2014; Orford et al. 2016). These misconceptions produce disagreements within the Muslim community, which, in addition to the lack of an overarching regulatory authority for halal food in member states and other countries in the developed world, opens the door for misinterpretation and potentially fraudulent activities in the halal food industry.

15.4 History of Religious Slaughter in the UK

The common method of slaughtering of large animals for food in the UK uses pre-stunning and a captive bolt pistol. In Britain, carbon dioxide (CO_2) gas stunning is the common method for slaughtering of poultry. In this method, a 65–70% gas and air mixture is given for 45 seconds and bleeding must begin within 30 seconds.

The Jewish community strictly refuses to accept stunning. Muslims have identical religious requirements and it is for this reason that both communities sought and obtained an exemption in 1911 when stunning was made compulsory in the UK.

In October 1985, the UK Ministry of Agriculture, Fisheries and Food invited the Muslim community and others for consultation on the proposal by the Farm Animal Welfare Councils (FAWC) on the welfare of livestock when slaughtered by religious methods, calling for a repeal of the legislative provisions permitting slaughter without stunning.

They cited pain and cruelty to animals when slaughtered without prior stunning as the main reason. The FAWC report also remarked that there was a widespread lack of understanding of why religious slaughter was considered necessary in Islam.

The Islamic Cultural Centre (ICC) spearheaded a strong representation to the government for the retention of exemption. Arguments were made justifying non-stunning on the grounds of compassion to animals, painless slaughter, and prohibition in the Holy Quran to consume blood and carrion, arguing that the strict procedure of slaughter allowed maximum bleeding. Other animal welfare concerns which are important in halal slaughter, such as ensuring that animals are fed and watered, and that they do not see the blood of carcasses or see other animals being slaughtered, were also addressed.

It is difficult to know what persuaded the government to retain the exemption, but good community relations, particularly as a common approach was made by Jewish and Muslim groups, probably prevailed above other considerations. Since then there has been consistent Muslim and Jewish pressure to ensure that the exemption is not removed or eroded.

Of all Muslim slaughter practices and aspects of Shariah law, the most fought over issue is stunning, which is seen to possibly render the animal or bird dead, although the primary aim is to make it unconscious so as slaughter the animal while it is alive. Scholars throughout the world have been consulted. The Food and Agriculture Organization (FAO) of the United Nations have published the Codex Alimentarius Commission, which specifies rules to be followed that particularly avoid the mention of stunning or otherwise and instead advise that the animal to be slaughtered should be alive or deemed to be alive at the time of slaughter.

European legislation provides laws, rules, and procedures regarding the slaughter of livestock (GWvD 1992; Council Directive 93/119/EC 1993; Council Regulation (EC) No 1099/2009). Article 4 of Council Regulation (EC) No. 1099/2009 describes mandatory pre-slaughter stunning, with exception of particular methods of slaughter prescribed by religious rites, to ensure unconsciousness and insensibility to prevent unnecessary suffering of animals. There is no consensus about the extent to which slaughter of conscious, meaning sensible and/or aware, animals causes them pain and distress (see Chapters 1, 3, 5, 6, and 7).

In a report by the House of Common in the UK in 2015, the Government said that it would prefer all animals to be pre-stunned before slaughter on welfare grounds but it observes the rights of religious communities. For animal slaughter to be lawful under Jewish law and Islamic Shariah Islamic law, Jewish (shechita) and Muslim (halal) conditions have to be met before an animal is cut and bled. These conditions also dictate how, and whether, pre-stunning of animals is acceptable. Around 80% of meat in the UK prepared by the halal method is pre-stunned, including supermarket own-brand meat.

15.5 Halal Certification in the UK

The emphasis on the importance of halal-compliant products demands more transparency in the halal supply chain. Recently the UK media was inundated with reports that suppliers, particularly restaurants and fast-food chains, were serving customers halal meat without prior notification. This caused consumer criticism and insistence on a demonstration of ethical responsibility by suppliers/producers to disclose halal supply chain information.

Conversely, the pressure is on producers to ensure that effective efforts are made towards halal verification and quality disclosure of their products.

A 2012 UK House of Commons Standard Note revealed that there was no legislative requirement in the UK for products to be labelled as halal. This meant that no effort had been made to collect data on halal products, the extent to which halal products are sold without being labelled as halal, and the quality of products labelled as halal.

The absence of UK legislative obligation required an alternative robust means to assure halal integrity and quality. Halal certification organizations (HCOs) may play a crucial role as not only a link between the halal producer and the halal-sensitive Muslim consumer but also as a supply chain partner in the halal producer's quality assurance to determine if a product is halal or not and to assure the quality of products represented as halal.

It cannot be visibly determined whether food is halal (as with other credence quality attributes relating to organic food and fair-trade products) so how does a consumer know which food is halal? There are basically three options:

- ask a religious leader which foods are permitted
- buy foods with a halal label
- buy from someone of known reputation (e.g. an Islamic butcher).

In traditional societies, a combination of the first two options is often applied. People living in a religious community that runs all political, economic, and communal matters internally rely on religious leaders and food suppliers of known reputation. In bygone days, this applied to many European Muslim communities, for whom the local Imam was often the final authority on halal issues. In the 1930s in the UK no food manufacturers were under halal certification supervision as most foods were prepared in the home.

Today, almost every supermarket in the UK stocks halal products. It is easier than ever before to eat halal foods as a Muslim living in a Western country.

It is hard to think of an era like the 1960s as the 'early days', but it was during this time that the Western world saw an influx of Muslim immigrants. Many chose to call the UK their home, but others found their way to countries like France and Spain.

This influx of people led to growing demand for halal food, which the Western markets of the time had previously not catered for. The idea of a halal label was alien to all involved. Newly-arrived Muslims were used to their food being halal because of where they had come from, while established Western businesses had not previously needed to meet a demand for halal food. This means that halal labelling on food is a phenomenon that is specific to Western countries, rather than Muslim ones.

First-generation Muslim immigrants were especially keen to maintain halal for their diet. Worries about the potential contamination of permitted meats with forbidden ones, such as pork, led to many buying their meats from Jewish butchers. The parallels between halal and kosher meats came to the fore here (Cutler 2007).

Initially reserved for special occasions, meat became a staple of the Muslim diet during this period, which again increased the demand for halal products. Throughout the 1970s, Islamic butcher shops began trading, especially in the UK and France. Usually owned by families, these shops certainly were not mainstream, but they offered Muslims more control over the food they put in their bodies. Often, the fact that such butcher shops were owned by Muslims was enough of a guarantee of halal that a label was not needed.

However, the 1990s also saw the rise of several food crises, the two most important of which were the rise of foot-and-mouth disease and Creutzfeldt–Jakob disease (CJD, mad cow disease).

This led to increasing public pressure to improve the quality of Western meat, from both Muslims and Western consumers. For the first time, Western abattoirs started equipping themselves with the equipment needed to create halal meat (Harvey 2010).

This added a level of industrialization to what had traditionally been a ritual slaughtering process, so there were missteps. However, the growing acceptance of the halal method led to the introduction of halal certification, which gave Muslims the confidence to purchase meats from mainstream providers, as well as the small butcher shops that had serviced them for so long.

Today, the prominence of halal in the food industry has led to more regulation that ensures products carrying halal certification actually meet the correct rules. Recently, Malaysia has secured a pledge from the FAO of the United Nations to create more stringent guidelines for halal products. As such, the halal label can be seen on jewellery, cosmetics, and anything else that comes into contact with the body (see Chapters 1 and 13).

15.5.1 Reliance on Local Suppliers and Religious Leaders

Due to the increase in industrially manufactured foods and the growing geographical distance between production and consumption (internationalization of the food market), reliance on local suppliers and religious leaders is often no longer enough. Traditional local arrangements are also disrupted by migration. Nowadays, consumers who seek halal foods are dependent on a label or trademark that identifies a product as halal. The consumer must trust the source and message of the labelling. These developments have resulted in a large number of certified products in US supermarkets and a growing number of halal labelled or certified products in Western European supermarkets.

The growth of halal certification in the UK fits into a general pattern of growing third-party certification and other regulatory arrangements involving a mix of private and public companies. Food safety regulation currently involves a large number of public and private organizations with complementary, overlapping, or competing roles. In other cases, private regulation is encouraged or enforced by governmental departments (such as many industrial hygiene codes).

According to the Muslim Council of Britain Muslims in the UK contributed an estimated £31 billion (US$46.5 billion) to the national economy in 2014. Of this £31 billion expenditure, Muslim household spend on food and beverages was an estimated £4.2 billion in 2014 with 5% compound annual growth rate compound annual growth rate (CAGR) to 2020. The BBC estimated the UK halal food market to be £3 billion in 2014 (BBC 2014). While UK Muslims make up around 5% of the UK population, around 25% of the country's 352 abattoirs now incorporate halal production. Halal food is sold in over 3000 independent food service outlets across the UK, and halal is increasingly offered in mainstream outlets.

The exponential increase in the population of Muslims across Europe in recent years may have contributed to the increase in the number of unregulated halal certification bodies (HCBs) (see Chapter 16), with the aim of independently verifying halal status and facilitating the trade in halal products. The UK does not have a national-level centralized

HCB. There are at least ten HCBs operating according to varying halal standards and competing with each other for a share of the halal certification market, raising doubts on ownership, control, prevalent practices, and even the authenticity of the halal (meat) industry in the UK. This has led to the emergence of a number of voluntary HCOs. There have been accusations and counter-accusations amongst the certifiers as to whose halal standard is genuine. The lack of government involvement or a central monitoring body of HCBs has meant that any individual can set up an HCB with little or no technical expertise about meat, potentially a lack of specific religious knowledge, and/or non-mainstream agreement about what makes meat halal. The development of these certification organizations is a response to the lack of relevant UK law and the existing structural factors. The following is a list of UK HCOs (in no particular order):

1) Institute of Islamic Jurisprudence
2) Halal Food Authority (HFA)
3) Islamic Foundation of Ireland
4) The Muslim Food Board (UK)
5) National Halal Food Group
6) Halal Monitoring Committee (HMC)
7) World Islamic Foundation
8) Halal Consultations Limited (HCL)
9) Universal Halal Agency
10) European Halal Development Agency
11) Halal Certification Europe (HCE)
12) Universal Halal Agency
13) The Halal Trust
14) Halal Food Council of Europe
15) Department of Halal Certification EU
16) Halal Correct

Of these, only the HFA and the Muslim Food Board (UK) are recognized by the Malaysian authorities, according to the Halal Hub Division, Department of Islamic Development Malaysia (JAKIM) (see Chapter 15).

All HCBs must comply with UK laws for food production as well as Islamic dietary compliance laws. The Food Standards Agency (FSA), an independent government agency, works with local authority enforcement officers to make sure food law is applied throughout the UK's food chain, including halal products. It has a Muslim Organisations Working Group that advices it on halal practices and policies. Licencing of Muslim slaughterers is done by the FSA and not by independent halal authentication and certification bodies. Under UK law the Meat Hygiene Services (MHS) has legal responsibility for animal welfare, health and hygiene issues, including for halal meat and livestock for halal slaughter. For instance, in the UK, although there are in excess of 12 HCBs, halal food products to be exported to Malaysia and Indonesia must be certified by either the HFA or HCE (formerly known as the Muslim Food Board). It must be noted that the HCE does not certify meat-based products; their main scope is the certification of non-meat processed foods and ingredients. The HMC is the UK's largest certifier of meat from animals slaughtered without stunning for the halal market, but certificates issued by the organization are only recognized by United Arab Emirate authorities, which means products certified by the HMC cannot officially be exported to either Malaysia or Indonesia. This is because the HMC is not accredited by either the Malaysian or Indonesian authorities (represented by JAKIM and MUI, respectively). Halal food business operators (HFBOs) are therefore limited in their choice of suitable HCBs, for instance a UK halal meat supplier wishing to export meat products to, say, Malaysia from the UK will have to use the HFA, unless the HFBO is willing to consider HCBs from other EU countries who hold accreditation for Malaysia. The HFA and HCL were co-operated to standardize their certification procedures, standards, and scope of certification. The scope of the HFA and HCL is similar: they both certify meat and processed food and their halal standards also permit the pre-slaughter stunning of animals for halal production. However, HCL approves the certification of kosher meat (Jewish or Shechita slaughter) as halal (see Chapter 21), whilst the HFA does not, and the concept of halal certification within the EU is a relatively new phenomenon. In the UK, the first HCB to certify meat as halal, the HFA, was established in 1994 (Halal Food Authority 2016). According to this organization, their main objective at inception was to monitor and authenticate the halal status of poultry and red meat because the halal status of up to 85% of halal meat in the UK at the time could not be verified (Ahmed 2008). Opponents of halal food certification have often questioned the motives of those ventures. It has been suggested that the exponential increase in the population of Muslims across Europe in recent years may have contributed to the increase in the number of unregulated HCBs, all with the aim of independently verifying halal status and facilitating the trade in halal products. In the UK alone, there are several HCBs who all compete to apply various halal standards in their quest to assure consumers about the authenticity of halal meat certified by themselves. There have been accusations and counter-accusations amongst the certifiers as to whose halal standard is genuine. Lack of technical understanding has led to a lack of proper monitoring and scrutiny of some halal certificated sites, which has resulted in the contamination of halal products at processing sites where halal and non-halal products are produced in tandem but with poor segregation and/or cleaning. Other HCBs have been accused of approving slaughter technologies that violate the basic requirements of halal slaughter. Although many Muslims would appear to prefer meat from animals slaughtered without stunning because the method guarantees a live, uninjured animal at the time of slaughter (EBLEX 2010), recent covert filming at non-stun halal slaughterhouses by animal welfare activists in Belgium and the UK have highlighted systemic abuse and suffering of animals destined for the halal food chain (PETA 2009; Newsweek 2015).

The high cost of halal certification to food business operators (FBOs) is another area that has attracted attention. The annual certification fee for an abattoir or food-processing site in the UK in 2016 ranged from a few hundred of pounds to tens of thousands of pounds, depending upon the HCB. The certification fee may be insignificant to the major players in the meat industry, but to small-throughput plants and food businesses this cost can be significant. In an attempt to avoid a fee, some FBOs have been found to intentionally mislabel non-halal meat as halal. Others have resorted to self-certification, where an FBO verifies its own halal status and communicates this to the halal consumer. This practice is common amongst Muslim-owned food businesses. Whilst some businesses have successfully implemented this model, others have struggled to effectively put it into operation. The practice of self-certification can only be used where the target consumers are local because most halal-importing countries require a halal certificate to accompany all halal meat products before such products can be given clearance.

15.5.2 Legislation

UK and EU law requires farm animals to be stunned prior to slaughter (British Veterinary Association 2015). Exemption for religious slaughter in UK law is contained in Schedule 12 of the Welfare of Animals (Slaughter or Killing) Regulations 1995 (SI 731) Jewish & Muslim.

Significant meat-producing countries such as Denmark and New Zealand legally mandate pre-stunning even for halal slaughter. The ability of the animal to feel pain following stunning is the subject of much debate and academic study, and exemptions to stunning are opposed by organizations such as the Royal Society for the Protection and Cruelty to Animals (RSPCA 2015).

Regulation exists to ensure meats labelled halal are authentic (trade description, food safety etc.) (see Chapters 17 and 18), but there is no legal requirement for non-halal meat to be labelled as such, or to be labelled stunned or not stunned at slaughter. It is estimated that 40% of poultry and 30% of lamb consumed in the UK meets halal specifications. Halal meat is therefore routinely served to non-Muslims and this can lead to objections from other faith groups. The Organization of Islamic Cooperation (OIC), an intergovernmental organization with 57 member states, is currently working on the development of a global halal standard (OIC 2015). There is lack of interest by local authorities due lack of experience about halal dietary law and resources, and therefore enforcement officers would find it difficult to protect halal consumers. Furthermore, consumer law does not currently give sufficient protection to halal consumers despite the Muslim authorities having made attempts to develop quality standards to instil some confidence in halal products.

15.6 Accreditation/Regulation

There is no single agreed standard for halal therefore there is no single authoritative UK accrediting body for validating authenticity. The World Halal Council (WHC) was established in December 1999 to implement an international halal certification standard. There has been an increasing trend for local government departments to take charge of halal certification as a result of the increasing demand for halal exports.

Some certification bodies purely address the religious authenticity of the products/production and not food safety, i.e. the British Retail Consortium (BRC). The UK HCBs such as the HFA and HCL permit the use of stunning while the HMC and the Muslim Council of Britain firmly reject such practice.

In Europe the concept of halal meat certification is a relatively new phenomenon. The ultimate goal of the certification of meat as halal by third-party HCBs is to provide assurance to halal consumers that some key requirements have been met during the slaughter, processing, packaging, storage, and transport of the meat. This is particularly important within the EU and other industrialized countries where many FBOs may be unfamiliar with the requirements of halal. Additionally, there is increased risk of cross-contamination of halal food with non-halal products such as pork during further processing, storage, and transport.

15.7 Halal Food Fraud

Some researchers have described food fraud as an ancient practice usually driven by profit (Manning et al. 2016). Fraudulent trading in meat that is claimed to be halal can take many forms, from the sale of illegally slaughtered animals generally associated with poor animal welfare, to the sale of unwholesome meat for human consumption, which poses a risk to human health. Manning and Soon (2014) argued that fraudulent trading in foods that poses a public health risk is often detected at a stage where the food is already at a point of sale or has potentially been consumed. In a recent food crime annual strategic assessment report in the UK, the FSA in conjunction with Food Standards Scotland (FSS) defined food crime as dishonesty on the part of the producer or supplier (FSA and FSS 2016). Food adulteration within the halal industry may fall under one of these categories. Food businesses (including halal producers) are expected to be familiar with these offences, and steps must be taken to prevent them. In the UK, organized criminal gangs are usually involved in these activities, a situation which because of the potential scale raises food safety concerns amongst regulatory and food safety professionals The most common offences included the sale of reprocessed chicken sludge bleached to improve its aesthetic value and the importation of unfit bush meat and smokies from parts of West Africa where such products are considered as delicacies. Blow-torching the skin or hide of an animal before splitting the carcass produces a smokie and is thought to improve the taste of the meat (Pointing and Teinaz 2004). Although banned within the EU (EC Regulation EC1662/2006), these products continue to be illegally imported into Europe for human consumption in some Member States. Aside from the dangers these products pose to public health, there are concerns for animal welfare and also conservation of endangered species. It has been reported that carcasses of endangered species such as giraffes, gorillas, and chimpanzees have all been imported from Africa and sold within the EU as bush meat. Some products may end up in the halal food chain with no proper documentation of their halal status (see Chapter 21).

15.7.1 Illegal Slaughter

Illegal slaughter is the slaughter of animals for human consumption outside a licenced establishment (European Council Regulation EC853/2004), except where permission is granted for the slaughter to be performed outside a licenced premise, such as the on-farm slaughter of animals for personnel use only (on-farm slaughtered meat cannot be sold to the general public, but must be used by the owner (usually the farmer) or his/her immediate family). In addition to meeting the EU legislative requirements specified in EC1099/2009 (see Chapter 19), animals destined for the halal food chain must also be slaughtered in the Islamic tradition. Any meat that falls short of the two requirements above is deemed unfit for Muslim consumption. Guidance issued to local enforcement authorities in the UK (FSA 2009) highlights the following scenarios that should arouse suspicion regarding the legality of meat:

- the slaughter of animals at unlicensed establishments
- the cutting of meat at unlicensed premises
- unlabelled or poorly labelled meat

- meat that may have been imported through illegal channels
- meat that is advertised, marketed and sold by itinerant vendors at car boot sales, pubs, clubs, and other public gatherings
- meat that was slaughtered legally but may become unfit to consume within the interpretation of European Council Regulation EC178/2002 due to storage under unsanitary conditions or at temperatures outside the legal limits.

Section 17.7.2 gives examples of fraudulent activities discovered in the halal meat industry in the UK. Muslims are instructed, according to the Islamic scriptures (the Holy Quran and Hadith; see Chapter 1), to consume only halal meat, therefore many Muslims regard the consumption of halal meat as a religious act so the significance of consuming halal meat to a Muslim cannot be underestimated. It is therefore surprising that some Muslims or food businesses owned by Muslims have been implicated directly or indirectly in the majority of the crimes committed regarding the falsification of the status of halal meat. According to experts in the halal certification and regulation industries, some production is motivated by greed and a desire to maximize profits by fair or foul means.

15.7.2 Mislabelling of Halal Meat

The accurate labelling, presentation, and advertisement of foodstuffs within the EU must comply with Article 2 of Council Directive 2000/13/EC. This specifically makes it an offence for the labelling of any food product to mislead consumers regarding the characteristics, identity, constituents, quantity, method of manufacture, durability, and geographical origin of the product (Pointing et al. 2008). Despite clarity regarding the meaning or definition of halal, non-halal products are sometimes intentionally mislabelled as halal in order to mislead consumers. The following is a list of some reported crimes involving halal meat in the UK:

- A halal meat company in Bradford used counterfeit halal labels on meat products and sold them as halal: £20000 fine (http://birminghamnewsroom.com/meat-supplier-ordered-topay-nearly-20000-for-fake-halal-labels).
- A halal meat wholesaler in Birmingham mislabelled and sold non-halal meat as halal: £62000 fine (https://www.bbc.com/news/uk-england-birmingham-29157457).
- A halal meat supplier in Birmingham misled halal consumers by claiming their products were certified by the HMC: £1100 fine (http://birminghamnewsroom.com/halal-meat-supplier-finedfor-misleading-customers).
- A halal meat butcher in Manchester mixed cheap minced beef with minced lamb and sold it as 100% lamb: £18000 fine (http://www.manchestereveningnews.co.uk/news/greater-manchester-news/halalbutchers-food-fraud-trafford-11178552).
- A meat processor in Newry, Northern Ireland, falsely labelled non-halal burgers as halal: £70000 fine (https://www.bbc.com/news/uk-northern-ireland-33317601).
- A meat supplier in Walton Summit, supplied halal chicken kebabs to a major retailers which were found to contain mainly connective tissue and beef: £10000 fine (http://www.lep.co.uk/news/business/food-firm-hit-by-kebab-fine-1-4670081).
- Four men illegally slaughtered sheep for the Eid festivities at unlicensed premises without prior stunning: £2075 fine (http://www.express.co.uk/news/uk/580745/Islam-Muslim-halalslaughter-farm-UK-Llechrydau-Farm-court).

A series of surveys regarding the labelling of chicken products in the UK concluded that there was widespread mislabelling of chicken (FSA 2001, 2003). With regard to chicken labelled as halal, the following conclusions were reached:

- The addition of water to 'pump up' chicken was a common practice. In some instances, over 50% of the chicken weight was added water.
- The water used in 'pumping up' the chicken was found to contain proteins to aid the retention of water in the carcasses which were of porcine and bovine origin. As mentioned above, it is forbidden for Muslims to consume pork or its derivatives. In addition, the bovine protein may have been derived from animals slaughtered contrary to Islamic slaughter requirements.
- Additives used in chicken were often not included on labels. Aldi had to apologize to its UK Muslim consumers after a product containing pig blood and pig skin was mislabelled as halal by its producer (International Business Times 2015). The retailer indicated that it was an isolated incident and that it did not pose any food safety risk to consumers.

The adulteration and mislabelling of halal meat is not restricted to the EU. Chuah et al. (2016) tested 143 processed halal meat (beef and poultry) products in Malaysia for the presence of halal-prohibited proteins from pigs, rats, cats, donkeys, dogs, and other undeclared halal-acceptable species. They found that 78% of the products were mislabelled. In addition, buffalo DNA was detected in 40 out of 58 products labelled as beef whilst 33 out of the 58 products contained undeclared chicken. However, none of the halal prohibited products were found in those tested.

Cross-contamination of halal meat with meat derived from pigs, carnivorous animals, birds of prey, and animals slaughtered contrary to Islamic principles will negate the halal status of the meat, so it will be regarded as non-halal. Pork appears to be the most important product of concern for the majority of Muslims living in the West because most halal meat-producing facilities in these countries also handle and process pork alongside halal meat products. This practice increases the risk of cross contamination between the halal products and pork, particularly where there is poor segregation and cleaning between the two products. The detection of even very low levels of pork or porcine genetic materials in halal products will render the products non-halal (see Chapters 10 and 19).

15.8 Halal Certification

15.8.1 The Aim of Certifying Products

The overall aim of certifying products, processes, or services (e.g. laboratory analyses) is to give confidence to all interested parties that a product, process, or service fulfils specified requirements. The value of certification is the degree of confidence and trust that is established by an impartial and competent demonstration of the fulfilment of specified requirements by a third party (i.e. an auditor) (Albersmeier et al. 2009).

15.8.2 The Halal Certificate

A halal certificate is an official document requiring food manufacturers to have observed and implemented Islamic requirements in producing halal foods in accordance with

Islamic law. A halal certificate is issued by an HCO after conducting an inspection to verify that the required standards have been met. When a halal certificate is issued by an HCO, it will be accredited by an international halal organization (such as MUI and JAKIM). JAKIM is the agency responsible for Islamic affairs, including halal certification, in Malaysia (Golnaz et al. 2012). It therefore plays a very important role in protecting Muslim consumers in Malaysia. JAKIM is responsible for assuring customers that products are halal, as urged by Shariah and Global Compliance Certification Pty Ltd (GCC), which is a third-party certification body accredited by the Joint Accreditation System of Australia and New Zealand (JAS-ANZ), GCC operates in compliance with internationally recognized standards such as ISO 17021, ISO 19011, International Accreditation Forum Mandatory Documents (IAF MD) and JAS-ANZ policies and procedures to ensure the integrity of the halal certification process and its continuing compliance with the requirements of international and national accreditation criteria. This provides trust and acceptability as a halal certification organization, which could lead to requests for services and supervision from halal food producers.

Halal food certification, although viewed by many halal consumers as a religious duty in helping consumers eat authentic halal food, has become big business. According to Salaam Gateway (SalaamGateway.com), the value of the global halal certification industry in 2015 was estimated to be US$12.5 billion. Their report indicated that there are over 400 unregulated global HCBs, and the amount of money spent by halal consumers on food and drinks annually is estimated to be US$1.2 trillion.

15.8.3 Halal Assurance System

The Halal Assurance System (HAS) is a management system intended to maintain the halal status of products which have already obtained halal certificates. It is compulsory for all halal-certified companies and must be applied within six months of the issue of a halal certificate. It seeks to ensure that a halal-certified company is producing halal food during the term of validity of the halal certificate.

The HAS requires a team of internal halal auditors, who are staff employed by each food company responsible for the implementation of the HAS. HAS documentation is needed, such as halal policy and standards, halal management organization, and the scope of the HAS. The HAS must be verified around an internal audit prescribed in the halal certification schedule and must include the corrective action required if errors occur. An audit report must be delivered to the HCB every six months. A HAS is essential if renewal of halal certification is being sought (see Chapter 15).

15.8.4 The Internal Halal Audit Team

The internal halal audit (IHA) team is a group of halal auditors who are staff from the food company that is seeking halal certification. The IHA team is mandatory according to the halal certification contract. This team is appointed by the company's management to coordinate implementation of halal production standards. The IHA team will be trained by halal certification auditors, and the team has direct contact with halal certification auditors.

15.8.5 Benefits of Halal Certification

Halal certification benefits food manufacturers and food premises wishing to increase the demand for and acceptance of their products. Moreover, the production of halal food products makes food companies global food companies (Power and Gatsiounis 2007). Halal certification helps in the organization of the global halal market to provide assurance to halal consumers. Furthermore, it provides a great opportunity to food manufacturers to increase their customers by exporting their products to over 1.6 billion halal consumers worldwide (Pew Research Center 2009, 2015). Halal certification of food products also provides opportunities for halal consumers, particularly Muslims, to increase their trust of halal food, to be sure that the ingredients used, and the process of production, are according to Islamic Shariah law, thus ensuring that they are always able to eat halal food without contravening Islamic law by eating non-halal food or using non halal products.

15.9 International Standards

15.9.1 Definition

International standards are available for consideration and use worldwide. The most prominent standards organization is the International Organization for Standardization (ISO). International standards cover a wide range of food and non-food products, for example:

- agricultural produce
- animal feed
- beverages
- chemicals and flavours
- ingredients and seasoning
- fats and oil products
- machinery and equipment
- meals and livestock
- packaging and containers
- prepared food
- processed food
- pharmaceuticals
- toiletries and cosmetics
- raw materials and ingredients
- rice and noodles
- snacks and confectionary
- nutraceutical, organic, and herbal food.

International halal standards are halal-oriented standards developed by international standards organizations. One example of an international halal standards organization is the Cooperation Council for the Arab States of the Gulf (GCC) Standardization Organization (GSO), which is a Regional Standardization Organization (RSO) that includes the State of the United Arab Emirates, the Kingdom of Bahrain, the Kingdom of Saudi Arabia, the Sultanate of Oman, the State of Qatar, and the State of Kuwait. The GSO started to operate in May 2004 and its headquarters are located in Riyadh, Kingdom of Saudi Arabia (Al-Mutairi et al. 2015).

The mission of the GSO is to assist the states of the GCC to achieve the objectives set forth in its Charter and in the GCC Economic Agreement by unifying the various standardization activities and following up implementation and compliance with the same in cooperation and coordination with the standardization bodies in the Member States of the National Standardization Bodies (NSBs) to develop the production and service sectors, foster the Intra-GCC trade, protect the consumer, environment and the public health, enhance the GCC economy and its competitiveness and meet the requirements of Gulf Custom Union and Gulf Common Market GSO Halal standards (GSO Standard 1998, 2009; GSO 2008):

- Halal food general requirements GSO: Halal food Part 1 – 2008
- FAD 04 = Supplementary accreditation requirements for product certification bodies
  FAD 12 Requirements for HCBs
  (FAD, Filed Application Document)

Malaysian halal standards:

- MS 1500:2009 Halal Food – Production, Preparation, Handling, and Storage General Guidance
- MS 2200:2008 Halal Cosmetic & Personal Care
- MS 1900:2005 Quality Management
- MS 2300:2009 Value-Based Management System
- MS 2400:2010 Logistic – Transportation
- MS 2400:2010 Logistic – Warehousing
- MS 2400:2010 Logistic - Retailing

Thailand halal standards:

- THS 1435-1-2557 Requirements for inspection of manufacturing process of halal products B.E. 2554
- THS 1435-2-2557 Requirements for halal certification on abattoir and animal dissection B.E. 2554
- THS 1435-3-2557 Halal manual
- TH 1435-4-2557 Guideline for halal certificate application on products and packaging
- THS 1435-5-2557 Regarding the operation of halal certification of factory, products, and fees of B.E. 2552

Brunei halal standard:

- Brunei Darussalam Standard For Halal Food

Philippine halal standard:
Halal Food – Production, Preparation, Handling, and Storage – General Guidelines PNS 2067:2007.

15.9.2 Positive Aspects of International Halal Standards

- Techno-economical studies have shown a positive relationship between application of international standards and trade (imports/exports) (Ortiz-Ospina et al. 2018).

- GSO. Main users of standards. Challenges faced by enterprises. An agreement on how governments can apply food safety and animal and plant health measures (sanitary and phytosanitary or SPS measures) sets out the basic rules in the World Trading Organisation covering a wide range of activities.

15.9.3 Negative Aspects of International Halal Standards

All halal standards are similar in their general requirements in promoting halal and avoiding haram.

However, if read carefully they are too loose with halal as prescribed by the Holy Quran and the Sunnah, for example:

- they allow the trade of products that resemble haram products, e.g. alcohol-free beer and lately alcohol-free whiskey
- they allow the use of alcohol (synthetic) for manufacturing certain ingredients (how can one be sure it is synthetic as chemically all alcohols are all the same; ethanol causes drunkenness, and all are toxic)
- they allow the presence or the use of najis (religiously prohibited ingredients) due to controversial religious edicts (fatwas) based on the concept of conformation (istihala).
- due to controversial religious ethics on halal issues, many Islamic countries have been forced to adopt religious ethics using the least constraints.

15.10 Common Mistakes Made by HCBs

With the increasing Muslim population, there has been a global increase in the demand for halal services and halal products (food and non-food). Muslim consumers look for the halal logo on products before purchase for consumption or use. This warrants, in some cases, the halal service or product to be genuinely halal.

However, the most important aspect of halal certification services is to ensure the integrity of the halal supply chain. Some common mistakes practiced by HBCs are:

- lack of a competent halal certification system
- not following halal procedures
- lack of transparency
- lack of Islamic behaviour
- lack of commitment from management
- lack of halal raw materials supply
- lack of halal technical training.

Not all halal-certified products/services or meat comply with halal requirements.

15.10.1 Lack of a Competent Halal Certification System

- Many HCBs do not comply with many governmental approved Halal standards or the food safety standard ISO 22000, e.g. the GSO Halal Standard, and Food Safety Management Standards ISO22000.

- GSO halal standards cover the six main Gulf countries: Saudi Arabia, Kuwait, the United Arab Emirates, Qatar, Bahrain, Oman. Lately Yemen has been accepted as a GSO member.
- Some HCBs do not have their own halal manual standard, e.g. written slaughtering requirements.
- The intention of all HCBs is to be accredited by halal organization like MUI of Indonesia so that their certificates are not rejected.

15.10.2 Not Following Halal Procedures

- Some HCBs do not follow halal procedures.
- Some HCBs may have a procedure manual or at least agree to follow a set of halal processes or standards but in practice often do not follow them, for example complete halal chain inspections and regular halal monitoring are not done.
- Some HCBs submit to non-halal laws.
- Some HCBs work on haram production lines.
- Many HCBs have a lack of commitment to consumers. This is due to ignorance in the professional skills of halal requirements. For this reason, many HCBs have been disqualified on religious grounds from carrying out the duties of halal services by halal experts.
- Many HCBs are non-ethical. Because of economical and animal welfare conflicts with non-Muslim owners of slaughterhouses or plants, in many cases HCBs find themselves forced to accept non-halal laws, without the knowledge of Muslim consumers.

15.10.3 Lack of Transparency

- Many HCBs do not disclose their standards, methods, and organization details.
- Many HCBs do not provide complete and clear enough information to consumers to allow them to make better choices, i.e. they give misleading halal certificates. For example, some HCBs certify stunned birds/animals as halal and advertise these as 'hand slaughtered'. This statement is misleading as it implies that the meat is purely halal with no stunning.
- Some HCBs are part of a manufacturing company that certifies its own services or products as halal.
- Transparency is a double-edged sword. Most HCBs keep the things to themselves lest they be criticized by a competing HCB. Muslim governments can play a role in forcing transparency amongst HCBs.
- The absence of any kind of coordination between HCBs occurs because other HCBs are considered to be enemies.
- Some HCBs that work closely with governmental agencies do not defending halal.
- Most certifying bodies are a one-man show, even when associated with a centre or a Mosque.

15.10.4 Lack of Islamic Behaviour

- Some HCBs give concessions on halal requirements for the sake of gaining acceptance from processing plants/slaughterhouses/laboratories.
- Some HCBs include non-Muslims.

- Many HCBs have no interest in investing part of their income to support halal activities (slaughtering techniques and genetically modified food issues etc.).
- Halal Muslim slaughterers sometimes deliberately do not perform their regular prayers five times, or at the time of slaughter do not utter the tasmiyah.
- Each halal organization is trying to protect its own territory only, which affects the infrastructure quality of the halal service.
- Most HCBs that certify meat from stunned birds/animals as halal do not check if the birds/animals at the time of slaughter were alive.
- HCBs who accept stunning say their slaughter men can detect dead chickens. But in an industrial context, when more than 5000 chickens are mechanically slaughtered every hour, it is impossible for a slaughterere to be alert and watchful all the time.
- Although it is halal, some HCBs accept certifying meats from very weak animals. This is not tayeb (poor quality meat). This is noticed by active halal groups in France.
- Many HCBs approve meat as halal without on-site halal audit/supervision of the meat. Halal audit/supervision is a procedure whereby halal meat/carcasses are checked to ensure they comply with halal requirements.
- Many HCBs approve meat as halal from birds/animals that have been inhumanely handled, e.g. captive bolt was used for stunning.
- Some HCBs certify processed meat that is manufactured by equipment normally used for haram/or non-halal processed meat as halal.
- Some HCBs accept non-approved slaughterers.
- Some HCBs accept mechanical slaughtering.
- Some HCBs accept stunning.
- Some HCBs accept istihala as an excuse to approve halal, e.g. istihala of fat in soap is thought to transformed it. Glycerine in soap can revert back to glycerine. i.e. the chemical reaction is reversible, the najis original nature of raw material still exist, and thus the istihala concept of najis glycerine in soap as proposed by Shariah scholars is not correct.

15.10.5 Lack of Commitment from Management

- To be halal accredited, a halal organization can receive high fees from the HCB and this can affect the credibility of accredited HCBs as they may be forced to bribe the halal accreditor organization to gain a place in the market.
- HCBs think they can trust the organizations they have certified, so they only inspect once or twice a year. Inspections must be done routinely daily or weekly.
- With few exceptions, some organizations have made an effort to reduce the credibility of the HCBs.
- Most HCBs are not internationally recognized.
- Most HCB do not have scientific and Shariah committees. These two committees are necessary to:
  - i) work hand in hand to investigate new technologies, solve emerging issues linked with certain technologies, and understand what actually happens during processing and the true nature of raw materials
  - ii) give guidance on how to change haram production lines to halal lines, i.e. how to apply acceptable cleansing on certain lines.

15.10.6 Lack of Halal Raw Materials Supply

- Some HCBs who certify processed food as halal do not directly obtain halal raw ingredients or halal casing from genuine halal suppliers so cannot guarantee the integrity of their halal supply chain.
- Some HCBs accept haram ingredients.

15.10.7 Lack of Halal Technical Training

- Halal technical operators, supervisors, and managers are not getting thorough training in what they need to know about halal.
- HCB staff lack training not only in Shariah and scientific knowledge but also in newly introduced food safety management systems such as ISO22000.
- Many HCBs do not have enough diverse professionals with the relevant competency and skills in halal.
- The halal industry needs to have halal professionals in every area to serve any targeted halal issue.
- Halal experience leading to a professional halal certificate confers added value.
- Being a scholar is not enough to be a halal auditor in this era of advanced technology processes.
- Scholars of HCB should learn from Muslim scientists and investigate emerging technologies and methods being used in the halal industry so that they can deliver the correct religious edict (fatwa).

15.10.8 Not All Halal-certified Products/Services or Meat Comply with Halal Requirements

Not all halal-certified products/services or meat are complied with halal requirements for the following reasons:

- halal laboratory analyses must be performed by Muslims
- some HCBs sign and seal a blank halal certificate
- many HCBs have no interest in forming a competent, well-trained halal team
- many HCBs do not certify their personnel as part the halal accreditation
- halal certification with integrity is very important
- HCBs do not inspect working premises.

15.11 Conclusion

The consumption of halal food is an important aspect of the Islamic faith. However, the desire of some halal food suppliers to maximize profits has led to a series of incidents involving the mislabelling of products that do not meet halal dietary laws as halal. Incidents of this nature can cause significant distress to practicing Muslims.

HCBs differ in their operational definitions of halal and their individual halal certification procedures. In the UK the development of these certification organizations is a response to the lack of relevant UK law, and the existing structural factors are all competing to apply various halal standards in their quest to assure consumers about the authenticity of their certified halal meat. There have been accusations and counter-accusations amongst the certifiers as to whose halal standard is genuine. The lack of government involvement or a central monitoring body of HCBs has meant that any individual can set up an HCB with little or no technical expertise regarding halal food, potentially a lack of specific religious knowledge, and/or non-mainstream understanding of what makes food halal.

In general, international halal standards claim to protect Muslim consumers, but if they are read carefully it can be seen that these standards mainly aim to protect the interests of the industries producing food and non-food items for Muslim consumers under the umbrella of halal. Being accredited as a halal service provider by an internationally approved halal accreditation body will open doors for halal activities. To go beyond the expectation of Muslim consumers organizations should follow strict halal regulations.

The lack of resources and expertise in Islamic dietary law and authentic halal requirements amongst some enforcement authorities also makes it difficult for law enforcement officers to identify and prosecute halal food fraud offenders. Nonetheless, in recent years there have been a number of successful prosecutions of halal meat suppliers who have intentionally sold non-halal meat as halal. The food safety aspects of halal meat have also been questioned as a significant proportion originates from smaller, potentially less well managed and regulated processors.

15.12   Recommendations

- HCBs should be accredited based on an internationally approved halal standard coupled with periodical audit by an approved internationally accredited Muslim body like the Gulf Accreditation Center GSO.
- HCBs should collaborate with other approved HCBs to serve halal consumers rather than serving individual interests.
- Mutual recognition of halal accredited HCBs will boost halal worldwide.
- There should be an international trustworthy Muslim organization to standardize halal standards, halal services, and licencing of HBCs globally.
- This international trustworthy Muslim organization could be the GSO or the Islamic International Fiqh Academy.
- Further research is needed to assist in the rational development and application of halal certification procedures in the UK.
- Further research into halal fraud in the UK should be conducted to identify its extent and scale.
- This study shows that regulatory procedures for halal food certification in the UK are necessary for halal consumer protection from misleading certification, with input from the FSA, and cooperation from halal certification organizations, which may re-organize and regulate halal certification gradually in the following way:
  - All halal certification organizations in the UK should be required to register with the FSA, an organization dealing with all other aspects of the food industry.
  - Each halal certification organization should register and document its halal requirements, standards, certificates, and labelling in order to retain its halal certification and organizational rights.

– All halal certification organizations should be required to place all relevant documented information, such as names, addresses, locations, specific requirements, halal standards, halal certificates, and halal labelling, on an official website that is easily accessible to food companies and food consumers alike, ensuring freedom of choice of halal certification organizations.

References

Ahmad, A.N., Rahman, A.A., and Rahman, S.A. (2015). Assessing knowledge and religiosity on consumer behaviour towards Halal food and cosmetic products. *International Journal of Social Science and Humanity* 5 (2): 10–14.

Ahmed, A. (2008). Marketing of halal meat in the United Kingdom. *British Food Journal* 110: 655–670.

Albersmeier, F., Schulze, H., Jahn, G., and Spiller, A. (2009). The reliability of third party certification in the food chain: from checklists to risk-oriented auditing. *Food Control* 20 (10): 927–935.

Al-Mutairi, S., Connerton, I., and Dingwall, R. (2015). Food safety organisations in Saudi Arabia – organisational, historical and future analysis. *Food Control* 47: 478–486.

Ambali, A.R. and Bakar, A.N. (2014). People's awareness on halal foods and products: potential issues for policy-makers. *Procedia – Social and Behavioral Sciences* 121: 3–25.15.

Anil, M.H., Yesildere, T., Aksu, H. et al. (2006). Comparison of halal slaughter with captive bolt stunning and neck cutting in cattle: exsanguination and quality parameters. *Animal Welfare* 15: 325–330.

BBC (2014). Birmingham Butcher to pay £62,000 for false halal sale. http://www.bbc.co.uk/news/uk-englandbirmingham-29157457 (accessed May 2018).

Chuah, L.O., He, X.B., Effarizah, M.E. et al. (2016). Mislabelling of beef and poultry products sold in Malaysia. *Food Control* 62: 157–164.

Codex (1997). General Guidelines for Use of the Term "Halal". CAC/GL 24-1997. http://www.fao.org/fao-who-codexalimentarius/sh-proxy/en/?lnk=1&url=https%253A%252F%252Fworkspace.fao.org%252Fsites%252Fcodex%252FStandards%252FCAC%2BGL%2B24-1997%252FCXG_024e.pdf (accessed March 2019).

Council Regulation (EC) No 1099/2009 (2009). Council Regulation on the protection of animals at the time of killing. http://eur-lex.europa.eu/LexUriServ/LexUriServ.do?uri=OJ:L:2009:303:0001:0030:EN:PDF (accessed June 2018).

Cutler, T.R. (2007). Food safety drives growth in kosher and halal foods. http://www.halalrc.org/images/Research%20Material/Literature/Food%20Safety%20Drives%20Growth%20in%20Kosher%20%26%20Halal%20Foods.pdf (accessed June 2018).

Demirhan, Y., Ulca, P., and Senyuva, H.Z. (2012). Detection of porcine DNA in gelatine and gelatine-containing processed food products – halal/kosher authentication. *Meat Science* 90: 686–689.

EBLEX (2010). Report on the Halal Meat Market: Specialist Supply Chain Structures and Consumer Purchase and Consumption Profiles in England by the English Beef & Lamb Executive. http://www.qsmbeefandlamb.co.uk/halal (accessed June 2018).

European Council Regulation EC853/2004 (2004). Laying down specific hygiene rules for on the hygiene of foodstuffs. http://eur-lex.europa.eu/LexUriServ/LexUriServ.do?uri=OJ:L:200 4:139:0055:0205:en:PDF (accessed June 2018).

Farouk, M.M., Al-Mazeedi, H.M., Sabow, A.B. et al. (2014). Halal and kosher slaughter methods and meat quality: a review. *Meat Science* 98: 505–519.

Fischer, J. (2009). The other side of the logo: the global halal market in London. In: *The New Cultures of Food: Marketing Opportunities from Ethnic, Religious and Cultural Diversity* (eds. A. Lindgreen and M.K. Hingley), 73–88. Gower Publishing.

Food Standards Agency UK (FSA) (2001). Water added to restaurant and take-away chicken, survey finds. http://tna.europarchive.org/20110116113217/www.food.gov.uk/news/ newsarchive/2001/dec/chickwater?view=printerfriendly (accessed June 2018).

Food Standards Agency UK (FSA) (2003). FSA water in chicken update. http://tna. europarchive.org/20110116113217/www.food.gov.uk/news/newsarchive/2003/may/waterch ickenupdate?view=printerfriendly (accessed June 2018).

Food Standards Agency UK (FSA) (2009). The food safety (Northern Ireland) order 1991. A guide for food businesses. https://www.food.gov.uk/sites/default/files/media/ document/nifood%20safety%20order%201991%20business%20guide.pdf (accessed April 2019).

Food Standards Agency UK (FSA) (2013). Horse DNA detected in canned beef from Romania. https://webarchive.nationalarchives.gov.uk/20150806005230/http://www.food.gov.uk/ news-updates/news/2013/5858/canned-beef (accessed March 2019).

FSA and FSS (2016). *Food Crime: Annual Strategic Assessment – A 2016 Baseline*. London: Food Standards Agency and Food Standards Scotland. http://www.food.gov.uk/sites/default/files/ fsa-food-crime-assessment-2016.pdf (accessed June 2018).

Golnaz, R., Zainalabidin, M., and Mad Nasir, S. (2012). Assessment of consumers' confidence on halal labelled manufactured food in Malaysia. *Pertanika Journal of Social Sciences & Humanities* 20 (1): 33–42.

GSO (2008). Halal Food, Part (2): The Requirements for Accreditation of Issuing the Halal Food Certification Bodies. http://www.classe-export.info/assistance/ANIA_ALLIANCE7/ TRANSV-ARABIE/GSO2055-Hallal-Products-P2.pdf (accessed March 2019).

GSO Standard (1998). Standard No. 993/1998 from GSO enacted on 15 September 1998, for Animal Slaughtering Requirements according to Islamic Law. https://www.halalcertifiering. se/newwebsiteimages/Gulf_standard.pdf (accessed 16 September 2019).

GSO Standard (2009) Standard No. 2055-1/2009 from GSO act on 24 May 2009, Halal Food Part (1): General Requirement. (available in Arabic). http://www.gso.org.sa/standards. (accessed March 2012).

GWvD (1992). *Gezondheids-en welzijnswet voor dieren*. The Hague, The Netherlands: Ministerie van landbouw, Natuurbeheer en Visserij.

Halal Food Authority (2016). Origin of Halal Food Authority. https://ucanr.edu/sites/ placernevadasmallfarms/files/103471.pdf (accessed March 2019).

Harvey, R. (2010). Certification of halal meat in the UK. Centre of Islamic Studies, University of Cambridge. https://www.academia.edu/1029422/Certification_of_Halal_Meat_in_the_ UK_Report_published_by_Cambridge_Centre_of_Islamic_Studies (accessed March 2019).

International Business Times (2015). American online news publication. *International Business Times*. www.ibtimes.com (accessed May 2015).

JAKIM (2010). Halal Malaysia. *JAKIM*. http://www.halal.gov.my/v2.

Khalid, R., Knowles, T.G., and Wotton, S.B. (2015). A comparison of blood loss during the Halal slaughter of lambs following Traditional Religious Slaughter without stunning, Electric Head-Only Stunning and Post-Cut Electric Head-Only Stunning. *Meat Science* 110: 15–23. https://doi.org/10.1016/j.meatsci.2015.06.008. (accessed April 2012)

Lever, J. and Miele, M. (2012). The growth of halal meat markets in Europe: an exploration of the supply side theory of religion. *Journal of Rural Studies* 28 (4): 528–537.

Manning, L. and Soon, J.M. (2014). Developing systems to control food adulteration. *Food Policy* 49: 23–32.

Manning, L., Smith, R., and Soon, J.M. (2016). Developing an organizational typography of criminals in the meat supply chain. *Food Policy* 59: 44–54.

Murugaiah, C., Noor, Z.M., Mastakim, M. et al. (2009). Meat species identification and halal authentication analysis using mitochondria DNA. *Meat Science* 83: 57–61.

Newsweek (2015). Animal cruelty at halal slaughterhouse highlights widespread abuse in the UK. *Newsweek*. http://europe.newsweek.com/animal-cruelty-halal-slaughterhouse-highlights-widespread-abuse-uk-304152 (accessed May 2018).

Organisation of Islamic Cooperation (OIC) (2015). Forum on unifying halal standards and procedures kicks off at OIC General Secretariat, Jedda. http://iinanews.org/page/public/news_details.aspx?id=137875#.XIj0N9IzYdU (accessed March 2019).

Orford, F., Ford, E.A., Brown, S.N. et al. (2016). The evaluation of two commercial electric sheep stunning systems: current applied and the effect on heart function. *Animal Welfare* 25 (3): 331–337.

Ortiz-Ospina, E., Diana Beltekian, D., and Roser, M. (2018). Trade and Globalization. Published online at OurWorldInData.org. https://ourworldindata.org/trade-and-globalization. https://ourworldindata.org/trade-and-globalization (accessed september 2019).

PETA (2009). The cruelty behind Muslim ritual slaughter. http://www.peta.org/blog/cruelty-behind-muslim-ritualslaughter (accessed May 2018).

Pew Research Center (2009). *Mapping the global Muslim Population: A Report on the Size and Distribution of the World's Muslim Population*. The Pew Forum on Religion & Public Life.

Pew Research Center (2015). The Future of World Religions: Population Growth Projections 2010–2050. Washington, DC: Pew Research Center, http://www.pewforum.org/2015/04/02/religious-projections-2010-2050 (accessed May 2018).

Pointing, J. and Teinaz, Y. (2004). Halal meat and food crime in the UK. Proceedings of the International Halal Food Seminar, Malaysia, Islamic University College of Malaysia (September 2004).

Pointing, J., Teinaz, Y., and Shafi, S. (2008). Illegal labelling and sales of halal meat and food products. *Journal of Criminal Law* 72: 206–213.

Power, C. and Gatsiounis, I. (2007). Meeting the Halal Test. *Forbes* 179 (8), dd. 16C04C2007.

Ratanamaneichat, C. and Rakkarn, S. (2013). Quality Assurance Development of Halal Food Products for Export to Indonesia. *Procedia - Social and Behavioral Sciences* 88 (10): 134–141.

Royal Society for the Protection and Cruelty to Animals (RSPCA) (2015). Slaughter without pre-stunning (for religious purposes): RSPCA Farm Animal Department Information Sheet. https://www.rspca.org.uk/adviceandwelfare/farm/slaughter/religiousslaughter (accessed March 2019).

The Guardian (2013a). Horsemeat scandal: timeline. *The Guardian*. https://www.theguardian.com/uk/2013/may/10/horsemeat-scandal-timeline-investigation (accessed March 2019).

The Guardian (2013b). 'Halal pork' supplier named. *The Guardian*. https://www.theguardian.com/world/2013/feb/03/supplier-halal-meat-pork-dna-named (accessed March 2019).

Wotton, S.B., Zhang, X., Mckinstry, J. et al. (2014). The effect of the required current/frequency combinations (EC 1099/2009) on the incidence of cardiac arrest in broilers stunned and slaughtered for the halal market. *PeerJ PrePrints* 2: e255v1. https://doi.org/10.7287/peerj.preprints.255v1. (accessed June 2018).

Part V

Food Law, Regulations, and Food Fraud

16

Legal Aspects of Halal Slaughter and Certification in the European Union and its Member States

Rossella Bottoni

Faculty of Political and Social Sciences, Università Cattolica del Sacro Cuore, Milan, Italy

16.1   Introduction

There is no legal definition of halal in the European Union (EU) law, nor are there any EU legal rules which have been envisaged, adopted, and applied specifically and exclusively to halal. The EU may act only within the limits of the powers conferred upon it by the Treaties, and 'the regulation of religion is primarily a matter for Member States at national level' (Doe 2011, p. 241).

When looking at the national legislation of the Member States, despite the differences existing amongst them, a European pattern of state–religion relations emerges, which is based on common features such as the state's recognition of the autonomy of religious organizations as regards their doctrine and internal structure (Ferrari 2010). Thus, it is left to Muslim communities to define what is halal ('lawful' or 'permitted') and what is not, and secular authorities may not interfere.

This does not mean that there are no legally binding rules applying also to halal in the EU and its Member States. Secular authorities have no direct explicit competence in regulating halal, but in fact they do so when related aspects concern areas that fall within their competence. This is the case for halal slaughter and certification, which will be specifically dealt with in this chapter. Halal slaughter as such is not regulated, but particular methods of slaughter prescribed by religious rites (including Islamic ones) are covered by the Acts adopted in order to define and respect the European standards on the protection of animals at the time of killing. Likewise, halal certification remains within the responsibility of Muslim organizations, but the issue of labelling (including halal products) has been hotly debated in the broader context of transparency and consumer protection.

The Halal Food Handbook, First Edition. Edited by Yunes Ramadan Al-Teinaz, Stuart Spear, and Ibrahim H. A. Abd El-Rahim.
© 2020 John Wiley & Sons Ltd. Published 2020 by John Wiley & Sons Ltd.

16.2 Legal Aspects of Halal Slaughter in the EU and its Member States

16.2.1 The EU

EU legal rules apply not so much to halal slaughter, as to any slaughter carried out according to a religious rite. Although in practice the only relevant rites are the Islamic and Jewish ones, any religious community having internal rules that regulate the ritual killing of an animal for the production of food for human consumption (and not, for instance, as an offering to a deity) would be covered by the same legal provisions.

Furthermore, EU legislation does not regulate religious slaughter as such, but it only takes into account specific aspects (stunning and inversion) that conflict with the general rules applying to conventional (non-religious) slaughter. Whereas conventional slaughter is preceded by stunning, religious slaughter tends to be characterized by the prohibition of previous stunning. In fact, religious slaughter is often defined as the method of slaughter where the animal is slaughtered without being previously stunned, although this definition is improper. First, the carrying out of religious slaughter, in whatever religious community, requires a greater number and more complex conditions than the mere prohibition of previous stunning. Second, as regards specifically Islam, there are Muslim communities that accept previous stunning, provided the animal is rendered unconscious but it is not killed by stunning and death occurs through the act of slaughter (Anil et al. 2010, p. 17). Because the latter method of halal slaughter complies with EU legal rules, which envisage compulsory previous stunning, here only religious slaughter without previous stunning is dealt with. The other debated aspect is the position of the animal at the time of killing, i.e. standing in conventional slaughter and inverted on its back with a 180° turn in religious slaughter.

Legally binding rules on religious slaughter are currently contained in Council Regulation 1099/2009 of 24 September 2009. When this regulation came into force in 2013, it repealed and replaced Council Directive 93/119/EC of 22 December 1993 (see Bottoni 2015).

The Directive lacked a definition of religious slaughter, and merely acknowledged the need 'to take account of the particular requirements of certain religious rites' as regards slaughter without previous stunning and the role of the religious authority in the carrying out of religious slaughter. Each EU Member State might authorize derogations from the provisions prescribing compulsory previous stunning 'in the case of animals subject to particular methods of slaughter required by certain religious rites' (Art. 5 § 2). The religious authority was regarded as competent 'for the application and monitoring of the special provisions which apply to slaughter according to certain religious rites', albeit under the responsibility of the official veterinarian (Art. 2 § 2).

These legal provisions aimed to guarantee the right to religious freedom in a twofold perspective: the right to manifest one's religion in practice and observance, and the respect for religious organizations' internal autonomy through the recognition of the religious authority's competence to apply and monitor the respective religious rules.

There was nonetheless a second interest promoted by EU legislation, i.e. the protection of animal welfare. This is why religious slaughter must be carried out under the responsibility of the official veterinarian, and why, according to Art. 1 § 2 of Annex B, bovine animals must be restrained before religious slaughter using a mechanical method.

The reform of the Directive, which had never been amended since its entry into force, was justified by the need to both adapt EU standards to technological developments and address EU citizens' increasing concerns about animal welfare (European Commission 2008a, pp. 5–6 and 11–15). The legal compromise on religious slaughter was not a determining factor and, thus, in its proposal of 18 September 2008 for the Council Regulation, the European Commission specified that the related legal provisions would not be subject to specific changes, in compliance with the principle of subsidiarity and the respect of the Members States' rules and customs concerning religious rites, cultural traditions, and regional heritage (European Commission 2008b, pp. 8 and 14). Art. 4 § 2 of the Commission's proposal confirmed the possibility for the Member States to grant a derogation from the compulsory requirement of prior stunning, provided that religious slaughter would take place in a slaughterhouse. It also specified that Member States may decide not to apply this derogation.

Although religious slaughter was not taken into account amongst the factors justifying the revision of the rules contained in the Directive, this issue was often mentioned in the debates developed during the procedure of approval of the Council Regulation.

On 18 December Rapporteur Wojciechowski submitted a draft report on the Commission's proposal to the Committee on Agriculture and Rural Development (Wojciechowski 2008), which approved it on 16 March 2009 after introducing some changes (Wojciechowski 2009). In its original version, Recital no. 37 stipulated that slaughter without previous stunning required 'an accurate cut of the throat to minimise suffering'. The Committee amended this clause to include the obligation to perform post-cut stunning (Wojciechowski 2009, amendment no. 11). In other words, animals used for religious slaughter had always to be stunned *after* the cut, where they had not been rendered unconscious previously. This was one of the amendments most conflicting with the principle of subsidiarity and the respect of the traditions and customs of the EU Member States, most of whom do not envisage any form of stunning in the case of religious slaughter.

This amendment was linked to that of Article 4 § 2, where the clause permitting Member States not to apply the derogation from the compulsory requirement of stunning was dropped (Wojciechowski 2009, amendment no. 27). This change, equally controversial, was meant to harmonize the various permissible means of slaughter (Wojciechowski 2009, amendment no. 28), but it was also liable to breach the principle of subsidiarity by forcing Member States prohibiting religious slaughter without previous stunning to prescribe instead post-cut stunning.

Finally, the clause contained in the Commission's proposal that prohibited inversion because it caused unnecessary stress and pain to the animals used for religious slaughter was dropped as inconsistent with the other Regulation provisions, which authorized the methods of slaughter according to a religious rite (including those prescribing inversion) (Wojciechowski 2009, amendment no. 81).

In the meantime, on 25 February the European Economic and Social Committee issued an opinion in which it considered the derogation from the compulsory requirement of previous stunning in the case of religious slaughter – envisaged by the original version of the Commission's proposal – as:

> inconsistent with the overall objective of the regulation which is to improve the protection of animals at slaughter. Innovative technology such as the Stun Assurance Monitor allows those who wish to slaughter with prior electrical stunning in

compliance with Halal rules to accurately monitor how much electrical charge is given to an animal. This ensures that it is properly stunned but still alive prior to slaughter. The monitor records each stun carried out and the voltage given to the animal. It has a real contribution to make to animal welfare (European Economic and Social Committee 2009, para. 4.4).

Interestingly, the Committee seemed to understand religious slaughter exclusively as halal slaughter, and to neglect the fact that Jewish communities also perform religious slaughter without previous stunning.

The same approach emerged in the debate in the European Parliament during the explanation of votes of 6 May 2009. According to MEP Carl Lang:

> [b]y affirming that animals must be slaughtered without unnecessary suffering, except in the case of religious rites, the majority of our House has demonstrated both its hypocrisy and its cowardice. 'Religious rites' mainly refer to the ritual slaughter practiced in particular during the Muslim festival of Eid-al-Adha, when hundreds of thousands of sheep have their throats cut.
>
> Legal recognition of such a practice is part of a much wider phenomenon, that of the Islamisation of our societies. Our laws and customs are changing progressively to accommodate Islamic Sharia law. In France, more and more local authorities are indirectly funding the construction of mosques. School menus are drawn up to meet Islamic dietary requirements. In some cities, such as Lille, the swimming pools have women-only sessions. By creating the Conseil Français du Culte Musulman in 2003, Mr Sarkozy, then Minister of the Interior, introduced Islam into France's institutions.
>
> To put an end to these developments, we have to reject the Islamically correct, reverse the flow of non-European migration and create a new Europe, a Europe of sovereign nations, without Turkey, affirming the Christian and humanist values of its civilization (see www.europarl.europa.eu at http://tinyurl.com/gotxcwa).

In this intervention – a remarkable example of Islamophobic discourse in the political arena – no mention was made of Jews, who also perform religious slaughter without previous stunning,[1] and little of animal welfare. Furthermore, the almost obsessive concern about the issue of stunning leads to neglect that improvement in standards of animal welfare needs to be pursued through the 'humanization' of all stages of animals' life – from transport to handling, from lairage to restraint. As stated by MEP Friedrich-Wilhelm Graefe zu Baringdorf:

> in this debate I think the issue of animal welfare has been narrowed down too much to the question of whether or not we should use stunning. It is not that I am against stunning, it is just that we must realise that stunning was originally introduced, not out of concern for animal welfare, but for economic reasons, to be able

---

1 It should not be neglected that anti-Semitism, too, has been one of the grounds to ban schechita. See Metcalf (1989) and Efron (2007).

to slaughter animals in factory farming, which involves mass killing in slaughterhouses, without affecting and lessening the quality of the meat through the anxiety caused by slaughter. Thus with religious slaughter it is about the high art of slaughtering without the animals experiencing additional suffering, and with the issue of stunning in slaughterhouses, it is not only about whether or not stunning should be used, but also about the handling of animals during transportation and during the waiting time in the slaughterhouses (see www.europarl.europa.eu at http://tinyurl.com/o2a87xe).

The European Commission accepted only a few of the changes proposed by the Committee on Agriculture and Rural Development and approved by the European Parliament. The others were dropped because the purpose of the reform was not the amendment of the legal rules concerning religious slaughter. Furthermore, a number of amendments breached the principle of subsidiarity and limited the Member States' right to respect their own national traditions concerning their relationships with religious communities and their regulation of the right to religious freedom. According to Recital no. 18 of the final text of the Regulation:

> Derogation from stunning in case of religious slaughter taking place in slaughterhouses was granted by Directive 93/119/EC. Since Community provisions applicable to religious slaughter have been transposed differently depending on national contexts and considering that national rules take into account dimensions that go beyond the purpose of this Regulation, it is important that derogation from stunning animals prior slaughter should be maintained, leaving, however, a certain level of subsidiarity to each Member State. As a consequence the present Regulation respects the freedom of religion and the right to manifest religion or belief in worship, teaching, practice and observance, as enshrined in Article 10 of the Charter of Fundamental Rights of the European Union.

Art. 4 § 4 of the final text further confirms the possibility for Member States to allow a derogation from the requirement of previous stunning in the case of religious slaughter. At the same time, Art. 26 § 2 c guarantees their right to 'adopt national rules aimed at ensuring more extensive protection of animals at the time of killing than those contained in this Regulation' in relation to, inter alia, religious slaughter.

Although the recognition of the possibility of performing religious slaughter without previous stunning is not a departure from the previous EU legal rules, there are indeed also novelties which should be highlighted. Unlike the 1993 Directive, the 2009 Regulation specifies that religious slaughter may only be carried out in a slaughterhouse (Art. 4 § 4) and it explicitly mentions methods of reversible or simple stunning, which are admitted provided that they are followed as quickly as possible by a procedure ensuring death (Art. 4 § 1 and Annex I). Further, it prohibits restraint systems by inversion as regards bovine animals, except in case of religious slaughter, provided that they can be adapted to their size and are fitted with a device restricting their head's lateral and vertical movements (Art. 15 § 2). Art. 27 § 2 commits the Commission to submit a report to the European Parliament and to the Council on restraint systems by inversion for bovine. It must be 'based on the results of a scientific study comparing these systems to the ones maintaining bovines in the upright position and shall take into account animal welfare aspects as well as the socioeconomic implications, including their acceptability by the religious communities and the safety of operators' and, if appropriate, it must be 'accompanied by legislative proposals with a view to amending this Regulation concerning the systems restraining bovine animals by inversion or any unnatural position'.

16.2.2 The EU Member States

The EU legal provisions on religious slaughter have allowed the adoption of a variety of normative solutions in the EU Member States, depending on the different outcome in each concerned country of the balance between the protection of the right to religious freedom, which includes the right to perform religious slaughter (Lerner and Rabello 2006/2007; Rovinski 2014), and the promotion of animal welfare (see Anil 2012).

The majority of EU Member States allow a derogation from the compulsory requirement of previous stunning. A small group require post-cut stunning (this is the case in Austria, Slovakia, and Estonia) and few other countries (e.g. Sweden) envisage only religious slaughter with previous stunning.[2] It should be noted that in the past few years a number of European countries have reformed their legal position on religious slaughter. Lithuania[3] and Poland[4] have allowed religious slaughter without previous stunning, which was previously prohibited; whereas Denmark (Delahunty 2015) and the Belgian regions of Wallonia and Flanders (Bottoni 2017; van der Schyff 2017) have prohibited this practice. Bills to prohibit religious slaughter without previous stunning have also been presented in the Netherlands and Finland (Zoethout 2013; Foblets and Velaers 2014).

A number of conditions must be respected in order to allow the carrying out of religious slaughter without previous stunning (or with post-cut stunning). Three of them are prescribed by the EU: religious slaughter may only be carried out (i) in a slaughterhouse, (ii) under the responsibility of the official veterinarian, and (iii) provided that bovine animals are mechanically restrained before slaughter. Further requirements, which have not been prescribed by the EU, have been introduced by a number of EU Member States. The two most important ones concern the notification or permission request to be submitted by either the concerned slaughterhouse or the religious community, and the slaughter men's licence. A few examples of their application in selected countries may help to illustrate how legal rules also applying to halal slaughter are implemented in practice.

---

2 The source of this paragraph, unless otherwise stated, is Ferrari and Bottoni (2010). For reasons of space, the normative references fully mentioned in the Dialrel Report will not be repeated here.

3 Law no. XII-1147 of 23 September 2014, entered into force on 1 January 2015. For the text in original language, see www.e-tar.lt at https://tinyurl.com/yb92qgnc. Last accessed on 20 September 2019.

4 By virtue of the Constitutional Court's judgement of 10 December 2014. For a summary in English, see www.trybunal.gov.plat https://tinyurl.com/y8gdzcq6. Last accessed on 20 September 2019. See also Szumigalska and Bazan (2014) and Gliszczyńska-Grabias and Sadurski (2015).

As mentioned, religious slaughter (just like conventional slaughter) may only be carried out in a slaughterhouse, but breaches of this rule have been reported, especially during the celebration of the Aid al-Kabir religious festival, when slaughterhouses cannot meet all the requests from local Muslim communities (for operational reasons or because of restrictions on the maximum number of animals that can be killed or for other reasons). Countries like France tried to accommodate such demands, with municipal authorities allowing exemption sites, that is, temporary structures outside slaughterhouses. At the end of the 1990s, France was warned by the EU Veterinary Office that it would be heavily fined if it continued doing this, and since then this practice has been stopped (see www.ec.europa.eu at https://tinyurl.com/y2fdpqaa).

As regards the permission request to be submitted in order to carry out religious slaughter, the German legal system provides for both a 'standard' and an 'exceptional' derogation. The former, which is easier to obtain, allows a religious community to modify stunning parameters in order to perform reversible stunning before religious slaughter and is typically granted to Muslim communities. The latter allows derogation from any form of previous stunning and can only be applied for by religious communities (*Religionsgemeinschaften*), whose rules require slaughter without stunning or prohibit consumption of the meat of animals slaughtered in another way. Because the existence of a religious commandment, which all believers must respect, is required, only Jewish communities have always been granted an 'exceptional' derogation. By contrast, Muslim communities' requests to perform religious slaughter without previous stunning have often been rejected (see also Lavi 2009). In 1995, examining an application lodged by a Muslim community who had not been granted the 'exceptional' derogation, the Federal High Administrative Court maintained that the applicant could not be regarded as a *Religionsgemeinschaft* and, further, there was no Islamic dogma forbidding the consumption of meat from animals stunned before slaughter. In 2002, the Federal Constitutional Court found unanimously that the applicant's religious freedom had been unreasonably restricted. It held first that a *Religionsgemeinschaft* is not a religious community hierarchically structured like Christian Churches, but it may be any 'group of persons with common beliefs organized in any form whatsoever'. Second, where there exist different conflicting opinions within Islam, state authorities are not competent to determine which one should prevail: an enquiry into Islamic doctrine breaches the religious communities' right to internal autonomy, as well as state neutrality (Rohe 2004, p. 329). In the same year the Parliament approved an amendment to the Constitution including animal welfare as a national objective, in order to give it constitutional protection and greater weight when balanced against religious freedom (Haupt 2007). Subsequent case law has remained divergent, with some courts granting 'exceptional' derogations and others not.

France offers an interesting example concerning the application of the additional requirement concerning slaughterers' licences. Religious slaughter may only be performed by slaughterers who are certified by religious bodies approved by the Ministry of Agriculture, upon proposal by the Minister of the Interior. In other words, French authorities have the discretional power to recognize the right to perform religious slaughter to some communities and to deny it to others. The exercise of this power has been challenged by an ultra-orthodox Jewish association, which had not been approved as such a religious body and which applied to the European Court of Human Rights (ECtHR) alleging a violation of Article 9 of the European Convention of Human Rights. The judges preliminarily maintained that:

> ritual slaughter must be considered to be covered by a right guaranteed by the Convention, namely the right to manifest one's religion in observance, within the meaning of Article 9 (*Cha'are Shalom ve Tsedek v. France*, application no. 27417/95, judgement of 27 June 2000, para. 74)

and that:

> by establishing an exception to the principle that animals must be stunned before slaughter, French law gave practical effect to a positive undertaking on the State's part intended to ensure effective respect for freedom of religion (para. 76)

At the same time, the ECtHR held that France was entitled to use its discretionary power to approve or not a religious body within its margin of appreciation and, by a majority of 12 votes to 5, it concluded that there had been no violation of Article 9. The right to perform religious slaughter was superseded by the right to the respect of religious dietary requirements. The applicants' right to religious freedom would have been violated only if there had been no access at all to meat from animals slaughtered according to their religious rite, but *glatt* meat could easily be obtained through import from Belgium (for a broader discussion of the judgment, see Cohen (2009) and Haupt (2007)).

Although this case is unrelated to halal slaughter, the principles established by the ECtHR may also apply to any contracting state and to any European Muslim community. It is irrelevant whether the latter comply with their religious rules by eating food produced internally or imported. As long as their need for halal meat is satisfied, no violation may be alleged. It may thus be concluded that European countries prohibiting halal slaughter without previous stunning do not breach the Convention, as long as they do not also prohibit the import of halal meat.

It should finally be noted that the legal recognition by an EU Member State of the derogation from the compulsory requirement of previous stunning should not lead to the conclusion that religious communities, in that state's territory, can actually and freely perform religious slaughter. The reality is indeed more complex. Luxembourg, despite the formal recognition of the derogation, applies a gentlemen's agreement with animal welfare associations, which commits the competent authorities not to grant the permission. The few requests that have so far been submitted have been rejected, and the concerned religious community has been encouraged to use the slaughterhouses located beyond the borders. In Slovakia the law stipulates that only religious communities having a specific legal status (as a registered religious Church or religious society) may request permission to perform religious slaughter without previous stunning. The Union of the Jewish Communities, which is a registered religious society, can apply for (and obtain) permission, whereas the Muslims, who have been so far denied the registration (Drobný 2014, p. 542), may only carry out religious slaughter with previous stunning.

16.3 Legal Aspects of Halal Certification in the EU and its Member States

According to the ECtHR, 'observing dietary rules can be considered a direct expression of beliefs in practice in the sense of Article 9' (*Jakóbsky v. Poland*, 18429/06, judgement of 7 December 2010, para. 45). Halal certification guarantees that Muslims may effectively exercise their right to religious freedom by eating meat produced consistently with their religious norms. However, the labelling of halal products concerns nowadays a realm going well beyond the respect of a human right, by also strongly affecting the economic dimension.

> Over the last 15 years dedicated markets for halal meat have emerged in a number of European countries. [....] Market expansion has also facilitated the rise of new certification bodies, each with their own marketing strategies and interpretations of what constitutes authentic 'halal', who question the reliability of certification policies that allow the practice of stunning before slaughter (Lever and Miele 2012, p. 528).

Further, products may be promoted as halal, although they are not certified as such by an Islamic authority. According to Dr. Yunes Teinaz, adviser to the London Central Mosque:

> the lack of a State body in Britain that is capable of inspecting the 'totally unregulated' halal market has left this market open to fraud, corruption, and without any kind of standards, uniform certification, or legislation (quoted by Fischer 2015, p. 355).

A study focused on the Netherlands also concludes that in this country 'public law does not protect consumers from misrepresentation or fraud involving food sold as kosher or halal' (Havinga 2010, p. 252). The British and Dutch cases exemplify a situation existing also in other EU Member States.[5] Unlike the USA, where several states have state-centred regulatory arrangements, and where law-makers, governmental enforcing agencies, and courts play an important role in labelling, EU countries tend to leave the issue of halal certification to commercial and religious organizations, that is, private actors, and do not regulate this issue by public law (Havinga 2010, pp. 241–244 and 250[6]). When EU Member States approve Muslim certification bodies, they do not do so to affirm state competence

---

5 'In most European countries several private and independent certification organizations are active' (van der Spiegel 2012).

6 This study compares the USA and the Netherlands, but its conclusions may extend to other EU Member States. As highlighted by the author, this reality is not consistent with the expectation one may have: 'In the United States, which is a liberal market economy, one would expect to find minimal governmental interference in the kosher and halal industry. In the Netherlands, which is a corporatist welfare state, one would expect to find a high level of state involvement in the regulation of the kosher and halal industry' (van der Spiegel 2012, p. 251).

in defining what is authentically halal,[7] as this decision ultimately belongs to Muslim religious authorities. This is an important difference with the situation existing in US states, such as New Jersey, where statutes have been enacted giving a legal definition of halal in order to 'prevent the fraudulent representation of food as being halal'. However, 'the constitutionality of these government-enacted definitions of halal is uncertain' (Havinga 2010, p. 249). In case of fraud, the EU Member States do not apply ad hoc rules specifically approved to regulate halal certification, but resort to general legislation aimed at preventing and punishing commercial fraud, irrespective of the religious orientation of the concerned products or actors. The same considerations apply to any compliance requirements stipulated by law. For instance, the Hazard Analysis and Critical Control Points (HACCP) is a system applied to the safety of food as a whole and, as such, 'it can also be applied for religious food safety' (van der Spiegel et al. 2012, p. 110), including that of halal food.

A different legal aspect of halal certification in Europe concerns the guarantee of the consumer's right to transparency and access to information. Law-makers in the EU and its Member States have in fact focused on initiatives to protect the right of ethically oriented consumers not to eat meat produced through a method inconsistent with their beliefs. For a discussion on labelling, 'ethically competent' and 'ethically non-competent' consumers, see Miele and Evans (2010).

Rapporteur Wojciechowski, in the draft report on the Commission's proposal for a Council Regulation submitted on 18 December 2008, amended Art. 4 § 2 to include a provision according to which products derived from religious rites had to be labelled in order 'to avoid confusion and give to the consumer the right information on all the products obtained under this derogation' (Wojciechowski 2008, amendment no. 5). As we know, not all meat from animals slaughtered according to a religious rite is labelled consistently.[8] Part of it may be and in fact is sold on the conventional market and unwittingly bought by consumers, including those opposing the religious method of slaughter. For this reason, according to Wojciechowski, an appropriate labelling of the meat coming from animals slaughtered without previous stunning according to a religious rite was needed in order to inform customers and put them in such a position as to choose knowingly whether or not to eat it (Wojciechowski 2008a, p. 29).

This amendment was dropped though, and the initiatives taken by some EU Member States to introduce compulsory labelling for halal (and kosher) products also failed. In France, two bills were withdrawn by their own proponents a few days after being

---

7 See, for instance, the agreement signed on 30 June 2010 by the Italian Ministries of Foreign Affairs, of Economic Development, of Health, and of Agricultural Food and Forestry policies to support the certification initiative of 'Halal Italia'. The reasons given for the signing of the agreement also include the need to promote the internationalization of the Italian productive system, the protection of 'made in Italy', the promotion of Italian interests abroad, and the consideration for Italian producers' interest in Islamic markets. See www.halalitalia.org at https://tinyurl.com/nnv73zo. Last accessed on 20 September 2019.

8 This problem arises in particular with *schechita,* but also affects halal products. The consumption of the sciatic nerve is forbidden in Judaism. Its removal is nonetheless too time-consuming in relation to the economic benefit gained from the sale of this part of the animal, which is thus sold on the conventional meat market.

presented.[9] In the UK, the Food Labelling (Halal and Kosher Meat) Bill, presented in the House of Commons, was rejected by 73 votes to 70 after a first reading on 24 April 2012. The text read as follows:

> Goods to be as described: meat products
>
> 1) All products containing halal and kosher meat shall be labelled as such at the point of sale by retail and food outlets.
> 2) A food outlet is anywhere where food is served to the public (Consumer Rights Bill, new clause 13; see www.publications.parliament.uk at http://tinyurl.com/o4yon9c and Gianfreda (2015)).

In 2014, these clauses were proposed again through an amendment to the Consumer Rights Bill, which was rejected by 381 votes to 17 (see http://hansard.parliament.uk at http://tinyurl.com/h49hayy). This legislation has been opposed by religious communities, on the basis inter alia of two legal arguments. The first is the violation of their right to internal autonomy, as would occur were religious authorities obliged to certify as halal (or kosher) a meat product which, despite coming from an animal slaughtered without previous stunning, is not consistent with other religious rules regulating the act of killing animals. The second is discrimination, which would take place insofar as compulsory labelling is imposed only upon religiously oriented meat products. More recently, in the heated debate sparked by the Brexit referendum, the UK government has been reminded that the EU has always resisted clearer labelling, and it has been urged to improve it following the exit from the EU. In this context, the issue of compulsory labelling for meat products from animals slaughtered without stunning has been raised again.

In fact, similar attempts by the EU have failed so far. Article 2 of the Council Directive 2000/13/EC only stipulated that labelling might not 'be such as could mislead the purchaser to a material degree', particularly as to a number of characteristics, which did not include method of slaughter.[10] This Act has been repealed and replaced by Regulation (EU) 1169/2011,[11] where the examined issue was raised and discussed.

---

9 Bill no. 2976 of 18 November 2010 aimed to improve the consumer's information as regards the method of killing animals (*visant à améliorer l'information du consommateur quant au mode d'abattage des animaux*) and bill no. 4379 of 21 February 2012 aimed to respect European regulation for the production of meat from animals slaughtered without stunning (*visant au respect de la réglementation européenne pour la production de viande provenant d'animaux abattus sans étourdissement*), presented respectively by Nicolas Dhuicq MP and Françoise Hostalier MP. See www.assemblee-nationale.fr at http://tinyurl.com/6kzm97f. Last accessed on 20 September 2019.

10 Directive 2000/13/EC of the European Parliament and of the Council of 20 March 2000 on the approximation of the laws of the Member States relating to the labelling, presentation and advertising of foodstuffs, published in the Official Journal of the European Union L 109 of 6 May 2000.

11 Regulation (EU) No. 1169/2011 of the European Parliament and of the Council of 25 October 2011 on the provision of food information to consumers, amending Regulations (EC) No. 1924/2006 and (EC) No. 1925/2006 of the European Parliament and of the Council, and repealing Commission Directive 87/250/EEC, Council Directive 90/496/EEC, Commission Directive 1999/10/EC, Directive 2000/13/EC of the European Parliament and of the Council, Commission Directives 2002/67/EC and 2008/5/EC and Commission Regulation (EC) No. 608/2004 Text with EEA relevance, published in the Official Journal of the European Union L 304 of 22 November 2011.

In its proposal, the European Commission stated that the new legislation is aimed at balancing the needs of consumers – demanding 'more and "better" information on labels' as well as 'clear, simple, comprehensive, standardised and authoritative information'–, and industry, which considers that 'there are too many labelling requirements which involve implementation of detailed, technical rules' and is concerned about the involved costs.[12] No provision on the labelling of meat from animals slaughtered according to a religious rite had been included though.

The inclusion of this norm was an amendment proposed by Rapporteur Sommer's draft report on the Commission proposal. 'Meat and meat products derived from animals that have not been stunned prior to slaughter, i.e. have ritually slaughtered' should be labelled 'meat from slaughter without stunning':

> EU legislation permits animals to be slaughtered without prior stunning to provide food for certain religious communities. A proportion of this meat is not sold to Muslims or Jews but is placed on the general market and can be unwittingly purchased by consumers who do not wish to buy meat derived from animals that have not been stunned. At the same time, however, adherents of certain religions specifically seek meat from animals which have been ritually slaughtered. Accordingly, consumers should be informed that certain meat is derived from animals which have not been stunned. This will enable them to make an informed choice in accordance with their ethical concerns. (see www.europarl.europa.eu at http://tinyurl.com/ojxqjlk)

The amendment, renumbered as no. 205, was approved by the European Parliament at first reading on 16 June 2010 and included in Annex III, point 2 under the heading 'meat products from special slaughter'. This wording, unlike the UK's abovementioned bill, was more neutral and did not suggest that religious authorities would be forced to certify as halal or kosher meat that did not comply with all their religious rules. Nonetheless, in its position of 21 February the Council of the EU did not accept the amendment, on the grounds that it 'did not intend to adopt a specific labelling for this meat' (see www.europarl.europa.eu at http://tinyurl.com/ojxqjlk).

European Parliament then restored the provision in the draft text (as amendment no. 359) in view of the second reading on 6 July. However, it later decided to withdraw it, after a compromise on the final text was reached on 14 June amongst the representatives of the European Parliament, the Council and the Commission (see www.europa.eu at https://tinyurl.com/oc8dsrh). In the end, only one provision has been included in the final text of the Regulation. According to Recital 50:

> Union consumers show an increasing interest in the implementation of the Union animal welfare rules at the time of slaughter, including whether the animal was stunned before slaughter. In this respect, a study on the opportunity to provide

---

12 European Commission, Proposal for a Regulation of the European Parliament and of the Council on the provision of food information to consumers, p. 4. See www.europarl.europa.eu at http://tinyurl.com/nckjgng. Last accessed on 20 September 2019.

consumers with the relevant information on the stunning of animals should be considered in the context of a future Union strategy for the protection and welfare of animals.

It should be stressed that this solution met the expectations not only of religious communities, but also of representatives of the meat industry. For example, the newsletter of June 2010 issued by Assocarni, Italy's National Association of Meat Industry and Trade, while reporting the discussion that was going on the Council Regulation, stated that:

> [a]fter a wide debate, the amendment envisaging the compulsory labeling of meat and meat products from animals slaughtered without previous stunning has been approved. This amendment, which Assocarni will oppose, hits also the meat produced through a religious rite but sold on the conventional market, as it is not requested by consumers of kosher or halal meat. (Assocarni, Notiziario 6/2010, p. 5, www.assocarni.it at http://tinyurl.com/qe9c8df. In Italian in the original text. The translation is mine.)

The latest development is the judgment delivered by the Court of Justice of the European Communities on the question of whether products from animals slaughtered according to a religious rite could be given the label 'organic farming'. In his opinion of 20 September 2018, Advocate General Wahl concluded that 'Council Regulation (EC) No 834/2007 of 28 June 2007 on organic production and labelling of organic products and Council Regulation (EC) No 1099/2009 of 24 September 2009 on the protection of animals at the time of killing', read in the light of Article 13 TFEU, must be interpreted as not prohibiting the issue of the European 'organic farming' label to products from animals which have been the subject of ritual slaughter without prior stunning carried out in the conditions laid down in Regulation No 1099/2009' (para 105).[13] However, the judgment delivered by the Court on 26 February 2019 reversed this conclusion on the grounds that religious slaughter 'does not allow the animal's suffering to be kept to 'a minimum' and is 'not tantamount, in terms of ensuring a high level of animal welfare at the time of killing, to slaughter with pre-stunning' (paras. 49–50). Because consumers must be 'reassured that products bearing the Organic logo of the EU have actually been obtained in observance of the highest standards, in particular in the area of animal welfare' (para. 51), the relevant EU legal norms may not be interpreted as authorizing the placing of 'organic farming' labels in products from animals that have been subject to religious slaughter without previous stunning.[14]

---

13 Text of the opinion: https://curia.europa.eu at https://tinyurl.com/y3pec8pe. Last accessed on 20 September 2019.
14 Text of the judgement: www.bailii.org at https://tinyurl.com/y29u3jkz. Last accessed 20 September 2019. See also Pocklington and Cranmer (2019).

References

Anil, H. (2012). Religious slaughter: a current controversial animal welfare issue. *Animal Frontiers* 2 (3): 64–67.

Anil H., Miele M., von Holleben K. et al. (2010). Religious Rules and Requirements. Halal Slaughter, p. 17. www.dialrel.eu.

Bottoni, R. (2013). The legal treatment of religious minorities: non-Muslims in Turkey and Muslims in Germany. In: *Religion, Identity and Politics: Germany and Turkey in Interaction* (eds. G. Gülalp and G. Seufert), 129–131. London: Routledge.

Bottoni, R. (2015). La disciplina giuridica della macellazione rituale nell'Unione europea e nei paesi membri [Legal regulation of religious slaughter in the European Union and its member states]. In: *Cibo, religione e diritto. Nutrimento per il corpo e per l'anima [Food, religion and law. Nourishment for the body and the soul]* (ed. A.G. Chizzoniti), 479–516. Tricase: Libellula Edizioni.

Bottoni, R. (2017). I recenti decreti delle Regioni vallona e fiamminga sulla macellazione rituale nel contesto dei dibattiti belga ed europeo in materia [The recent Walloon and Flemish decrees on religious slaughter and the related debate in Belgium and Europe]. *Quaderni di diritto e politica ecclesiastica* 2: 523–558.

Cohen, J. (2009). Kosher slaughter, state regulations of religious organizations, and the European Court of Human Rights. *Intercultural Human Rights Law Review* 4: 355–386.

Delahunty, R.J. (2015). Does animal welfare Trump religious liberty? The Danish ban on Kosher and Halal butchering. *San Diego International Law Journal* 16: 341–379.

Doe, N. (2011). *Law and Religion in Europe. A Comparative Introduction*. New York: Oxford University Press.

Drobný, J. (2014). Slovakia. In: *Yearbook of Muslims in Europe* (ed. J.S. Nielsen), 540–546. Leiden: Brill.

Efron, J.M. (2007). The most cruel cut of all? The campaign against Jewish ritual slaughter in Fin-de-Siècle Switzerland and Germany. *Leo Baeck Institute Yearbook* 52: 167–184.

European Commission (2008a). Commission Staff Working Document. Impact Assessment Report Accompanying the Proposal for a Council Regulation on the Protection of Animals at the Time of Killing, 18 September 2008. www.ipex.eu at http://tinyurl.com/o8fklct. Last accessed on 20 September 2019.

European Commission (2008b). Proposal for a Council Regulation on the protection of animals at the time of killing, 18 September 2008. www.ipex.eu at http://tinyurl.com/pj3pvrv. Last accessed on 20 September 2019.

European Economic and Social Committee (2009). Opinion on the protection of animals at the time of killing, 4.4. *Official Journal of the European Union* C/218 of 11 September 2009: 55.

European Union (1993). Council Directive 93/119/EC of 22 December 1993 on the protection of animals at the time of slaughter or killing. *Official Journal of the European Union* L 36 (340 of 31 December 1993): 21.

European Union (2009). Council Regulation (EC) No. 1099/2009 of 24 September 2009 on the protection of animals at the time of killing. *Official Journal of the European Union* L 303 of 18 November 2009: 1.

Ferrari, S. and Bottoni, R. (2010). Legislation regarding religious slaughter in the EU member, candidate and associated countries. www.dialrel.eu. at https://tinyurl.com/qc2zfss. Last accessed on 20 September 2019.

Ferrari, S. (2010). Islam and the European system of state-religions relations throughout Europe. In: *Cultural Diversity and the Law: State Responses from around the World* (eds. M.-C. Foblets, J.-F. Gaudreault-DesBiens and A.D. Renteln), 477–502. Brussels: Bruylant.

Fischer, J. (2014). Muslim material culture in the Western world. In: *Routledge Handbook of Islam in the West* (ed. R. Tottoli), 348–362. New York: Routledge.

Foblets, M.-C. and Velaers, J. (2014). In search of the right balance. Recent discussions in Belgium and the Netherlands on religious freedom and the slaughter of animals without prior stunning. In: *Recht, Religion, Kultur. Festschrift für Richard Potz zum 70. Geburtstag* (eds. B. Schinkele, R. Kuppe, S. Schima, et al.), 67–85. Wien: Facultas Verlags- und Buchhandels AG.

Gianfreda, A. (2015). Alimentazione e religione nel Regno Unito: ambiti normativi e questioni aperte. In: *Cibo, religione e diritto. Nutrimento per il corpo e per l'anima* (ed. A.G. Chizzoniti), 392–395. Tricase: Libellula Edizioni.

Gliszczyńska-Grabias, A. and Sadurski, W. (2015). Freedom of religion versus humane treatment of animals: Polish constitutional Tribunal's judgment on permissibility of religious slaughter. *Judgment of the Constitutional Tribunal of Poland of 10 December 2014, K 52/13. European Constitutional Law Review* 11: 596–608.

Haupt, C.E. (2007). Free exercise of religion and animal protection: a comparative perspective on ritual slaughter. *George Washington International Law Review* 39: 839–886.

Havinga, T. (2010). Regulating halal and kosher foods: different arrangements between state, industry and religious actors. *Erasmus Law Review* 3 (4): 241–255.

Lavi, S. (2009). Unequal rites – Jews, Muslims and the history of ritual slaughter in Germany. In: *Juden und Muslime in Deutschland. Recht, Religion, Identität* (eds. J. Brunner and S. Lavi), 164–184. Göttingen: Wallstein Verlag.

Lerner, P. and Mordechai Rabello, A. (2006/2007) The prohibition of ritual slaughtering (Kosher Shechita and Halal) and freedom of religion of minorities. *Journal of Law and Religion* 22 (1): 1–62.

Lever, J. and Miele, M. (2012). The growth of halal meat markets in Europe: an exploration of the supply side theory of religion. *Journal of Rural Studies* 28 (4): 528–537.

Metcalf, M.M. (1989). *Regulating slaughter: animal protection and antisemitism in Scandinavia, 1880–1941. Patterns of Prejudice* 23 (3): 32–48.

Miele, M. and Evans, A. (2010). When foods become animals: ruminations on ethics and responsibility in care-full practices of consumption. *Ethics, Place and Environment* 13 (2): 171–190.

Pocklington, D. and Cranmer, F. (2019). CJEU rules non-stun slaughter incompatible with organic labelling, 1 March 2019. www.lawandreligionuk.com at https://tinyurl.com/y2yceh8f. Last accessed on 20 September 2019.

Rohe, M. (2004). Application of Sharî'a rules in Europe: scope and limits. *Die Welt des Islams* 44 (3): 323–350.

Rovinski, J.A. (2014). The cutting edge: the debate over regulation of ritual slaughter in the Western world. *California Western International Law Journal* 45: 79–107.

Szumigalska, A. and Bazan, M. (2014). Ritual slaughter issue in Poland: between religious freedom, legal order and economic-political interests. *Religion and Society in Central and Eastern Europe* 7 (1): 53–69.

van der Schyff, G. (2017). Reviewing the Recent Ban on Ritual Slaughter in Flanders, 16 August 2017. https://verfassungsblog.de at https://tinyurl.com/y9ljc8ty. Last accessed on 20 September 2019.

van der Spiegel, M., van der Fels-Klerx, H.J., Sterrenburg, P. et al. (2012). Halal assurance in food supply chains: verification of halal certificates using audits and laboratory analysis. *Trends in Food Science & Technology* 27: 109–119.

Wojciechowski, J. (2008). Committee on Agriculture and Rural Development. Draft Report on the proposal for a Council regulation on the protection of animals at the time of killing, 18 December 2008. www.europarl.europa.eu at http://tinyurl.com/hgv2rm8. Last accessed on 20 September 2019.

Wojciechowski, J. (2009). Committee on Agriculture and Rural Development. *Report on the proposal for a Council regulation on the protection of animals at the time of killing*, 24 March 2009. www.europarl.europa.eu at http://tinyurl.com/hgv2rm8. Last accessed on 20 September 2019.

Zoethout, C.M. (2013). Ritual slaughter and the freedom of religion: some reflections on a stunning matter. *Human Rights Quarterly* 35 (3): 651–672.

17

The Legal Framework of General Food Law and the Stunning of Animals Prior to Slaughter

John Pointing

Barrister, London, UK

17.1 Background to the General Food Law

Food law in the UK is governed by EU law, the framework of which is set out by the General Food Law: Regulation (EC) 178/2002. This legislation, which came fully into force on 1 January 2006, lays down the general principles and requirements for food law for all Member States of the EU. Should the UK leave the EU it is likely that the General Food Law will continue to provide the legislative basis for domestic law in the UK.

Besides aiming to achieve free movement of food and feed between Member States, Article 5 of Regulation (EC) 178/2002 states:

> Food law shall pursue one or more of the general objectives of a high level of protection of human life and health and the protection of consumers' interests, including fair practices in food trade, taking account of the protection of animal health and welfare, plant health and the environment.

Regulation (EC) 178/2002 also established the European Food Safety Authority (EFSA) to restore and maintain public confidence in food safety, provide scientific advice, coordinate research, and promote consumer protection. A related batch of European Regulations came into force on 1 January 2006 dealing with the hygiene of foodstuffs, specific hygiene rules for food of animal origin, and specific rules governing official controls on products of animal origin intended for human consumption.

Regulation (EC) 178/2002 is drafted in the form of Articles setting out the framework of food law for EU Member States. Some Articles are quite detailed and some are expressed in the form of general principles. This means that subsidiary legislation is often required in order to flesh out how the Articles can be fulfilled. Such subsidiary legislation – both European and domestic measures – must be consistent with Regulation (EC) 178/2002, whether enacted before or after this Regulation came into force. European Regulations are directly applicable within each Member State and cannot be modified or repealed by domestic legislatures.

The Halal Food Handbook, First Edition. Edited by Yunes Ramadan Al-Teinaz, Stuart Spear, and Ibrahim H. A. Abd El-Rahim.
© 2020 John Wiley & Sons Ltd. Published 2020 by John Wiley & Sons Ltd.

Subsidiary domestic legislation is permitted in order to provide the enforcement mechanism for bringing a European Regulation into operation. Further, Member States can (and should) amend or repeal existing domestic legislation should it conflict with a European Regulation. Thus, in the UK, the Food Safety Act 1990 (Amendment) Regulations 2004 were enacted in order to bring Regulation (EC) 178/2002 into force. Offences involving injury to health provided under section 7 of the Food Safety Act 1990 and selling food in breach of the food safety requirements under section 8 were amended by regulations 9 and 10 of the General Food Regulations 2004 in order to ensure consistency with European law.

17.1.1 The Precautionary Principle and Risk

The precautionary principle with regard to environmental and health risks is generally promoted by EU policy. It is introduced into food law by Article 7 of Regulation (EC) 178/2002. Article 6 establishes that the enforcement of food law shall be risk-based. Risk assessment relies on a scientific analysis of the risks and hazards posed by a particular operation or process. It allows action to be taken where a risk is identified and the hazard consequential on that risk is sufficiently serious to warrant action. The precautionary principle thus permits action to be taken where a risk assessment cannot, or does not, indicate a high standard of scientific certainty. Article 7 also provides that provisional measures may be based on an initial assessment pending a more comprehensive risk assessment. It emphasizes that these measures must be proportionate, since the likelihood is that they will damage the trade opportunities of particular persons or industries.

17.2 Consumer Protection

Detection of regulatory breaches is difficult and to an increasing extent enforcers rely on producers to implement robust systems of consumer protection. Article 17 of Regulation (EC) 178/2002 places primary responsibility on food and feed business operators for ensuring that the requirements of food law are met. Article 8 of Regulation (EC) 178/2002 prescribes that food law should aim to prevent fraudulent or deceptive practices, food adulteration, and any other misleading practices being carried out by the operators of food businesses. Food law also needs to protect consumers by enabling them to make informed choices concerning the foods they consume. The principles of transparency set down in Articles 9 and 10 reinforce consumer protection. Article 9 requires that open and transparent consultation shall take place during the preparation, evaluation, and revision of food law. Article 10 requires public authorities to inform the general public 'where there are reasonable grounds to suspect that a food or feed may present a risk to human or animal health'. Public authorities must identify the risk and inform the public as to what measures are being taken to prevent, reduce, or eliminate it.

17.3 Article 14: Food Safety Requirements

The food safety requirements are set down in Regulation (EC) 178/2002. Failure to observe these or to put them into effect is the basis for a number of offences. Under Article 14 it is a criminal offence to place food on the market if it is unsafe. Food is deemed to be unsafe if it is injurious to health or unfit for human consumption. This

Article outlines the factors to be taken into account when deciding whether food is unsafe, injurious to health, or unfit for human consumption.

17.3.1   Definition of 'Food'

One of the objectives of the General Food Law is to harmonize food law across the EU, both in the interests of promoting freedom of internal trade and in order to achieve a high standard of public health and consumer protection. Such harmonization is based on a common definition of 'food' and 'feed'. The definition of 'food' contained in Article 2 of Regulation (EC) 178/2002 is a wide one which includes:

- drink, including water
- chewing gum
- any substance, including water, intentionally incorporated into the food during its manufacture, preparation, or treatment.

Expressly excluded are matters regulated in other legislation, namely:

- feed
- live animals unless they are prepared for placing on the market for human consumption
- plants prior to harvesting
- medicinal products
- cosmetics
- tobacco and tobacco products
- narcotic or psychotropic substances
- residues or contaminants.

17.4   'Placing on the Market'

Article 14 does not refer to the 'sale' of food, but instead refers to food which is 'placed on the market'. 'Placing on the market' is defined in Article 3 of Regulation (EC) 178/2002, as the holding of food or feed for the purpose of sale. It includes offering food for sale or any other form of transfer and applies whether free of charge or not. Sale, distribution, and other forms of transfer are included.

Article 14 is not confined to food which is placed on the market by food business operators. This means that offences can be committed where a one-off sale or transfer of food is made and where food is supplied free of charge. So, all possible occasions when public health can be affected by the supply of food are covered, except private domestic consumption, which is exempted under Article 1(3). Article 1 prescribes that Regulation (EC) 178/2002 shall apply to all stages of production, processing, and distribution of food and feed, but shall not apply to primary production for private domestic use, or to the domestic preparation, handling, or storage of food for private domestic consumption.

17.4.1   Food That is Unsafe

Article 14 prescribes that 'food shall not be placed on the market if it is unsafe'. Food is deemed to be unsafe if it is injurious to health or unfit for human consumption. The factors to be taken into account here include the normal conditions of use by the consumer at each stage of production, processing, and distribution. Also relevant are the normal practices of preparation, such as using sufficient cooking times. Thus, the information provided to the consumer is a relevant factor to be taken into account when determining whether the food is safe to eat. Careful labelling by the processor as to the correct way of preparing the food to avoid any ill effects should satisfy the food safety requirements.

17.4.2 Food That is Injurious to Health

Factors to be taken into account when determining whether food is injurious to health include the probable long-term effects as well as immediate and short-term effects. These are probable ill effects, although reference to the precautionary principle in Article 7 suggests that where there is a lack of scientific certainty, the risk of possible ill effects might be sufficient in evaluating risks to health. Article 14 includes probable cumulative ill effects. So, if single or occasional consumption of the food would not be injurious to health, there is still an offence committed if cumulative consumption could cause injury.

The effect on the sensitivities of a specific category of consumers for whom the food is intended is a relevant factor in determining whether food is injurious to health. Food intended for diabetics, for example, must take account of their particular health sensitivities. Issues such as the problems of obesity in children might fall within this provision where foods are specifically targeted at children, thus requiring food processors to consider the health effects of the ingredients used in the food they produce. Just as public campaigns have exerted pressure on food producers to reduce the amount of salt contained in food, concerns about the amounts of sugar and fat raise the issue as to whether such foods may be injurious to health within the meaning of Article 14.

17.4.3 Food That is Unfit for Human Consumption

This provision covers the adulteration, contamination, or decay of food by extraneous matter or otherwise so as to render it injurious to health or unfit for human consumption. Unfitness for human consumption is a complex concept. Article 14 defines the conditions for unfitness as occurring where it is 'unacceptable for human consumption [of the food] according to its intended use, for reasons of contamination, whether by extraneous matter or otherwise, or through putrefaction, deterioration or decay'. Unfitness encompasses both injury to health and anything falling below that level which causes the food to be unacceptable for human consumption. Article 14 links the acceptability of the food for human consumption to its intended use. Therefore, the determination of whether food is fit for human consumption must take place at the final stage of its production or preparation, i.e. just before when consumption is intended.

17.4.4 Batch, Lot or Consignment

Article 14 provides that where food which is unsafe forms part of a lot, batch or consignment, then there is a presumption that all of the food in that batch is unsafe. It follows that the power to seize food under section 9 of the Food Safety Act 1990 will apply to the whole batch in such circumstances. This presumption can be rebutted by the food business operator where a properly carried out and detailed assessment shows that the rest of the batch is not unsafe.

17.5 Food Safety Offences

Food safety offences can be divided between those involving the adulteration of food and those which fail the food safety requirements. Misdescription, inferior quality, and mislabelling of food are consumer protection offences considered later in this chapter. Food fraud can involve any of these types of criminal behaviour and is a more serious offence involving dishonesty. Charges of fraud are not often brought by prosecutors, but would be appropriate in cases where there is evidence of organized crime, particularly when this occurs on a large scale (Pointing 2005). With fraud, unlike statutory food safety offences, the prosecution is required to prove *mens rea*, or guilty intent, and this complicates matters for prosecutors. Food crimes are mostly classified as strict liability offences, for which limited statutory defences are available, such as the due diligence defence under section 21 of the Food Safety Act 1990.

17.5.1 Food Adulteration: Section 7 Food Safety Act 1990

Section 7 of the Food Safety Act 1990 creates offences in respect of food which has been rendered so as to be injurious to health. The meaning of 'injurious to health' is defined by Article 14(4) of Regulation (EC) 178/2002. Section 7 is concerned with food adulteration and dates back to legislation originally enacted in the nineteenth century to prohibit the adding of harmful substances to food.

The section 7 Food Safety Act offence applies to:

Any person who renders any food injurious to health by means of any of the following operations, namely-

a) adding any article or substance to the food
b) using any article or substance as an ingredient in the preparation of the food
c) abstracting any constituent from the food
d) subjecting the food to any other process or treatment,

with intent that it shall be sold for human consumption.

Section 7 indicates that adulteration of the food can be effected in various ways, including by adding something to it or by taking something away. Not all forms of adulteration come within the scope of section 7. It does not include the dilution of the food, such as by adding water that is not contaminated. Neither does the practice of adding water boosted by proteins derived from pigs to improve the palatability of chicken come within the scope of section 7. This is because the product has not been rendered injurious to health. Depending on the circumstances, a more appropriate charge based on a consumer protection offence might be brought, or a charge for breaching food labelling regulations. A charge of fraud might be more appropriate where the chicken to which porcine proteins had been added to the water was being passed it off as halal (Pointing et al. 2008, pp. 210–211).

A section 7 offence is committed when the food is adulterated with intent that it be sold (that is placed on the market) for human consumption. The intent required pertains to the sale of food for human consumption: it is not necessary for the prosecutor to prove intention to adulterate the food, which may have occurred accidentally. There is a rebuttable presumption that where food is found in the possession of a person involved in its processing that that person intends it to be sold, or otherwise placed on the market, for human consumption. This presumption can be rebutted if an accused person can prove – on the balance of probabilities – that the food did not form part of the food chain for human consumption or had been withdrawn from the market.

17.5.2 Selling Food Not Complying with the Food Safety Requirements: Section 8 Food Safety Act

Selling food that does not comply with the food safety requirements forms the grounds for the usual charge brought by food authorities when prosecuting those involved in placing unfit food on the market. On 1 January 2005, when Article 14 of (EC) 178/2002 came into force, a new section 8 Food Safety Act offence came into effect through the operation of the General Food Regulations 2004. Section 8 of the Food Safety Act 1990 now provides:

> food fails to comply with food safety requirements if it is unsafe within the meaning of Article 14 of Regulation (EC) No. 178/2002 and references to food safety requirements or to food complying with such requirements shall be construed accordingly.

17.6 Breaches of Food Safety and Hygiene Regulations

Section 16 of the Food Safety Act 1990 gives wide powers to the Secretary of State to make food safety and consumer protection regulations, which must be fully compliant with EU Law. The Food Safety and Hygiene Regulations form a complex and detailed body of secondary legislation that is regularly updated to respond to minor changes and requirements of food law (Lawless 2012). They comprise detailed provisions that flesh out the general principles enshrined in Regulation (EC) 178/2002.

The current regulations in force in England are the Food Safety and Hygiene (England) Regulations 2013. Similar regulations are in force in other parts of the UK, made by the devolved legislative bodies of Scotland, Wales, and Northern Ireland. Regulation 19 of the Food Safety and Hygiene (England) Regulations 2013 provides that 'any person who contravenes or fails to comply with any of the specified EU provisions commits an offence'. Offences under these Regulations are triable either way, though prosecutions generally takes place in the magistrates' court. The penalty for each charge tried before magistrates is an unlimited fine. A conviction in the Crown Court attracts a maximum sentence of an unlimited fine and/or imprisonment for up to two years.

17.7 Consumer Protection Offences

Sections 14 and 15 of the Food Safety Act 1990 set out the basis for consumer protection offences in respect of food. These are long-standing provisions that pre-date the entry of EU legislation into UK food law. Nevertheless, they are supplemented by Article 16 of Regulation (EC) 178/2002, which creates a further offence in addition to those remaining under sections 14 and 15 of the Food Safety Act 1990.

17.7.1 Article 16: Labelling, Presentation, and Advertising

Article 16 of Regulation (EC) 178/2002 deals with the labelling, presentation, and advertising of food (or feed) and prescribes that these shall not be undertaken so as to mislead consumers. It includes such matters as the shape, appearance, type of packaging and packaging materials used, the setting in which the food is displayed, and the information which is made available about the food. This Article could be invoked where consumers are likely to be misled by packaging. Examples of this could include where the meat was falsely described as halal. It could also apply where meat was labelled as the product of an animal that had been slaughtered without stunning when in fact the animal had been stunned prior to slaughter.

17.7.2 Section 15: Falsely Describing or Presenting Food

Section 15 of the Food Safety Act 1990 provides additional consumer protection offences in relation to the presentation of food. Under section 15(1) an offence is made out where a label falsely describes the food or is likely to mislead the consumer as to its nature, or substance, or quality. A further offence is made out, under section 15(2), where a person publishes an advertisement that falsely describes the food or misleads in the above ways. A person who sells food, or who offers or exposes it for sale, or who possesses it for the purpose of sale also commits an offence, under section 15(3), where the presentation of the food is likely to mislead as to its nature, substance, or quality. These offences could apply to false descriptions and representations made about food purporting to be halal.

17.7.3 Section 14: Selling Food Not of the Nature or Substance or Quality Demanded

Consumer protection legislation also seeks to regulate the conditions surrounding the selling of food. Section 14 of the Food Safety Act 1990 makes it an offence to sell to the purchaser's prejudice any food which is not of the nature, or substance, or quality demanded by the purchaser. Section 14 establishes an offence for what would otherwise be a breach of contract at common law. It is not necessary to prove that the food is unsafe, which, additionally, it may be and thereby fail the food safety requirements.

17.7.4 Nature, Substance, or Quality of the Food

Sections 14 and 15 of the Food Safety Act both refer to the nature, substance, or quality of the food. These are separate elements that correspond to separate offences. In drafting an information – the document that sets out the legal and factual basis for the prosecution – these separate ways of committing the offence should not be rolled up by the prosecutor into a single charge. Use of more than one of these three words in a single charge would constitute duplicity, which is a serious technical fault. An information found to be at fault in this way could not properly be used as the basis for a prosecution since it would not be clear which of the three offences is being alleged.

It would be useful, therefore, to examine these separate elements further.

1) Nature

An offence based on this provision would cover cases where the item being sold was of a different type from that demanded by the purchaser. This would include, for example, a case of caustic soda being sold instead of lemonade. It might be considered where food that was not halal was being sold as halal under section 14, though generally this would fit better as a false description offence under section 15.

2) Substance

This provision addresses problems with the composition of the food and includes cases of adulteration. In such cases – and others where the food fails to comply with the food safety requirements – there will be an overlap with section 8 of the Food Safety Act. Section 8 would normally form the basis as the more appropriate charge, where a foreign body present in the food posed a risk to health.

3) Quality

Where the purchaser demanded food of a higher quality than that provided by the supplier, it would fall under this provision. This concept of quality in sections 14 and 15 of the Food Safety Act arguably corresponds with the concept of tayyib in Shariah law: positive requirements for wholesomeness, quality, and nourishment, going 'a step beyond halal' (Kamali 2013, p. 6).

17.8 Offences by Suppliers

Article 18 of Regulation (EC) 178/2002 creates an offence in relation to the traceability of food, feed, food-producing animals, and any other substance intended, or expected, to be incorporated into food or feed. Article 18 applies to food and feed business operators, and requires them to be able to identify any person from whom they have been supplied and the other businesses to which their products have been supplied. They must be able to trace forwards and backwards in the food supply chain. A food (or feed) business operator is defined under Article 3 of Regulation (EC) 178/2002 as the 'natural or legal persons responsible for ensuring that the requirements of food law are met within the food (or feed) business under their control'.

Food business operators must have in place effective systems and procedures to identify persons from whom they have been supplied as well as persons to whom they supply. Records must be kept of supplies from outside the EU. Food labelling should ensure that traceability is effective. Such procedures to keep track of supplies will assist enforcement officers when tracing food that does not comply with food law and regulations. Enforcement officers have powers under section 9 of the Food Safety Act 1990 to seize documents and records revealing the source and destination of supplies, and will be able to prosecute where no records have been kept that comply with Article 18.

17.9   Penalties

Most prosecutions take place in the magistrates' courts, which have more limited sentencing powers than the Crown Court. For conviction for an offence under the Food Safety Act 1990 heard in the magistrates' court, maximum penalties comprise an unlimited fine and/or six months' imprisonment. Penalties for breaches of the Food Safety and Hygiene (England) Regulations 2013 are limited to a fine (also without a maximum). Very serious or more complex cases can be tried before a jury in the Crown Court. The maximum penalties for conviction on indictment in the Crown Court for all food law offences are an unlimited fine and/or a maximum of two years' imprisonment.

17.10   Halal Slaughter and Food Law

The preparation for sale of halal food must conform to state law and regulation. State law – whether grounded in UK law or in the law of the EU – is as binding for halal food as for non-halal. State law in the UK is not specifically concerned with halal foods, so how does halal fit into the system of state law? How does Shariah law relate to state law? It may be imagined that these different systems of law are irreconcilable or that religious law conflicts with state law. This may be true when these systems compete or try to occupy the same legislative space, or where religious sources of law try and oust state law, but it may be more useful to see Shariah law pertaining to halal as forming part of the network provided by state law. This way of seeing things offers the possibility that Shariah and state law are compatible, or at least do not occupy the same legislative space. However, primacy must be accorded to state law, which determines the structure of the net and provides the space allowing the entry of Shariah concepts, including the determination of what is permitted and comprises halal (Pointing 2014).

This position is not unusual. This process has been described by Tamanaha (2001, p. 228) in respect of religious law generally, which 'often has its own independent institutional existence' and which is 'often intertwined with and incorporated into the state legal system'. Thus, Islamic or Shariah law can have effect only to the extent permitted by state law. It means that within the scope of the permitted exceptions state law does not interfere. Any exception which is provided on religious or other grounds can be repealed or abolished because state law has supremacy in law-making. Moreover, the implications of withdrawing a permitted exception founded in Shariah law would be politically damaging. Such withdrawal may amount to a breach of human rights and would be justiciable in the European Court.

The exemption of Muslim (and Jewish) methods of slaughtering animals from the general provisions of state food law is provided by European Regulations. Much of UK domestic law on the slaughtering of animals and on animal welfare law arises from membership of the EU. The Treaty of Amsterdam 1997 requires the European Commission and

Member States to take animal sentience into consideration whenever legislation is being drafted, i.e. to recognize that animals can suffer and that their welfare should be of prime concern. The way that Islamic law applies to animal slaughter is that state law exempts certain requirements for Muslims (and Jews) in order to take religious beliefs and sensibilities into account. The exemption in the requirement to stun animals prior to slaughter first arose for the UK with the Welfare of Animals (Slaughter or Killing) Regulations 1995/731. This legislation has been superseded by Regulation (EC) 1099/2009, which came into effect in January 2013.

Islamic law thus makes rules regulating the slaughter of permitted animals for human consumption in a humane way. Such rules are backed by general legislation which provides a derogation, on grounds of religion, with regard to the use of stunning of animals prior to their slaughtering for food. The derogation is limited so that state law remains applicable to all forms of slaughter, including religious slaughter. Thus recital (2) of Regulation (EC) 1099/2009 provides that:

> Killing animals may induce pain, distress, fear, or other forms of suffering to the animals even under the best available technical conditions. Certain operations related to the killing may be stressful and any stunning technique presents certain drawbacks. Business operators or any person involved in the killing of animals should take the necessary measures to avoid pain and minimize the distress and suffering of animals during the slaughtering or killing process, taking into account the best practices in the field and the methods permitted under this Regulation. Therefore, pain, distress, or suffering should be considered as avoidable when business operators or any person involved in the killing of animals breach one of the requirements of this Regulation or use permitted practices without reflecting the state of the art, thereby inducing by negligence or intention, pain, distress or suffering to the animals.

The reference to best practice in this recital is important. It encourages the development of rules and practices that achieve the highest, 'state-of-the-art' standards to ensure that animals are protected from undue distress and suffering. A failure to achieve sufficiently high standards has the effect that the derogation does not apply. Slaughtering – however it is carried out – that falls below the required standard would be unlawful therefore.

The animal welfare principles that form a prerequisite for the exemption from the stunning of animals prior to slaughter to apply are set down in Regulation (EC) 1099/2009. Article 3 states that:

Business operators shall, in particular, take the necessary measures to ensure that animals:

a) are provided with physical comfort and protection, in particular by being kept clean in adequate thermal conditions and prevented from falling or slipping
b) are protected from injury
c) are handled and housed taking into consideration their normal behaviour
d) do not show signs of avoidable pain or fear or exhibit abnormal behaviour
e) do not suffer from prolonged withdrawal of feed or water
f) are prevented from avoidable interaction with other animals that could harm their welfare.

Regulation (EC) 1099/2009 does not, however, provide detailed rules or procedures regarding the non-stunning of animals prior to slaughtering. These are provided by religious law. As with the domestic UK legislation that preceded Regulation (EC) 1099/2009, the basic position is that the exemption provides a derogation, on religious grounds, from the general requirement that stunning must precede the slaughtering of animals for food. Member States are allowed a margin of appreciation in interpreting the scope of religious slaughter when implementing the Regulation. Thus recital (18) stipulates:

> Since Community provisions applicable to religious slaughter have been transposed differently depending on national contexts and considering that national rules take into account dimensions that go beyond the purpose of this Regulation, it is important that derogation from stunning animals prior to slaughter should be maintained, leaving, however, a certain level of subsidiarity to each Member State. As a consequence, this Regulation respects the freedom of religion and the right to manifest religion or belief in worship, teaching, practice, and observance, as enshrined in Article 10 of the Charter of Fundamental Rights of the EU.

Besides giving effect to the principle that differences between Member States should allow for some variations in slaughtering practices, Regulation (EC) 1099/2009 affirms that the derogation from stunning is based on the protection of religious rights by the institutions of the EU. Subsidiarity thus has the effect that differences are allowed to remain between Member States with respect to religious rights, so retaining the right to slaughter without stunning is under threat in some EU countries.

References

Kamali, M.H. (2013). *The Parameters of Halal and Haram in Shariah and the Halal Industry*. International Institute of Islamic Thought: London & Washington.

Lawless, J. (2012). The complexity of flexibility in EU food hygiene regulation. *European Food and Feed Law* 5: 220–231.

Pointing, J. (2005). Food crime and food safety: trading in bushmeat – is new legislation needed? *Journal of Criminal Law* 69 (1).

Pointing, J. (2014). Strict liability food law and halal slaughter. *Journal of Criminal Law* 72 (3).

Pointing, J., Teinaz, Y., and Shafi, S. (2008). Illegal labelling and sales of halal meat and food products. *Journal of Criminal Law* 78 (5).

Tamanaha, B. (2001). *A General Jurisprudence of Law and Society*. Oxford: Oxford University Press.

18

Detecting Adulteration in Halal Foods

M. Diaa El-Din H. Farag

National Center for Radiation Research and Technology, Atomic Energy Authority, Nasr City, Cairo, Egypt

18.1 Introduction

The world Muslim population today is estimated to be around 1.6–1.8 billion in over 112 countries, about one-fifth of the total world population (Ratanamaneichat and Rakkarn 2013). The majority of Muslims are from the Asia Pacific region (61.9%) and the Middle-East (20.1%). These numbers are expected to increase to 2.2 billion by 2030 (Jamaludin et al. 2011). This means that the demand for halal (permitted by Islamic law) products will increase, creating substantial marketing opportunities for halal food products (Table 18.1).

Almost 50 years ago, the terms halal and haram (prohibited by Islamic law) were not popular and were not commonly used in the world food industry. Today, they form an important element of the food industry because of halal represents the hygiene, cleanliness, and quality of the food consumed. The halal marketplace is emerging as one of the most profitable and influential market arenas in the world food business today. The global halal food market will be worth US$1.6 trillion by 2018, up from US$1.1 trillion in 2013, according to a report commissioned by Dubai Chamber of Commerce. The halal market is growing fast and has been increasing at an estimated 25% per year. Halal food is becoming a lucrative business not only amongst Muslim, but also non-Muslim countries (Mariam 2010; Ahmad et al. 2013). Halal food made up 16.6% of the total world food market as of 2013, according to Thompson Reuters report, and up to 2018 the halal food sectors are set to grow at an average rate of 6.9% a year, faster than the food sector in general. By 2018 halal was forecast to make up 17.4% of the world food market (http://www.foodnavigator.com/Regions/Middle-East/Global-Halal-market-to-hit-1.6tn-by-2018).

In accordance with religious requirements, all Muslims must eat, drink, and take medicines that are halal. Halal food should be clean and safe without any traces of dirt as well as free from prohibited ingredients (haram), which are lawfully proscribed by Islamic law. Halal food encompasses meats, seafood, frozen foods, pasta and noodles, canned food, biscuits and cookies, fruits, chocolate and candy, snack food, seasoning and spices, sauce/spread/vinegar, packaged cooking sauces, cereals, and beverages. Halal food and drinks

The Halal Food Handbook, First Edition. Edited by Yunes Ramadan Al-Teinaz, Stuart Spear, and Ibrahim H. A. Abd El-Rahim.
© 2020 John Wiley & Sons Ltd. Published 2020 by John Wiley & Sons Ltd.

Table 18.1   Global halal food market size classified by region in US$

Region	2004	2005	2009	2010
Asia[a]	369.6	375.8	400.1	416.1
Africa	136.9	139.5	150.3	153.4
Indonesia	72.9	73.9	77.6	78.5
Europe	64.3	64.4	66.6	67.0
American	15.3	15.5	16.1	16.2
Australasia	1.1	1.1	1.5	1.6

Source: World Halal Forum (2009), post event report 4-5 May 2009, p. 46 (after: Ratanamaneichat and Rakkarn , 2013).
[a] Includes Indonesia, Gulf Cooperation Council counties, Malaysia, China, and India.

can be described as anything that humans can eat or drink as long as there is no legal evidence prohibiting it and its constituents are free from any unlawful or impure elements. It is good and pure, and its consumption brings no harm. Animals' meat must be from animals slaughtered in the Islamic manner, must not be dedicated to anyone but Allah, and must be obtained in a lawful manner (Saida et al. 2014).

Halal animals can be divided into two groups: halal animals properly slaughtered according to Shariah law, such as cows, goats, sheep, chickens, ducks, rabbits, and others that are allowed in Islam. If these animals have not been slaughtered according to Islamic law, they and their derivatives are considered as non-halal or haram. Pigs and any food derived from pigs, dogs, rats, and others which are forbidden in Islam are non-halal or haram animals.

With the increasing price of commercial meats, the prevalence of incidents of adulteration has increased. Adulteration is the act of intentionally substituting one species for another whereby the food products from one species are mixed intentionally with either a similar substitute ingredient or a cheaper species. This is easy in the case of meat products based on comminuted meats such as sausage goods, burgers, patties, etc. Many cases have been reported worldwide involving adulteration of haram or mushbooh ingredients in food production.

In the food industry, pork and its derivatives are amongst the most widely used materials, for example gelatin, sodium stearoyl lactylate, shortening, collagen, whey, calcium stearate, capric acid, myristic acid, oleic acid, pancreatic extract, bone ash, and lard. This has necessitated the development of methods for the detection and quantitation of such admixtures. The determination of food authenticity and detection of adulteration are major concerns not only to consumers, but also to the industry and policy makers at all levels of the production process. The development of reliable analytical methods that can sensitively identify and quantify unknown species in processed and composite mixtures has thus become more important and, at the same time, more complicated. Moreover, identification of the animal origin is important for fair trade, for ensuring compliance with labelling regulations, and for other issues of health.

Several analytical methods used for the detection of the adulteration of meat products, oils, and fats, including lard, are based on the differences in the nature and the composition of the minor and major components of the adulterant and those of the unadulterated meats, oils, or fats. These methods usually depend on their physical–chemical constants or chemical and biological measurements (Kowalski 1989). The methods developed for species identification were mainly intended to facilitate the legal enforcement of regulations governing the import of raw meats at ports of entry, since species substitutions, such as the substitution of horse or kangaroo meat for beef (Martin 1981; Whittaker et al. 1983) and pork for beef or sheep meat, has been reported in several countries. Moreover, as processed meat and prepared ready-to-eat products have become increasingly available to consumers, the possibility of fraudulent adulteration and substitution of the expected species with other meats has also increased. It has been observed that the adulteration problem occurs more frequently in precooked meats than in fresh meat, possibly due to the lack of reliable and economical analytical methods for cooked meats.

In most countries, food manufacturers choose to use lard as a substitute ingredient for oil because it is cheaper and easily available. Pork, lard, and gelatin are serious matters in the view of some religions such as Islam and Judaism that forbid then (Venien and Levieux 2005) and for health reasons (Rashood et al. 1996). In Islam, foods containing pig sources are haram (unlawful or prohibited) for Muslims to consume. Hence, it is important for food control laboratories to carry out species differentiation of raw materials to be used for industrial food preparation and the detection of animal species in food products. This is especially crucial for halal (lawful or permitted) authentication of food products. In order to protect consumers from fraud and adulteration several analytical approaches have been developed to identify animal species in food products. On top of that verification of pork and/or other unlawful meat adulteration in commercial meat products is increasingly important for the authentication of halal labels in processed foods. The recent explosion of interest in the development of identification techniques for the adulteration, species identification, and detection of pork in meat and meat products requires simple, specific, sensitive, accurate means for the identification and quantification of each declared or undeclared component in finished commercial products and reliable analytical and authentication techniques (Darling and Blum 2007; Ali et al. 2012a; Karabasanavar et al. 2014). Hence, detection methods for meat species and lard adulteration in products are very important criteria for identifying their origins. Researchers have developed and employed detection techniques such as polymerase chain reaction (PCR), enzyme linked immunosorbent assay (ELISA), and electronic nose (EN), among others. The application and characteristics of various meat adulteration detection technologies are reviewed and discussed in this chapter.

18.2 Deoxyribonucleic Acid Techniques

Nucleic acids are involved in the storage and transfer of genetic information in all living organisms, including the simplest viruses. There are two types of nucleic acid in cells: deoxyribonucleic acid (DNA) and ribonucleic acid (RNA). Nucleic acids are so named because DNA was first isolated from nuclei, but both DNA and RNA also occur in other parts of the cell, e.g. DNA is also found in mitochondria and chloroplasts, whilst RNA is also found in the cytoplasm, particularly at the ribosomes. Both DNA and RNA are polymers, the monomeric units being called nucleotides. DNA and RNA are therefore polynucleotides. There are five different nitrogen-containing organic bases present in nucleic acids: *adenine* (A), *guanine* (G), *cytosine* (C), *thymine* (T), and *uracil* (U). A, T, C, and G are found in DNA, while RNA contains A, G, and C, but has U instead of T (uracil replaces thymine). The order of the nitrogen bases is a genetic code to produce specific proteins when the living organism needs them.

The most important nucleic acid used for the detection of adulteration and identification of animal species is DNA. DNA techniques have become very important and are widely used nowadays. It cannot be denied that there are several advantages of DNA analysis techniques. DNA-based methods have already been proven to be effective in deciphering minute levels of adulteration in highly processed meat products, heralding good prospects of transparency and fair trade in the food industry (Fajardo et al. 2010; Ali et al. 2011a,b; Doosti et al. 2011) due to the following principles:

1) DNA is the hereditary material responsible for all the characteristics of an organism and it controls all the activities of a cell.
2) DNA is a remarkably stable molecule at high temperature and is relatively resistant to degradation.
3) DNA potentially enables analysis in raw, cooked, and/or processed food materials due to its conserved structure (Beneke and Hagen 1998).
4) DNA carries an organism's genetic information and is ubiquitous: copies of the DNA target sequence will be present throughout all tissues of an organism/food, due to the degeneracy of the genetic code as one goes from DNA to protein (Wolf et al. 2000).
5) DNA of interested species can be recovered and amplified from a few copies to easily detectable quantities, even from a complex background of highly degraded samples (Ali et al. 2011a,b; Sakai et al. 2011).
6) DNA can be extracted from all kinds of tissue due to its ubiquity in every type of cell (Wolf and Lüthy 2001; Wolf et al. 2000).

18.3 DNA Extraction and Sampling Effects

All food products contain sufficient levels of DNA from the tissue of origin to allow detection and quantification. DNA is a very stable and long-lived biological molecule present in all tissues of all organisms. The quantity and purity of extracted nucleic acid are important factors in PCR-based detection and the presence of PCR inhibitors in food samples, including complex polysaccharides, haemoglobin, urea, RNA, and DNases (Amagliani et al. 2007; Levin 2008), affects the efficiency and the validity of the assay and leads to false-negative results.

Other important factors which should be taken in to account when selecting a DNA extraction method are the time need to complete the extraction and the toxicity and cost of the chemical products employed in the extraction (Chapela et al. 2007). Moreover, raw and cooked meat can contain several chemical compounds, including products of the Maillard reaction, milk proteins, glycogen, fat, collagen, fulvic acids, and iron, that may be co-purified with the target DNA. All these substances play an important role in the inhibition of nucleic acid amplification (Wilson 1997; Scholz, et al. 1998) and the use of an internal amplification control can reveal false negatives (Hoorfar et al. 2004). For most applications, DNA is purified by utilizing one of several commercially available kits, which are predicated on the adsorption of DNA to special resins.

The major advantage of these procedures is the efficacious abstraction of sundry inhibitors of the PCR reaction that often are present in food samples. However, the relative performance of the kits depends on the food commodity (Di Bernardo et al. 2007). The removal of inhibitory substances can be achieved during preparation of food samples where the samples are incubated in the presence of the detergent hexadecyltrimethyl-ammonium bromide (CTAB), then extracted with chloroform, and the DNA is precipitated with isopropanol. In another method the food sample (usually meat) is treated with proteinase K and SDS followed by the use of a DNA-binding silica resin to purify the released DNA (Levin 2008). The application of PCR-based techniques is therefore closely linked to the selection of suitable methods for DNA extraction.

18.4   PCR-based Techniques

18.4.1   Polymerase Chain Reaction

The PCR technique was developed in 1985 by the American biochemist Kary Mullis (US patent 4683202) who received the Nobel Prize and the Japan Prize for developing PCR in 1993 (Bartlett and Stirling 2003). PCR is an extremely powerful biochemical tool that has allowed the genetic discrimination of related food species in addition to allowing the quantitative determination of the presence of a single species at low levels in a given food.

PCR techniques are capable of differentiating meat from males and females. The other benefits of PCR over conventional methods include detection of wide variety of meat samples (Kesmen et al. 2010). Fresh or processed meat can be easily detected by this technique. It is very reliable and very low levels of adulteration (up to 1%) can be easily identified.

PCR, the repetitive bi-directional DNA synthesis via primer extension of a region of nucleic acid, is simple in design and can be applied in seemingly endless ways. In addition, it is highly specific and can be exquisitely sensitive for target DNA, but is dependent on appropriate experimental design. The progress in PCR technologies has increased throughput, brought higher assay sensitivity and more reliable data analysis, and reduced cost. PCR has been used in many different applications because it has great flexibility in the field of molecular biology. Its principal use is to generate a large amount of a desired DNA product starting from a given template, but it can also be used to amplify very long fragments of DNA in such a way as to synthesize whole genes, to amplify and quantify specific RNA species, to produce RNA fingerprinting, or PCR-mediated cloning, to screen DNA libraries, and to produce DNA sequences.

Hence, the PCR assay has an important contribution to make in detecting adulteration in halal food or commercial products (Ali et al. 2012c). A number of studies have demonstrated that PCR appears to be the ideal assay for confirming the identity of and quantifying the species present in meat and other food products, and could be easily used for successful analysis of DNA molecules that are exposed to heat and begin to breakdown into smaller and smaller fragments due to bacterial, biochemical, and oxidative processes operating under natural environmental conditions (Butler 2005, 2006). Heating for protracted periods ravages DNA, which especially obstructs the DNA-based species identification of extremely heated meat and bone meal. However, bovine DNA could be amplified from meat subjected to the most common cooking procedures, with the exception of pan-frying for 80 min (Arslan et al. 2006). A promising approach is the binding and subsequent sequence analysis of highly fragmented DNA to globules, followed by emulsion PCR and high-throughput sequencing (Kesmen et al. 2010).

Methods based on DNA techniques in which multiple copies of specific piece of DNA sequences *in vitro* can be obtained include species-specific polymerase chain reaction (PCR-SSP) (Karabasanavar et al. 2014), PCR-restriction fragment length polymorphism (PCR-RFLP), PCR product sequencing (Ali et al. 2011a, 2013), real-time PCR (Kesmen et al. 2013), and DNA barcoding (Di Pinto et al. 2013; Ali et al. 2014). PCR techniques are highly selective and specific tests to determine the species of meat in a mixture of meat sample. They are highly sensitive techniques in which even a single copy sequence from a single cell sample can be found out. PCR is a qualitative test and the quality of the mixture is easily determined. These methods have been applied successfully in detection, identification, and quantification of adulteration in foods, and have proved to be very efficient and applicable in monitoring adulterated species in meat and meat products.

Summing up, PCR techniques involve three principles steps: denaturation, annealing, and extension. Any PCR-based detection strategy depends on the selection of oligonucleotide primers and detailed knowledge of the molecular structure and DNA sequences used. The PCR reaction allows the million-fold amplification of a specific target DNA fragment framed by two primers (synthetic oligonucleotides, complementary to either one of the two strands of the target sequence).

A summary of a typical DNA analysis procedure is as follows:

1) Sampling effects and sample preparation.
2) DNA extraction and purification.
3) DNA quantification.
4) PCR setup.
5) Equipment operation.
6) Software analysis.
7) Manual analysis.
8) User interpretation.

PCR amplification is based on the hybridization of specific oligonucleotides to a target DNA and synthesis, *in vitro*, of millions of DNA copies flanked by these primers. The amplification of DNA fragments, followed by agarose gel electrophoresis for fragment size verification, is the simplest PCR strategy applied to evaluate the presence of a species in a meat product.

Additional confirmation methods and/or examination of PCR products can be accomplished by (i) sequencing of DNA amplicons (PCR-sequencing) (Karlsson and Holmlund 2007), (ii) analysis of PCR-single strand conformation polymorphism (Ripoli et al. 2006), (iii) simultaneous amplification of two or more fragments with different primer pairs (multiplex PCR) (Tobe and Linacre 2008), (iv) analysis of PCR-RFLP (Park et al. 2007),

(v) analysis of random amplified polymorphic DNA (PCR-RAPD) (Arslan et al. 2005) or (vi) real-time fluorescence PCR assays (Jonker et al. 2008). Other strategies focused on detection of genetic variability within closely related populations like short sequence repeat markers (microsatellites) or DNA chips (microarrays) can also be applied for meat speciation purposes (Felmer et al. 2008; Teletchea et al. 2008).

A primary requirement for successfully detecting a species by PCR is to choose adequate genetic markers to develop the assay. Both nuclear and mitochondrial genes have been broadly targeted for the identification of game and domestic meat species (Fajardo et al. 2008a). The detection of nuclear DNA might be limited as a result of the generally low copy number of sequences. Utilization of mtDNA increases PCR amplification sensitivity because there are several copies of mtDNA per cell. In addition, mitochondrial genes evolve much faster than nuclear ones and, thus, contain more sequence diversity facilitating the identification of phylogenetically related species (Girish et al. 2005). There are several advantages to the use of mt genes, one of them being that mammalian cells normally harbour 800–1000 mitochondria per cell. In addition, the nucleotide sequences of the mitochondrial *cytochrome b* (mt *cytb*) gene from a large number of animal species are available from GenBank.

The use of universal primer systems for mt *cytb* PCR facilitate the identification of several species and the enhanced sensitivity of the PCR assays within a single analysis. The use of mitochondrial DNA (mtDNA) offers a series of advantages over cell nucleus DNA. Mitochondrial DNA facilitates PCR amplification even in cases where the availability of DNA template after extraction is insufficient for detection (Murugaiah et al. 2009). This is attributed to the fact that mtDNA is several times more abundant than that of the nuclear genome; each mitochondrion is estimated to contain 2–10 mtDNA (Murugaiah et al. 2009).

Mitochondrial genes such as the ATPase subunit 8 and subunit 6, D-loop and *cytb* genes, satellite DNA, actin genes, Art2 short, and CR1 long interspersed repetitive elements in genomic genes have been studied for the purpose of identifying animal species, with most of the methods concentrating on the *cytb* gene as a target sequence (Colombo et al. 2000; Brodmann and Moor 2003). However, the *cytb* gene, which localizes on the mitochondrial genome, has been determined to be a powerful marker for identifying species with DNA analytical techniques (Kocher et al. 1991; Chikune et al. 1994; Forrest and Carnegie 1994; Abdulmawjood and Büelte 2002). Moreover, mtDNA is present in a much higher copy number compared to nuclear DNA, which makes it a useful tool in forensic casework (Wilson et al. 1995; Lutz et al. 1996) and used in species and stock identification studies for the construction of species-specific primers.

A PCR protocol using species-specific mtDNA primer pairs has been developed for the identification of three commercial Russian sturgeon (caviar) species (Desalle and Birstein 1996) and for species differentiation of meat from snails (Abdulmawjood and Büelte 2001). Another reason for using PCR-SSP includes more stability of mtDNA and strength in comparison to nuclear DNA. mtDNA is protected from degradation, even when exposed to harsh environmental conditions. Besides mitochondrial genes, nuclear markers are also described for meat species discrimination exploiting the existence of introns of different sizes which allow the amplification of species-specific DNA fragments. Some examples are the growth hormone gene (Brodmann and Moor 2003), the actin gene (Hopwood et al. 1999), and the melanocortin receptor 1 (MC1R) gene (Fajardo et al. 2008a).

Sequencing of amplified fragments results in a large amount of information without the need to use enzymes or post-analysis. By means of a universal primer pair, single-band amplification products from a wide range of animals can be obtained (Kocher et al. 1989). Further sequence analysis of the generated PCR amplicons can be used for interspecific and intraspecific identification of animal DNA in food products, allowing discrimination of even very closely related species. Due to their adequate level of mutation and great availability of sequences in the databases, mt *cytb*, 12S, and 16S rRNA genes are the most extended genetic markers for species discrimination by PCR-sequencing (Karlsson and Holmlund 2007). Nowadays, gene sequences in mitochondrial DNA (mtDNA), particularly the mt *cytb* gene, are commonly targeted for species identification.

18.4.2 PCR Product Detection

Any PCR response discriminatingly relies on upon the primers. With just a couple of exceptions, primers for animal species identification target variable regions in mtDNA. Mitochondrial DNA is more variable than nuclear DNA, but its high copy number increases sensitivity relative to the PCR of single-copy nuclear sequences. However, because of its maternal origin, mtDNA may not be representative if samples originate from hybrids between species (Lenstra 2003).

Kocher et al. (1989) described a number of universal mtDNA primers that allow the sequencing or detection of various mtDNA segments from known or unknown species (Lenstra 2003; Rostogi et al. 2007). For purposes of detecting all animal DNA in foodstuffs, primers specific for the 16S mtDNA gene were designed that matched completely to species from all mammalian orders (Bottero et al. 2003a,b). The nucleotide sequences of the mt *cytb* gene from a large number of animal species are available from GenBank.

18.4.3 PCR Using Species-Specific Primers

PCR analysis of species-specific mtDNA sequences is the most common method currently used for identification of meat species in food (Bellagamba et al. 2003; Bottero et al. 2003a). PCR-SSP is a unique technique that offers simplicity, specificity, and high sensitivity for meat authentication studies. It is used to identify specific meat species from a mixture of meat samples and to control food authenticity. PCR-SSP targeting muticopy mitochondrial genes has received enormous attention in recent years (Che Man et al. 2010; Fajardo et al. 2010; Yusop et al. 2012). The method involves the amplification of a segment of mt-gene using a pair of species-specific primers followed by electrophoretic detection on gel agarose or polyacrylamide by ethidium bromide staining. Both the single plex (Che Man et al. 2010; Ali et al. 2011a,b) and multiplex (Köppel et al. 2009, 2011; Sakai et al. 2011) PCR assays have been documented.

The advantage of mitochondrial-based DNA analyses derives from the fact that there are many mitochondria per cell and many mtDNA molecules within each mitochondrion, making mtDNA a naturally amplified source of genetic variation. These methods can therefore be used for routine analysis of large numbers of samples, even when aggressive processing treatments have been applied to the food (Mafra et al. 2008; Rojas et al. 2009).

The identification of the species origin of such heated meat remains with the detection of particular mtDNA, which retains its specificity during processing. The specific identification of a target species in matrices containing a pool of heterogeneous genomic DNA sequences is possible and it is a well-adapted technique for the analysis of thermally or otherwise processed products with highly damaged DNA. The main drawback, however, is the need for accurate data on the species target sequences in order to design the corresponding specific primers (Rojas et al. 2009). Compared to single-species PCR systems, multiplex PCR, in which many primers are used together for the amplification of more than one target region, is a promising technique that may reduce cost and enhance the speed, efficiency, and reliability of analysis for the simultaneous identification of various meat species (Tobe and Linacre 2008).

According to studies done by Carney et al. (1997), Rigaa et al. (1997), Yoshizaki et al. (1997), and Abdulmawjood and Büelte (2002), mtDNA analysis requires the isolation of the mtDNA molecule and digestion of the mtDNA with a variety of restriction endonucleases. The resulting fragment patterns are then examined for polymorphisms within and amongst the populations examined. A specific PCR amplification of mtDNA followed by RFLP analysis has already been used in species and stock identification studies (Wilson et al. 1995; Ram et al. 1996; Yoshizaki et al. 1997; Carrera et al. 1999; Cespedes et al. 1999; Abdulmawjood and Büelte 2001, 2002). For example, Fajardo et al. (2007a,b) describe the use of specific oligonucleotides designed on the mitochondrial 12S rRNA and D-loop genes for the identification of various cervid and wild ruminant meats.

Analysis of binary meat mixtures including each target species allowed the detection of 0.1% of all game species. Similarly, Rojas et al. (2009), Rojas et al. (2010a,b) developed PCR techniques with species-specific primers targeting the mitochondrial 12S rRNA and D-loop genes for the identification of various game bird species. The detection level of the PCR assays was set at 0.1% either on raw and sterilized muscular binary mixtures for each of the targeted species. Ha et al. (2006) achieved the unequivocal identification of deer and other ruminant species in animal feedstuffs by using species-specific primers targeting the mitochondrial 12S and 16S rRNA genes. The detection limit of the assay was set at 0.05% for the four developed primer pairs.

Based on PCR, Tartaglia et al. (1998) developed a bovine-specific PCR assay using a bovine-specific mtDNA sequence, which allowed the detection of bovine meat and bone meal in feedstuffs at the 0.125% level. By amplifying the satellite DNA using a bovine-specific primer pair, Guoli et al. (1999) were able to identify raw, cooked (100 °C, 30 min), and autoclaved (120 °C, 30 min) beef without cross-amplification with other animal species tested. In 2007, Che Man and his colleagues developed a technique for pork identification in four types of food products, sausages and their casings, bread, and biscuits, using PCR-SSP detection of a conserved region in the mitochondrial 12S ribosomal RNA (rRNA) gene. The genomic DNA of the food products was successfully extracted, except for the casing samples, where no genomic DNA was detected.

This could be the result of degraded DNA, insufficient target DNA or DNA contaminated by inhibitors such as organic and phenolic compounds (Cardarelli et al. 2005). The result corresponds with the findings reported by Kuiper (1999) which showed that there is no mtDNA detected in highly heat-treated food products, hydrolyzed plant proteins, purified lecithin, starch derivatives, and refined oils.

However, the extracted DNA from other samples was sufficient to be used as a template for PCR amplification and produced clear PCR products on the amplification of the 12S rRNA gene of 387 base pairs (bp) from pork species. PCR-SSP has a number of advantages, for example it offers simple, fast, specific, and highly sensitive species identification. It can be used to analyse cooked or processed products despite the highly damaged DNA. PCR-SSP is a simple and promising method for the detection of pig derivatives that can be adopted by research bodies and quality control laboratories for halal authentication and verification (Che Man et al. 2007). The specificity of PCR primer pairs that encode the *cytb* gene sequence has permitted the amplification of degraded tuna DNA for the identification of cooked and canned tuna fish species in commercial preparations (Unseld et al. 1995).

Carefully designed PCR-SSP under optimized conditions is conclusive in detecting and identifying species, eliminating the need for restriction digestion and/or sequencing of PCR products (Rodríguez et al. 2004; Karabasanavar et al. 2014). Recently, Ali et al. (2015) reported that PCR-SSP is considered a robust method in comparison with other methods such as single nucleotide polymorphism (SNP) analysis, PCR-RFLP, PCR-RAPD, and DNA barcoding (Ballin 2010; Bottero and Dalmasso 2011; Ali et al. 2014; Karabasanavar et al. 2014).

18.4.4 Species-Specific Multiplex Polymerase Chain Reaction

The inherent complexity, low and unequal amplification efficiency, and variable sensitivity for different-length templates have made them unsuitable for target quantification (Bai et al. 2009; Iwobi et al. 2011; Köppel et al. 2011). Compared to conventional single-species PCR systems, multiplex PCR assays with species-specific primers are very promising since they offer multiple target detection in a single assay platform, reducing both cost and time implications (Bottero and Dalmasso 2011; Koppel et al. 2011; Zha et al. 2011; Ali et al. 2014).

Multiplex PCR assays allow simultaneous identification of several species with one PCR by using one universal primer from a conserved DNA sequence in the gene paired with multiple primers targeted to hybridize on species-specific sequences for each species (Rodríguez et al. 2003). Thus, a number of amplicons with different lengths are amplified and detected using a single PCR assay. Multiplex PCR assays use comparatively longer and variable-length DNA templates (Koppel et al. 2011; Sakai et al. 2011) which are not stable in the harsh conditions of food processing (Bielikova et al. 2010) and also entail different sensitivities for different species, making them unsuitable for the analysis of processed commercial goods (Ali et al. 2011a). Additionally, an optimized real-time PCR assay under a background of complex matrices found in processed commercial foods has yet to be described.

Several multiplex PCR assays have been documented for the identification of various animal species (Dalmasso et al. 2004; Di Pinto et al. 2005; Zhang 2013), but none has been aimed at the authentication of prohibited species in Islamic foods. However, Abdulmawjood et al. (2003) developed a multiplex PCR assay for the detection of five haram meat species, namely, pig, dog, cat, monkey and rat, in raw and processed meats and commercial meat products such as meatballs. Later, Abdulmawjood et al. (2012) investigated the efficiency of multiplex real-time PCR for the determination of pig sex in meat and meat products. They reported that the real-time PCR assay allowed the detection of male pig meat at a concentration of 1%, yielding a detection probability of 100% while the detection probability for investigating meat samples containing 0.1% male pig meat was 44.4%. The analytic sensitivity of this system was assessed to be <5 pg DNA per PCR reaction. The assessment of the accuracy of the real-time PCR assay to correctly identify sex individuals was investigated with 62 pigs including males ($n = 29$) and females ($n = 33$) belonging to different breeds/lines.

18.4.5 PCR-RFLP

PCR-RFLP has a special interest for meat species identification. This technique involves the amplification of a large amplicon of several species using a pair of consensus sequence-specific primers followed by digestion of the amplified products using a single or a set of restriction enzymes to produce a species-specific pattern of the restriction fragments (Doosti et al. 2011; Sait et al. 2011). The method is inherently more accurate than PCR-SSP and can distinguish between closely related species by authenticating real products (Ali et al. 2011a; Doosti et al. 2011; Sait et al. 2011). However, PCR-RFLP is clumsier and costlier than PCR-SSP. Moreover, the method uses a comparatively longer DNA template which is prone to degradation in the harsh conditions of food processing (Bielikova et al. 2010).

Ali et al. (2011a) have developed a PCR-RFLP assay with a very short length (109 bp) of amplicon for the authentication of pork in commercial meat products. The assay involves a single restriction digestion and attains the sensitivity of real-time PCR.

Compared to other techniques for species identification by DNA-based methods, PCR-RFLP analysis of mtDNA has offered the greatest advantage (Bellagamba et al. 2001). The advantages of PCR-RFLP are many, for example one universal PCR-primer system in combination with a few restriction enzymes (REs) can be sufficient for species identification (Meyer et al. 1995), no references are necessary once the restriction patterns of the species of interest have been determined, and a careful selection of RE prevents ambiguous results caused by intraspecies polymorphisms (Wolf ct al. 1999). PCR-RFLP constitutes a simpler alternative to sequencing for the identification of genetic variation between and within species (Borgo et al. 1996).

The analysis of PCR-RFLP of *cytb* fragments has already been successfully applied in species differentiation in heated and processed meat products, e.g. sausages (Meyer et al. 1995).

The genomic DNA of pork and lard was extracted using commercial kits and subjected to PCR amplification targeting the mt *cytb* gene. Aida et al. (2005) employed a three-step analysis to determine the identity of meat and fat samples. First, they isolated the genomic and mtDNA from fat and meat and subjected mtDNA to PCR amplification of the *cytb* gene then the *cytb* amplicon I was cut with restriction enzyme to reveal the restriction enzyme cutting pattern so that the identity of the meat and fat source were revealed. Additionally, they reported that the genomic DNA from lard was found to be of good quality and produced clear PCR products on the amplification of the mt *cytb* gene of approximately 360 bp. They concluded that the mtDNA is good enough for routine detection of *cytb*, and the PCR-RFLP analysis of *cytb* represents a powerful and easy method for identification of species. It is a potentially reliable technique for detection of pig meat and fat from other animals for halal authentication assay yielded.

Amongst such DNA-based methods for species identification, PCR-RFLP studies aimed at investigating mitochondrial sequences, such as the variable regions of the *cytb* gene (Herman 2001; Verkaar et al. 2002), offer two main advantages:

i) mtDNA is present in thousands of copies per cell (as many as 2500 copies), especially in the case of post-mitotic tissues such as skeletal muscle (Sorenson and Quinn 1998; Greenwood and Paabo 1999). This increases the probability of achieving a positive result even in the case of samples suffering severe DNA fragmentation due to intense processing conditions (Schmerer et al. 1999; Bellagamba et al. 2001).

ii) The large variability of mtDNA targets as compared with nuclear sequences facilitates the discrimination of closely related animal species even in the case of mixtures of species (Prado et al. 2002). Partis et al. (2000) developed a PCR-RFLP technique as a routine analytical tool for the identification of 22 animal species, including five fish species. The primer pair *cytb* 1/*cytb* 2 amplified a 359-bp sequence of the mt *cytb* gene. Resulting amplicons were digested with *Hae*III and *Hinf*I. All 22 species were discriminated, using the two restriction nucleases, except for kangaroo and buffalo. Each of the two nucleases generated from one to four DNA bands. The PCR amplicons and PCR-RFLP products were separated on 9% polyacrylamide gels and visualized by staining with ethidium bromide. Cooking (0.5 g in a microwave oven set on high for 30 s) did not affect the DNA extractions or banding profiles. The method, however, was found to be unsuitable for analysing mixtures.

Meanwhile, Abdulmawjood et al. (2003) developed a PCR-mediated method for the detection of dog and cat meat in meat mixtures and animal feed. The method was based on PCR-RFLP analyses as they designed species-specific primers for dog and cat by using a conserved mitochondrial DNA region of the *cybt* gene. They analysed the *cytb* gene sequence of both species by PCR-RFLP analysis. The use of the restriction enzymes *Alu* I and *Hae* III yielded specific restriction profiles characteristic for each species. They claim that the meat of both species could additionally be differentiated with species-specific oligonucleotide primers based on specific parts of the *cytb* gene sequences characteristic for dog and cat. The use of these oligonucleotide primers allowed a direct identification of dog and cat meat in meat mixtures even after heat treatment.

Pascoal et al. (2004) conducted a survey of the authenticity of meat species in food products subjected to different technological processes to evaluate the usefulness of a PCR-RFLP method, based on the universal primers *cytb* 1/*cytb* 2, initially designed by Kocher et al. (1989), complemented by RFLP analyses with endonucleases *Pal*I, *Mbo*I, *Hinf*I and *Alu*I (the restriction enzymes), to achieve species identification in 50 raw or processed food products containing one or more meat species. These foods included products subjected to one or more technological processes inherent to the meat sector, such as mincing, flavouring, precooking and freezing, curing, cooking and smoking, dehydration, and sterilization.

Twenty of the products declared mixtures of meat species on their labels. The reference samples of beef, pig, wild-boar, chicken, turkey, duck, quail, deer, roe-deer, ostrich, rabbit, sheep, goat, and horse were considered in their study. Their result exhibited that 15 (30%) of the 50 food samples investigated displayed incorrect qualitative labelling. While this affected only one (11.1%) of the nine raw/cured products, 14 (34.2%) of the 41 products subjected to some type of heat-processing were not correctly labelled. The undeclared presence of turkey was the most frequent concern, since it was detected in seven food products.

The complete absence of a declared species of high commercial value – such as beef or roe-deer – was observed in another four cases.

However, the authors declared that the PCR-RFLP method, involving the use of universal primers aimed at amplifying mitochondrial sequences of the *cytb* gene, used allowed the detection of three different meat species in minced, flavoured, pre-cooked and frozen, cured, cooked and smoked, dehydrated, and sterilized food products. The method proved to be a rapid and easy-to-perform two-step analytical approach to achieve qualitative meat species identification in raw and cooked food products containing one or more different species (Pascoal et al. 2004).

Aida et al. (2007) developed an approach based on the use of PCR-RFLP in mitochondrial genes for the presence of pig derivatives in three types of food products – sausages and casings, bread and biscuits – using PCR-RFLP analysis of a conserved region in the mt *cytb* gene. The genomic DNA of sausages and casings, bread and biscuits were extracted. The genomic DNA from the food products was found to be of good quality for the sausages and produced clear PCR products on the amplification of the mt *cytb* gene of approximately 360 bp. However, no genomic DNA was detected from the casing samples and poor quality genomic DNA was extracted from bread and biscuits. No amplification of the mt *cytb* gene was produced from bread and biscuit samples. To differentiate between samples, the amplified PCR products were digested with RE BsaJI, resulting in species-specific RFLP. The *cytb* PCR-RFLP species identification assay gave excellent results for detection of pork adulteration in food products and is a potentially reliable technique to tackle species adulteration or fraudulent species substitution for halal authentication.

Similarly, Murugaiah et al. (2009) used PCR-RFLP analysis of the *cytb* gene of mitochondrial DNA to trace adulteration present in mixed meat. PCR products of 359-bp were successfully obtained from the *cytb* gene of beef, pork, buffalo, quail, chicken, goat, and rabbit meats for the purpose of species identification and halal authentication. *AluI, BsaJI, RsaI, MseI,* and *BstUI* enzymes were identified as potential restriction endonucleases to differentiate the meats. The genetic differences within the *cytb* gene amongst the meat were successfully confirmed by PCR-RFLP. Aida et al. (2005) used PCR-RFLP of *cytb* gene to detect pork adulteration in raw meats. The PCR-RFLP technique presents the advantage of being cost-effective, simple and especially adaptable for routine large-scale studies like those required in inspection programmes (Pfeiffer et al. 2004). However, PCR-RFLP has the disadvantage of not being applicable in processed foods due to DNA destruction as amplification of the large DNA fragments which are required for enzymatic restriction is impeded by thermal DNA degradation (Fajardo et al. 2010).

In a large-scale study Doosti et al. (2011) conducted extensive work on the identification of fraud and adulteration in industrial meat products in Iran. They applied the PCR-RFLP technique in the detection of beef, sheep, pork, chicken, donkey, and horse meats in food products. They collected 224 meat products, including 68 sausages, 48 frankfurters, 55 hamburgers, 33 hams, and 20 cold-cut meats, from different companies and the local food markets. In this study, genomic DNA was extracted and PCR was performed for gene amplification of meat species using specific oligonucleotide primers. Raw meat samples are served as the positive control.

For differentiation between donkey's and horse's meat, the mtDNA segment (*cytb* gene) was amplified and products were digested with *AluI* restriction enzyme. The results of an investigation by Doosti et al. (2011) showed that 6 of 68 fermented sausage (8.82%), 4 of 48 frankfurter (8.33%), 4 of 55 hamburger (7.27%), 2 of 33 ham (6.6%), and 1 of 20 cold-cut meat (5%) samples were found to contain haram (unlawful or prohibited) meat. These results indicated that 7.58% of the total samples did not containing halal (lawful or permitted) meat. The authors concluded that molecular methods such as PCR and PCR-RFLP are potentially reliable techniques for detection of meat type in meat products for halal authentication.

Ali et al. (2012a) combined the use of species-specific primers and RFLP analysis to develop a PCR-RFLP assay for the authentic and sensitive detection of swine DNA in 11 different meat-providing animals (fresh raw meat such as pork/pig, beef/cow, mutton/sheep, chevon/goat, venison/deer, and chicken/hen) and fish species (cichlid, shad, shrimp, tuna, and cuttlefish). They simulated pork adulteration in processed meat products, and pork–beef and pork–chicken binary mixtures were prepared by spiking 10, 5, 1, 0.1, and 0.01% (w/w) pork with beef and chicken in 100-g mixtures on three different days by three independent analysts. The mixtures were mashed with a blender and autoclaved at 120 °C under 45 psi pressure for 2.5 h. They also checked four types of the most popular finished meat products (meatballs, streaky beacon, frankfurters, and burgers) in the Malaysian food market in order to verify the PCR-RFLP assay performance and sensitivity.

The assay targeted a 109-bp fragment of swine mt *cytb* gene which is present in multiple copies per cell and highly conserved in evolution. The amplicon contained two *Alu*I sites within it, offering opportunities for product authentication by RFLP analysis. The authors claimed that the assay was accurate, sensitive enough to detect 0.0001 ng of swine DNA in pure formats and to trace 0.01% (w/w) spiked pork in an extensively processed ternary mixture of pork, beef, and wheat flour. Variations of analysts, days, and samples did not significantly affect the assay performance, showing good reproducibility and precision. Hence, they concluded that the use of PCR-RFLP analysis further confirmed the authenticity issues, eliminating doubts from all possible sources of the tested food mixture. In another study, Haider et al. (2012) investigated the efficiency of the mitochondrial cytochrome c oxidase subunit 1 (*COI*) gene of mtDNA for the identification of the species origin of raw meat samples of cow, chicken, turkey, sheep, pig, buffalo, camel, and donkey using PCR-RFLP. PCR yielded a 710-bp fragment in all species. Haider et al. (2012) digested the amplicons with seven restriction endonucleases (Hind II, Ava II, Rsa I, Taq I, Hpa II, Tru 1I, and Xba I) that were selected based on preliminary *in silico* analysis. Different levels of polymorphism were detected amongst samples. They found that the level of *COI* variation revealed using only Hpa II was sufficient to generate easily analysable species-specific restriction profiles that could distinguish unambiguously all targeted species. The authors concluded that meats of cow, buffalo, sheep, camel, turkey, chicken and donkey can be qualitatively identified and differentiated by robust and reliable PCR-RFLP of the mitochondrial COI gene, and this technique could be used as a routine control method in food control laboratories.

It worth mentioning that PCR-RFLP can be successfully applied to distinguish and authenticate similar-sized PCR products obtained from the meat products of multiple species (Doosti et al. 2011), demonstrating its superiority over PCR-SSP. Due to its inherent ability to authenticate and differentiate PCR products of identical molecular size, PCR-RFLP is increasingly being used in clinical laboratories to differentiate closely related bacterial strains, which is often not possible by PCR-SSP (Sait et al. 2011). The techniques can qualitatively identify all types in mixed meat products, showing good reproducibility and precision, therefore, it could be used for detection of pork adulteration in food products or fraudulent species substitution for Halal authentication.

18.4.6 PCR-RAPD

The random amplified polymorphic DNA technique consists of the amplification of DNA fragments using a short arbitrary primer that ties multiple locations on the genomic DNA followed by separation of amplified fragments based on their sizes using gel electrophoresis. Samples are identified by comparing the DNA bands of the fingerprints, which are expected to be consistent for the same primer, DNA and experimental conditions used. The PCR-RAPD method has the advantage that little or no information on the gene fragments to be amplified is needed to generate species-specific patterns. PCR-RAPD is a simple and fast method that can be used for halal authentication of meat without complex analytical steps like DNA restriction, sequencing or hybridization (Wu et al. 2006; Ballin 2010). Known standards must be run together each time a sample is tested (Koh et al. 1998). Martınez and Yman (1998) and Koh et al. (1998) carried out a PCR-RAPD study of game meats from elk, kangaroo, reindeer, buffalo and ostrich, wild boar, and red deer, as well as some domestic meat species, identifying meats species successfully.

Species-specific profiles where obtained in fresh, frozen, and canned samples. Martınez and Danielsdottir (2000) correctly identified different seal and whale meat products (frozen, smoked, salted, dried, etc.) by RAPD and PCR-SSCP techniques using consensus primers designed on the mitochondrial cytochrome gene. Yau et al. (2002) authenticated three snake species by PCR-RAPD to avoid illegal trade of threatened populations. The PCR-RAPD method has also been successfully used for species identification in meat, fish and vegetable foodstuffs (Arslan et al. 2005; Koveza et al. 2005; Mohindra et al. 2007).

Huang et al. (2003) performed authentication of meats samples from ostrich, quail, dove, emu, and pheasant using PCR-RAPD fingerprinting. Arslan et al. (2005) differentiated meats from wild boar, bear, camel and domestic species by PCR-RAPD using a unique 10 bp oligonucleotide. Wu et al. (2006), Rastogi et al. (2007), and El-Jaafari et al. (2008) successfully used by PCR-RAPD to identify different species belonging to the family Cervidae (sika deer, sambar deer, rusa deer, tufted deer, black muntjac and Reeve's muntjac).

Rastogi et al. (2007) successfully applied the PCR-RAPD technology to identify snake and buffalo, amongst other species, targeting the mitochondrial 16S rDNA and NADH dehydrogenase subunit 4 (ND4) genes and the nuclear actin gene. The main disadvantage of PCR-RAPD is the difficulty of obtaining reproducible results, as PCR amplifications have to be developed under strictly controlled and standardized conditions such as temperature, number of cycles, and reagent concentration. It also requires high-quality starting DNA in order to achieve reproducible RAPD profiles, so the application of the technique is limited in highly processed meats with extensively degraded nucleic acids. In addition, RAPD analysis is not suitable for identification of a target species in admixed meats consisting of more than one species due to the non-specific nature of the PCR reaction (Fajardo et al. 2010).

18.4.7 Real-time PCR

Real-time PCR or quantitative PCR refers to the detection of PCR-amplified target DNA (amplicons) usually after each PCR cycle. The PCR process generally consists of a series of temperature changes that are repeated 25–50 times. These cycles normally consist of three stages: the first, around 95 °C, allows the separation of the nucleic acid double chain; the second, around 50–60 °C, allows the binding of the primers with the DNA template, the third, between 68 and 72 °C, facilitates the polymerization carried out by the DNA polymerase. Conventional thermocyclers often require 2–3 h to complete 35–40 'thermal' cycles. The reactions can be monitored at the early stages and the reactions taking place can be monitored at every step. The early detection or prediction of results can be achieved at an early stage of the reactions where the signal is readily followed on a computer screen. Each point is automatically plotted and the extent of amplification is followed as an ongoing continual direct graphical plot. Computer software handles all of the preprogrammed calculations and plotting of data.

However, real-time PCR depends on the emission of an ultraviolet (UV)-induced fluorescent signal that is proportional to the quantity of DNA that has been synthesized. Several fluorescent systems have been developed for this purpose and there are two types of dye chemistries adapted to real-time PCR: (i) nonspecific fluorescent dye that intercalates with double-stranded (ds) DNA in a blind fashion and (ii) sequence-specific DNA probes consisting of oligonucleotides that are labelled with a fluorescent reporter dye at one end and a quencher at the other. Reporter dye fluoresces only when the probe is hybridized with the DNA target, signalling the amplification of specific targets (Fajardo et al. 2010; Yusop et al. 2012). The first category of fluorescence chemistry sometimes provides nonspecific detection as it intercalates with both the amplified and non-amplified ds-DNA and includes SYBR green I (Farrokhi and Joozani 2011) and EvaGreen (Ihrig et al. 2006) chemistries.

The second category of fluorescence chemistry provides additional target screening through the probe hybridization and thus is more specific than PCR-SSP. The two categories of fluorescence-based chemistries include the TaqMan hydrolysis probe (Fajardo et al. 2010; Rojas et al. 2010c; Köppel et al. 2011) and the molecular beacon probe (Yusop et al. 2012). Further, TaqMan chemistries are simpler and more reliable than those of the molecular beacon probe. The simplest, least expensive, and most direct fluorescent system for real-time PCR involves the incorporation of SYBR Green I dye (Symmetrical Syanine dye) whose fluorescence under UV greatly increases when bound to the minor groove of double-helical DNA. The resulting DNA–dye complex absorbs blue light ($\lambda_{max} = 497 \, nm$) and emits green light ($\lambda_{max} = 520 \, nm$). SYBR Green I lacks the specificity of fluorescent DNA probes, but has the advantage of allowing a DNA melting curve to be generated and software calculation of the thermal denaturation temperature (*Tm*) of the amplicon after the PCR. This allows identification of the amplified product and its differentiation from primer dimers, which also result in a fluorescent signal with SYBR Green I but usually have a lower *Tm* value. Use of SYBR Green I is a more flexible method without the need for individual probe design (Farrokhi and Joozani 2011). Some of the advantages of real-time PCR-based fluorescence technology are (i) the potential for quantitative measurements at an early stage in the PCR process, which makes it more precise than end-point analysis, (ii) the discrimination of the origin of DNA without the need for any additional time-consuming and laborious steps such as sequencing, enzyme digestion or conformational analysis, (iii) fluorescence data can be collected directly from a real-time PCR instrument or a fluorescence spectrophotometer, avoiding the need for electrophoresis, (iv) the rapidity of assays, allowing the routine high-throughput screening of multiple samples, and (v) the great reduction in potential contamination of the PCR mixture with target DNA because the reaction tubes remain closed throughout the assay (Chisholm et al. 2008; Fajardo et al. 2010). Real-time PCR assays can be automated and are sensitive and rapid. They can quantify PCR products with greater reproducibility while eliminating the need for post-PCR processing, thus preventing carryover contamination (Jothikumard and Griffiths 2002).

Real-time PCR assays have been described for the detection of a number of meat species, such as beef, pork, lamb, horse, chicken, turkey and duck (Laube et al. 2007; Jonker et al. 2008). Fajardo et al. (2008b,c) accomplished a SYBR Green I real-time PCR assay to detect red deer, fallow deer, roe deer, chamois and pyrenean ibex in meat mixtures using species-specific primers targeting the mitochondrial 12S rRNA and D-loop genes. In a further study, these authors reported an improvement of the assay in terms of specificity, sensitivity, efficiency, and accuracy by using specific TaqMan probes instead of the SYBR Green I fluorescent intercalator (Fajardo et al. 2009). Rojas et al. (2010b) developed real-time PCR assays using TaqMan probes for verifying the authenticity of meat and commercial meat products from game birds including quail, pheasant, partridge, guinea fowl, pigeon, Eurasian woodcock, and song thrush. The assay is based on specific primers and probes designed for each target species on the mt 12S rRNA gene.

Real-time PCR approaches allow the detection of even minute traces of different animal species in products of complex composition and are considered one of the most promising molecular tools for meat authentication. However, most reports lack validation and applicability assays to commercial food samples. Application of real-time PCR techniques for the specific detection of pork meat has been reported, but some groups have only used the technique for the specific detection of pork species without achieving quantification (Farrokhi and Joozani 2011; Ali et al. 2012a; Cammà et al. 2012). However, a real-time PCR-based method has been developed to quantify bovine contamination in buffalo products (Drummond et al. 2013). The quantification procedure proposed involves amplifying a sample with both primer sets and then normalizing the total DNA using the total non-normalized bovine and buffalo DNA. To correct for the potential deviations between the real and measured DNA quantity caused by biological differences between species, the use of calibration curves generated from each analysed matrix is proposed. A recent method proposed by Ballin et al. (2012) is based on the use of PCR LUX primers to amplify repetitive and single-copy sequences to establish the species-dependent number of amplified repetitive sequences per genome. Though the approach gave promising results regarding relative pork meat quantification, further improvement is required to increase specificity. With the advent of real-time PCR, the technique is simplified. Real-time PCR does not only depict amplification in real time (on line or live), but is also quantitative. Detection of meat adulteration simultaneously creates the need for the assessment of adulteration level. Andreo et al. (2006) assessed the level of adulteration in a series of DNA mixtures containing (percentages) 1/99, 5/95, 10/90, 40/60, 50/50, 60/40, 90/10, 95/5, and 99/1 ratios of cattle/horse, cattle/wallaroo, pork/horse, and pork/wallaroo by using duplex real-time PCR, allowing identification of the peaks in double-species duplex reactions were established as follows: 5% (cattle or wallaroo) in cattle/wallaroo mixtures, 5% pork and 1% horse in porcine/horse mixtures, 60% pork and 1% wallaroo in porcine/wallaroo mixtures, and 1% cattle and 5% horse in cattle/horse mixtures. In all cases, 1% corresponded to 0.4 ng of DNA.

Soares et al. (2013) proposed the SYBR Green I real-time PCR as a simple, fast, sensitive, and reliable method for the quantitative detection of pork meat in processed meat products. For the development of the method, binary meat mixtures containing known amounts of pork meat in poultry meat were used to obtain a normalized calibration model from 0.1 to 25% with high linear correlation and PCR efficiency. The method revealed high specificity by melting curve analysis. The method was validated using blind mixtures and subsequently applied to quantify the presence of pork meat in poultry processed meat products available commercially. The proposed SYBR Green I real-time PCR method proved to be a powerful and simple technique, highly specific and sensitive for pork species identification without requiring any post-PCR treatment and the use of more expensive fluorescent probes (Soares et al. 2013).

Recently, Kumari et al. (2015) detected meat adulteration by using real-time PCR for the identification of cattle and buffalo meat from a mixed meat sample using *cytb* gene variability by real-time PCR. In real-time PCR, the common forward primer with cattle-specific reverse primer showed a melting temperature peak 76.2 °C on cattle DNA while the common forward primer with buffalo-specific reverse primers showed a melting peak at 78.2 °C on buffalo DNA. Even in duplex PCR it showed only species-specific melting peaks in respective species DNA, but when duplex PCR was evaluated on a cattle-buffalo mixed DNA template in equal proportion it exhibited two peaks, a major buffalo specific and a minor cattle specific, merging into one broader peak at 78.2 °C. It was possible to detect presence of mixed DNA by real-time PCR duplex primers. The duplex real-time PCR showed only a single broader peak at 78.2 °C at 1 : 10 and all further ratios. Hence all independent cattle-specific real-time PCR was run on mixed DNA which produced a cattle-specific melting peak at 76.2 °C up to 1 : 1000 dilutions. Thus, it was possible to detect and differentiate cattle meat mixed in buffalo meat up to 1 : 1000 fraction, i.e. 9 pg of cattle DNA adulterated in buffalo DNA, by running a duplex PCR followed by a cattle-specific real-time PCR. Duplex real-time PCR did not produce any amplification and melting peaks on DNA templates from sheep, goat, and chicken. Thus, real-time PCR was found to be successful in differentiating cattle and buffalo mixed-meat samples.

18.4.8 Species-Specific Real-time PCR (TaqMan)

Conventional PCR techniques allow the qualitative detection of different animal species in an admixture, but they are not appropriate for the quantitation of species in a product. DNA-based quantitative methods for the identification of species are based on quantitative competitive PCR (Frezza et al. 2003), densitometry (Calvo et al. 2002) or real-time PCR (Rodríguez et al. 2004). The TaqMan assay differs from conventional PCR methods in its use of fluorescent dyes for detection purposes. Real-time quantitative PCR (Q-PCR) is a recent technological advance allowing the real-time visualization of amplicon production and as a consequence allows PCR to be used quantitatively for the first time. Q-PCR may be performed using a number of platforms, the most popular of which is the TaqMan system. The TaqMan fluorogenic probe, labelled with a reporter and a quencher dye, binds to a target DNA between the flanking primers. During PCR amplification the 5′–3′ exonuclease activity of the Taq DNA polymerase cleaves the probe hybridized to the template, releasing the 5′ reporter from the quenching effects of the 3′ quencher. Fluorescence emission is measured during the reaction and is directly proportional to the amount of specific PCR products.

The measurement of fluorescence throughout the reaction eliminates the need for post-PCR processing steps, such as gel electrophoresis and ethidium bromide staining of target DNA, easing automation of the technique and large-scale sample processing. Moreover, there is reduced potential for contamination of the PCR mixture with target DNA because the reaction tubes remain closed throughout the assay. Restrictions on the PCR amplicon size (maximum 150 bp) lend this system to detection of DNA in highly processed food products where the likelihood of DNA degradation is high. TaqMan PCR has been used for authentication of food products (Vatilingom et al. 1999; Wiseman 1999).

TaqMan technology is at the forefront of analytical techniques for food authentication. However, there are some factors that can affect the accurate quantitation of DNA, including species and sample preparation (temperature, processing), DNA target and extraction approach, level of degradation, DNA recovery, tissue type, PCR efficiency, matrix background, and DNA template amount. Species identification by PCR in meat samples could be complicated because of the presence of such inhibitory substances in products with a complex composition, as mentioned earlier. The risk of amplification inhibition is higher when a large amount of DNA (>50 ng) is tested (Dooley et al. 2004), therefore several attempts have been made to use species-specific real-time PCR (TaqMan) analytical methods to overcome such difficulties and significantly enhance the specificity and reliability of the assay (Koppel et al. 2011; Sakai et al. 2011). Hird et al. (2004) achieved the identification of deer and some domestic species by the optimization of real-time TaqMan technology with truncated primers located on the mt *cytb* gene. Lopez-Andreo et al. (2006) developed TaqMan real-time PCR systems on the mt *cytb* gene for the detection and quantification of DNA from ostrich and other meat species. This research group also evaluated the usefulness of post-PCR melting temperature analysis for the identification of kangaroo, horse, bovine, and porcine species in mixed samples using mt *cytb* sequences and the SYBR Green I fluorescent molecule (Lopez-Andreo et al. 2006). Chisholm et al. (2008) identified DNA from pheasant and quail in commercial food products using species-specific primers and TaqMan signed on the mt *cytb* gene. In this context, Dooley et al. (2004) developed DNA detection assays that would enable sensitive identification of beef, pork, lamb, chicken, and turkey in raw and cooked meat products. Assays were developed around small (amplicons <150 bp) regions of the mt *cytb* gene. For detection purposes, Dooley et al. (2004) developed two TaqMan probes: the first specific to the mammalian species (beef, lamb, and pork) and the second to the poultry species (chicken and turkey). The normal end-point TaqMan PCR conditions were applied and they limited PCR to 30 cycles to ensure that assays were reliable and sensitive enough to detect their respective target meat species. Applying the assays to DNA extracts from raw meat admixtures, it was possible to detect each species when spiked in any other species at a 0.5% level. The absolute level of detection, for each species, was not determined, but experimentally determined limits for beef, lamb, and turkey were below 0.1% (Dooley et al. 2004).

Based on the amplification of a fragment of the mitochondrial 12S ribosomal RNA gene (rRNA), Rodríguez et al. (2005) developed specific Q-PCR for the quantitation of pork in binary pork/beef muscle mixtures. The method combined the use of pork-specific primers that amplify a 411 bp fragment from pork DNA and mammalian-specific primers that amplify a 425–428 bp fragment from mammalian species DNA, which are used as endogenous control. They used the internal fluorogenic probe (TaqMan), which hybridizes in the 'pork-specific' and also in the 'mammalian' DNA fragments, to monitor the amplification of the target gene. They cited that the comparison of the cycle number at which mammalian and pork-specific PCR products are first detected, in combination with the use of reference standards of known pork content, allows the determination of the percentage of pork in the pork/beef muscle binary mixtures. They evaluated a linearity test, regression line parameters, and sensitivity parameters for the pork-specific TaqMan PCR system. Their significant data demonstrated the specificity and sensitivity of the assay for detection and quantitation of pork in the range 0.5–5%. In another study, analysis of pork adulteration in commercial meatballs targeting porcine-specific mt *cytb* gene by TaqMan probe real-time PCR was conducted by Ali et al. (2012a). They mentioned that the assay combined porcine-specific primers and the TaqMan probe for the detection of a 109 bp fragment of porcine *cytb* gene showed 94–106% recovery of pork DNA in beef meatballs with 0.01–100% of spiked pork, and showing a strong agreement with the actual values of spiked pork in tested samples. The detection limit of the assay was 0.01% of pork in beef meatballs with a very low relative error ($\leq6\%$). High precision was confirmed by residual analysis and also by enough replicates in each measurement. Analysis of 45 commercial meatballs of various meat species revealed high specificity and reproducibility of the assay in quantifying pork in meatball formulations. The authenticity of the assay was verified by PCR-RFLP analysis. In their work, Ali et al. (2012a) not only considered all the typical factors found in a meatball formulation during the process optimization but also limited the application of the developed assay to quantify pork only in meatball formulations. Hence, they claimed that the applied method showed the suitability of the assay to determine pork in commercial meatballs with a high accuracy and precision, therefore it has strong potential to be used by regulatory and enforcement bodies and quality control laboratories for halal and kosher food verification and certification (Ali et al. 2012a).

18.4.9 Immunological Techniques (ELISA)

The first immunoassay was developed by Yalow and Berson in 1959 for the determination of insulin in blood. Immunological techniques are based on the ability of antibodies (immunoglobulins) to recognize three-dimensional structures and play a major role in biochemical research and routine control (e.g. pregnancy testing). As an analytical tool, the techniques are discussed by Billett et al. (1996).

Immunological techniques require no expensive equipment or elaborate protocols and are still in use. The ideal immuno-chemical reactions, such as ELISA, are cost-effective, rapid and highly sensitive, and specific for the determination of a specific adulterant in extracts of meat products and to verify food authenticity for regulatory purposes. ELISA results can be obtained within two to three hours. ELISA is well suited to market survey studies where a large number of samples are collected and handled at a time.

Most applications in this field focus on identification of meat and milk of numerous animal species, either directly in the raw material or in processed food. It is a good technique for closely related species identification, and adulteration up to 2% can be easily detected. Pressure-cooked meat at 133 °C for 20 min can be identified by this technique.

Today, ELISA is used because of its sensitivity, selectivity, and versatility to detect animal proteins and a number of commercial immunoassays are available where species-specific proteins, or epitopes, have been developed for most animals used for meat production, including pig, cattle, sheep, and poultry (Giovannacci et al. 2004; Asensio et al. 2008).

Chen and Hsieh (2000) were the first to develop ELISA using a monoclonal antibody to a porcine thermal-stable muscle protein for detection of pork in cooked meat products. Although heating decreases the sensitivity and specificity of the antisera, adequate performance of a species-specific ELISA with commercial antisera (Giovannacci et al. 2004) and of a pork-specific indirect ELISA (Jha et al. 2003) has been reported. However, the ELISA procedures are able to detect porcine skeletal muscle (the most abundant type of muscle in the animal body), but not cardiac muscle (located only in the heart), smooth muscle (a critical component of numerous animal tissues providing elasticity to the walls of arteries, linings of the gastrointestinal tract, reproductive, urinary and respiratory tracts, and the lymphatic system), blood, and non-muscle organs, It is also not adequate for sensitive detection of ruminant material in feed (Myers et al. 2007).

ELISA employs an antiserum with antibodies to a specific protein of that species. It is involves the immobilization of either an antigen or an antibody onto a solid surface followed by the detection of antigen–antibody interactions by virtue of a labelled enzyme (a protein that catalyses a biochemical reaction) that converts a suitable substrate into a colour product or releases an ion which reacts with another reactant to generate a detectable change in colour (Bonwick and Smith 2004), i.e. detects the presence of an antibody or an antigen in a sample. The two ELISA formats commonly used for meat authentication and speciation and in food component analysis are indirect ELISA and double-sandwich ELISA. Indirect ELISA utilizes two antibodies, one specific to the antigen and the other coupled to an enzyme. This second antibody gives the assay its 'enzyme-linked' name, and will cause a chromogenic or fluorogenic substrate to produce a signal. In other words, indirect ELISA involves the adsorption of a fixed amount of meat proteins that may or may not contain the antigen molecules of a target species onto a microtiter plate. An antispecies antibody is then used to bind the antigen to the solid phase. The immunoreaction between the bound antigen and the antibody can be detected by a secondary anti-immunoglobulin enzyme conjugate, and subsequently colour is developed by the addition of the enzyme substrate.

The colour intensity is proportional to the amount of antigen present in the protein extract. Indirect ELISA is often used to determine the titers and specificity of an antibody reagent. However, it is neither a convenient nor a reliable assay format for the routine analysis of unknown samples because the protein concentration for each meat extract must be determined before the sample can be properly diluted and coated on the microtiter plate. The dilution and composition of the meat extract affect the antibody binding, and thus the ELISA results (Chen et al. 1998; Hsieh et al. 2002). On the other hand, in double-sandwich ELISA the antigen is bound between two antibodies: the capture antibody and the detection antibody. This means that this technique employs an antibody immobilized on the solid phase as the capture antibody and another antibody as the detecting antibody (detection antibody) to form a sandwich with the antigen in the middle and two antibodies attached on the different epitopes of the antigen molecule (Liu et al. 2006). The detection antibody is coupled to an enzyme to realize a detectable change in colour (Berg et al. 2002). This is a more user-friendly format that is widely used for commercial test kits because properly diluted sample extracts can be directly added to the assay plates, which are pre-coated with the first capturing antibody reagent.

ELISA tests can be obtained in either qualitative or quantitative formats. While qualitative ELISA provides either positive or negative results, quantitative ELISA determines antigen concentration by interpolating optical or fluorescence intensity into a standard curve generated by a serial dilution of targets (Goldsby et al. 2003; Asensio et al. 2008). Nevertheless, the prerequisite for an ELISA test is the availability of a sufficient amount of antibodies to detect analytes of interests. Both monoclonal (Chen et al. 2004; Liu et al. 2006) and polyclonal (Hsu et al. 1999) antibodies can be raised against an antigen for food component identifications. Polyclonal antibodies (PAbs) are preferred for the detection of denatured proteins because they offer broad recognition of different epitopes and more tolerance to small changes, such as denaturation and polymerization of antigens (Asensio et al. 2008).

However, PAbs are associated with a number of problems, such as variable affinity, limited production, and requirements of extensive purification to eliminate cross-species reaction. On the other hand, monoclonal antibodies (MAbs) can be homogeneously produced by hybridoma technologies and have well-defined biological function and consistent specificity (Asensio et al. 2008). Both MAbs and PAbs can be raised against soluble and structural proteins of the muscle cell for species detection. However, soluble proteins are more susceptible to denaturation and hence are not suitable for authentication of species in processed food (Smith et al. 1996; Hsu et al. 1999). On the other hand, MAbs raised against heat-stable structural proteins, such as troponin I, perform well to distinguish species in raw meats as well as processed foods and feeds (Hsieh et al. 2002; Chen et al. 2004).

Commercial MAbs in the form of strips are currently available for the detection of animal, mammalian, and ruminant proteins in foods and feedstuffs. Although ELISA is useful to discriminate the tissue source and type of animal proteins, cross-species reactions between closely related species are reported in several cases (Chen et al. 2004; Asensio et al. 2008). Moreover, the sensitivity of ELISA assays is not equivalent to that of DNA-based methods (Fajardo et al. 2010; Ali et al. 2011a; Yusop et al. 2012) and also varies significantly in a mixed background of multiple species (Liu et al. 2006). Certain important factors must be taken into consideration to identify an animal species using an immunoassay, for example a suitable species marker must be chosen as the antigen for antibody development. Once a suitable antigen has been selected, the most important criterion for the successful development of enzyme immunoassays for meat speciation is the production of species-specific antibodies raised against the antigen. The antibodies are the key immunoreagents that differentiate individual species and for food authentication consideration should be given to the selection of the antigen bound by the antibody, the accuracy, and validation and matrix effects.

18.5   Advantage and Disadvantage of Immunochemical Techniques

Immunochemical technology has several advantages: it is robust, rapid, highly sensitive, selective, has high working capacity, simplicity of working protocols, few instrumental requirements, little sample treatment, and it is capable of detecting low levels of significant variations in food matrices. It needs a minimum sample size (microlitre volumes), reduces the consumption of chemicals due to their use in the form of a test kit, works as a screening method for *in situ* applications, has a low cost per sample analysis, and is reliable and easy to use. Its disadvantages include the need for the widespread availability of appropriate antibodies and standards because the ELISA approach includes the initial difficulty of producing an antibody specific to a particular target (protein). The presence of interfering substances decreases the selectivity and the capacity of an assay to discriminate between similar substances. However, this is a relatively minor difficulty to overcome when the selectivity of the technique is taken into account. In addition, accurate measurement is only possible if (i) sample matrices are identical to the reference material or (ii) matched standard materials or standards that have been validated for the matrix are available. As the heat lability of proteins is the main obstacle to the general application of these methods, the drawbacks of the immunochemical techniques may result from the fact that reagents or sample components behave as specific reagents (antigen or antibody), therefore the availability of adequate immuno-reagents is a very important criterion, as well as changes in pH or use of denaturing conditions, immuno-reagent stability, single-analyte determination, and matrix interferences due to cross-reactivity might occur with the secondary antibody, resulting in non-specific signal, difficulty in non-aqueous media such as the oily matrix, and qualitative information that is difficult to interpret.

Moreover, the ELISA assay sometimes needs confirmation of screening results, therefore careful design of standard sample results in calibration curve analysis alongside spike samples can help to evaluate the performance of the assay. In general, the methodology is still not well accepted in the food industry, and it needs working protocols, reference materials and standardization between protocols, validation and in-house prevalidation tests, and collaborative trials to ensure proficiency testing and the elaboration of reference materials for testing targets. It also requires standardization between protocols, reference materials, and accreditation of analysis by independent organizations. ELISA techniques that are designed to identify the species of origin of meats in cooked products usually involve a complex extraction of thermostable antigens, and antisera must then be purified to achieve species specificity.

An early study conducted by Patterson and Jones (1989) described two ELISA methods designed to detect heated lean meats, particularly pork, in a variety of meat products. An indirect ELISA produced a linear response of optical density versus percentage of pig meat in lean-meat mixtures with beef, lamb or chicken. The assay permitted the detection of 5% pork lean meat mixed with other lean meats. A significant variation in the response to individual pig muscles meant that the assay could not be used to evaluate absolutely the lean meat content of products. Liu et al. (2006) used the double-sandwich ELISA method for the detection of pork in processed meats. They stated that the technique was able to detect 0.05% (w/w) of laboratory-adulterated pork in chicken, 0.1% (w/w) pork in beef mixtures, 0.05% (w/w) pork meal in soy-based feed, and 1% commercial meat and bone meal, containing an unknown amount of pork, in soy-based feed. The specificity of the assay was 100%, and no false-positive results were found. The heat treatment of meat samples (132 °C, 2 h) did not affect the assay performance. Thermostability is not absolute and assay response and specificity diminish with prolonged cooking or heating, and these assays cannot be used quantitatively. Moreover, for food authentication, consideration should be given to the selection of the antigen bound by the antibody, the accuracy, and validation and matrix effects.

18.6 Electronic Nose

EN technology includes various types of electronic chemical semi-selective gas sensors (sensor array) with broad and partly overlapping selectivity for the measurement of the volatile compounds present in the headspace of a food sample. EN has several advantages over other techniques for analysing food aroma. It is a non-destructive, rapid, accurate, relatively low cost, and reliable method. Only a small amount of sample preparation is involved, it provides fast analysis and is an environmentally friendly tool for the detection of porcine-based ingredients in foods. It also has a good correlation with data from sensory analyses.

However, this technique employs sensors that are not very selective for particular types of compounds, thus preventing any real identification or quantitation of individual compounds present in a food sample. Such a drawback has obvious implications for food authentication, as an adulterant could not be definitively identified. In spite of this, EN technology has been used to develop a rapid method for halal screening. Combination with suitable statistical methods, EN is able to recognize complex odours and thus provide an odour fingerprint of the sample (Peris and Escuder-Gilabert 2009; Nurjuliana et al. 2011a,b). Principle component analysis (PCA) has been employed to explore the performance of EN in classification of the adulteration, and stepwise linear discriminant analysis (step-LDA) was employed to optimize the data matrix. The results were evaluated by discriminant analysis methods, and step-LDA was found to be the most effective method. Canonical discriminant analysis (CDA) was used as a pattern recognition technique to detect the presence of pork in minced mutton. Partial least square analysis (PLS), multiple linear regression (MLR) and back-propagation neural net-work (BPNN) techniques were used to build a predictive model for the prediction of pork content in minced mutton. The model built by BPNN was the most effective method for the prediction of pork content, predicting the adulteration more precisely than PLS and MLR. The applications of EN in the food industry are well-known in monitoring processes, shelf-life, freshness evaluation, authenticity assessment (Peris and Escuder-Gilabert 2009), and other quality controls in a wide range of food products, including meat products (Tian et al. 2013) as well as the originality of food component.

The working rule of EN is that it is guaranteed to mimic the human nose. The sensory array represents the sensors in the human nose (Win 2005). The circuitry represents the transformation of the chemical reaction on the human sensors to electrical signals into the brain, and the software analysis represents the brain itself. The EN is accordingly intended to be analogous to the human olfactory system. An EN instrument is composed of three basic elements: (i) a sample handling unit, (ii) a detection unit, and (iii) a data-processing unit. A good review of EN and various sample handling systems can be found elsewhere (Tisan and Oniga 2003; Win 2005; Peris and Escuder-Gilabert 2009).

Using the characteristic fingerprint aroma, it is possible to distinguish and detect any contaminated or adulteration practices, and lard adulteration can be viewed and recognized as part of a previously learned image set (Gan et al. 2005a,b). EN technology has been successfully used for the differentiation of olive oils on the basis of geographical origin (Guadarrama et al. 2001) and adulteration with either sunflower oil or olive-pomace oil (Martin et al. 2001). Oliveros et al. (2002) successfully applied EN and chemometric analysis for the detection of adulteration of olive oil samples with sunflower and olive-pomace oil at levels as low as 5%. EN analysis of sunflower oil and different grades of olive oil demonstrated that it was also possible to differentiate extra virgin olive oil, non-virgin olive oil, and sunflower oil (James et al. 2004).

In addition, EN has been used for detection of lard adulteration in RBD-palm olein (RBD-POl) by Che Man et al. (2005), who developed a surface acoustic wave sensing EN (zNose™) method for lard detection. Samples of RBD-POl were spiked with lard at levels of 1–20% (w/w). The zNose™ produced a Vapour Print™, a two-dimensional olfactory image that could be used qualitatively for immediate detection of lard in the sample mixtures. The lard adulteration could be determined by a few distinct peaks with the chromatogram in the zNose™. The best relationship between percentage of lard in the adulterated RBD-POl and SAWdetector response was observed in one of the adulterant peaks (peak E), with R^2 of 0.906.

Several authors have reported that EN was able to discriminate adulteration of oil (Haddi et al. 2011), wines (Penza and Cassano 2004), and meat. For meat discrimination, studies were done on freshness evaluation (Musatov et al. 2010), processing methods evaluation (Limbo et al. 2010), meat product differentiation, and authenticity assessment (Nurjuliana et al. 2011b; García et al. 2006).

The lard adulteration was identified by distinctive peaks in the zNose™ chromatogram as well as by two-dimensional olfactory images called VaporPrint™. The VaporPrint™ image is a polar plot of the odour amplitudes from the surface acoustic wave (SAW) detector frequency. The VaporPrint™ images for the aroma pattern of lard adulterated RBD palm olein were different from those of unadulterated olein (Che Man et al. 2005). The zNose™ system was also developed as a fast method to analyse the aroma of raw and processed meat for halal authentication (Nurjuliana et al. 2009). Using the VaporPrint™, derived from the frequency of the SAW detector, the zNose™ was successfully exploited to provide an individual fingerprint aroma of pork and different types of sausage (beef, chicken, and pork sausages). The chemometrics technique of PCA was used to structure the data matrix. The first PCA described 94% of the total variance and showed four well-defined classes (chicken sausage, pork sausage, beef sausage, and pork). The high negative score of pork was determined by it having the highest amounts of C8, C13, and C16 of all the samples. Nurjuliana et al. (2011a) successfully combined zNose™ chromatograms with PCA for the differentiation and discrimination of various animal fats such as lard, beef, mutton, and chicken fats.

PC provided a good grouping of samples: PC1 and PC2 accounted for 61% and 29% variations of the data. The volatile compounds of pork, other meats, and meat products were studied using an EN and gas chromatography mass spectrometer with headspace analyser (GCMS-HS) for halal verification. Nurjuliana et al. (2011b) successfully employed the EN technology for rapid qualitative detection and discrimination of pork from other types of meat and meat products alongside a gas chromatography-mass spectrometry with headspace analyser (GCMS-HS) to distinguish the aroma profiling of pork and sheep, cow, and chicken meats for halal verification. Measurements of the volatile compounds by GCMS-HS were also employed and indicated that the EN has adequate selectivity and sensitivity to perform flavour detection in meats. For data interpretation, they applied PCA as an unsupervised classification method to visualize the similarities and differences amongst different measurements in the data sets.

Analysis by PCA was able to cluster and discriminate pork from other types of meats and sausages. It was shown that PCA could provide a good separation of samples with 67% of the total variance accounted for by the first principle component (Nurjuliana et al. 2011b), establishing a very good model for meats and meat product discriminations. However, application of EN to discriminate between meat species in a complex matrix that is frequently encountered in commercial meat products has yet to be defined.

The main advantages of EN technology for pig derivatives (i.e. any substances or compounds resulting from or derived from a pig, such as lard obtained from pig fat) analysis include the relatively small amount of sample preparation needed and its rapidity. However, this technique employs sensors that are not very selective for particular types of compounds in pig derivatives (PD) such as lard, pork, and gelatin, hence preventing any real qualitative and quantitative analyses of individual pig derivatives present in the food samples (Reid et al. 2006; Rohman and Cheman 2012). Recently, Tian et al. (2013) successfully analysed and detected pork adulteration in minced mutton by combining traditional methods (pH and colour evaluation) and EN of metal oxide sensors.

References

Abdulmawjood, A. and Büelte, M. (2002). Identification of ostrich meat by restriction fragment length polymorphism (RFLP) analysis of cytochrome b gene. *Journal of Food Science* 67 (5): 1688–1691.

Abdulmawjood, A. and Büelte, M. (2001). Snail species identification by RFLP-PCR and designing of species specific oligonucleotide primers. *Journal of Food Science* 66 (9): 1287–1293.

Abdulmawjood, A., Krischek, C., Wicke, M., and Klein, G. (2012). Determination of pig sex in meat and meat products using multiplex real time-PCR. *Meat Science* 91: 272–276.

Abdulmawjood, A., Schonenbrucher, H., and Bulte, M. (2003). Development of a polymerase chain reaction system for the detection of dog and cat meat in meat mixtures and animal feed. *Journal of Food Science* 68 (5): 1757–1761.

Ahmad, N.A.B., Tunku Abaidah, T.N., and Abu Yahya, M.H. (2013). A study on halal food awareness among Muslim customers in Klang Valley. *Proceedings of the International Conference on Business and Economic Research (4th ICBER 2013)*, Golden Flower Hotel, Bandung, Indonesia (4–5 March 2013). ISBN: 978-967-5705-10-6. www.internationalconference.com.my. (accessed: 21 August 2013).

Aida, A.A., Che Man, Y.B., Raha, A.R., and Son, R. (2007). Detection of pig derivatives in food products for halal authentication by polymerase chain reaction–restriction fragment length polymorphism. *Journal of the Science of Food and Agriculture* 87: 569–572.

Aida, A.A., Che Man, Y.B., Wong, C.M.V.L. et al. (2005). Analysis of raw meats and fats of pigs using polymerase chain reaction for halal authentication. *Meat Science* 69: 47–52.

Ali, M.E., Hashim, U., Mustafa, S., and Che Man, Y.B. (2011a). Swine-specific PCR-RFLP assay targeting mitochondrial cytochrome b gene for semiquantitative detection of pork in commercial meat products. *Food Analytical Methods* https://doi.org/10.1007/s12161-12011-19290-12165.

Ali, M.E., Hashim, U., Dhahi, T.S. et al. (2011b). Analysis of pork adulteration in commercial burgers targeting porcine-specific mitochondrial cytochrome b gene by TaqMan probe real-time polymerase chain reaction. *Food Analytical Methods* https://doi.org/10.1007/s12161-011-9311-4.

Ali, M.E., Hashim, U., Mustafa, S. et al. (2011c). Nanoparticle sensor for label free detection of swine DNA in mixed biological samples. *Nanotechnology* 22: 195503.

Ali, M.E., Hashim, U., Mustafa, S. et al. (2012a). Analysis of pork adulteration in commercial meatballs targeting porcine-specific mitochondrial cytochrome b gene by TaqMan probe real-time polymerase chain reaction. *Meat Science* 91: 454–459.

Ali, M.E., Hashim, U., Mustafa, S. et al. (2012b). Gold nanoparticle sensor for the visual detection of pork adulteration in meatball formulation. *Journal of Nanomaterials* 2012: 103607.

Ali, M.E., Hashim, U., Mustafa, S. et al. (2012c). Analysis of pork adulteration in commercial meatballs targeting porcine specific mitochondrial cytochrome b gene by TaqMan probe real-time polymerase chain reaction. *Meat Science* 91 (4): 454–459.

Ali, M.E., Hashim, U., Mustafa, S., and Che Man, Y.B. (2012d). Swine-specific PCR-RFLP assay targeting mitochondrial cytochrome B gene for semi quantitative detection of pork in commercial meat products. *Food Analytical Methods* 5: 613–623.

Ali, M.E., Rahman, M.M., Hamid, S.B.A. et al. (2013). Canine-specific PCR assay targeting cytochrome b gene for the detection of dog meat adulteration in commercial frankfurters. *Food Analytical Methods* 7 (1): 234–241.

Ali, M.E., Razzak, M.A., and Hamid, S.B.A. (2014). Multiplex PCR in species authentication: probability and prospects – a review. *Food Analytical Methods*.

Ali, M.E., Razzak, M.A., Hamid, S.B. et al. (2015). Multiplex PCR assay for the detection of five meat species forbidden in Islamic foods. *Food Chemistry* 177: 214–224.

Amagliani, G., Giammarini, C., Omiccioli, E. et al. (2007). Detection of *Listeria monocytogenes* using a commercial PCR kit and different DNA extraction methods. *Food Control* 18: 1137–1142.

Andreo, L.M., Lugo, L., Pertierra, A.G., and Puyet, A. (2006). Evaluation of post-polymerase chain reaction melting temperature analysis for meat species identification in mixed DNA samples. *Journal of Agricultural and Food Chemistry* 54: 7973–7978.

Arslan, A., Ilhak, O.I., and Calicioglu, M. (2006). Effect of method of cooking on identification of heat processed beef using polymerase chain reaction (PCR) technique. *Meat Science* 72: 326–330.

Arslan, A., Ilhak, I., Calicioglu, M., and Karahan, M. (2005). Identification of meats using random amplified polymorphic DNA (RAPD) technique. *Journal of Muscle Foods* 16 (1): 37–45.

Asensio, L., Gonzaiez, I., Garcia, T., and Martin, R. (2008). Determination of food authenticity by enzyme-linked immunosorbent assay (ELISA). *Food Control* 19 (1): 1–8.

Bai, W., XuW, H.K., Yuan, Y. et al. (2009). A novel common primer multiplex PCR (CP-M-PCR) method for the simultaneous detection of meat species. *Food Control* 20 (4): 366–370.

Ballin, N.Z. (2010). Authentication of meat and meat products. *Meat Science* 86 (3): 577–587.

Ballin, N.Z., Vogensen, F.K., and Karlsson, A.H. (2012). PCR amplification of repetitive sequences as a possible approach in relative species quantification. *Meat Science* 90: 438–443.

Bartlett, J.M.S. and Stirling, D. (2003). A short history of the polymerase chain reaction. *Methods in Molecular Biology* 226: 3–6.

Bellagamba, F., Moretti, V.M., Comincini, S., and Valfr, F. (2001). Identification of species in animal feedstuffs by polymerase chain reaction-restriction fragment length polymorphism analysis of mitochondrial DNA. *Journal of Agricultural and Food Chemistry* 49: 3775–3781.

Bellagamba, F., Valfre, F., Panseri, S., and Moretti, V.M. (2003). Polymerase chain reaction-based analysis to detect terrestrial animal protein in fish meal. *Journal of Food Protection* 66: 682–685.

Beneke, B. and Hagen, M. (1998). Applicability of PCR (polymerase chain reaction) for the detection of animal species in heated meat products. *Fleischwirtschaft* 78: 1016–1019.

Berg, J.M., Tymoczko, E.L., and Stryer, L. (2002). Proteins can be detected and quantitated by using an enzyme-linked immunosorbent assay. In: *Biochemistry*, 5e. New York: W H Freeman, Chapter 4, Section 4.3.3.

Bielikova, M., Pangallo, D., and Turna, J. (2010). Polymerase chain reaction-restriction fragment length polymorphism (PCR-RFLP) as a molecular discrimination tool for raw and heat-treated game and domestic animal meats. *Journal of Food and Nutrition Research* 49 (3): 134–139.

Billett, E.E., Bevan, R., Scanlon, B. et al. (1996). The use of a poultry specific murine monoclonal antibody directed to the insoluble muscle protein desmin in meta speciation. *Journal of the Science of Food and Agriculture* 70: 396–404.

Bonwick, G.A. and Smith, C.J. (2004). Immunoassays: their history, development and current place in food science and technology. *International Journal of Food Science & Technology* 39 (8): 817–827.

Borgo, R., Souly-Crosset, C., Bouchon, D., and Gomot, L. (1996). PCR-RFLP analysis of mitochondrial DNA for identification of snail meat species. *Journal of Food Science* 61: 1–4.

Bottero, M.T., Civera, T., Nucera, D. et al. (2003). A multiplex polymerase chain reaction for the dentification of cows', goats' and sheep's milk in dairy products. *International Dairy Journal* 13: 277–282.

Bottero, M.T., Civera, T., Nucera, D., and Turi, R.M. (2003a). Design of universal primers for the detection of animal tissues in feedstuff. *Veterinary Research Communications* 27 (Suppl. 1): 667.

Bottero, M.T. and Dalmasso, A. (2011). Animal species identification in food products: evolution of biomolecular methods. *The Veterinary Journal* 190 (1): 34–38.

Bottero, M.T., Dalmasso, I.A., Nucera, D. et al. (2003b). Development of a PCR assay for the detection of animal tissues in ruminant feeds. *Journal of Food Protection* 66: 2307.

Brodmann, P.D. and Moor, D. (2003). Sensitive and semi-quantitative TaqMan real-time polymerase chain reaction systems for the detection of beef (*Bos taurus*) and the detection of the family Mammalia in food and feed. *Meat Science* 65 (1): 599–607.

Butler, J.M. (2005). *Forensic DNA Typing-Biology, Technology and Genetics of STR markers*, 2e. New York: Elsevier.

Butler, J.M. (2006). *MiniSTRs: Past, Present, and Future*. Forensic News. Foster City: Applied Biosystems.

Calvo, J.H., Osta, R., and Zaragoza, P. (2002). Quantitative PCR detection of pork in raw and heated ground beef and pate. *Journal of Agricultural and Food Chemistry* 50: 5265–5267.

Cammà, C., Di Domenico, M., and Monaco, F. (2012). Development and validation of fast real-time PCR assays for species identification in raw and cooked meat mixtures. *Food Control* 23: 400–404.

Cardarelli, P., Branquinho, M.R., Ferreira, R.T.B. et al. (2005). Detection of GMO in food products in Brazil: the INCQS experience. *Food Control* 16: 859–866.

Carney, B.L., Gray, A.K., and Gharrett, A.J. (1997). Mitochondrial DNA restriction site variation within and among five populations of Alaskan coho salmon (*Oncorhynchus kisutch*). *Canadian Journal of Fisheries and Aquatic Sciences* 54 (7): 940–949.

Carrera, E., Garcia, T., Caspedes, A. et al. (1999). PCR-RFLP for the identification of eggs of Atlantic salmon (*Satmo salar*) and rainbow trout (*Oncorhynchus mykiss*). *Arch Lebensmittelhyg* 50 (4): 67–70.

Cespedes, A., Garcia, T., Carrera, E. et al. (1999). Genetic discrimination among *Solea solea* and *Microchirus azevia* by RFLP analysis of PCR amplified mitochondrial DNA fragments. *Archiv Für Lebensmittelhygiene* 50 (4): 49–72.

Chapela, M.J., Sotelo, C.G., Perez-Martin, R.I. et al. (2007). Comparison of DNA extraction methods from muscle of canned tuna for species identification. *Food Control* 18: 1211–1215.

Che Man, Y.B., Aida, A.A., Raha, A.R., and Son, R. (2007). Identification of pork derivatives in food products by species-specific polymerase chain reaction (PCR) for halal verification. *Food Control* 18: 885–889.

Che Man, Y.B., Gan, H.L., NorAini, I. et al. (2005). Detection of lard adulteration in RBD palm olein using an electronic nose. *Food Chemistry* 90 (4): 829–835.

Che Man, Y.B., Mustafa, S., Khairil Mokhtar, N.F. et al. (2010). Porcine specific polymerase chain reaction assay based on mitochondrial D-loop gene for the identification of pork in raw meat. *International Journal of Food Properties* https://doi.org/10.1080/10942911003754692.

Chen, F.C. and Hsieh, Y.H.P. (2000). Detection of pork in heat-processed meat products by monoclonal antibody-based ELISA. *Journal of AOAC International* 83 (1): 79–85.

Chen, F.C., Hsieh, Y.H.P., and Bridgman, R.C. (1998). Monoclonal antibodies to porcine thermal-stable muscle protein for detection of pork in raw and cooked meats. *Journal of Food Science* 63 (2): 201–205.

Chen, F.-C., Hsieh, Y.-H.P., and Bridgman, R.C. (2004). Monoclonal antibody-based sandwich enzyme-linked immunosorbent assay for sensitive detection of prohibited ruminant proteins in feedstuffs. *Journal of Food Protection* 67: 544–549.

Chikune, K., Tabata, T., Saito, M., and Monma, M. (1994). Sequencing of mitochondrial cytochrome b genes for the identification of meat species. *Animal Science and Technology* 65 (6): 571–579.

Chisholm, J., Sanchez, A., Brown, J., and Hird, H. (2008). The development of species-specific real-time PCR assays for the detection of pheasant and quail in food. *Food Analytical Methods* 1 (3): 190–194.

Colombo, F., Viacava, R., and Giaretti, M. (2000). Differentiation of the species ostrich (*Struthio camelus*) and emu (*Dromaius novaehollandiae*) by polymerase chain reaction using an ostrich-specific primer pair. *Meat Science* 56 (1): 15–17.

Dalmasso, A., Fontanella, E., Piatti, P. et al. (2004). A multiplex PCR assay for the identification of animal species in feedstuffs. *Molecular and Cellular Probes* 18 (2): 81–87.

Darling, J. and Blum, M. (2007). DNA-based methods for monitoring invasive species: a review and prospectus. *Biological Invasions* 9 (7): 751–765.

Desalle, R. and Birstein, V.J. (1996). PCR identification of black caviar. *Nature* 381: 197–198.

Di Bernardo, G., Del Gaudio, S., Galderisi, U. et al. (2007). Comparative evaluation of different DNA extraction procedures from food samples. *Biotechnology Progress* 23: 297–301.

Di Pinto, A., Di Pinto, P., Terio, V. et al. (2013). DNAbarcoding for detecting market substitution in salted cod fillets and battered cod chunks. *Food Chemistry* 141 (3): 1757–1762.

Di Pinto, A., Forte, V.T., Conversano, M.C., and Tantillo, G.M. (2005). Duplex polymerase chain reaction for detection of pork meat in horse meat fresh sausages from Italian retail sources. *Food Control* 16 (5): 391–394.

Dooley, J.J., Paine, K.E., Garrett, S.D., and Brown, H.M. (2004). Detection of meat species using TaqMan real-time PCR assays. *Meat Science* 68: 431–438.

Doosti, A., Ghasemi Dehkordi, P., and Rahimi, E. (2011). Molecular assay to fraud identification of meat products. *Journal of Food Science and Technology* https://doi.org/10.1007/s13197-011-0456-3.

Drummond, M.G., Brasil, B.S.A.F., Dalsecio, L.S. et al. (2013). A verasatile real time PCR method to quantify bovine contamination in buffalo products. *Food Control* 29: 131–137.

El-Jaafari, H.A.A., Panandam, J.M., Idris, I., and Siraj, S.S. (2008). RAPD analysis of three deer species in Malaysia. *Asian-Australasian Journal of Animal Sciences* 21 (9): 1233–1237.

Fajardo, V., Gonzalez, I., Lopez-Calleja, I. et al. (2007b). PCR identification of meats from chamois (*Rupicapra rupicapra*), pyrenean ibex (*Capra pyrenaica*) and mouflon (*Ovis ammon*) targeting specific sequences from the mitochondrial D-loop. *Meat Science* 76 (4): 644–652.

Fajardo, V., Gonzalez, I., Lopez-Calleja, I. et al. (2007a). Identification of meats from red deer (*Cervus elaphus*), fallow deer (*Dama dama*) and roe deer (*Capreolus capreolus*) using polymerase chain reaction targeting specific sequences from the mitochondrial 12S rRNA gene. *Meat Science* 76 (2): 234–240.

Fajardo, V., Gonzalez, I., Martı'n, I. et al. (2008c). Real-time PCR for quantitative detection of chamois (*Rupicapra rupicapra*) and pyrenean ibex (*Capra pyrenaica*) in meat mixtures. *Journal of AOAC International* 91 (1): 103–111.

Fajardo, V., Gonzalez, I., Martı'n, I. et al. (2008a). Differentiation of European wild boar (*Sus scrofa scrofa*) and domestic swine (*Sus scrofa domestica*) meats by PCR analysis targeting the mitochondrial D-loop and the nuclear melanocortin receptor 1 (MC1R)genes. *Meat Science* 78 (3): 314–322.

Fajardo, V., Gonzalez, I., Martı'n, I. et al. (2008b). Real-time PCR for detection and quantification of red deer (*Cervus elaphus*), fallow deer (*Dama dama*), and roe deer (*Capreolus capreolus*) in meat mixtures. *Meat Science* 79 (2): 289–298.

Fajardo, V., Gonzalez, I., Dooley, J. et al. (2009). Application of PCR-RFLP analysis and labon-a-chip capillary electrophoresis for the specific identification of game and domestic meats. *Journal of the Science of Food and Agriculture* 89 (5): 843–847.

Fajardo, V., Gonzᴊlez, I., Rojas, M. et al. (2010). A review of current PCR-based methodologies for the authentication of meats from game animal species. *Trends in Food Science and Technology* 21 (8): 408–421.

Farrokhi, R. and Joozani, R.J. (2011). Identification of pork genome in commercial meat extracts for halal authentication by SYBR green I real-time PCR. *International Journal of Food Science and Technology* 46: 951–955.

Felmer, R., Sagredo, B., Chavez, R. et al. (2008). Implementation of a molecular system for traceability of beef based on microsatellite markers. *Chilean Journal of Agricultural Research* 68 (4): 342–351.

Forrest, A.R. and Carnegie, P.R. (1994). Identification of gourmet meat using FINS (forensically informative nucleotide sequencing). *Biotechniques* 17 (1): 24–26.

Frezza, D., Favaro, M., Vaccari, G. et al. (2003). A competitive polymerase chain reaction-based approach for the identification and semiquantification of mitochondrial DNA in differently heat-treated bovine meat and bone meal. *Journal of Food Protection* 66: 103–109.

Gan, H.L., Che Man, Y.B., Tan, C.P. et al. (2005a). Monitoring the storage stability of RBD palm olein using the electronic nose. *Food Chemistry* 89: 271–282.

Gan, H.L., Che Man, Y.B., Tan, C.P. et al. (2005b). Characterisation of vegetable oils by surface acoustic wave sensing electronic nose. *Food Chemistry* 89: 507–518.

García, M., Aleixandre, M., Gutiérrez, J., and Horrillo, M.C. (2006). Electronic nose for ham discrimination. *Sensors and Actuators B: Chemical* 114 (1): 418–422.

Giovannacci, I., Guizard, C., Carlier, M. et al. (2004). Species identification of meat products by ELISA. *International Journal of Food Science & Technology* 39: 863–867.

Girish, P.S., Anjaneyulu, A.S.R., Viswas, K.N. et al. (2005). Meat species identification bypolymerase chain reaction-restriction fragment length polymorphism (PCR-RFLP) of mitochondrial 12S rRNA gene. *Meat Science* 70 (1): 107–112.

Goldsby, R.A., Kindt, T.J., Osborne, B.A., and Kuby, J. (2003). Enzyme-linked immunosorbent assay. In: *Immunology*, 148–150. New York: W.H. Freeman, & Company.

Greenwood, A. and Paabo, S. (1999). Nuclear insertion sequences of mitochondrial DNA predominate in hair but not in blood of elephants. *Molecular Ecology* 8: 133–137.

Guadarrama, A., Rodríguez-Mendez, M.L., Sanz, C. et al. (2001). Electronic nose based on conducting polymers for the quality control of the olive oil aroma – Discrimination of quality, variety of olive and geographic origin. *Analytica Chimica Acta* 432: 283–292.

Guoli, Z., Mingguang, Z., Zhijiang, Z. et al. (1999). Establishment and application of a polymerase chain reaction for the identification of beef. *Meat Science* 51: 233–236.

Ha, J.C., Jung, W.T., Nam, Y.S., and Moon, T.W. (2006). PCR identification of ruminant tissue in raw and heat-treated meat meals. *Journal of Food Protection* 69 (9): 2241–2247.

Haddi, Z., Amari, A., Ali, A.O. et al. (2011). Discrimination and identification of geographical origin virgin olive oil by an e-nose based on MOS sensors and pattern recognition techniques. *Procedia Engineering* 25: 1137–1140.

Haider, N., Nabulsi, I., and Al-Safadi, B. (2012). Identification of meat species by PCR-RFLP of the mitochondrial *COI* gene. *Meat Science* 90 (2012): 490–493.

Herman, L. (2001). Determination of the animal origin of raw food by species-specific PCR. *Journal of Dairy Research* 68: 429–436.

Hird, H., Goodier, R., Schneede, K. et al. (2004). Truncation of oligonucleotide primers confers specificity on real-time polymerase chain reaction assays for food authentication. *Food Additives and Contaminants* 21 (11): 1035–1040.

Hoorfar, J., Malorny, B., Abdulmawjood, A. et al. (2004). Practical consideration in design of internal amplification controls for diagnostic PCR assays. *Journal of Clinical Microbiology* 42 (5): 1863–1868.

Hopwood, A.J., Fairbrother, K.S., Lockley, A.K., and Bardsley, R.G. (1999). An actin gene-related polymerase chain reaction (PCR) test for identification of chicken in meat mixtures. *Meat Science* 53: 227–231.

Hsieh, Y.H.P., Zhang, S., Chen, F.C., and Sheu, S.C. (2002). Monoclonal antibody based ELISA for assessment of endpoint heating temperature of ground pork and beef. *Journal of Food Science* 67 (3): 1149–1154.

Hsu, Y.C., Pestka, J.J., and Smith, D.M. (1999). ELISA to quantify triose phosphate isomerase to potentially determine processing adequacy in ground beef. *Journal of Food Science* 64 (4): 623–628.

Huang, M.C., Horng, Y.M., Huang, H.L. et al. (2003). RAPD fingerprinting for the species identification of animals. *Asian Australasian Journal of Animal Sciences* 16 (10): 1406–1410.

Ihrig, J., Lill, R., and Mühlenhoff, U. (2006). Application of the DNA-specific dye Eva Green for the routine quantification of DNA in microplates. *Analytical Biochemistry* 359 (2): 265–267.

Iwobi, A., Huber, I., Hauner, G. et al. (2011). Biochip technology for the detection of animal species in meat products. *Food Analytical Method* 4 (3): 389–398.

Jamaludin, M.A., Zaki, N.N.M., Ramli, M.A. et al. (2011). Istihalah: analysis on the utilization of gelatin in food products. *Proceedings of the 2nd International Conference on Humanities, Historical and Social Sciences*, Cairo, Egypt (21–22 October 2011). IPEDR vol. 17 (2011). Singapore: IACSIT Press.

James, D., Scott, S.M., O'Hare, W.T. et al. (2004). Classification of fresh edible oils using a coated piezoelectric sensor array-based electronic nose with soft computing approach for pattern recognition. *Transactions of the Institute of Measurement and Control* 26: 3–18.

Jha, V.K., Kumar, A., and Mandokhot, U.V. (2003). Indirect enzyme-linked immunosorbent assay in detection and differentiation of cooked and raw pork from meats of other species. *Journal of Food Science and Technology* 40: 254–256.

Jonker, K.M., Tilburg, J.J.H.C., Hagele, G.H., and De Boer, E. (2008). Species identification in meat products using real-time PCR. *Food Additives and Contaminants* 25 (5): 527–533.

Jothikumard, N. and Griffiths, M. (2002). Rapid detection of Escherichia coli O157:H7 with multiplex real-time PCR assays. *Applied and Environmental Microbiology* 68: 3169–3171.

Köppel, R., Ruf, J., and Rentsch, J. (2011). Multiplex real-time PCR for the detection andquantification of DNA from beef, pork, horse and sheep. *European Food Research and Technology* 232 (1): 151–155.

Köppel, R., Zimmerli, F., and Breitenmoser, A. (2009). Heptaplex realtime PCR for the identification and quantification of DNA from beef, pork, chicken, turkey, horse meat, sheep (mutton) and goat. *European Food Research and Technology* 230 (1): 125–133.

Karabasanavar, N.S., Singh, S.P., Kumar, D., and Shebannavar, S.N. (2014). Detection of pork adulteration by highly-specific PCR assay of mitochondrial D-loop. *Food Chemistry* 145: 530–534.

Karlsson, A.O. and Holmlund, G. (2007). Identification of mammal species using species-specific DNA pyrosequencing. *Forensic Science International* 173 (1): 16–20.

Kesmen, Z., Celebi, Y., Güllüce, A., and Yetim, H. (2013). Detection of seagull meat in meat mixtures using real-time PCR analysis. *Food Control* 34 (1): 47–49. https://doi.org/10.1016/j.foodcont.2013.04.006.

Kesmen, Z., Yetim, H., and Şahin, F. (2010). Identification of different meat species used in sucuk production by PCR assay. *GIDA* 35 (2): 81–87.

Kocher, T.D., Irwin, D.M., and Wilson, A.C. (1991). Evolution of the cytochrome b gene of mammals. *Journal of Molecular Evolution* 32 (8): 128–144.

Kocher, T.D., Thomas, W.K., Meyer, A. et al. (1989). Dynamics of mitochondrial DNA evolution in animals: amplification and sequencing with conserved primers. *Proceedings of the National Academy of Sciences of the United States of America* 86: 6196–6200.

Koh, M.C., Lim, C.H., Chua, S.B. et al. (1998). Random amplified polymorphic DNA (RAPD) fingerprints for identification of red meat animal species. *Meat Science* 48 (3–4): 275–285.

Koveza, O.V., Kokaeva, Z.G., Konovalov, F.A., and Gostimsky, S.A. (2005). Identification and mapping of polymorphic RAPD markers of pea (*Pisum sativum* L.) genome. *Genetika* 41 (3): 341–348.

Kowalski, B. (1989). Sub-ambient differential scanning calorimetry of lard and lard contaminated by tallow. *International Journal of Food Science and Technology* 24: 415–420.

Kuiper, H.A. (1999). Summary report of the ILSI Europe workshop on detection methods for novel foods derived from genetically modified organisms. *Food Control* 10: 339–349.

Laube, I., Zagon, J., and Broll, H. (2007). Quantitative determination of commercially relevant species in foods by real-time PCR. *International Journal of Food Science and Technology* 42 (3): 336–341.

Lenstra, J.A. (2003). DNA methods for identifying plant and animals species in foods. In: *Food Authenticity and Traceability* (ed. M. Lees), 51–70. Cambridge: Woodhead Publishing.

Levin, R.E. (2008). DNA-based technique: polymerase chain reaction (PCR). In: *Modern Techniques for Food Authentication* (ed. D.-W. Sun), 411–476. Academic Press, Elsevier.

Limbo, S., Torri, L., Sinelli, N. et al. (2010). Evaluation and predictive modeling of shelf life of minced beef stored in high-oxygen modified atmosphere packaging at different temperatures. *Meat Science* 84 (1): 129–136.

Liu, L., Chen, F.-C., Dorsey, J., and Hsieh, Y.-H.P. (2006). Sensitive monoclonal antibody-based sandwich ELISA for the detection of porcine skeletal muscle in meat and feed products. *Journal of Food Science* 71 (1): M1–M6.

Lopez-Andreo, M., Garrido-Pertierra, A., and Puyet, A. (2006). Evaluation of post-polymerase chain reaction melting temperature analysis for meat species identification in mixed DNA samples. *Journal of Agricultural and Food Chemistry* 54 (21): 7973–7978.

Lutz, S., Weisser, H.J., Heizmann, J., and Pollak, S. (1996). mtDNA as a tool for identification of human remains. *International Journal of Legal Medicine* 109 (4): 205–209.

Mafra, I., Ferreira, I.M.P.L.V.O., and Oliveira, M.B.P.P. (2008). Food authentication by PCR-based methods. *European Food Research and Technology* 227 (3): 649–665.

Mariam, A.L. (2010). Halal integrity-Malaysian perspective. *Workshop on Halal Awareness Programme ASEAN WG on Halal Food and IMT-GT*. Retrieved 15 December 2010www.hdc.com.my. (accessed: 15 December 2010).

Martin, C.R.-A. (1981). Sale of horse flesh. *British Food Journal* 83: 101–102.

Martin, Y.G., Oliveros, M.C.C., Pavon, J.L.P. et al. (2001). Electronic nose based on metal oxide semiconductor sensors and pattern recognition techniques: Characterisation of vegetable oils. *Analytica Chimica Acta* 449: 69–80.

Martinez, I. and Danielsdottir, A.K. (2000). Identification of marine mammal species in food products. *Journal of the Science of Food and Agriculture* 80 (4): 527–533.

Martinez, I. and Yman, I.M. (1998). Species identification in meat products by RAPD analysis. *Food Research International* 31 (6e7): 459–466.

Meyer, R., Hofelein, C., Luthy, J., and Candrain, U. (1995). Polymerase chain reaction-restriction fragment length polymorphism analysis: a simple method for species identification in food. *Journal Association of Official Analytical Chemists International* 78: 1542–1551.

Mohindra, V., Khare, P., Lal, K.K. et al. (2007). Molecular discrimination of five Mahseer species from Indian peninsula using RAPD analysis. *Acta Zoologica Sinica* 53 (4): 725–732.

Murugaiah, C., Noor, Z.M., Mastakim, M. et al. (2009). Meat species identification and halal authentication analysis using mitochondrial DNA. *Meat Science* 83 (1): 57–61.

Musatov, V.Y., Sysoev, V.V., Sommer, M., and Kiselev, I. (2010). Assessment of meat freshness with metal oxide sensor microarray electronic nose: a practical approach. *Sensors and Actuators B – Chemical* 144 (1): 99–103.

Myers, M.J., Yancy, H.F., Farrell, D.E. et al. (2007). Assessment of two enzyme-linked immunosorbent assay tests marketed for detection of ruminant proteins in finished feed. *Journal of Food Protection* 70: 692.

Nurjuliana, M., Che Man, Y.B., and Hashim, D.M. (2011a). Analysis of lard's aroma by an electronic nose for rapid halal authentication. *Journal of the American Oil Chemists' Society* 88: 75–82.

Nurjuliana, M., Che Man, Y.B., Hashim, D.M., and Mohamed, A.K.S. (2009). Rapid detection of porkin food products by electronic nose for halal authentication. In: *Proceedings of 3rd IMT-GT International Symposium on Halal Science and Management 2009, Kuala Lumpur, Malaysia (21–22 December 2009)* (eds. P. Hashim et al.), 7–13. Halal Products Research Institute.

Nurjuliana, M., Che Man, Y.B., Hashim, D.M., and Mohamed, A.K.S. (2011b). Rapid identification of pork for halal authentication using the electronic nose and gas chromatography mass spectrometer with headspace analyzer. *Meat Science* 88 (4): 638–644.

Oliveros, M.C.C., Pavon, J.L.P., Pinto, C.G. et al. (2002). Electronic nose based on metal oxide semiconductor sensors as a fast alternative for the detection of adulteration of virgin olive oils. *Analytica Chimica Acta* 459: 219–228.

Park, J.K., Shin, K.H., Shin, S.C. et al. (2007). Identification of meat species using species-specific PCR-RFLP fingerprint of mitochondrial 12S rRNA gene. *Korean Journal for Food Science of Animal Resources* 27 (2): 209–215.

Partis, L., Croan, D., Guo, Z. et al. (2000). Evaluation of DNA fingerprinting method for determining the species origin of meat. *Meat Science* 54 (4): 369–376.

Pascoal, A., Prado, M., Castro, J. et al. (2004). Survey of authenticity of meat species in food products subjected to different technological processes, by means of PCR-RFLP analysis. *European Food Research and Technology* 218: 306–312.

Patterson, R.L.S. and Jones, S.L. (1989). Species identification in heat processed meat products. Proceedings of 35th International Congress on Meat Science and Technology. August 20–25, 1989, 2, 529–536. Copenhagen, Denmark.

Penza, M. and Cassano, G. (2004). Chemometric characterization of Italian wines by thin-film multisensors array and artificial neural networks. *Food Chemistry* 86: 283–296.

Peris, M. and Escuder-Gilabert, L. (2009). A 21 st century technice for food control: electronic noses. *Analytica Chimica Acta* 638 (1): 1–15. https://doi.org/10.1016/j.aca.2009.02.009 PMid:19298873.

Pfeiffer, I., Burger, J., and Brenig, B. (2004). Diagnostic polymorphisms in the mitochondrial cytochrome b gene allow discrimination between cattle, sheep, goat, roe buck and deer by PCR–RFLP. *BMC Genetics* 5: 30.

Prado, M., Franco, C.M., Fente, C.A. et al. (2002). Comparison of extraction methods for the recovery, amplification and species-specific analysis of DNA from bone and bone meals. *Electrophoresis* 23: 1005–1012.

Ram, J.L., Ram, M.L., and Baidoun, F. (1996). Authentication of canned tuna and bonito by sequence and restriction site analysis of polymerase chain reaction products of mitochondrial DNA. *Journal of Agricultural and Food Chemistry* 44 (8): 2460–2467.

Rashood, K.A., Abou-Shaaban, R.R.A., Abdel-Moety, E.M., and Rauf, A. (1996). Compositional and thermal characterization of genuine and randomized lard: a comparative study. *Journal of the American Oil Chemists' Society* 73: 303–309.

Rastogi, G., Dharne, M.S., Walujkar, S. et al. (2007). Species identification and authentication of tissues of animal origin using mitochondrial and nuclear markers. *Meat Science* 76 (4): 666–674.

Ratanamaneichat, C. and Rakkarn, S. (2013). Quality assurance development of halal food products for export to Indonesia. *Procedia – Social and Behavioral Sciences* 88 (2013): 134–141.

Reid, L.M., O'Donnell, C.P., and Downey, G. (2006). Recent technological advances for the determination of food authenticity. *Trends in Food Science and Technology* 17: 344–353.

Rigaa, A., Cellos, D., and Monnerot, M. (1997). Mitochondrial NA from the scallop *Pecten maximum*: an unusual polymorphism detected by restriction fragment length polymorphism analysis. *Heredity* 79 (4): 380–387.

Ripoli, M.V., Corva, P., and Giovambattista, G. (2006). Analysis of a polymorphism in the DGAT1 gene in 14 cattle breeds through PCRSSCP methods. *Research in Veterinary Science* 80 (3): 287–290.

Rodríguez, M.A., Garcia, T., Gonzalez, I. et al. (2004). Quantitation of mule duck in goose foiegras using TaqMan real-time PCR. *Journal of Agricultural and Food Chemistry* 52: 1478–1483.

Rodríguez, M.A., Garcia, T., Gonzalez, I. et al. (2003). Identification of goose, mule duck, turkey, and swine in foie grass by species-specific polymerase chain reaction. *Journal of Agricultural and Food Chemistry* 51: 1524–1529.

Rodríguez, M.A., García, T., Gonzalez, I. et al. (2005). TaqMan real-time PCR for the detection and quantification of pork in meat mixtures. *Meat Science* 70 (1): 113–120.

Rohman, A. and Cheman, Y.B. (2012). Analysis of pig derivatives for halal authentication studies. *Food Reviews International* 28: 97–112.

Rojas, M., Gonzalez, I., Fajardo, V. et al. (2009). Authentication of meats from quail (*Coturnix coturnix*), pheasant (*Phasianus colchicus*), partridge (*Alectoris* spp.), and guinea fowl (*Numida meleagris*) using polymerase chain reaction targeting specific sequences from the mitochondrial 12S rRNA gene. *Food Control* 20 (10): 896–902.

Rojas, M., Gonzalez, I., Pavon, M.A. et al. (2010a). Polymerase chain reaction assay for verifying the labeling of meat and commercial meat products from game birds targeting specific sequences from the mitochondrial Dloop region. *Poultry Science* 89 (5): 1021–1032.

Rojas, M., Gonzalez, I., Pavon, M.A. et al. (2010b). Novel TaqMan real-time polymerase chain reaction assay for verifying the authenticity of meat and commercial meat products from game birds. *Food Additives and Contaminants* 27 (6): 749–763.

Rojas, M., Gonzalez, I., Pavon, M.A. et al. (2010c). Novel TaqMan realtime polymerase chain reaction assay for verifying the authenticity of meat and commercial meat products from game birds. *Food Additives and Contaminants* 27 (6): 749–763.

Saida, M., Hassan, F., Musa, R., and Rahman, N.A. (2014). Assessing consumers' perception, knowledge and religiosity on Malaysia's halal food products. *Procedia – Social and Behavioral Sciences* 130: 120–128.

Sait, M., Clark, E.M., Wheelhouse, N. et al. (2011). *Veterinary Microbiology* 151: 284.

Sakai, Y., Kotoura, S., Yano, T. et al. (2011). Quantification of pork, chicken and beef by using a novel reference molecule. *Bioscience, Biotechnology, and Biochemistry* 75 (9): 1639–1643.

Schmerer, W.M., Hummel, S., and Herrmann, B. (1999). Optimized DNA extraction to improve reproducibility of short tandem repeat genotyping with highly degraded DNA as target. *Electrophoresis* 20: 1712–1716.

Scholz, M., Giddings, I., and Pusch, C.M. (1998). A polymerase chain reaction inhibitor of ancient hard and soft tissue DNA extract is determined as human collagen type I. *Analytical Biochemistry* 259: 283–286.

Smith, D.M., Desrocher, L.D., Booren, A.M. et al. (1996). Cooking temperature of turkey ham affects lactate dehydrogenase, serum albumin and immunoglobulin G as determined by ELISA. *Journal of Food Science* 61 (1): 209–212.

Soares, S., Amaral, J.S., Oliveira, M.B.P.P., and Mafra, I. (2013). A SYBR Green real-time PCR assay to detect and quantify pork meat in processed poultry meat products. *Meat Science* 94 (2013): 115–120.

Sorenson, M.D. and Quinn, T.W. (1998). Numts: a challenge for avian systematics and population biology. *The Auk* 115: 214–221.

Tartaglia, M., Saulle, E., Pestalozza, S. et al. (1998). Detection of bovine mmitochondrial DNA in ruminant feeds: a molecular approach to test for the presence of bovine derived materials. *Journal of Food Protection* 61 (5): 513–518.

Teletchea, F., Bernillon, J., Duffraisse, M. et al. (2008). Molecular identification of vertebrate species by oligonucleotide microarray in food and forensic samples. *Journal of Applied Ecology* 45 (3): 967–975.

Tian, X., Wanga, J., and Cui, S. (2013). Analysis of pork adulteration in minced mutton using electronic nose of metal oxide sensors. *Journal of Food Engineering* 119: 744–749.

Tisan, A. and Oniga, S. (2003). Current status of electronic nose: the sensing system. *Proceedings of the International Multidisciplinary Conference 5th edition*, Baia Mare (23–24 May 2003). Scientific Bulletin, Series C 17, p. 517.

Tobe, S.S. and Linacre, A.M.T. (2008). A multiplex assay to identify 18 European mammal species from mixtures using the mitochondrial cytochrome b gene. *Electrophoresis* 29 (2): 340–347.

Unseld, M., Beyermann, B., Brandt, P., and Hiesel, R. (1995). Identification of the species origin of highly processed meat products by mitochondrial DNA sequences. *PCR Met Application* 4: 241–243.

Vatilingom, M., Pijnenburg, H., Gendre, F., and Brignon, P. (1999). Real-time quantitative PCR detection of genetically modified maximizer maize and roundup ready soybean in some representative foods. *Journal of Agricultural and Food Chemistry* 47: 261–5266.

Venien, A. and Levieux, D. (2005). Differentiation of bovine from porcine gelatines using polyclonal anti-peptide antibodies in indirect and competitive indirect ELISA. *Journal of Pharmaceutical and Biomedical Analysis* 39: 418–424.

Verkaar, E.L.C., Nijman, I.J., Boutaga, K., and Lenstra, J.A. (2002). Differentiation of cattle species in beef by PCR-RFLP of mitochondrial and satellite DNA. *Meat Science* 60: 365–369.

Whittaker, R.G., Spencer, T.L., and Copland, J.W. (1983). An enzyme linked immunosorbent assay for species identification of raw meat. *Journal of the Science of Food and Agriculture* 34: 1143–1148.

Wilson, I.G. (1997). Inhibition and facilitation of nucleic acid amplification. *Applied and Environmental Microbiology* 63: 3741–3751.

Wilson, M.R., DiZinno, J.A., Polanskey, D. et al. (1995). Validation of mitochondrial DNA sequencing for forensic casework analysis. *International Journal of Legal Medicine* 108 (2): 68–74.

Win, D.T. (2005). The electronic nose – a big part of our future. *AU Journal of Technology* 9 (1): 1–8.

Wiseman, G. (1999). Quantitative PCR Detection of *T. aestivum* Adulteration in Commercial *T. durum* Pasta Using PSR 128 Primers: Optimisation. Final Report for MAFF Project ANO667.

Wolf, C., Rentsch, J., and Hubner, P. (1999). PCR-RFLP analysis of mitochondrial DNA: a reliable method for species identification. *Journal of Agricultural and Food Chemistry* 47 (4): 1350–1355.

Wolf, C., Burgener, M., Hübner, P., and Luthy, J. (2000). PCR-RLFP analysis of mitochondrial DNA: differentiation of fish species. *Lebensmittel-Wissenschaft und – Technologie* 33: 144–150.

Wolf, C. and Lüthy, J. (2001). Quantitative competitive (QC) PCR for quantification of porcine DNA. *Meat Science* 57: 161–168.

Wu, X.B., Liu, H., and Jiang, Z.G. (2006). Identification primers for sika deer (*Cervus nippon*) from a sequence-characterised amplified region (SCAR). *New Zealand Journal of Zoology* 33 (1): 65–71.

Yau, F.C.F., Wong, K.L., Shaw, P.C. et al. (2002). Authentication of snakes used in Chinese medicine by sequence characterized amplified region (SCAR). *Biodiversity and Conservation* 11 (9): 1653–1662.

Yoshizaki, G., Yamaguchi, K., Strossmann, O.T., and CA, T.F. (1997). Cloning and characterization of pejerrey mitochondrial DNA and its application for RFLP analysis. *Journal of Fish Biology* 51 (1): 193–203.

Yusop, M., Mustafa, S., Che Man, Y. et al. (2012). Detection of raw pork targeting porcine-specific mitochondrial cytochrome B gene by molecular beacon probe real-time polymerase chain reaction. *Food Analysis Methods* 5: 422–429.

Zha, D.-M., Xing, X.-M., and Yang, F.-H. (2011). Rapid identification of deer products by multiplex PCR assay. *Food Chemistry* 129 (4): 1904–1908.

Zhang, C. (2013). Semi-nested multiplex PCR enhanced method sensitivity of species detection in further-processed meats. *Food Control* 31 (2): 326–330.

19

Food Fraud

John Pointing[1], Yunes Ramadan Al-Teinaz[2], John Lever[3],
Mary Critchley[4] and Stuart Spear[5]

[1] *Barrister, London, UK*
[2] *Independent Public Health & Environment Consultant, London, UK*
[3] *University of Huddersfield Business School, Huddersfield, UK*
[4] *University of London, (Warmwell) based south-west France*
[5] *Freelance Journalist, London, UK*

19.1   Introduction

Halal food is essential for people of Islamic faith, and is, perhaps, particularly vulnerable to fraud and adulteration compared to non-halal (Fuseini et al. 2017; McElwee et al. 2017). While there are specific problems for Muslims around the authenticity of halal meat and associated certification practices (Lever and Miele 2012), the wider implications of food fraud are global and can affect people of all religions and beliefs.

One definition of food fraud is that it is the act of knowingly defrauding, for economic gain, the buyers of food or food ingredients, be they consumers, food manufacturers, retailers or importers in the food industry. For England, Wales and Northern Ireland, section 2 of the Fraud Act 2006 defines the crime of false representation as occurring when:

1) A person ...
   a) dishonestly makes a false representation, and
   b) intends, by making the representation –
      i) to make a gain for himself or another, or
      ii) to cause loss to another or to expose another to a risk of loss.

Section 2 goes on to define false representation, as follows:

2) A representation is false if –
   a) it is untrue or misleading, and
   b) the person making it knows that it is, or might be, untrue or misleading.

The various types of food fraud include substituting a cheaper ingredient for a more expensive one, adulteration, dilution, counterfeiting, the inclusion of unapproved enhancements, mislabelling, the diversion of unfit meat from the pet food trade into the human food chain, freezing out-of-date meat and re-dating it, the sale of meat that has become unfit due to poor transport and storage, the sale and handling of condemned meat by disguising signs of uncleanliness or disease, the use of unlicensed and unhygienic premises, the fraudulent use of health and identification marks, and the deliberate selling of non-halal meat (haram) as if it were halal.

Some of the earliest reported cases of food fraud involved products such as olive oil, milk, honey, spices, tea, and wine. The global rise of packaged and processed foods has undoubtedly increased concerns about the safety and quality of food, as well providing fresh opportunities for food fraud and adulteration. This led the *New York Times* to parody such developments in the following way (cited in Pilcher 2017, p. 68):

> Mary had a little lamb,
> And when she saw it sicken,
> She shipped it off to Packingtown,
> And now it's labelled chicken.

Food fraud is big business (Gallagher and Thomas 2010) and globally is estimated to cost about £40 billion annually (PwC 2015). It is an international trade that makes large profits for criminals, who are becoming more sophisticated and reckless about the consequences for consumers. *The Guardian* newspaper reported, in May 2013, that 900 individuals had been arrested in China for a series of meat frauds totalling 20 000 t of unfit meat, which included meat from fox, mink, and rat being passed off as lamb.

The UK Consumer magazine *Which?* campaigns against food fraud and regularly publishes food test results. In April 2014 an investigation into takeaway food in the UK revealed that 40% of lamb served at takeaways contained meat other than lamb (*Which?* 2014). The public health effects of food fraud may be disgust, illness, disease, and even death. In the UK, the bovine spongiform encephalopathy (BSE) outbreak in the 1990s – in addition to the public health effects – had far-reaching implications in terms of economic losses for the economy as a whole, as well as particular problems for farmers and the tourist industry. New legislation entered the statute book quickly amid real fears and concerns that past consumption of BSE infected beef could result in the invariably fatal human form of BSE – the brain infection known as 'new variant CJD'. More recently, the horsemeat scandal of 2013 caused great public alarm, with equine DNA being found in numerous beef products in UK supermarkets; the scandal eventually affecting all but one EU Member States (Marsden and Morley 2014).

19.2 Food Ingredients and False Labelling

False labelling is one way that fraudsters target consumers (Pointing et al. 2008). At particular risk are people with specific food allergies who depend on reassurances that gluten, peanuts, lactose, and so on are absent in the foods they buy. Manufacturers' claims may be fraudulent or food labels may simply omit listing problematic ingredients. Spices are a particularly high-risk sector because they are expensive and very easy to adulterate. In 2013, the Food Standards Agency (FSA) issued an urgent alert to people in England and Wales with nut allergies about a batch of ground cumin that had been found to contain traces of almond protein not listed on the label. An article in *The Guardian*, in February 2005, revealed how a 'chain of transactions' resulted in an illegal dye, normally used in wax and floor polish, being added to chilli powder. Over 400 well-known, reputable processed foods including three brands of sausages sold in Sainsbury's, ready meals made by major manufacturers, and a seafood sauce made by Unilever had been adulterated by dyes known generally as Sudan I, III, and IV. Although not strictly classifiable as a direct cause of cancer in humans, these dyes are classified as category three carcinogens by the International Agency for Research on Cancer and are thus banned from inclusion in foods in many countries.

The 1985 the diethylene glycol (DEG) wine scandal involved a limited number of Austrian wine producers deliberately adulterating their wines using DEG (a primary ingredient in some brands of antifreeze) to make the wines appear sweeter and more full-bodied. Antifreeze is often implicated in deaths due to poisoning for humans, and its sweet taste makes it seem innocuous. There are documented cases of more than a dozen DEG poisoning incidents where it was substituted for glycerine in pharmaceutical preparations. These poisonings are particularly evident in developing countries where intensive medical care and quality control procedures tend to be substandard (Schier et al. 2013).

Another recent case occurred in 2007–2008, when Chinese manufacturers were discovered to be deliberately adding melamine to high-protein feed and milk-based products, including powdered baby milk. Where products had been illegally diluted, the addition of melamine increased the nitrogen content of the milk and therefore its apparent protein content. Wheat gluten and rice protein concentrate from China, adulterated with melamine, were unwittingly used in the manufacture of pet food in the USA, which caused kidney failure and death for a large number of dogs and cats. Until this point, the toxic nature of melamine and its related compounds had not been fully appreciated by food authorities.

Sections of the public have grown to distrust food from sources considered unhealthy and turn away from products produced through large-scale factory farming. There has been a surge in demand for organic and free-range meat, and eggs from hens allowed to lead as 'natural' a life as possible (Miele and Lever 2013). That free-range eggs can cost significantly more than battery-cage eggs can lead to swindles, and as the demand for free-range grows, so too does the size of so-called free-range farms. Despite the existence in the UK of the British Lion and Freedom Foods quality assurance schemes, they fail to prevent free-range farms from housing as many as 16 000 hens in a building from which only a few birds can find their way into the open air. In 2010, about 100 million mislabelled eggs were sold by the UK supermarkets Sainsbury and Tesco. As many as 150 German and 36 Dutch poultry farms were investigated for fraud concerning the intentional mislabelling of eggs as free range in 2013.

19.3 Types of Meat Fraud

Meat fraud occurs in many forms and with respect to many types of meat that should never enter the human food chain. We consider some important examples in this chapter. Meat fraud is not restricted to the halal sector, but is widespread and extends to all stages of the production process. It includes the sale of condemned meat, which fraudsters trim, treat, and bleach before selling on as 'fresh' at markets and from the backs of vans. Such meat will often have forged health marks stamped on carcasses. Following the public outcry and widespread panic about variant Creutzfeldt–Jakob disease (vCJD) in the late twentieth century, UK laws still forbid the sale of meat from cattle or sheep which have not had the spinal cord or specified risk material (SRM) removed, or where the skin remains attached to the animal. The ban also takes into account fears about *Escherichia coli*, salmonella and other bacteria that can be passed on to consumers. Non-SRM meat by-products, which were once saleable, now require payment for disposal, and it is thus tempting for criminals to hide such parts in meat products offered for sale.

'Bushmeat' is the name given to wild animals that have been slaughtered illegally and then exported to communities which cannot obtain such meat from legal sources. It is an illegal trade that spreads from Africa into Europe, and sales occur clandestinely in the street markets and back street shops in parts of cities where there is a continuing strong demand for bushmeat. It has been estimated that 7500 t of illegal bushmeat enters Britain each year (*Daily Mail* 2014). The trade is worrying from a conservationist point of view, not least because endangered species may be involved (Pointing 2005). There are also risks to public health owing to the viruses and bacteria such meat will often contain (*Daily Mail* 2014).

Another form of meat fraud concerns 'smokies'. Smokies are prepared by blowtorching the fleece of the carcass of an elderly sheep or goat, which leaves a much-prized smoky flavour in the burnt-on skin (Pointing and Teinaz 2004). The process of slaughtering old sheep and goats can be callous, and the unlicensed premises where production takes place are often filthy and dangerous (McElwee et al. 2017). The production of smokies was banned in the UK in 1987 mainly because of the associated health risks. Following this ban, the trade in smokies became the province of underground criminal meat producers. The UK National Farmers Union has lobbied for smokies to be produced legally, as happens in Australia. This, they claim, would make the meat safer for consumers to eat, noting the difficulties in policing the current underground trade (McElwee et al. 2017).

It has been estimated that 200 000 smokies are bought and sold each year in London by butchers, restaurants, and other retail outlets in areas where demand is high (Teinaz 2004). A recent documentary reported that in parts of south London with large west African communities 'every second customer' requests 'smoky meat', with one butcher stating that he could sell around 60 smokies a week if production were legalized (Vice 2014). Where demand is high, it has been estimated that a van full of smokies could be worth between £5000 and £10000. An old cull ewe, purchased for a few pounds, can retail at £100 and £150 after being turned into a 'smoky' (Teinaz 2004).

Over a decade ago, the Chartered Institute of Environmental Health (CIEH) warned that rural Wales was becoming a centre for the illegal production of smokies. At the 2003 Cracking Down on Meat Crime conference, the CIEH suggested that criminal gangs could put some farming communities at risk if they were not stopped. However, as the above documentary showed, the trade is still very much alive and concerns about health, hygiene, and animal cruelty are still not adequately being addressed. The ongoing threat of criminal gangs stealing sheep in Wales and other rural locations across the UK to produce smokies for the halal market was recently highlighted by McElwee et al. (2017). This research focused on the activities of just one criminal gang over the course of a 20-year period. The authors discuss how the gang members went about stealing sheep from farms to produce smokies for the halal market, and also how they avoided detection because of the lack of understanding and concern for such crimes amongst regulatory agencies.

Some types of meat fraud occur because of regulatory failures in the food industry rather than through organized crime. A type of fraud particularly affecting the Muslim community concerns 'hot meat' (Pointing and Teinaz 2012). Hot meat is derived from old cull ewes – from thin, emaciated, often oedematous animals. These are slaughtered as 'halal' at Meat Hygiene Service licenced plants for Muslim markets throughout the UK. The halal certification process is deeply flawed. Hot meat – which is often dark in colour – is transported without chilling immediately after slaughter. It is loaded whilst warm for immediate despatch under the wrong presumption that this is a halal requirement. Meat from old ewes does not have the same keeping qualities as meat from young lambs, especially when not chilled. Because of their poor quality and the defective production processes, carcasses will often be delivered at the point of sale in a condition that is unfit for human consumption.

19.4 Fraud Involving Chicken

Muslims consume a large percentage of chicken produced in the UK at fast food restaurants, such as Kentucky Fried Chicken, and bought from supermarket chains, such as Tesco. Surveys conducted by the FSA during 2001 and 2003 suggested that there were widespread problems concerning the addition of water to frozen chickens, once defrosted, and regarding the mislabelling of chicken. The FSA also became concerned around this time about widespread confusion in the industry concerning Islamic requirements for halal production, including whether pre-slaughter stunning had taken place during processing. Such confusion partly resulted from changes in the running of slaughterhouses. In the early 1990s the processing of halal red meat and chicken was segregated from non-halal production in slaughterhouses and meat-cutting facilities. However, as the market expanded this practice has become expensive and segregation more difficult to maintain, so halal meat production has become entangled with non-halal production (Lever and Fischer 2018).

This problem has not been resolved with the entry of halal certifiers playing an important role in the industry. Over the last two decades, the Halal Food Authority's (HFA) percentage share of the UK halal certification market has been estimated at between 80 and 90% (Lever and Fischer 2018). During this period, many mainstream poultry manufacturers and distributors, including most of the UK's major slaughter facilities, have been under their control at some point, notably the 2 Sisters Food Group. The production practices involved are increasingly complex and difficult to untangle, with stunned halal chicken often being produced in the same factories as stunned non-halal chicken.

During 2014, a poultry scandal broke at plants owned by the 2 Sisters Food Group and by Faccenda indicating that some deep-seated problems have damaged the integrity of the chicken processing industry. *The Guardian* reported, in July 2014, that serious breaches of food hygiene regulations had been revealed by whistleblowers and that food safety records, including slaughter dates, had been altered. It emerged that workers had been asked to stretch the commercial life of poultry to allow their employer to dupe consumers into buying chicken that was unfit for human consumption.

A major reason for poor practices in the chicken processing industry in the UK is that profit margins are squeezed and products are sold very cheaply – a policy made possible by the huge volume of production of both halal and non-halal chickens. A FSA survey found in 2003 that there were issues over the mislabelling of chickens. The survey also revealed that both halal and non-halal chickens contained proteins derived from beef and pork, but their labelling gave no indication of such adulteration. The presence of undeclared mammalian proteins in chicken was also found in a subsequent FSA survey, published in 2009.

19.5   Problems of Halal Regulation

Meat purporting to be halal which is sold to Muslims by the unscrupulous can turn out to be a far cry from what Muslims believe it to be. Meat that is unfit for human consumption cannot be considered halal, whether or not it is derived from a halal-slaughtered animal. Many fraudsters come from the Muslim community, and some are able to justify their criminal activity by the lack of consensus about what constitutes halal. Closer involvement between halal regulation and state law is also required, as argued in by Pointing (2011, p. 43) in the *Halal Journal*:

> This sort of crime is not simply a case of wrongs committed by non-Muslims on Muslims. It involves Muslim businessmen operating as middlemen, Muslim butchers and Muslim retailers. At the consumer end, it is a crime committed by Muslim against Muslim, as well as by and against non-Muslims. So for a number of reasons the 'Halalness' of food needs to engage with state law.

The EU General Food Law provides the basis for a rigorous system of enforcement, though it is less effective in curtailing organized food crime (Pointing 2005). EU legislation requires that all animals are stunned prior to slaughter. It is only through a derogation in European law (by Regulation [EC] No. 1099/2009) that the rights of Muslim and Jewish consumers are protected, so enabling them to consume meat from animals that are not stunned prior to slaughtering. Derogation allows these religious minorities to practise non-stun slaughter in line with religious freedoms granted through Article 10 of the Charter of Fundamental Rights of the EU. The legislation accommodates rights: there is no requirement in European law for Muslims to consume non-stunned meat or to eschew any non-halal product.

The Islamic requirement for halal meat – combined with the complexity and variety of methods used in halal meat production – has left the Muslim community in Britain open to hostility and suspicion from those who do not understand what is actually meant by 'halal' (Lever and Fischer 2018). This creates a problem more of misinformation rather than one resulting from deliberate fraud. A FSA survey of slaughterhouses, carried out in 2013, indicated that over 80% of all halal meat produced in the UK comes from stunned animals. It also found that 31 million poultry animals, 2.5 million sheep and goats, and 44 000 cattle were not stunned. Combine this with increasing supermarket-led vertical integration in the meat industry (Lever and Milbourne 2015) and the difficulty of distinguishing mainstream (stunned) meat from stunned halal meat produced in the same industrial-scale slaughterhouses, and we start to understand the rise of anti-halal sentiment in the UK.

The widespread public perception – driven by anti-Muslim media-driven political rhetoric and discourse – is that supermarkets, restaurants, and public institutions are selling or providing non-stunned halal meat in increasing volumes (Lever and Fischer 2018). Richard North, a qualified meat inspector and former technical adviser to the Small Abattoirs Association, argues that not only are consumers deliberately not being informed that their meat is produced according to religious requirements, but many retailers themselves do not know the true facts. To complicate matters further, retailers have become reticent about displaying the logos of their certification partners (Lever and Fischer 2018).

19.6 Conclusion

There is no lack of goodwill to overcome the problems discussed in this chapter and specific governmental bodies have been set up to tackle them. For England and Wales, the FSA was set up in 1999 to protect consumers, and for Scotland, Food Standards Scotland. The European Commission's Directorate General for Health and Consumer Protection (DG SANCO) was also established in 1999, which was followed, in 2002, by the European Food Safety Authority (EFSA). At a conference hosted by the European Commission in 2015, the deputy head of the Commission's food fraud unit spoke of the need for vigilance at an international level to prevent further scandals: 'We live in a globalized world. This is the internationalization of food. More and more we will have this wide-scale fraud over the world. We have to be vigilant and exchange information'.

In the UK, first the National Audit Office Review (2008) and then the Elliott Review (2014) have provided the impetus to bring about change. The need for better leadership and cooperation on food fraud and enforcement matters was highlighted in both reviews. This was necessary not least because the FSA had fought shy over the years in getting involved in matters concerning the rights of Muslim consumers, having recoiled from the regulatory chaos over food fraud in the early 2000s. The FSA has a statutory duty under section 1 of the Food Standards Act 1999 to protect the interests of all consumers, including Muslims. It needs to take a proactive role as the government body responsible for seeing the full implementation of EU food law, including religious freedoms protected by Article 10 of the Charter of Fundamental Rights of the EU.

More recently, the spectre of Brexit has heightened concerns about food safety and the increased threat of food fraud once the UK leaves the EU (Lever, unpublished). If prices rise quickly or unexpectedly post Brexit, opportunities for food fraud may increase, as producers look to cut corners or substitute cheaper ingredients for more expensive ones. New trade relations with the USA could result in a lowering of standards and allow the importation of low-grade products, such as bleached chicken, that are currently prevented from entering the EU. Such exposure to US sources could result in damage to the already shrinking agricultural and farming sectors in the UK. There would also be the possibility for a general weakening of regulatory controls, perhaps igniting a 'bonfire' of regulations thought to be the epitome of excessive Brussels bureaucracy (Pointing 2009). On the other hand, some argue that Brexit creates opportunities for the UK government to provide more effective food regulation protecting consumers. As far as the halal market is concerned, there has been talk of Brexit providing the opportunity to ban non-stun slaughter in the UK, as has occurred in Denmark. The political pressures to do so are likely to be strong, with the resurgence of right-wing populism and lobbying of powerful groups such as the British Veterinary Association (BVA) (see p.1 of Introduction).

References

Daily Mail (2014), A. Malone. Secret trade in monkey meat that could unleash Ebola in UK: How an appetite for African delicacies at British market stalls may spread killer virus. *Daily Mail* (1 August 2014). https://www.dailymail.co.uk/news/article-2713707/Secret-trade-monkey-meat-unleash-Ebola-UK-How-appetite-African-delicacies-British-markets-stalls-spread-killer-virus.htm (accessed 15 August 2018).

Elliott Review (2014). Elliott Review into the Integrity and Assurance of Food Supply Networks – Final Report. A National Food Crime Prevention Framework. HM Government: July 2014.

Food Standards Agency (2001). Water added to restaurant and take-away chicken, survey finds. *FSA Press Release* (11 December 2001).

Food Standards Agency (2003). FSA water in chicken update.

Food Standards Agency (2009). Illegal meat: guidance for local enforcement authorities in Wales.

Fuseini, A., Wotton, S., Knowles, T., and Hadley, P.J. (2017). Halal meat fraud and safety issues in the UK: a review in the context of the European Union. *Food Ethics* 1: 127–142.

Lever, J. and Fischer, J. (2018). *Religion, Regulation, Consumption: Globalising Kosher and Halal Markets*. Manchester: Manchester University Press.

Lever, J. and Miele, M. (2012). The growth of the halal meat markets in Europe: an exploration of the supply side theory of religion. *Journal of Rural Studies* 28: 528–553.

Lever, J. and Milbourne, P. (2015). The structural invisibility of outsiders: the role of migrant labour in the meat-processing industry. *Sociology* 51 (2): 306–322.

Marsden, T. and Morley, A. (eds.) (2014). *Sustainable Food Systems: Building a New Paradigm*. Abingdon: Routledge.

McElwee, G., Smith, R., and Lever, J. (2017). Illegal activity in the UK halal (sheep) supply chain: towards greater understanding. *Food Policy* 69: 166–175.

Miele, M. and Lever, J. (2013). Civilizing the market for welfare friendly products in Europe? The techno-ethics of the Welfare Quality (R) assessment. *Geoforum* 48: 63–72.

National Audit Office (2008). *Effective Inspection and Enforcement: Implementing the Hampton Vision in the Food Standards Agency*. London: Better Regulation Executive and National Audit Office.

Pilcher, J.M. (2017). *Food in World History*, 2e. Abingdon: Routledge.

Pointing, J. (2005). Food crime and food safety: trading in bushmeat – is new legislation needed? *Journal of Criminal Law* 69 (1).

Pointing, J. (2009). Food law and the strange case of the missing regulation. *Journal of Business Law* 6: 592–605.

Pointing, J. (2011). Should halal conform to state food law? *Halal Journal* 40: 42–45.

Pointing, J. and Teinaz, Y. (2004). Halal meat and food crime in the UK. *Proceedings of the International Halal Food Seminar*, Islamic University College of Malaysia, Kuala Lumpur. https://www. researchgate.net/publication/317033542_HALAL_MEAT_AND_FOOD_CRIME_IN_THE_UK (accessed September 2004).

Pointing, J. and Teinaz, Y. (2012). Meat crimes in the UK. *The Meat Hygienist: Journal of the Association of Meat Inspectors* 155: 7–9.

Pointing, J., Teinaz, Y., and Shafi, S. (2008). Illegal labelling and sales of halal meat and food products. *Journal of Criminal Law* 72 (3).

PwC (2015). A recipe for food trust. PwC's Food and Integrity Services. https://www.pwc.com/ foodtrust (accessed 15 August 2018).

Schier, J.G., Hunt, D.R., Perala, A. et al. (2013). Characterizing concentrations of diethylene glycol and suspected metabolites in human serum, urine, and cerebrospinal fluid samples from the Panama DEG mass poisoning. *Clinical Toxicology (Philadelphia)* 51 (10): 923–929. Available from: http://www.ncbi.nlm.nih.gov/pubmed/24266434 (accessed 15/08/18).

Teinaz, Y. (2004). Meat Crimes in the UK. *Speech given to The Royal Society for the Promotion of Health, Fellows Lunch*, The Royal Overseas League, London (19 May 2004). https://www. researchgate.net/search.Search.html?type=publication&query=Yunes%20Teinaz (accessed 15 August 2018).

The Guardian (2005), R. Ramesh, S. Jha, F. Lawrence, V. Dodd. From Mumbai to your supermarket: on the murky trail of Britain's biggest food scandal. *The Guardian* (23 February 2005).

The Guardian (2013), J.Kaiman. China arrests 900 in fake meat scandal. *The Guardian* (3 May 2013). https://www.theguardian.com/world/2013/may/03/china-arrests-fake-meat- scandal (accessed 15 August 2018).

The Guardian (2014), F. Lawrence, A. Wasley, R. Ciorniciuc. Revealed: the dirty secret of the UK's poultry industry. *The Guardian* (23 July 2014). https://www.theguardian.com/ world/2014/jul/23/-sp-revealed-dirty-secret-uk-poultry-industry-chicken-campylobacter (accessed 15 August 2018).

Vice (2014), The Politics of Food: Smokies. https://munchies.vice.com/videos/the-politics-of- food-smokies (accessed 15 August 2018).

Which? (2014) Food fraud: What's in your takeaway? http://www.which.co.uk/news/2014/04/ food-fraud-whats-in-your-takeaway-362886 (accessed 15 August 2018).

Part VI

Halal vs Kosher

The Dialogue Between Science, Society and Religion about
Religious Slaughter

20

The Halal and Kosher Food Experience in the UK

Yunes Ramadan Al-Teinaz[1], Joe M. Regenstein[2], John Lever[3],
A. Majid Katme[4] and Sol Unsdorfer[5]

[1] Independent Public Health & Environment Consultant, London, UK
[2] Department of Food Science, College of Agriculture and Life Sciences, Cornell University, Ithaca, NY, USA
[3] University of Huddersfield Business School, Huddersfield, UK
[4] Islamic Medical Association, UK
[5] Active member of the orthodox Jewish community, London, UK

20.1   Introduction

Islamic dietary laws and Jewish dietary laws are both very detailed and demonstrate some similarities. They share a common root – a code of laws found in Leviticus and in the Holy Quran and prophetic sayings and behaviour (Sunnah). Halal and kosher are terms often heard in the context of meat and dairy foods, and it is common knowledge that the terms refer to what is permitted by Islamic and Jewish religious laws, respectively. Halal is an Islamic term that means lawful or permitted. Although halal refers to anything that is permitted by Islamic Shariah law, it is most often used in the context of permissible dietary and meat consumption (Shafi et al. 2013; see Chapter 1). Kosher is a similar term used to describe food that is proper or fit for consumption according to Kashrut, the Jewish dietary laws, part of the bigger set of Jewish laws or Halacha (Lever and Fischer, 2018). This chapter will restrict itself to only the religious dietary laws.

Although it may appear that kosher and halal foods deal with similar products, there are differences in slaughter and inspection methods (Regenstein et al., 2003). In linguistic terms, both the terms halal and kosher are almost similar. Halal is Arabic meaning permissible whilst kosher is a Hebrew word that means proper or fit.

However, halal and kosher are two different legal systems with differences, although similar, in their meaning and spirit. Halal and kosher are mainly associated with the foods of Muslims and Jews, they also have relevance to other rituals that people of both religions follow in their lives.

Halal and kosher have their roots in their respective scriptures, halal is mentioned in the Holy Quran and kosher is identified in the Hebrew Scriptures, especially the Torah, the first five books of Scripture.

In this chapter, the focus is on the commonality of the slaughter of animals. Other similarities and differences are presented in Table 20.1.

20.2   Halal and Shechita: The Muslim and Jewish Religious Humane Methods

Halal and shechita are the Jewish and Muslim religious-humane methods of slaughtering animals and birds for food. Halal and shechita may only be used for permitted (see below) animals and birds, and are the only methods that can be used to provide Jews and Muslims with permissible meat, poultry, and their products. All other methods, including stunning pre- or post-slaughter, even if done in conjunction with shechita or halal, renders the meat non-halal (see Chapter 1) or non-kosher according to most authorities and are thus forbidden for Jews and Muslims (see Chapter 1). A ban on halal and shechita or imposition of other methods may lead to a situation where Muslims and Jews would be unable to buy and consume acceptable meat, poultry or meat products.

The source for the method of Shechita originated in the Torah in Deuteronomy 12:21:

> 'And you shall slaughter of your herd and your flock (for food) in the manner I have commanded you'.

Therefore, for Jews the permissible animals must be slaughtered in the manner as instructed to Moses on Mount Sinai by the Almighty Himself. Since it is believed that the physiology of animals is designed by God, Jews hold the view that by adhering to the laws of the Torah, man's spiritual status is improved. This is fundamental in Judaism. The actual details are a part of the Oral Law handed down at Sinai that were eventually written down and codified in the Talmud early in the Common Era.

Only ruminants having cloven hooves are permitted by Torah law as Kosher meat. This excludes swine, horses, camels (which are ruminants but have padding in their hooves), and rabbit amongst other animals. Domestic fowl, including turkey, chickens, ducks, geese, pheasant (with some disagreement amongst scholars), partridge, and quail (and their eggs), are permitted as kosher birds. Halacha specifies those species of birds which may not be eaten. Birds of prey are not permitted as food. Kosher fish, from the sea or freshwater, are identified by having both fins and removable scales. Molluscs, crustaceans, and marine mammals are not permitted as they are fish without the requisite scales. Fish, however, do not require shechita, and are actually pareve, i.e. they are not meat or milk, which in halacha must be kept separate.

20.3   Legislation

In the UK the practice of shechita and halal is authorized and regulated by the current Welfare of Animals at the Time of Killing (WATOK) Regulations (formerly known as The Welfare of Animals [Slaughter or Killing] Regulations 1995 [Statutory Instrument 1995 No. 731]).

Table 20.1 Comparison chart – halal vs kosher, differences and similarities

	Halal	Kosher
Introduction	Halal is anything that is permissible according to Islamic law. The term covers and designates not only food and drink as permissible according to Islamic law, but also all matters of daily life.	Kosher foods are those that conform to the regulations of kashrut, the Jewish dietary law.
Guidelines	Follow Islamic Shariah dietary laws.	Follow Jewish Halachic dietary laws.
Etymology	'Halal' in Arabic means permissible or lawful.	The noun 'kashrut' is derived from the Hebrew word 'kasher' (adjective), which means proper or fit.
Roots	The Holy Quran	Torah
How to slaughter	Quick and swift at single point on the throat; blood has to be completely drained.	Quick and swift at single point on the throat; blood has to be completely drained.
Slaughterer	Any trained Muslim having reached puberty is allowed to slaughter after saying the name of Allah and facing Qiblah (Makkah). The Holy Quran permits meat slaughtered by people of the book, i.e. those who follow Hebrew and/or Christian scriptures, if they follow the same original religious method of animal slaughter. Prophetic Islamic instructions for animal welfare at time of slaughter: – offer food/drink to the animal before slaughter – avoid sharpening the knife in front of the animal – avoid slaughtering the animal in front of other animals – use a very sharp suitable knife. The animal to slaughter: – should be healthy with healthy body, not damaged, alive and conscious, fed before his/her own natural vegetarian diet – should be handled gently by the slaughterer, slaughtered by hand not by machine and put in a comfortable position	Animals must be slaughtered by a Jew specifically trained (shochet) using a special knife (chalaf) that is twice the width of the neck. If the chalaf has the slightest nick or irregularity it must not be used. If the chalaf is found to be imperfect after it has been used for shechita the animal is non-kosher and may not be eaten by Jews but is sold in the secular market so as to not waste it.

(Continued)

Table 20.1 (Continued)

	Halal	Kosher
Dead meat	Both the Islam and Jewish religions prohibit the eating of meat from a dead animal. It is forbidden to eat the meat from an animal that died of illness, a blow, by trapping, or a natural death, or any animal killed in any manner other than by slaughter or the hunt.	Both the Islam and Jewish religions prohibit the eating of meat from a dead animal. It is forbidden to eat the meat from an animal that died of illness, a blow, by trapping, or a natural death, or any animal killed in any manner other than by slaughter or the hunt (hunting is not permissible in Judaism).
Prayer	Before zabiha occurs, the slaughterer recites a blessing, saying: 'Bism Allah, Allah Akbar (In the Name of Allah, Allah the Greatest)'. Requires a prayer to Allah before every slaughter.	Requires a blessing (bracha) before commencing a slaughter session.
Animals	The animal should be one of those permitted to be eaten (as halal) by Muslims, such as sheep, goats, cattle, poultry, camels, rabbits, non-predatory animals, and birds, etc. Animals should be alive and healthy at the time of slaughter. In Islam, meat should be halal and also tayyib (meaning natural, organic, free from harmful chemicals).	According to kosher dietary law only ruminants with cloven hooves are permitted as food. Chickens, ducks, and turkeys and their eggs are permitted, but not birds of prey. For those that may be eaten there are certain rules to be followed and a specific way to be slaughtered. Permitted fish, both sea and freshwater, must have fins and removable scales.
Not permitted animals	Animals who have died by unlawful treatment, such as pre-slaughter stunning. Animals which have been dedicated to any purpose other than God. Pigs, dogs, domestic donkeys, elephants, mules. Predatory carnivorous animals and birds such as lions, tigers, bears, eagles, and falcons. Carrion, strangled or suffocated (as in electric water bath for poultry), and fatally beaten animals.	The following animals and meat products are not considered kosher according to Jewish dietary law: • animals not slaughtered according to Jewish law • camels and pigs • rabbits and hares • predatory and scavenger birds • shellfish • rodents • reptiles and amphibians.
Blood	In Islam the eating or drinking of blood that has flowed from flesh is prohibited. A slaughtered animal must be drained of all flowing harmful blood. Accordingly, while preparing meat for cooking, many Muslims wash raw meat to remove the surface blood from it.	Blood is forbidden to be consumed by Jews. For meat to be kosher, all blood and large blood vessels must be removed from the meat. The residual blood is commonly removed by soaking for a half hour and salting for one hour, but it can also be done under some circumstances by a special broiling process. Liver of both animal and fowl can only be 'koshered' by this process. The hindquarters of a mammal are not kosher until the sciatic nerve and certain fat surrounding the internal organs are removed.

Fruit and vegetables	Considered halal if there are no whole bugs or chemicals in them, which can be removed after inspection, but they should be natural (tayyib), as in the Holy Quran. Many Muslims are concerned about the risk to health when insecticides and pesticides are used in the production of fruits and vegetables.	Considered kosher only if there are no whole bugs in them, which can be removed after inspection.
Alcohol	In Islam, alcohol and alcohol-derived products and ingredients, including (most) wine vinegars, are prohibited, even when used in cooking. A Muslim is not allowed to eat any meat or food or drink if it contains alcohol.	Kosher wine in moderation; grape-based alcohols must be prepared under Orthodox Rabbinical supervision. For a wine to be kosher it must be produced under the supervision of an authorized Rabbinical authority. All other (non-grape) alcoholic drinks can be certified kosher, so long as they contain no non-kosher ingredients and the equipment used to prepare them is inspected under the supervision of an authorized Rabbinical authority (Lever and Miele 2012).
Cheeses and dairy products	Cheese is allowed in the Muslim diet. However, Muslims are concerned with the slaughter of the calf from which the rennet came – it should come from halal slaughter. Muslims may choose to eat cheeses using vegetarian clotting enzymes to avoid doubt. Genetically modified and microbial rennet are generally permitted. In Islam there is no prohibition against mixing meat and dairy products, they can be consumed together.	For any cheese, cream or butter to be kosher, it must be produced from the milk of a kosher animal, i.e. cow, sheep, goat, whether or not it is made with genetic modification, microbial or animal rennet from a kosher-killed animal (Gordimer, 2011). It must be 'Jewish cheese' (gevinat Yisrael), i.e. cheese produced under authorized Rabbinical supervision with participation by an observant Jew. (this was not explained by Cholev Yisroel). Meat and dairy products cannot be consumed together. Jewish dietary law forbids: 1) cooking meat and milk together in any form 2) eating such cooked products or 3) deriving benefit from them. As a safeguard, the Rabbis extended this prohibition to disallow the eating of meat and dairy products at the same meal or preparing them with the same utensils. Milk products cannot be consumed after eating meat for a period of time. Most customs say to wait six hours (ranges from one to six hours). Prior to eating meat after dairy, one must eat a solid food, either drink a liquid or thoroughly rinse one's mouth, and check the cleanliness of hands. Meat and fish may not be cooked together in the same utensil as this is considered injurious to health.

(Continued)

Table 20.1 (Continued)

	Halal	Kosher
Pig	In Islam, the eating or application (e.g. cosmetics) of pig products (pork, gelatine, etc.) is forbidden.	In Judaism, the eating of pig or its products is forbidden, but injection (e.g. injectable collagen) is acceptable. However, there have been recent rabbinical rulings that suggest a trend away from the religious permissibility of topical and injectable pig-derived products. (I think this is too broad a statement.)
Insects	Islam prohibits eating insects, except for the Arabian locust. Vegetables must be washed to remove insects. Vegetables and fruits, especially leafy vegetables and berries, must be checked for insect infestation.	Eating whole insects is a major concern. However, as in Islam, certain sub-species of locust are kosher. To be kosher, vegetables and fruits, especially leafy vegetables and berries, must be checked for insect infestation.
Utensils	Islam does not mandate separate utensils. Utensils previously used to prepare pork or another haram substance must be thoroughly washed. Muslims are concerned about the source of enzymes before using them. If the enzyme comes from a non-halal animal, it is prohibited for a Muslim.	A kosher kitchen must have two different sets of utensils, one for meat and poultry and the other for dairy foods. There must be separate, distinct sets of pots, pans, plates, and silverware. There are complex procedures for changing the status of kosher equipment although some items cannot be made kosher or have their use changed. Most homes also have additional sets of utensils for meat and/or dairy on Passover.
Ingredients, intoxicants	According to halal law, all intoxicating alcohols, wines, liquors and drugs are prohibited. Halal permits the mixing of dairy and meat. See Chapter 9.	Animal enzymes are generally considered non-kosher. Plant enzymes and extracts of plant tissue usually pose no special Kashrus concerns. Once the enzymes are separated from the organism, they need to be diluted and standardized. In kosher foods, dairy and meat cannot be mixed (pet food with dairy and pork is okay but not dairy and beef).
Accreditation	In the UK, there are at least 12 halal certification bodies all operating according to varying halal standards and competing with each other for a share of the halal certification market.	According to the KosherFest website there are 900 kosher symbols used throughout the world (*Kashrus Magazine* lists almost 1400!). A kosher certification mark is referred to as a hechsher. The main accreditation bodies within the UK are the London Beth Din, the London Board for Shechita, Sephardi Beth Din, the Joint Kashrus Committee of England, the Federation of Synagogues Kashrus, and the Manchester Beth Din. Some Scottish companies have also opted for US accreditation, including the New York based Orthodox Union, the largest kosher certifying agency in the world. Of the UK bodies, the London Beth Din is the largest. Indeed, it is one of the largest kosher accreditation bodies in Europe and Asia.

Halal and shechita in the UK may only take place in a licenced abattoir certified by the Food Standards Authority (FSA) and under the inspection of at least one of their veterinary officers.

To meet the Jewish religious requirements, shechita is supervised and licenced by the rabbinical authorities, who are responsible for training, supervision, and certification (licencing) of competent shochtim. The Rabbinical Commission for the Licensing of Shochetim was established by Parliament by the Slaughter of Animals Act 1933, an Act to 'provide for the humane and scientific slaughter of animals' in England, effective 1 January 1934.

Halal is certified by a few organizations which are not regulated and not externally audited or inspected by the local authorities or any official religious body, and are in competition with each other.

The rabbinical authorities aim to ensure that meat and poultry sold as kosher conforms to the strict requirements of halacha. All kosher meat and poultry must bear the seal of the Rabbinical Authority authorizing the shechita when it leaves the abattoir to be delivered to premises from where it is distributed to the public. Without authorized identification (a hechsher) as to its kosher status all meat, poultry, and their products, are considered to be non-kosher and may not be eaten by Jews.

British legislation, now implementing EU legislation, requires the pre-stunning of animals before slaughter in normal circumstances, so that in the absence of mis-stuns death should be painless. Religious slaughter, on the other hand, is a controversial issue because the animals are not stunned (Regenstein, 2011). The requirement in British legislation for the pre-stunning of animals in slaughterhouses has always provided exemptions for the Jewish and Muslim methods of slaughter (Bergeaud-Blackler, 2007). The exemption dates back to the Slaughter of Animals (Scotland) Act 1928 and the Slaughter of Animals Act 1933 (which applied to England and Wales). Schedule 12 of The Welfare of Animals (Slaughter or Killing) Regulations 1995 (SI 731) lays down provisions for slaughter by a religious method, additional to EU law.

1) The exemption for religious slaughter in Schedule 12 of The Welfare of Animals (Slaughter or Killing) Regulations 1995 (SI 731) makes clear that it relates to a method of slaughter for people of that religion, not for everybody (i.e. slaughter by a religious method)
2) In this Schedule references to slaughter by a religious method are references to slaughter without the infliction of unnecessary suffering
   a) by the Jewish method for the food of Jews by a Jew who holds a licence in accordance with Schedule 1 (which relates to the licencing of slaughtermen) and who is duly licenced –
      i) in England and Wales by the Rabbinical Commission referred to in Part IV of this Schedule; or
      ii) in Scotland by the Chief Rabbi; or
   b) by the Muslim method for the food of Muslims by a Muslim who holds a licence in accordance with Schedule 1 (see Chapters 16 and 17).

The act of shechita is carried out by the shochet. His official title in Jewish usage is 'Shochet U'Bodek', which means 'Shochet and Examiner' because he must be competent in inspecting the animal or bird for any anatomical irregularities which might cause the animal or bird to be non-kosher. The expertise of the shochet in matters of halacha and anatomy takes many years to acquire (Rosen, 2004).

Before the actual shechita takes place, the shochet is required to pronounce a 'bracha' (blessing) attesting to the fact that he is fulfilling a divinely ordained precept and that it is only through the commandments of the Almighty that the performance of shechita is permitted.

Shechita involves a swift incision using a surgically sharp instrument free of nicks, i.e. the chalaf, severing the major structures and blood vessels in the neck of the animal or bird. Shechita conforms to the principles of surgery which require that it must be performed *cito, tuto et jucunde* (with certainty, quickly and painlessly). The shochet is also required to examine the shechita incision immediately afterwards to ascertain that the shechita was done correctly, i.e. that the trachea, oesophagus, and the four blood vessels (two carotids and two jugulars) have been fully cut. The chalaf must be extremely sharp and smooth, and must be examined by the shochet before and after shechita to ensure that it does not have even the minutest nick or irregularity. If one is found, the animal is not kosher and may not be eaten by Jews.

The act of shechita and the halal slaughter of animals are believed to be painless. They take a few seconds to complete, resulting in an immediate drop in blood pressure and lack of oxygen to the brain and the loss of consciousness. They also provide for the maximum exsanguination (blood outflow from the carcass). Shechita is claimed to avoid the suffering of animals caused by stunning, which may not always be effective, sometimes resulting in re-stunning.

Those who are opposed to shechita and halal, i.e. those who want to see it banned, are continually pressuring the UK and EU governments. For many others, however, the Jewish and Muslim humane methods are practices that conform in every way with the dictates of hygiene, compassion, and humaneness (Regenstein, 2000).

Many people confuse halal and kosher, and believe that they are equivalent. There are indeed similarities and differences. Table 20.1, in a synthesized and basic view, provides an overview of the comparisons that can be made between these two sacred food codes.

20.4 Conclusion

Islamic dietary laws and Jewish dietary laws are both very detailed and demonstrate some similarities. They share a common root – a code of laws found in Leviticus and in the Holy Quran. However, kosher and halal are two different entities that have differences in their meaning and spirit (see table 21.1. Kosher and halal are mainly associated with the food of Muslims and Jewish people.

This chapter compares halal with kosher as it understood and practiced in the UK. It focusses mainly on the commonality of animal slaughter practices, but other similarities and differences are considered, including permissible and prohibited animal species and foods. While halal is most often used in the context of permissible dietary and meat consumption practices, kosher used similarly to describe food that is 'proper' or 'fit' for consumption.

There have been persistent calls to ban shechita and halal slaughter (without stunning) for almost two centuries in the UK, with those who want to see it banned continually pressuring the UK and EU governments. Jewish and Muslim humane methods are practices that conform in every particular with the dictates of hygiene, compassion, and humaneness. However, there is no overall scientific consensus about what constitutes the most 'humane' way of slaughtering animals for food: as well as being scientific, the issue is also ethical and subject to interpretation (Bergeaud-Blackler, 2007; Bergeaud-Blackler et al., 2015). These debates are likely to continue.

References

Bergeaud-Blackler, F. (2007). New challenges for Islamic ritual slaughter: a European perspective. *Journal of Ethnic and Migration Studies* 33: 965–980.

Bergeaud-Blackler, F., Fischer, J., and Lever, J. (eds.) (2015). *Halal Matters: Islam, Politics and Markets in Global Perspective*. Routledge.

Gordimer, R.A. (2011). How is cheese made kosher? *Conference of Halal Industry and Its Services*, Salmiyah, Kuwait (24–26 January 2011). http://AskMoses.com reprinted from https://oukosher.org (Retrieved June 2018).

Lever, J. and Fischer, J. (2018). *Religion, Regulation, Consumption: Globalising Kosher and Halal Markets*. Manchester University Press.

Lever, J. and Miele, M. (2012). The growth of halal meat markets in Europe: an exploration of the supply side theory of religion. *Journal of Rural Studies* 28: 528–537.

Regenstein, J.M. (2000). Humane (halal) on-farm slaughter of sheep and goats. Northeast Sheep and Goat Marketing Program, Department of Animal Science, Cornell University, Ithaca, NY. http://sheepgoatmarketing.info/images/HumaneSlaughter.pdf (accessed 16 september 2019).

Regenstein, J.M. (2011). Expert Opinion on Considerations When Evaluating All Types of Slaughter: Mechanical, Electrical, Gas and Religious Slaughter and A Critical Scientific Review of Report 161: Ritual Slaughter and Animal Welfare (September 2008); Report 398: Report on Restraining and Neck Cutting or Stunning and Neck Cutting in Pink Veal Calves (September, 2010) by the Animal Sciences Group, Wageningen UR; and the 2009 New Zealand Papers by Gibson et al. Preliminary Report. Prepared for the Netherlands Parliamentary Debate on Religious Slaughter. Issued by the Netherland Jewish Community, 49 pages.

Regenstein, J.M., Chaudry, M.M., and Regenstein, C.E. (2003). The kosher and halal food laws. *Comprehensive Reviews in Food Science and Food Safety* 2: 111–127.

Rosen, S.D. (2004). Physiological insights into Shechita. *Veterinary Record* 154: 759–765.

Shafi, S., Teinaz, Y., Haluk, A., and Pointing, J. (2013). Halal making religious slaughter authentic. *The Meat Hygienist* 156: 25–30.

21

Establishing a Dialogue Between Science, Society and Religion About Religious Slaughter

The Experience of the European Funded Project Dialrel

Mara Miele[1], John Lever[2] and Adrian Evans[3]

[1] *School of Geography and Planning, Cardiff University*
[2] *University of Huddersfield Business School, Huddersfield, UK*
[3] *Centre for Agroecology, Water and Resilience, Coventry University, Coventry, UK*

21.1 The Work With the Advisory Board

As part of the activities aimed at promoting a dialogue between the different stakeholders in the halal and kosher supply chains an advisory board was set up at the beginning of the Dialrel project. The activities of the advisory board increased when the research team started to produce the first results of the investigation. The members of the advisory board included representatives of the Muslim Council of Britain, the Halal Food Authority and Shechita UK Boards, Shechita France, the Federation of Veterinarians of Europe (FVE), and Eurogroup for Animals, amongst others. They showed great interest in the results produced and made a valuable contribution to the debate, underlining both the different religious requirements and their commitment to explore possible strategies to improve the welfare of animals at time of slaughter that would be consistent or acceptable with their religious requirements. Many members of the advisory board also pointed out the need for more transparency in the meat supply chains and better information for the general public about religious slaughter and certification processes. Particular attention was given to the risk of the adulteration and/or mislabelling of halal meat and other halal animal food products.[2]

The activities of the advisory board ended with the final Dialrel Workshop on 15 and 16 March 2010 in Istanbul, Turkey, where the final recommendations of the project were presented and the majority of the members of the advisory board added letters of support or comments to the final recommendation (see Velarde et al. 2010).

From our study carried out in 2007/2008 of the media coverage of issues related to religious slaughter and halal and kosher food in a sample of European countries (Germany, France, the UK, and Norway) it emerged that whilst there was a lack of coverage of these issues within traditional media sources (e.g. newspapers), this was to some extent compensated by the availability of information through new media sources (in particular via the internet). However, in the last year of the project there has been a significant increase in media attention given to the issue of religious slaughter. We can see that there has been a significant change in the issues and topics covered in recent years. Questions and issues concerning diversity and integration were part of the media debate in all study countries in 2007/2008. The expanding market for halal was a very prominent theme, especially within the UK (see Lever and Miele 2012). Debates regarding kosher food were, for various reasons, largely omitted from media debates within the four study countries.

A significant contribution to the debate about religious slaughter was given by a growing number of halal certifying bodies (see Lever and Miele 2012). In the following years the debate about religious slaughter became more prominent in a number of European countries. A key element affecting the development of this debate has been the increased demand for halal meat in developing countries. As Delgado pointed out 'by 2020, developing countries will consume 107 million metric tons (mmt) more meat and 177 mmt more milk than they did in 1996/1998, dwarfing developed-country increases of 19 mmt for meat and 32 mmt for milk' (Delgado 2003 in Miele 2016, p. 47).

These changes in the global demand for meat are linked to the expected growth in the populations of Muslim background that will affect the demand for halal meat. Recent studies (Grim and Karim 2011)[3] have forecasted an annual growth in the population of Muslim background of 1.5%, which will lead to an increase of about 35% in the next 20 years, rising from 1.6 billion in 2010 to 2.2 billion by 2030.[4] If this trend is confirmed, Muslims will make up 26.4% of the world's total population of 8.3 billion in 2030, with a median age of 24 globally (Miele 2016).

The rapid expansion of halal meat markets world-wide has been coupled with a growth in the export of halal meat from non-Muslim countries and the emergence of a growing number of certifying bodies to attract Muslim consumers with a guarantee of the halal status of the meat. For halal certification the central issue is the definition of what constitute 'halal slaughter', with various and often diverse interpretations of the religious rules regarding the acceptability of the practice of stunning animals before cutting the throat (Lever and Miele 2012; Wilson 2014; Bergaud-Blacker et al. 2015). In Europe the number of animals slaughtered without stunning is not systematically recorded in most countries and there are certifying bodies that grant the halal certification of meat obtained from stunned as well as non-stunned animals (see Lever and Miele 2012 for a discussion about this issue). These trends have raised significant concerns about the welfare of farm animals at time of killing and several non-governmental organizations (NGOs) have asked for the ban of religious slaughter without stunning in the UK, Spain and other European countries (Miele and Parisi 2001; Mukherjee 2014; White 2014; Miele and Rucinska 2015). The activities of the certifying bodies started to be addressed by the Dialrel project, which pointed out the increased lack of transparency of the market for halal meat as well as for those parts of the animal carcasses that would not be circulated in the religious market but originated from animals slaughtered without stunning.

21.2 The Activities of the Certifying Bodies

In the five years prior to the Dialrel project, markets for certified halal meat products emerged across a number of European countries and this gave rise to numerous controversies (Lever and Miele 2012; Mukherjee 2014). The market for halal meat expanded rapidly in the UK in this period and led to discussion and debate about the authenticity of halal products sold in supermarkets. There was little transparency in the market at this time and it soon became clear that many consumers displayed trust only in face-to-face interactions with local retailers and neighbourhood butchers. The major UK halal certification body is the Halal Food Authority (HFA), which accepts the practice of stunning animals before throat cutting in its slaughter standard and provides halal meat from stunned animals in line with the welfare claims of mainstream science and mechanical stunning practices. Interviews with corporate retailers and supermarkets quickly illustrated the relevance of this situation, with many supermarkets employing local Muslims to run in-store halal meat counters. Drawing on the advice and expertise of local of Muslim scholars, the Halal Monitoring Committee (HMC) started to make small inroads in the market at this time by refusing the practice of stunning in their halal standard and claiming that only halal meat from non-stunned animals was 'authentic' halal (see Miele 2016). This was controversial, and fierce competition soon emerged between the HFA and the HMC as the fast-food chain KFC started selling HFA-certified halal chicken (Lever and Miele 2012).[5]

In France, the market for halal products experienced growth and greater visibility in the years prior to Dialrel, but certification and stunning only emerged as major contested issues in the immediate years before the project. The French halal market was the largest in Europe and most meat was declared halal through 'auto-certification' practices, moreover there were few independent certification bodies. Differently from the UK, most halal meat was from non-stunned animals and debate about the provision of halal meat from stunned animals only began to emerge as mainstream actors in the food supply chains started to encroach on the market. However, while halal meat was starting to appear in mainstream supermarkets and fast food chains, it was mostly still distributed through dedicated retailers and local neighbourhood butchers. There was no consensus or overall definition of halal and the processes through which meat travelled from the slaughterhouse to the consumer were not clear (Lever and Miele 2012; Mukherjee 2014, Evans and Miele 2017).

While the demand for halal certification was increasing in Germany as well around this time, debate and controversy around stunning and slaughter practices were not at the same level as they were in the UK and in France. Concerns about the authenticity of commercial forms of halal meat had only recently emerged and the Aldi and Lidl supermarket chains started to consider the implications of selling halal meat (Schröder 2009). Much like the HMC in the UK, the European Halal Certification Institute (EHZ) certified in line with the opinion of Muslim scholars. Unlike the HMC, they worked closely with the meat industry and permitted stunning and mechanical slaughter in different areas of production to a greater or lesser extent. Pre-stunning was generally becoming more acceptable amongst German Muslims, while halal poultry registered the biggest market growth because the religious rules for slaughtering poultry were less stringent than for other farmed animals.

Unlike many EU countries, where derogation legislation allows minority groups to practice stunning without slaughter in line with religious freedoms granted by human rights legislation, under Norwegian law it was and still is a requirement that all animals are stunned prior to slaughter. The Norwegian market for halal meat, however, at this time, was still emerging steadily as a dialogue between religious communities and the meat industry facilitated high levels of trust and transparency. The Norwegian Islamic Council worked closely with the Norwegian Food Safety Authority to find a slaughtering method that different Muslim groups could agree on and validate.

In Turkey, an associate member of the EU and a secular Muslim state, the mainstream meat market was based on implicit and taken-for-granted everyday Islamic practices at this time. For the majority of the Turkish population all meat marketed within the country was taken to be halal at face value and most production companies only needed a letter from the local mufti (or cleric) to operate. Some meat was arriving at supermarkets with a label saying *this is halal*, but concerns over halal standards and certification had only just started to emerge in the meat industry (Lever and Anil 2015). The independent certification body GİMDES was the first organization to issue halal certificates to Turkish companies exporting to markets in Asia and the Middle East in 2009, which coincided with the rise of the politics of halal across Europe. In 2008, the chairman of GİMDES claimed that fewer than 40% of all products imported into Turkey were investigated. He argued that this situation had had created widespread public mistrust and cynicism and that halal standardization and certification were necessary to overcome such problems. Others, conversely, argued that the government of a secular state should not tell people what they should and should not eat on religious grounds (Cagatay and Yegenoglu 2006 in Lever and Anil 2015).

21.3 Muslim and Jewish Consumers' Attitudes to Halal and Kosher Foods and Religious Slaughter

An important part of the Dialrel project involved gaining a better understanding of the views and concerns of Muslim and Jewish consumers across Europe. In particular, the project sought to expand knowledge about the range of requirements and expectations that Muslim and Jewish consumers had about halal and kosher foods, and how these varied across countries and between different socio-cultural groups. The research also explored consumers' knowledge and views of religious slaughtering practices, with specific attention given to the issue of stunning (see Bergeaud-Blackler et al. 2010 for the original Dialrel report upon which this section is based). This research added to broader academic interest in both the public understanding of farm animal rearing and slaughter, and ethical food consumption (e.g. see Miele and Evans 2010; Miele et al. 2011).

In order to research the views of Muslim consumers, six in-depth qualitative focus groups were undertaken. The focus groups took place in Renaix in Belgium, Berlin in Germany, Bordeaux in France, Cardiff in the UK, Amsterdam in the Netherlands, and Istanbul in Turkey. Each focus group consisted of between seven and eight male and female consumers, all of whom were regular halal eaters and permanent residents in the country where the research was conducted. A common discussion guide was used across the five countries.

Results from the focus groups shed light on the perceived availability of halal foods and on halal consumers' trust in different networks of provision. Focus group participants from Amsterdam, Bordeaux, Cardiff, and Istanbul believed that halal foods were widely available in butcher shops, whereas participants from Berlin and Renaix indicated that the supply of halal food was only 'average' or 'low' in these outlets. In contrast, the perceived availability of halal-certified foods in supermarkets was low to average in all places, except for Istanbul. In all countries, focus group participants expressed a preference for purchasing halal meat in Islamic butcher shops rather than in supermarkets. This was because butchers were believed to provide a good balance of hygiene, quality, price, variety, and proximity. Furthermore, many participants had developed a personal relationship of trust with their butcher and they felt that this was the best way to guarantee that the meat they purchased was genuinely halal. In contrast, participants were less positive about purchasing halal meat in supermarkets. Halal food labels and certification schemes were present in all study countries, but participants often questioned their reliability. Certain participants highlighted the complexity of food chains in industrialized societies and hence the difficulty of guaranteeing that 'halal' requirements are followed at all stages of production. Participants were also suspicious about the proliferation of different halal labels on both food and non-food products, and many were unwilling to trust a halal label without additional assurances (e.g. authentication by trustworthy religious institutions).

During the focus groups participants were asked about their views regarding religious slaughter, animal welfare, and stunning prior to slaughter. The majority of focus group participants expressed a strong preference for the Islamic way of killing and they were keen to emphasize the animal welfare benefits of halal methods. Most male participants believed that they had a good knowledge of the Islamic duty of care for animals at the time of slaughter, including good feeding, stroking, not showing the knife, and not seeing the slaughter of other animals. However, these views were often based on their personal or family experience of sacrifice during Aid el Kebir/Kurban bayrami, rather than on contemporary large-scale halal slaughtering practices. Some focus group participants were eager to stress that the concept of halal should go beyond slaughtering requirements to encompass a general principle of care for animals, including on the farm and during transport. One participant even suggested that this duty of care could extend to reducing our overall amount of meat consumption. There was also some debate as to the appropriateness of using the halal label to certify certain ethically contentious food items, such as foie gras. Several of the focus group participants were also sceptical about contemporary Western secular methods of animal slaughter, as they perceived them to be highly intensive and driven by profit at the expense of animal welfare. All participants supported the aim of ensuring that animals suffered as little as possible during slaughter, but there were mixed views concerning the use of stunning as a means to achieve this aim. Those who rejected stunning cited incompatibility with religious requirements or concerns that stunning might cause harm to the animals and damage the meat. In contrast, certain participants believed that stunning methods were not forbidden by Islam and drew comparisons with the use of anaesthetics to eliminate or minimize pain in humans. These participants focused on the effectiveness of stunning methods, especially on their capacity to induce insensibility, rather than on the religious permissibility of stunning.

In order to research the views of kosher consumers, a further six in-depth qualitative focus groups were undertaken. The focus groups took place in Brussels in Belgium, Berlin in Germany, Bordeaux in France, Cardiff in the UK, Amsterdam in the Netherlands. and in Tel Aviv in Israel. Each focus group consisted of between seven and ten male and female consumers, all of whom were regular kosher eaters and permanent residents in the country where the research was conducted (see Bergeaud-Blackler et al. 2010, 2013).

Results from the focus groups suggested that whilst eating kosher was important for the focus group participants, and indeed considered to be an obligation for Jews, the level of commitment to eating kosher foods seemed to be less than found in the Muslim focus groups in relation to halal. This can be explained in part by the low availability and high price of kosher foods, and by the growth in secular practices amongst certain Jews. Focus group participants tended to rely on centralized rabbinic determination of whether or not a particular product was kosher and many were content to allow experts to make these types of determinations. However, certain participants expressed frustration with the shear amount of different kosher standards and there was a perceived lack of standardization in kosher certification processes.

The vast majority of focus group participants expressed the view that shechita was the best method of animal slaughter and was preferable to conventional and halal methods. Many different reasons were given for the preference for shechita slaughter, including the fact that the permission to kill an animal can only be granted by God and must therefore be carried out using the methods prescribed by God. Other reasons focused on animal welfare and there was a belief that the shechita method reduced animal suffering at death due to the skill of the shochet, the nature of the incision technique, and the sharpness of the knife ensuring a quick loss of consciousness in the animal. Views regarding the permissibility and effectiveness of stunning at slaughter differed between focus group participants in Europe and Israel. In Israel there was some debate about the effectiveness of stunning as a technique, with certain participants arguing that it was ineffective and should not be used, even putting religious beliefs aside. There was also a strong feeling that the shechita process should not be interfered with. In Europe many participants questioned the motivation for banning stunning and some linked it with anti-Semitism. The effectiveness of stunning techniques was also challenged and participants expressed a belief that religious practices should be upheld.

21.4 Conclusions

Over recent decades the acceptability of practices of religious slaughter that do not allow the stunning of animals prior to throat cutting have been increasingly questioned by NGOs in Europe and other industrialized countries. At the same time the status and qualification of halal meat has been questioned by Muslims in a number of European countries, especially those with large Muslim immigrant populations (Lever and Fischer 2017). The Dialrel project represented an opportunity for a dialogue between the Muslim and Jewish religious authorities, the scientific authorities (FVE), the representatives of many animal welfare organizations, and representatives of the meat supply chain to address the welfare of farm animals at time of killing and the transparency of the meat markets. During the lifetime of the project, a plethora of new certifying bodies and schemes emerged to propose different standards for Halal MEAT with the aim of reassuring the growing number of Muslim consumers in the study countries about the religious quality of the meat they buy. Lever and Miele (2012), reflecting upon the experience of Dialrel, have argued that this segmentation of the market, and the increasing differentiation emerging from it, have contributed to a debate about what is 'authentic' halal meat and assisted the growth in the demand for halal meat largely based on the lack of transparency of the market, but they have not explicitly engaged with the issue of animal suffering at time of killing.

The findings of the Dialrel project addressed some fundamental issues that in the following years have become more pressing with the implementation of European Regulation 1099/2009 (Protection of animals at the time of killing), directly applied from 1 January 2013. National inspection data and FVO reports indicate that not all requirements are realized. Regulatory problems at the time of killing concern the implementation of the standard operating procedures, the training of staff, the development of new stunning and killing techniques, lairage facilities, the handling and restraining of animals, training of staff, veterinary supervision and presence of a trained animal welfare officer, documentation, and quality of inspection, as well as (training for) emergency killing on farms. The attempt to establish a dialogue about the suffering of animals at time of killing started with the Dialrel project is still very much needed.

References

Bergaud-Blacker, F., Fischer, J., and Lever, J. (2015). *Halal Matters: Islam, Politics and Markets in Global Perspective*. Routledge.

Bergeaud-Blackler, F., Evans, A., and Zivotofski, A. (2010). *Final Report Consumer and Consumption Issues: Halal and Kosher Focus Group Results*, Deliverable D3.1 for DIALREL EU SSA-43075. Cardiff: Cardiff University Press.

Bergeaud-Blackler, F., Zivotofsky, A.Z., and Miele, M. (2013). Knowledge and attitudes of European kosher consumers as revealed through focus groups. *Society and Animals* 21 (5): 425–442.

Cagatay, S. and Yegenoglu, Y. (2006). *Halal Turkey*. Bitter Lemons International http://www.bitterlemons-international.org/inside.php?id=503 (accessed 9 March 2010).

Evans, A. and Miele, M. (2017). Food labelling as a response to political consumption: effects and contradictions. In: *The Routledge Handbook in Consumption* (eds. M. Keller, B. Halkier, T.-A. Wilska and M. Truninger). London & New York: Routledge.

FSA (2013). Horse DNA detected in canned beef from Romania. Food Standards Agency UK. http://webarchive.nationalarchives.gov.uk/20150624093026/http://food.gov.uk/news-updates/news/2013/5858/canned-beef (accessed 7 June 2016).

FSA (2014). FSA publishes latest report on industry beef product test results. Food Standards Agency UK. http://webarchive.nationalarchives.gov.uk/20150624093026/http://food.gov.uk/news-updates/news/2014/5950/test-results (accessed 7 June 2016).

Fuseini, A., Wotton, S.B., Knowles, T.G., and Hadley, P.J. (2017). Halal meat fraud and safety issues in the UK: a review in the context of the European Union. *Food Ethics* 1: 127–142.

Grim, B.J. and Karim, M.S. (2011). *The Future of the Global Muslim Population: Projections for 2010–2030*. Pew Research Center http://www.npdata.be/Data/Godsdienst/PEW/FutureGlobalMuslimPopulation-WebPDF.pdf.

ITV News (2013). Pork found in Halal meat in Birmingham. http://www.itv.com/news/central/update/2013-03-27/pork-found-in-halal-meat-in-birmingham (accessed 6 June 2016).

Lever, J. and Anil, H. (2015). From an implicit to an explicit understanding; new definitions of Halal in Turkey. In: *Halal Matters: Islam, politics and markets in global perspective* (eds. F. Bergeaud-Blackler, J. Fischer and J. Lever), 38–54. London: Routledge.

Lever, J., & Fischer, J. (2017). *Religion, Regulation, Consumption: Globalising Kosher and Halal Markets,* Manchester: University Press, Manchester.

Lever, J. and Miele, M. (2012). The growth of the halal meat markets in Europe: an exploration of the supply side theory of religion. *Journal of Rural Studies* 28 (4): 528–537.

Miele, M. (2016). Killing animals for food: how religion, science and technology affect the debate about and halal and conventional slaughter. *Food Ethics* 1 (1): 47–60.

Miele, M. and Evans, A. (2010). When foods become animals, ruminations on ethics and responsibility in care-full spaces of consumption. *Ethics, Place and Environment* 13 (2): 171–190.

Miele, M. and Parisi, V. (2001). L'Etica del Mangiare, i valori e le preoccupazioni dei consumatori per il benessere animale negli allevamenti: un'applicazione dell'analisi means-end chain. *Rivista di Economia Agraria* LVI (1): 81–103.

Miele, M. and Rucinska, K. (2015). Producing halal meat: the case of halal slaughter practices in Wales, UK. In: *The Political Ecologies of Meat Production* (eds. J. Emel and H. Neo), 253–277. London: Earthscan.

Miele, M., Veissier, I., Evans, A., and Botreau, R. (2011). Establishing a dialogue between science and society about animal welfare. *Animal Welfare* 20 (1): 103–117.

Mukherjee, S.R. (2014). Global halal: meat, money, and religion. *Religion* 5: 22–75.

Schröder, D. (2009). Halal is big business: Germany Waking up to growing market for Muslim food. Spiegel Online International. http://www.spiegel.de/international/germany/halal-is-big-business-germany-waking-up-to-growing-market-for-muslim-food-a-653585.html (accessed 10 October 2010).

The Guardian (2013a). Horsemeat scandal: timeline. *The Guardian* 10 May 2013. https://www.theguardian.com/uk/2013/may/10/horsemeat-scandal-timeline-investigation (accessed 7 June 2016).

The Guardian (2013b) 'Halal pork' supplier named. *The Guardian,* 10 May 2013. http://www.theguardian.com/world/2013/feb/03/supplier-halal-meat-pork-dna-named (accessed 7 June 2016).

Velarde, A., Rodriguez, P., Fuentes, C. et al. (2010). *Improving Animal Welfare During Religious Slaughter Recommendations for Good Practice*, Dialrel Report N.2.4, DIALREL EU SSA-43075. Cardiff: Cardiff University Press http://www.dialrel.eu/images/recom-light.pdf. (accessed June 2013)

Wilson, A.J. (2014). The halal phenomenon: an extension or a new paradigm? *Social Business* 4 (3).

White, K. (2014). Halal meat furore: What does the food industry say? *The Grocer* 255–271, 8 May.

Further Reading

Miele, M., Bock, B., and Horling, L. (2015). Animal welfare: the challenges of implementing a common legislation in Europe. In: *The Handbook of International Political Economy of Agriculture and Food* (eds. A. Bonanno and L. Busch), 295–321. New York and Cheltenham: Edward Elgar Publishing Ltd.

Pointing, J., Teinaz, Y., and Shafi, S. (2008). Illegal labelling and sales of halal meat and food products. *Journal of Criminal Law* 72 (3): 206–213.

Notes

1 Religious slaughter, improving knowledge and expertise through dialogue and debate on issues of welfare, legislation and socio-economic aspects. Contract no. 43075, European Union, FP6 Priority 5 'Food Quality and Safety', FP6-2005-FOOD-4-C.

2 In the following years, this concern was proved to be founded. As argued by Fuseini et al. (2017) the uncovering in the UK of undeclared horsemeat (and horse DNA) in some products (*The Guardian* 2013a; FSA 2013) led to more rigorous tests of processed foods for undeclared materials (FSA 2014). The subsequent increase in product testing led to the detection of pork meat and porcine DNA in various halal products destined for Muslim consumption (*The Guardian* 2013b; ITV News 2013, p. 1).

3 See the 2011 Pew Research Center's Forum on Religion & Public Life report available at http://www.pewforum.org/files/2011/01/FutureGlobalMuslimPopulation-WebPDF-Feb10.pdf.

4 Muslim population growth (1.5%) is about twice the rate of growth of the non-Muslim population (0.7%).

Part VII

Halal in Different Countries

22

Halal Food Production in the Arab World

Majed Alhariri[1] and Hani Mansour M. Al-Mazeedi[2]

[1] Independent Halal Researcher in Food Science and Technology, Cairo University, Cairo, Egypt
[2] Kuwait Institute for Scientific Research, Kuwait City, Kuwait

22.1   Introduction

The Arab world, also known as the Arab Nation, consists of the 22 Arabic-speaking countries of the Arab League. The Arab countries are Algeria, Bahrain, Comoros, Djibouti, Egypt, Iraq, Jordan, Kuwait, Lebanon, Libya, Mauritania, Morocco, Oman, Palestine, Qatar, Saudi Arabia, Somalia, Sudan, Syria, Tunisia, the United Arab Emirates (UAE), and Yemen. They are members of the Organization of Islamic Cooperation (OIC). Islam is the fastest growing and second largest religion in the world. The Muslim population has grown exponentially in the last 50 years and by 2030 the Muslim population is projected to be 26.4% of the world's population (Table 22.1). The rural population is about 156.74 million.

Arab countries are considered to be one of the most important halal markets in the Muslim world as the population of these countries comprises about 20% of the total Muslim population of the world with good disposable income.

The basic guidance about halal food laws is revealed in the Holy Quran from Allah the Almighty to the Prophet Mohammed (peace be upon him) for all people. The food laws are explained and put into practice through the Sunnah (the life, actions, and teachings of Mohammed (pbuh)) as recorded in the Hadith (the compilation of the traditions of Mohammed (pbuh)). In general, everything is permitted for human use and benefit. Nothing is forbidden except what is prohibited either by a verse of the Holy Quran or an authentic and explicit Sunnah of Mohammed (peace be upon him). These rules of Shariah (Islamic) law bring freedom for people to eat and drink anything they like as long as it is not haram (prohibited).

Food is considered one of the most important factors for interaction amongst various ethnic, social, and religious groups. All people are concerned about what they eat: Muslims want halal assurance for their food, Jews require kosher food, Hindus, Buddhists, and certain other groups seek vegetarian foods.

The concept of halal products or foods is now gaining a worldwide discussion due to its recognition as an alternative benchmark for the safety, hygiene, and quality assurance of

Table 22.1 Total population and Muslim population of Arab countries (2014)

Country	Total population (million)	Muslim percentage (%)	Muslim population (million)
Algeria	38.3	99.7%	38.19
Bahrain	1.1	90.4%	0.94
Comoros	0.8	98%	0.78
Egypt	84.7	95%	80.47
Eretria	5.8	50%	2.9
Iraq	35.1	97%	34.05
Jordan	7.3	94%	6.86
Kuwait	3.5	85%	2.98
Lebanon	4.8	59.7%	2.87
Libya	6.5	99%	6.44
Mauritania	3.7	100%	3.7
Morocco	33.6	99%	32.26
Oman	4	99%	3.96
Palestine	4.4	98%	4.31
Qatar	2.2	77.5%	1.71
Saudi Arabia	30.1	100%	30.1
Somalia	10.4	100%	10.4
Sudan	34.2	97%	33.17
Syria	21.9	90%	19.71
Tunisia	10.9	99%	10.79
UAE	9.3	76%	7.07
Yemen	25.2	100%	25.2
Total	377.8	95%	358.86

what we consume or drink daily. For a Muslim consumer, halal food and drink means products that have met the requirements laid down by Shariah law, whereas for a non-Muslim consumer halal products represent hygiene, quality, and safety when produced strictly under the halal assurance management system. The awareness of Muslim and non-Muslim consumers demonstrates their perception and cognitive reaction to products or foods in the market. As such, their awareness is an internal state or a visceral feeling by way of sensory perception towards the products/foods they use or consume. The findings of Ambali and Bakar (2014) show that the religious belief, exposure, certification logo, and health reasons are potential sources of Muslim awareness about halal consumption. Halal is the key to life for Muslims, and the meaning of halal is no longer limited to meat or meat products, it is not even limited to food. The halal concept extends to include, pharmaceuticals, cosmetics, tourism, etc. Halal in brief is the Muslim lifestyle.

22.2   The Reality of Halal Food Production in the Arab World

Arab countries, collectively, are the biggest importers of food in the world with an annual food import bill reaching close to US$80 billion and projected to rise to US$90 billion within the next few years. This situation is exacerbated by rising populations, climate change, and global food commodity prices. Arab countries are heavy consumers of livestock products, with most of their needs being met through imports (AFED 2014). However, consumption and import patterns are not uniform, with higher values noted in the Gulf (livestock product imports estimated at US$8.6 billion in the Gulf countries versus US$7.8 billion in all other Arab countries, in 2011). This consumption is projected to increase in the future driven by three forces: population, wealth, and urban growth. The population of the Arab countries is projected to increase from about 378 million today to 460 million by 2025 and to 621 million by 2050. This represents twice the world average population growth rate, and an extra 243 million people needing food in the region by 2050.

The livestock sector is considered to be one of the key economic sectors that underpins economic and social development in a number of Arab countries. It contributes significantly to the gross domestic product, providing employment opportunities and a decent livelihood for a large sector of the rural population. Moreover, it provides the raw materials for many industries, and supports export and hence foreign-currency earnings.

The livestock sector productivity in Arab countries is challenged by the scarcity of natural resources in terms of feed and water, lack of supporting infrastructure and services. and a history of arbitrary policies that affected the sector negatively. Arab countries face a heavy reliance on feed imports, estimated at US$10.4 billion, after attempts to grow feed locally resulted in drainage of non-renewable water reserves. In addition, the natural feed resource for mixed and pastoral systems has been largely degraded, leading to loss of biodiversity, soil erosion, and a decrease in carrying capacity and thus livestock productivity.

The estimated number of livestock in the Arab countries in 2012 was about 341 million head; cattle and buffalo constitute about 17%, sheep about 52%, goats about 26%, and camels about 5% (Table 22.2). In spite of the relative increases in meat production, Arab countries still depend on imported meat to fill the gap between production and consumption of meat and meat products. Red meat production increased to 12% in 2013 compared to the average production during the period 2002–2011. Milk production has increased to 9.1% compared to 1.9% in 2012. Sudan is considered the largest Arab country by number of cattle, sheep, and goats: it has about 55%, 22% and 35%, respectively, of the total livestock numbers in the Arab countries.

The production of poultry, meat, and eggs increased at high rates in 2013 compared to an average period from 2002 to 2011, and the rates of annual growth exceed the growth of other animal product rates.

Sudan is the biggest producer of red meat, milk, and milk products of the Arab countries, and is capable of increasing this percentage due to their large livestock numbers. Sudan contributes about 40% of the total Arab production of red meat, milk, and milk products. Other important sources of meat in the Arab world are Somalia and Mauritania, but these countries need a strong strategy and plan to further develop their meat industry.

Table 22.2 Livestock numbers, slaughtered animal numbers, and animal production in the Arab countries (2012)

Country	Livestock numbers (1000 head)					Slaughtered animals (1000 head)		Production of animal products (1000MT)		
	Buffalo	Cattle	Sheep	Goats	Camel	Cattle and buffalo	Sheep and goats	Red meat	Poultry meat	Milk
Jordan	0.1	64.4	2389.6	885	11.05	24	703	44.52	244.17	356.4
Emirates	–	64	1400	1900	364	7.08	334.27	47.6	78.79	152
Bahrain	–	10	41	18.5	1.03	8.1	699.06	18.34	6.45	10.4
Tunisia	–	654	6802	1272	73	261.72	4569.23	116.3	127.5	1124
Algeria	–	1843.94	25194	4595	340	533.54	5921.63	240.87	365.4	3063.84
Djibouti E	–	298	470	514	71	55	409	11.39	–	–
Saudi Arabia	–	501	10129	3408	810	133.05	4062.11	116.09	588	1712.42
Sudan	–	29840	39483	30837	4751	7338	47604	1962.2	45	6240.57
Syria	–	1108.47	18062.84	2292.75	55.01	459.06	7434.24	255.86	138.2	2452.25
Somalia	–	4870	12300	11600	7000	544	3520	198.6	4.5	1053.9
Iraq	312	2788	8428	1612	64	236	531	63	90	292.67
Oman	–	358.74	404.11	1788.8	134	50.72	298.69	14.13	41.8	76.8
Palestine	–	33.67	730.89	215.34	2.06	32.39	46.31	9.49	53.58	261.55
Qatar	–	10.6	308	215	64	4	105.42	3.65	10.46	41.46
Kuwait	–	29.84	489.37	162.27	9.09	16.9	1588.57	42.41	48.42	65.45
Lebanon	–	78	400	450	0.2	211.7	349.5	114.26	100	117.23
Libya	–	198	7150	2550	170	45.5	4916.67	160.81	124	232.1
Egypt	3985	4946	5430	4306	142	2620	4887	780.65	859	5719
Morocco	–	3029	19006	5602	182	1077	9731	373.96	620	2500
Mauritania	–	1750	9000	5600	1425	230	9468	227.8	4.56	394.7
Yemen	–	1688.87	9419	9158	443	236.1	3288	98.58	155.5	280.8
Total	4304.51	54164.53	177036.8	88981.66	16112.24	14123.86	110466.7	4900.506	3705.33	26147.54

MT, metric ton.

In spite of the growth in the production of white meats and eggs, the poultry industry in the Arab world still depends on materials imported largely from abroad, and this is considered a real threat to the halal industry in Arab countries. This threat requires serious alternatives to be found that will reduce imports of materials used in the poultry industry. The Arab world production of poultry meat in 2013 was about 4 million tons, accounting for 3.7% of global production, which stood at 107 million tons. In 2012, poultry meat production in Egypt was estimated to be about 23.2% of the total production of poultry in the Arab world, followed by Morocco at 16.7%, Saudi Arabia at 15.9%, Algeria at 10.0%, Jordan at 6.7%, and Yemen at 4.2%.

In 2012, Arab countries imported about 564.72 thousand head of live cattle, 14 million sheep and goats, 1.3 million tons of red meat, 2 million tons of white meat, and 11.6 million tons of milk and dairy products. The total value of these imports reached about US$18 billion. Imports of red meat constituted about 24.1%, poultry meat imports about 19.5%, milk and dairy products about 33.3%, fish imports about 11.3%, live sheep and goats' imports about 6.5%, and live cattle imports about 3.8% of the total imports of live animals, meat, and milk products. Red and poultry meat imports into the Arab countries in 2012 were 1293.08 and 1995.23 metric tonnes, respectively, with values of US$4318.37 million and US$3754.54 million, respectively (Table 22.3).

22.3 The Potential Value of the Halal Market in the Arab World

The size of the halal product market has grown considerably during recent decades due to the increase in demand for halal products and services. The volume of the halal market stands at US$635 billion per annum. This is close to 17% of the whole food industry around the world. The global halal market grew to 12.6% between 2004 and 2010. An interesting characteristic of the halal market is that it has not been affected by the global financial crisis.

According to the Global Islamic Economy Report 2013, Muslim consumer expenditure globally in the food and lifestyle sectors was estimated to be US$1.62 trillion in 2012 and was expected to reach US$2.47 trillion by 2018. This forms the potential core market for the halal food and lifestyle sectors. Muslim consumers globally spent US$1088 billion on food and beverages in 2012, which is 16.6% of global expenditure. This expenditure is expected to grow to US$1626 billion by 2018. This shows the potential of the halal food market worldwide.

Despite the size of the halal market, it is characterized by chaos and the illegal and unlawful exploitation of the halal brand by many producers and importers for their own benefit.

Arab and Middle Eastern countries are considered to be net importers of processed foods for both the food service and retail markets. Saudi Arabia, the UAE, and other Middle Eastern countries have been importing food for decades. North Africa and other African countries also offer opportunities for the export of processed food as their economies and political conditions improve (Table 22.4).

The World Bank, FAO and IFAD (2009) reported that meat and milk consumption in the Arab countries will continue to rise. Meat consumption is predicted to increase by 104% and milk consumption by 82% from 2000 to 2030. Increases in consumption of animal products will be more pronounced in oil-rich Arab countries, nearly doubling from 2000 to 2030, driven by surging income and population growth. This will lead to substantially

Table 22.3 Total live animal, meat, and milk product imports to Arab countries (2012)

Country	Live cattle		Live sheep and goats		Red meat		Poultry meat		Milk and milk products	
	Value (US$ million)	Quantity (1000 head)	Value (US$ million)	Quantity (1000 head)	Value (US$ million)	Quantity (1000 MT)	Value (US$ million)	Quantity (1000 MT)	Value (US$ million)	Quantity (1000 MT)
Jordan	0.52	0.99	56.58	539.21	86.39	85.21	101.45	62.35	335.54	651.34
Emirates	17.93	31.77	114.00	1850.00	1350.82	490.75	537.44	297.64	790.55	434.27
Bahrain	0.20	0.65	22.83	475.10	110.35	19.97	72.47	35.83	111.52	71.88
Tunisia	20.49	14.44	–	–	23.43	5.48	6.53	3.21	36.66	127.63
Algeria	44.66	19.90	0.00	0.00	172.21	62.30	1.22	0.47	861.68	2831.71
Djibouti	–	–	–	–	20.21	11.24	2.52	1.35	15.25	24.35
Saudi Arabia	46.50	39.20	690.52	7855.32	801.15	192.72	1754.92	786.37	775.64	1581.01
Sudan	1.48	0.26	5.28	1.66	0.13	0.52	0.00	0.00	34.47	319.56
Syria	18.99	11.88	1.29	0.99	68.88	24.07	12.05	10.02	80.35	228.70
Iraq	–	–	0.46	4.39	2.36	0.76	409.22	262.54	44.15	319.06
Oman	31.92	10.53	55.59	1048.14	84.76	43.82	96.00	48.00	399.81	1033.61
Palestine	59.08	40.00	1.51	26.40	18.42	9.50	18.22	12.00	24.96	22.14
Qatar	2.65	0.66	115.65	28.39	204.30	41.58	56.57	165.97	281.19	123.24
Kuwait	8.57	4.36	128.23	1518.96	180.45	41.09	216.22	115.18	201.72	647.30
Lebanon	285.25	94.67	12.87	5.78	105.90	15.59	30.58	10.83	366.51	550.18
Libya	19.78	34.04	11.50	121.80	56.03	19.68	0.31	0.07	193.81	335.99
Egypt	43.18	39.08	0.00	0.00	995.96	218.43	224.26	60.91	539.98	976.28
Morocco	36.18	39.94	0.083	0.47	26.30	4.83	1.06	0.84	249.57	613.96
Mauritania	–	–	–	–	0.01	0.01	7.12	7.31	48.30	149.75
Yemen	35.92	182.36	20.30	565.43	10.31	5.53	206.38	114.34	302.69	736.78
Total	673.29	564.72	1236.61	14041.6	4318.37	1293.08	3754.54	1995.23	5703.44	11785.27

MT, metric ton.

Table 22.4 Total and food imports of Arab countries, 2012

Country	Total imports (US$ million)	Food imports (US$ million)	Food imports (%)
Jordan	20804.05	3217.53	15.47
Emirates	162744.48	6134.31	3.77
Bahrain	11797.61	552.35	4.68
Tunisia	22127.49	1535.99	6.94
Algeria	53782.41	5515.01	10.25
Djibouti	410.00	48.42	11.81
Saudi Arabia	155787.29	21684.14	13.92
Sudan	2653.01	202.50	7.63
Syria	17392.05	2652.71	15.25
Somalia	944.00	437.35	46.33
Iraq	25791.53	1174.31	4.55
Oman	28001.27	986.01	3.52
Palestine	4697.30	373.65	7.95
Qatar	30755.20	2056.48	6.69
Kuwait	22963.78	1221.31	5.32
Lebanon	22462.99	3472.10	15.46
Libya	20460.00	1828.80	8.94
Egypt	69537.80	14991.24	21.56
Morocco	46898.22	5064.10	10.80
Mauritania	369.79	84.01	22.72
Yemen	112657.63	3468.02	3.08
Total	833037.90	76700.34	9.21

greater dependence on imports of these products in the future. Non-oil-producing Arab countries will also increase meat and milk consumption, but have the natural resources necessary to produce enough to keep imports near current levels.

The growing popularity of beef reflects social and economic changes in the Middle East, in particular the growing financial clout of the oil-rich nations in the region.

The halal industry in the Arab countries and Middle East is estimated to be worth more than US$20 billion. The Middle East and Arab regions are strong markets for halal products and services as they import 80% of their food requirements. The Gulf Cooperation Council (GCC) member countries include the wealthy nations of Saudi Arabia, Kuwait, Bahrain, Qatar, the UAE, and Oman. The region's annual food imports are expected to double from US$25.8 billion in 2010 to US$53 billion in 2020 with total imports of halal meat (chicken and beef) exceeding 1 million metric tonnes on an annual basis.

In tandem with the expected rise in the region's population of 40% by 2030, the consumption of food in the Middle East may well reach 51 million tonnes by 2020 to record an annual average growth of 4.6%.

The region imports about 90% of its beef and beef products from countries such as Brazil, India, New Zealand, and Australia. The total imports of these products were estimated at US$5.1 billion in 2013, with Saudi Arabia, Egypt, and the UAE emerging as the top importers in the region. In recent years, meat consumption has shown an upward trend (GIFR 2013).

Another booming halal sector is the halal cosmetics and beauty product market. This sector is said to be worth about US$2 billion in the Middle East and Arab countries. In many of these countries the market for natural and organic cosmetics is growing by over 20% a year, whilst the halal cosmetics market is booming at a 12% growth rate per annum. Rising consumer affluence and growing knowledge about cosmetic ingredients is leading to consumers paying a premium for cosmetics that meet their ethical and religious beliefs.

In the Middle Eastern countries, per capita incomes and consumption rates are higher than in other countries. There is a higher concentration of wealth in countries like the UAE, Saudi Arabia, Qatar, Oman, and Kuwait. Over the last few years there has been rapid economic activity in these countries, resulting in larger household incomes. In this region, the UAE and Saudi Arab top the list with higher demand for halal products. North Africa is also a major market for halal food. This region fulfils most of its needs through imports. Egypt is the biggest market player in this region, with a Muslim population of more than 80 million (Global Pathfinder Report 2011). Figure 22.1 illustrates the estimated size of the halal food market in some Arab countries.

22.4 Halal Organization and Halal Certification Bodies in Arab Countries

All countries in the Arab world and Muslim majority countries in the Middle East, the Gulf, and other Arab countries require that imported products be accompanied by a halal certificate issued by a recognized Islamic organization in the country of export (Appendix B).

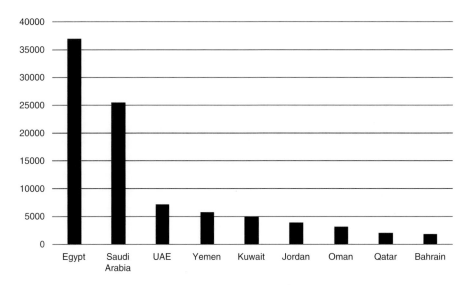

Figure 22.1 Estimated size of the halal food market (US$ million).

Several governments and non-governmental organizations (NGOs) in Arab world are involved in halal evaluation, halal monitoring services, halal enforcement, disseminating information, issuing halal certificates, and authorizing the use of halal logos and markings to ensure that foods are imported or produced domestically, for example:

- Egyptian Organization for Standardization and Quality (EOS), Egypt
- Central Organization for Quality Control (COSQC), Iraq
- Jordan Institution for Standards and Methodology (JISM), Jordan
- Public Authority for Industry Standards and Industrial Services Affairs (KOWSMD), Kuwait
- Kuwait Institute for Scientific Research, Biotechnology Department (KISR), Kuwait
- Sudanese Standards and Metrology Organization (SSMO), Sudan
- Syrian Arab Organization for Standardization and Metrology (SASMO), Syria
- Tunisian National Institute for Standardization and Industrial Property (INNORPI), Tunisia
- Emirates Authority for Standardization and Metrology (ESMA), the UAE
- Yemen Standardization, Metrology, and Quality Control Organization, Yemen
- Saudi Arabian Standard Organization (SASO), Saudi Arabia
- Saudi Food and Drug Authority, Saudi Arabia
- Qatar General Organization for Standards and Metrology (QS), Qatar
- The Moroccan Industry Normalization Service (SNIMA), Morocco.

Gulf countries under the leadership of Saudi Arabia have formulated quite elaborate standards for meat and prepared foods to be used by the member countries. One standard is known as the Gulf Standard. It includes guidelines and requirements for the import of a variety of food and food products and meat and meat products. The Gulf Standardization Organization draft standard is the draft Gulf standard specification for halal food prepared by the technical committee for the sector of foods and agriculture products. The Gulf Standardization Organization is a regional forum that has among its members the national departments for standards and specifications in the Gulf Arab states. The organization's objective is to prepare the Gulf Standards specifications through technical specialized committees (Appendix A).

22.5 The Obstacles and Challenges Facing Halal Production in the Arab World

International trade of halal food products and the halal food industry are facing many challenges due to obstacles that can be encountered in Arab countries.

In any Muslim majority country the concept of halal is an important part of daily Muslim life and a key to consumption of any consumer products. Consuming halal is an order of Allah and an essential part of the Islamic faith. Allah the Almighty has repeatedly emphasized the consumption of halal in His Book. The meaning of the Arabic word halal, i.e. 'permitted' or 'lawful', is very clear. The permission means that of the lord of all existence, Allah the Almighty, who commands Muslims and all people to consume halal products in the Holy Quran.

'O mankind! Eat of that which is lawful and wholesome in the earth'. *(2/168)*.

Halal food is no longer limited to the slaughtering method or being free from pork derivatives and alcohol. The halal product concept has become more complex, especially for imported foods, due to huge developments in food processing technology and food handling, such as using biotechnology, nanotechnology techniques, food additives, etc.

The main challenge in the halal market is to understand what Muslims consider halal. There is a dissimilarity across Muslim consumers of different ethnicities (Anwar-UlHaq et al. 2014). The level of halal food perception amongst non-Arab Muslims is higher than amongst Arab Muslims. Non-Arab Muslims are more careful than Arab Muslims when purchasing food products.

Hashim and Othman (2011) conducted a comparative study of Arab Muslims and non-Arab Muslims consumers in Malaysia to understand Muslim consumer's perception towards halal-certified food across different ethnicities. The research examined the difference between Arab Muslims and non-Arab Muslims in terms of the level of their religiosity and halal perception. The result showed that the level of religiosity and halal perception amongst non-Arab Muslims is higher compared to Arab Muslims, and that non-Arab Muslims are more careful when purchasing food products compared to Arab Muslims. The survey was conducted for three main food groups: meat-based products (sausages, nuggets, meatballs, pizza, and meat-based processed food), non-meat-based food products (pasta, noodles, vegetable-based canned food, fish-based canned food, cookies, fruit-based canned food, chocolate, candy, snack food, seasoning, soy sauce, tomato sauce, chilli sauce, spices, butter, margarine, cheese, cake, muffin, vinegar, and cereals) and drinks (non-alcoholic drinks, fruit juices, and mineral water). The result also showed that non-Arab Muslims place significantly higher importance on the halal logo when purchasing studied food items compared to Arab Muslims.

The lack of awareness of halal-certified products is one of the biggest obstacles that faces the halal food industry in Arab countries. Halal has one correct meaning but consumers are often confused about its meaning. Some believe that if a product is allowed to enter a Muslim country, it is halal, and some do not know what is considered as haram and should be avoided. Arab countries have sought to increase the level of Muslim awareness by organizing training, conferences, workshops, etc. to promote the importance of the halal concept among Muslims as well as seeking to increase awareness about the risk of consuming non-halal certified products. They have embarked on a mass and holistic awareness program alerting Arab communities to the gross violation of their religious observances to try and improve the confused understanding about halal products.

The confused meaning of halal may result in a lack of necessary information regarding the integrity of the complete halal chain, i.e. that the full manufacturing chain of raw materials must be halal and there must not be any contamination with haram or forbidden raw material. This confusion mainly arises due to lack of awareness among consumers and differing fatwa from scholars belonging to different schools of thought.

Lack of awareness about the importance of consuming halal in Arab countries compared to other Muslim countries may be due to the lack of information about food production and the lack of promotion of halal standards, which leads to a misleading concept about halal and misunderstanding of the truth of importance of halal and tayyib food consumption. Misunderstandings about halal food include the following:

- Products made in Muslim countries are not necessarily halal.
- Products sold in markets owned by Muslims are not necessarily halal.
- Products that have their ingredients written in Arabic are not necessarily halal (Al-Mazeedi 2011).
- Halal is not just limited to meat and meat products but extends to all food technology sectors, in addition to cosmetics and skincare products, pharmaceuticals, medicine, tourism, etc.
- Products that have some components from vegetable sources are not necessarily halal.
- Products that have labels stating they do not contain fat or meat from pigs are not necessarily halal.
- Vegan meals (100% vegetable sources) are not necessarily halal.
- Jewish kosher meals are not necessarily halal.
- Carcasses printed with the words 'slaughtered by hand' are not necessarily halal because slaughtering by hand does not negate stunning before slaughter.
- Products that have the word halal printed on their packaging are not necessarily halal.

Most Arab countries import meat and critical raw materials from non-Muslim countries. Arab governments endeavour to enhance food security for their countries and seek ways to increase the productivity of both livestock and plants. One challenge they face is that most critical raw materials used in the food, cosmetics and pharmaceuticals industries come from non-Muslim countries. In the production of these raw materials, many critical additives come from animal sources, such as enzymes, gelatin, fatty acids, amino acids, emulsifiers, etc. In addition, ethanol may be used in solvent extraction or as a carrier material.

The halal industry in the Middle East and Arab countries lacks unified regulation, particularly in labelling and listing standards, therefore Arab countries should develop a unified halal standard in order to facilitate food trade and to be in a position of strength to impose their requirements and conditions for halal food and cosmetics according to global industry requirements. Arab countries also have to work for adoption of a unified rule for accrediting Islamic societies and meat exporters abroad to ensure sanitary conditions and adherence to halal methods will be a given priority by the authority.

Additionally, they have to develop a mechanism to control the import of ingredients and food products to Arab countries. Most Arab countries have no control over the halal quality of meat imported from Western countries. Most of the meat imported from the West to the Middle East and the Arabian Gulf countries has been stunned. This includes meat that comes from Western countries such as all European countries, the USA, Canada, Brazil, all South American countries, Australia, and New Zealand, in addition to most meat that comes from Malaysia, Indonesia, Turkey, and the Far East. All products imported or manufactured locally are required to have a valid halal certificate issued by a reputable and recognized Islamic body. Halal certificates issued by non-trusted halal certification bodies should not be accepted.

The productivity of the livestock sector in Arab countries is challenged by the scarcity of natural resources in terms of feed and water, lack of supporting infrastructure and services, and a history of arbitrary policies that affected the sector negatively. Arab countries have a heavy reliance on feed imports estimated at US$10.4 billion, after attempts to grow feed locally resulted in drainage of non-renewable water reserves. In addition, the natural feed resources for mixed and pastorals systems have been largely degraded, leading to loss of biodiversity, soil erosion, and a decrease in their carrying capacity, and thus livestock productivity.

Despite the importance of the halal market segment and its growing revenue, research on halal food consumption in the Muslim market segment has been largely ignored (Bonne and Verbeke 2007; Fischer 2008; Bonne et al. 2009). Not much has been done to understand Muslim consumers' perception of halal food and many questions relating to their attitudes and food choices are still being researched. In addition, the halal food industry is not a 'single' market. It is a fragmented market where every country or region has its own characteristics (Sungkar 2010). However, research across countries and ethnicity on consumer behaviour and halal food consumption is lacking (Lada et al. 2010). Furthermore, there is almost no research being done in comparing ethnicity and religiosity in halal food consumption, which is a challenge faced by the halal industry.

The standardization authorities of some countries and unions have embarked on developing standards for halal without regard to the Islamic religious sanctity of the entire halal certification process. These standardization authorities have established technical committees for the development of a standard in total disregard to the religious concerns being safeguarded by recognized halal certification and accreditation authorities worldwide. Halal is a divine obligation based on the Holy Quran and Prophetic teachings. The interpretation, standardization, regulation, governance, and application of halal is regarded as an act of worship. In addition to these challenges, there are many other obstacles facing the halal food industry, including industry scale and inefficiency, lack of leadership, raw material/supply chain challenges, human capital deficiency, low consumer confidence, global perception challenges, global halal market, non-unified interpretation of the meaning of halal, and majority of halal products being imported from non-Muslim countries, and lack of legislation for the halal logo in exporting countries.

References

AFED (2014). Arab environment: food security. In: *Annual Report of the Arab Forum for Environment and Development* (eds. A. Sadik, M. El-Solh and N. Saab). Beirut, Lebanon: Technical Publications.

Al-Mazeedi, H.M. (2011). Halal services: obstacles over the past 30 years. *Proceedings of The First Gulf Conference on Halal Industry and Its Services*, Holiday Inn Hotel, Al-Salmiyah, State of Kuwait, 24–26 January 2011.

Ambali, A. and Bakar, A.N. (2014). People's awareness on halal foods and products: potential issues for policy-makers. *Procedia – Social and Behavioral Sciences* 121: 3–25.

Anwar-UlHaq, M., Zafar-uz Zaman, and Usman, M. (2014). Global halal food market and opportunities for Pakistan. *International Journal of Education and Research.* 2 (3): 1–8.

AOAD (2013a). *Arab Agricultural Statistical Yearbook*. Khartoum, Sudan: Arab Organization for Agricultural Development.

AOAD (2013b). *Report on the Arab Food Security Situation*. Khartoum, Sudan: Arab Organization for Agricultural Development ISSN 1811–5020.

Bonne, K. and Verbeke, W. (2007). Muslim consumer trust in halal meat status and control in Belgium. *Meat Science* 79 (1): 113–123.

Bonne, K., Vermeir, I., and Verbeke, W. (2009). Impact of religion on halal meat consumption decision making in Belgium. *Journal of International Food and Agribusiness Marketing*: 5–26.

Fischer, J. (2008). Religion, science and markets: modern halal production, trade and consumption. *EMBO Reports* 9: 828–831.

GIFR (2013). Halal Industry and Islamic Finance. *The Global Islamic Finance Report*.

Global Pathfinder Report (2011). *Halal Food Trends* (ed. I.M. Bureau). Canada: Agriculture and Agri-Food Canada.

Hashim, A.H. and Othman, M.N. (2011). Halal food consumption: a comparative study between Arab Muslims and NonArab Muslims consumers in Malaysia. *Proceedings of the Australian and New Zealand Marketing Academy Conferencey*, pp. 1–8, Perth, Australia, 28–30 November 2011.

Lada, S., Geoffrey, H.T., and Hanudin, A. (2010). Predicting intention to choose halal products using theory of reasoned action. *International Journal of Islamic and Middle Eastern Finance and Management* 3 (4): 66–76. ISSN 1753-8394.

Muslim Population (2014). Muslim Population in the World. http://www.muslimpopulation.com.

Sungkar, I. (2010). Consumer awareness: thoughts and trends across the globe. *The Halal Journal* 2 (1): 22–28 (Last accessed: January 4, 2016).

USDA (2015). US exports of agricultural products. http://www.fsis.usda.govwps/portal/fsis/topics/international-affairs/exporting-products/export-library-requirements-by-country.

World Bank, FAO, and IFAD (2009). *Improving Food Security in Arab Countries*. Washington, DC: World Bank, Food and Agriculture Organization (FAO), İnternational Fund for Agricultural Development (Last accessed: November 12, 2017).

23

Halal Food in Egypt

M. Diaa El-Din H. Farag

National Center for Radiation Research and Technology, Atomic Energy Authority, Nasr City, Cairo, Egypt

23.1   Introduction

Food is a basic necessity of human beings. However, food habits differ across various regions, religions, and tribes. The selection and processing of food also depend on factors like ethnicity, religion, and culture, with religion being the most significant factor to influence the patterns and habits of diet (Qureshi et al. 2012). The concept of halal refers to the production goods and services in a manner approved by Islamic law or Shariah. This involves not only food and functional products and food preparation but also pharmaceutical products and financial practices. The halal food industry is thus crucial for Muslims all over the world as it serves to assure them that the food items they consume daily are Shariah compliant. It is an industry which is set for tremendous growth.

The food habits of Muslims are unique and different to other dietary traditions around the globe. Muslims consume halal food and the growing population of Muslims has attracted the attention of food marketing companies towards this very lucrative market segment (Zakaria 2008). Moreover, the concept of halal is now gaining worldwide interest due to its recognition as an alternative benchmark for safety, hygiene, and quality assurance of what we consume or drink daily. Thus products or foods that are produced in line with halal prescriptions are readily acceptable by Muslim consumers as well as consumers from other religions. For a non-Muslim consumer, halal represents hygiene, quality, and safety when products are produced strictly under the Holistic Halal Assurance Management System. The awareness of the Muslim and non-Muslim consumers reflects their perception and cognitive reaction to products or foods in the market. As such, their awareness is an internal state or a visceral feeling by way of sensory perception towards the products/foods they use or consume.

Egypt (officially the Arab Republic of Egypt) has a population of about 90.3880 million (March 2016 estimate), with an annual growth of 2.03%. The great majority of its people live near the banks of the Nile River. Egypt is a transcontinental country occupying the north-east corner of the African continent that borders the Mediterranean Sea, between Libya and the Gaza Strip, and the Red Sea north of Sudan, and includes the Asian Sinai Peninsula. Egypt lies primarily between latitudes 22°and 32°N, and longitudes 25° and 35°E. it is approximately bisected by the highly fertile Nile valley, where most economic activity takes place. The total area of Egypt is $1\,001\,450\,km^2$ (land $995450\,km^2$ and water $6000\,km^2$). The official language is Arabic, and the main foreign languages taught in schools, in order of popularity, are English, French, German, and Italian.

Egyptian agricultural products include cotton, rice, corn, wheat, beans, fruits, vegetables; cattle, water buffalo, sheep, goats, and poultry. On the other hand, the main industrial products are textiles, food processing, tourism, chemicals, pharmaceuticals, hydrocarbons, construction, cement, metals and other light manufacturing. Industrial production growth rate is around 1.4% (2013 estimate).

Egypt is a predominantly Sunni Muslim country with Islam as its state religion. The percentage of adherents of various religions is a controversial topic in Egypt. An estimated 90% of the population is identified as Muslim, and around 10% is Christian (majority Coptic Orthodox, other Christians include Armenian Apostolic, Catholic, Maronite, Orthodox, and Anglican) (2012 estimate).

All Egyptian consumers are increasingly becoming aware of the quality and variety of imported consumer-oriented products, therefore their buying habits are changing dramatically. In the past, most Egyptian consumers bought products such as meat, fresh fruits, and vegetables from small neighbourhood shops. However, with an increasing number of supermarkets and hypermarkets in different Egyptian cities and offering all these services in one place, a large number of middle- to high-income consumers have started to purchase most of their requirements from supermarkets. Also, with changing eating habits and the expansion of fast-food chains, local restaurants, and resorts, there is great potential for imported products. Egyptian authorities are stricter in enforcing product standards on imported food products than on locally produced food products. However, the local food producers are still under mandatory government standards. Import regulations are routinely developed and enforced. Furthermore, Egypt has mandatory standards for a number of quality characteristics, moving beyond restrictions imposed for health and safety reasons. The Egyptian food-processing industry has many advantages, such as very good agro-climatic conditions, low labour costs and increased availability of skilled labour, proximity to Europe and Gulf markets, and an improving business environment, which are reflected in its recent very encouraging performance.

23.2 Global Halal Market

Global food processing and marketing value chains are now organized more tightly than before and entry into these chains has become an important source of dynamic competitive advantage that is crucial for the Egyptian food processing industry. It is vital for Egypt to tap into these global value chains, both by building competitive domestic enterprises and by attracting international companies.

It is estimated that the world's halal food trade averages nearly US$60 billion per annum and growing. The existence of such a big market naturally opens the door of economic opportunity for those engaged in the business, directly or indirectly. Global Muslim spending on food and beverages increased 10.8% to reach $1292 billion in 2013. This makes the potential core halal food market to 17.7% of the global expenditure in 2013 compared to 16.6% the year before. This expenditure is expected to grow to $2537 billion by 2019 and will account for 21.2% of the global expenditure. Thomson Reuters in collaboration with DinarStandard™ produced the *State of the Global Islamic Economy Report 2014–2015*, which showed that the top Muslim food consumption countries are Indonesia ($190 billion), Turkey ($168 billion), Pakistan ($108 billion), Iran ($97 billion), and Egypt ($94.8 billion) based on 2013 data (GIEI, 2014–2015; Table 23.1). Meanwhile, Malaysia, the United

Table 23.1 Top Muslim food consumption market by size (2013)

Country	Size (US$ billions)
Indonesia	190.4
Turkey	168.5
Pakistan	108.4
Iran, Islamic Republic	97.0
Egypt, Arab Republic	94.8
Bangladesh	59.9
Saudi Arabia	52.7
Russian Federation	43.7
India	41.1
Nigeria	37.7
Iraq	35.4
Algeria	35.4
Sudan	27.0
Morocco	24.5
UAE	21.3
Malaysia	16.6
Kazakhstan	14.5
USA	12.8
Azerbaijan	12.5
France	11.9
Yemen	11.5
China	10.1
Germany	9.9
Kuwait	8.7
Tunisia	8.4

Source: 2014–2015 GIEI report.

Arab Emirates (UAE), and Australia lead the Halal Food Indicator, which focuses on the health of the halal food ecosystem of a country relative to the size of the halal products market. A special focus report on halal food logistics estimated the logistic costs for the potential global halal food market to be $151 billion in 2013. The *State of the Global Islamic Economy Report 2018–2019* stated that global Muslim spending on food and beverages is expected to grow to 6.1% and forecast to reach US$1863 billion by 2023. Such growth will create a significant opportunity for investment and the creation of global halal food brands (GIEI, 2018–2019).

23.3 Halal Definitions and Requirements for Food Products and Ingredients

Halal food is prepared following a set of Islamic dietary laws and rules that determine what is permissible, lawful, and clean. Traditionally in Egypt, halal meat was always prepared and usually sold by Egyptian Muslims. Halal foods were made from scratch at home and complex processed ingredients were not used. Today, Muslims continue to require food products that conform to acceptable halal standards. Permissible food categories include meat, poultry, fish, seafood, milk, eggs, fruits, and vegetables. Halal food is defined as safe and not harmfully prepared, it does not contain non-halal and unclean ingredients, and is processed and manufactured using equipment that is not contaminated with things that are unclean. The term haram (prohibited) describes, in contrast, that which is forbidden and unacceptable.

Islamic law specifies halal requirements for food products and ingredients. Meat and poultry are permissible under halal regulations, but animals must be of halal species and must be slaughtered using Islamic methods. It is required that animals be slaughtered by a Muslim and that the name of Allah be pronounced at the time of slaughter. Humane handling must also be practiced and animal suffering minimized. All blood must be removed from the carcass.

The following raw materials and foodstuffs are haram, in other words forbidden, substances and may not be used for food production. Halal requirements are not met in the following examples:

1) Meat from dead animals and its by-products or from animals improperly slaughtered according to Islamic law.
2) Pork and its by-products pork and pork products – ham, bacon, lard, and hydrolyzed porcine collagen.
3) Gelatin, animal shortening, and hydrolyzed animal protein (if from a pig source).
4) Blood and its by-products.
5) Meat from carnivorous animals with claws and fangs such as lions, tigers, bears, and other similar animals.
6) Meat from birds of prey with claws such as eagles, vultures, and other similar birds.
7) Meat from rats, centipedes, scorpions, and other similar animals.

8) Meat from animals which are considered repulsive generally like lice, flies, maggots, and other similar animals.
9) Meat from animals that live both on land and in water such as frogs, crocodiles, and other similar animals.
10) Meat from mules and domestic donkeys, dogs, snakes and monkeys.
11) All poisonous and hazardous aquatic animals.
12) Meat from any other animals not slaughtered according to Islamic law.
13) Alcohol (ethyl alcohol) is prohibited in Islam.
14) Food items with alcohol as an ingredient or used in cooking are not allowed. This includes fruit essences (with alcohol naturally present) and alcohol used for technical reasons (e.g. extraction of flavours like vanilla).
15) Alcohol obtained from fermentation and then used in soups or sweets (intoxicants). Flavourings, colouring, and compound mixtures which contain alcohol (ethanol) in the end product.
16) Enzymes from the stomach lining of pig or calves from cattle which were not slaughtered according to Islamic rules (such as rennet enzymes in cheese and other products).
17) Dairy ingredients, whey, yoghurt, and cheese (these depend on the enzyme used, which is normally not indicated).
18) Pancreatin (pancreatic extract), pepsin, protein, and fat from non-halal animals.
19) Antioxidants from animals that were not slaughtered according to Islamic rules.
20) Dyes that use animal glycerin as a carrier, if the animals were not slaughtered according to Islamic rules.
21) Amino acids, gelatin, collagen, and glycerin (from pork or animals not slaughtered according to Shariah law).
22) Emulsifiers (from lard or fat from animals not slaughtered according to Shariah law) such as sodium stearoyl lactylate and calcium stearate.
23) Foods contaminated by hazardous chemicals/ingredients.
24) Foods containing questionable ingredients.
25) Placenta from human and non-halal animals or animals not slaughtered according to Shariah law (used for wound dressings and other medical applications).
26) Surgical powder (bone powder).
27) Plasma substitute.
28) Animal feed ingredients such as blood, bone, meat, meat-and-bone meal, and other animal feed additives.

Basically, it is recommended that:

- mixing any product with pork or lard is incompatible with halal requirements
- separate production lines should be operated for halal products rule out any possible impurities and contamination
- during the production of halal products, any materials used should not contain non-permitted or haram substances, and all equipment, utensils, machines, and processing aids that are used must be clearly marked and stored separately to avoid possible contamination.

23.3.1 Questionable Products

A number of food products and ingredients are in questionable or prohibited (haram) as their source is not normally indicated.

- Fish with scales are acceptable by all Muslim groups. Fish without scales, shellfish, and crustaceans are only accepted by some Muslim groups.
- Animal-derived ingredients should come from animals slaughtered by Muslims or from fish.
- Milk and eggs of all acceptable animal species are permitted if they are not fed rations contain questionable or prohibited ingredients.
- All vegetable ingredients are halal except intoxicating ones.
- Preservatives are also questionable food ingredients, as well as other products used in the production of food, including processing aids, lubricants, cleaning agents, sanitizers, and edible packaging material.
- Genetically modified organisms and biotechnology raise new challenges for halal certification.
- Regarding transgenic foods, plant-to-plant gene transfer is acceptable, but animal-to-plant and animal-to-animal gene transfer are questionable and may or may not be acceptable.

23.4 Relationship between Halal, Hygiene, Safety Food, and Phytosanitary Measures in Egypt

The concept of halal totally encompasses all aspects of human life. Consumption of halal food as ordained by Allah must be viewed in the overall scope and in perspective. Thus, to Muslims, food must not just be of good quality, safe, and hygienic but also be halal (Hayati et al. 2008). Islam only permits its followers to consume lawful, hygienic, safe, and good food, drinks and products as stated in the Holy Quran and Shariah. Food hygiene is important to provide foods that are safe and suitable for consumption. Food producers also need to ensure that consumers are provided with clear and easily understood information on labelling or otherwise. This prevents food from being contaminated with food-borne pathogens. Food hygiene practice should apply throughout the food supply chain from primary production through to the final stage of consumption, setting out the key hygienic controls and conditions at each stage of production. Hygiene, safety, and cleanliness are strongly emphasized in Islam via halal, includes the various aspects of the person, clothing, equipment, and working premises for the processing or manufacture of foods, drinks, and products. In fact the basis of halal itself is hygiene and health (Hayati 2008). The objective is to ascertain that the food (whatever kind) produced is safe, hygienic, and not hazardous to human health. In the context of halal, hygienic food, drinks, and products can be defined as free from najis or contamination and harmful germs. It is obvious that halal is very important in food matters, especially in the practice of keeping ourselves and the things around us clean in order to prevent diseases. Hence a safe food, drink, or product is one that does not cause harm to consumers, Muslim or non-Muslim, when it is prepared and/or eaten or in accordance to its intended usage. It therefore worth noting that in Islam the consumption of halal foods and use of halal products are obligatory in serving Allah. In this context, Muslim communities must be mindful of food or drink ingredients, handling processes, and packaging. Processed foods and drinks as well as products are only halal if the raw materials and ingredients used are halal and are fully compatible with Islamic guidelines. As such Muslims must be aware of the halal aspect of what they are consuming. In order to ensure we are safe, producers should take the necessary steps to comply with good manufacturing practice (GMP) and good hygiene practice (GHP). GMP is where the producers apply a combination of manufacturing and quality control procedures to ensure products are consistently manufactured to the specifications and halal prescriptions given by the halal certification body.

Food imports are sometimes subject to quality standards that appear to lack technical and scientific justification. The sanitary and phytosanitary measures of fruits and vegetables vary frequently. In addition, importers have to fulfil complicated labelling and packaging requirements, which increases the processing costs of the imported goods. Imports of poultry and meat products, for instance, must be preserved in sealed packages and shipped directly from the country of origin to Egypt, with content details given in Arabic on both the inside and outside of the package. These details should include the country of origin, the title of the commodity as well as the registered trade mark, the name of the slaughterhouse, the date of slaughter, the name of the importer, and the name of the supervising institution (by Islamic law, these institutions must be accredited by the commerce authorities of the country of origin). However, Egyptian companies use HACCP and others follow a number of measures that ensure safety, and they achieve other targets for the cleanliness of halal foods through the identification of cross-contaminants and allergens, as well as the declaration of all ingredients in non-halal items. Additional safeguards include careful employee training and halal food laws and regulatory requirements.

23.5 Standards, Testing, Labelling, and Certification

Food standards are established by the Egyptian Organization for Standardization and Quality Control (EOS) in the Ministry of Trade and Industry (MTI). Verification of compliance is the responsibility of agencies affiliated with various ministries, including the Ministry of Health and Population, the Ministry of Agriculture and, for imported goods, the General Organization for Export and Import Control (GOEIC) in the MTI (https://www.google.com.eg/url?sa=t&rct=j&q=&esrc=s&source=web&cd=1&cad=rja&uact=8&ved=0ahUKEwiy9uHogbvMAhXBvhQKHY39A5QQFggdMAA&url=http%3A%2F%2Fwww.goeic.gov.eg%2Fen%2Findex_r.asp&usg=AFQjCNGFFs_OCaWeY9UiVaMXyXB-yZh9YQ). Egypt has increased efforts to bring mandatory regulations into line with international standards.

Of Egypt's 8500 standards, 5000 are Egyptian technical regulations or mandatory standards. The EOS reports that it has harmonized mandatory standards with international standards and that about 80% of its mandatory standards are based on standards issued by international institutions such as the Geneva-based International Organization for Standardization (ISO). In the absence of a mandatory Egyptian standard, Ministerial Decree Number 180/1996 allows importers to choose a relevant standard from seven international systems: the ISO and European, American, Japanese, British, German, and, for food, Codex standards. Most of these specifications are optional, except for those related to general health, public security, and consumer protection. A ministerial decision issued by the MTI is needed to require compliance to these specifications. Obligatory standards constitute around 15% of the total number of Egyptian specifications. The EOS also issues quality and conformity marks. The conformity marks are mandatory for certain goods that can affect health and safety. The quality mark is issued by the EOS upon request by a producer and is valid for two years. Goods carrying the mark are subject to random testing. Egyptian food is by definition halal. However, lacking halal certification has up till now prevented certain Egyptian foods from being exported to countries with strict halal certification requirements. The International Trade Centre (ITC), through its Enhancing Arab Capacity for Trade (EnACT) program, funded by the Government of Canada, has helped many Egyptian food companies acquire the required knowledge to obtain the certificates needed to export halal products. As a result, the number of halal-certified companies more than doubled from 21 to 52 and the presence of the Egyptian food-processing sector in the Malaysian market has increased by 30% (see Table 23.2). Egypt traditionally exported halal products to the North American and European markets, which – while growing – are much smaller than markets in South Asia, Southeast Asia, and the Middle East. It is estimated that there are over 2.2 billion customers for halal products worldwide, an increasing number of them middle class with growing purchasing power. Yet Egyptian exporters have secured only a small share of the Asian halal market so far, due in part to a lack of halal certification, branding, and packaging.

23.6 The Demand for Halal Product Certification

The temptation to follow a certain strategy is of course facilitated if consumers cannot easily judge product quality. Typically, transactions involving food are characterized by information asymmetries. The seller tends to know more about the quality than the buyer. Where, when, and under what conditions was it harvested, preserved, processed, stored, and mixed? The asymmetries increase as the distance from farm to fork increases because consumers cannot themselves trace food products (any more) to the original producers, with more intermediaries and processers in the food value chain this becomes practically impossible. For consumers, instead of tracing their food products they can trust the halal certification system as it guarantees the purity and cleanliness of the sources from which products are derived, the processes by which they are made, and the rights of consumers regarding what they eat. Halal certificate will therefore allow them to meet their religious compliance. Furthermore, as the food chain gets longer, original foodstuffs are cut up, processed, mixed, etc. in a great many combinations, resulting eventually in the readymade foodstuffs that we find in supermarkets. Chewing gum, candy, and chocolate contain gelatin as a thickening agent that may be derived from pork skins and bones, which are non-halal. In many cases, every major and minor ingredient is not listed on the label of such products, therefore, many Muslims do not have

Table 23.2 List of halal-certified Egyptian companies

No.	Company	Address	Certificate no.	Date of issue	Validity	Product(s) certified
1	Misr Cafe Co.	Industrial Zone A1, 10th Ramadan City, Egypt	001/2005	1/2/2013	2 years	Instant granulated coffee(Misr café, Coffee Break, Lorzo)
2	Mass Food Co.	6th of October City, Egypt	002/2005	1/4/2014	2 years	Corn flakes, Sweet Flakes (Temmy's)
3	Hashem Brother for Essential Oils	Kafr El-Sohaby, El Kanater, El Kalyoubya, Egypt	003/2005	1/6/2014	2 years	Essential oils (11 products), concretes (4 products), absolute (4 products) (Note: Once the solvent is removed after extraction of volatile aroma, the resulting product is known as a concretes. Absolutes are a further derivation of concretes that result when concrete is subjected to a secondary solvent extraction process with ethanol.)
4	Machalico Co. for Essential Oils and Concentrates	58 Kasr El Eini St Garden City, Egypt	004/2005	1/9/2012	2 years	Essential oils, concretes
5	The International Company for Agricultural Development (Farm Frites)	10th Ramadan City, Industrial Zone A2 (Lot C), Egypt	007/2005	2/12/2013	2 years	Pommes frites (13 products)
6	Dream Co.	First Industrial Zone New Borg El-Arab- Alexandria, Egypt	008/2066	1/1/2014	2 years	Food powder products (50 products)

(Continued)

Table 23.2 (Continued)

No.	Company	Address	Certificate no.	Date of issue	Validity	Product(s) certified
7	New Benisuef Co. for Preservation, Dehydration and Vegetables	12,13 Light Industrial Zone, Benisuef El-Gadida, Egypt	014/2007	1/1/2014	2 years	Dehydrated onion
8	El-Nasr Co. for Gelatin	67 Shaiakhet – Ghiet El Enabqesm, Karmouz, Alexandria, Egypt	015/2007	15/4/2014	2 years	Edible beef gelatin
9	Nestle Egypt	6th of October City, Industrial Zone Plot 5,6,7,8, Egypt	017/2007	1/5/2014	2 years	Movenpick Ice Cream, Movenpick Waffle Mix for Ice
10	Mass Food International Co.	Plot.44,3rd Industrial Zone, 6th October City, Egypt	023/2009060/2013	1/7/20141/10/2013	2 years	Corn Flakes-Sweet Flakes (18 Products), Kicker flaks, Cookies (Temmy's)
11	Nile Co. for Food Industries (Enjoy)	Combera, Embaba, Giza, Egypt	024/2009	1/11/2012	2 years	Enjoy Nectar (6 products)
12	El Amin Co. for Gelatin	3rd Industrial Zone, 6th October City, Egypt	026/2010	1/7/2014	2 years	Gelatin (2 products)
13	Givaudan Egypt	Piece 37, Zone 3, 6th October City, Egypt	028/2010050/2013053/2013	1/4/20141/5/20131/6/2013	2 years	Flavours (72 products, 45 products, 9 products)
14	Egy Swiss Food Co.	Plot 167, Industrial Zone, Belbes, El Asher Road, Sharquia, Egypt	029/2012030/2012	1/10/20121/10/2012	2 years	Poultry (23 products), meat (50 products)
15	G.S.F. Egypt	Lot 321, 1st Industrial Zone, 6th October City, Giza, Egypt	031/2011	1/7/2013	2 years	Ketchup, mayonnaise (133 products)
16	National Food Company (Meat Division)	Jeddah, Industrial City Fourth Stage, KSA	032/2011	1/9/2013	2 years	Burger Arabic spices, breaded chicken burger, breaded fish fillet (Americana)

17	Unilever Mashrek Foods	Block 9, 1st Industrial Zone, New Borg, El-Arab City, Alexandria, Egypt	033/2012	1/2/2014	2 years	Chicken pullet, beef boeuf, chicken stock, mushroom cream soup, creamy chicken soup, chicken noodle soup
18	Arma Food Industries	10th of Ramadan, Industrial Zone b2, Egypt	034/2012	1/3/2014	2 years	Vegetable oils (Samna Crstal, Al Hanim Taza, Koot Al Klob, Al Asil, Sakia, El Montaz), shortening (Arma Short- Scing), margarine, coca butter substitute, coconut oil, Arcon
19	Arma Oils	10th of Ramadan, Industrial Zone b2, Egypt	035/2012	1/3/2014	2 years	Corn oil (Crstal-Hala, Asil), sunflower oil (Crstal-Hala, Asil) soya bean oil (Crstal) vegetable oil for frying (Qaleya-Hadewa, Asil, El Montaz)
20	Misr Food Additives	Plot 154 El Roubeky St, Badr City, Egypt	037/2012	1/7/2014	2 years	Lacta (501–555) (emulsifier and stabilizer), yoghurt Lacta (501–555) (emulsifier and stabilizer), cheese
21	Farco Pharmaceuticals	Kilo 31 Alex, Cairo Desert Road, El Amreya, Alexandria, Egypt	038/2012	15/7/2014	2 years	Urinex (soft gelatin capsules), Lady 4 (soft gelatin capsules) Baraka 100 mg (soft gelatin capsules)Baraka 450 mg (soft gelatin capsules)
22	Cairo Poultry Processing	3rd Industrial Zone, Al Robaiky St, 10th of Ramadan City, Egypt	039/2012	1/11/2012	2 years	Koki (Americana) (37 products)
23	The Egyptian Co. for Development	New Borg El Arab City, Industrial Zone, Zone 3, 4 – Block 28, Alexandria, Egypt	040/2012	1/12/2012	2 years	Nectar Faragello (mango, guava, orange, cocktail)Drink Faragello (mango, guava, orange, cocktail)

(*Continued*)

Table 23.2 (Continued)

No.	Company	Address	Certificate no.	Date of issue	Validity	Product(s) certified
24	Olio for Industry	Obour City, Part No. 9, Block 13039 Industrial Zone One, Kaliobeia, Egypt	041/2012	1/2/2013	2 years	Shamy kebab Olio, meat spring rolls, cheese spring rolls, fried Kofta mozzarella finger, vine leaves with oil (22 products)
25	Kemet for Natural Food. Co.	Lot no. 48/B, 5th Industrial Zone, 6th of October City, Egypt	042/2013	1/1/2013	2 years	Pretzo, mixy, pretzo Dr. (15 products)
26	The Egyptian Food Co.	New Borg El Arab City, First Industrial Zone, Block 28, Zone 2, Alexandria, Egypt	044/2013	1/1/2013	2 years	Nectar Faragello (mango, guava, cocktail, orange)
27	The Egyptian Food Co.	New Borg El Arab City, First Industrial Zone, Block 26, Zone 3, Alexandria, Egypt	043/2013	1/1/2013	2 years	Chicken stock (Faragello, Elwady, quicky-hilwa) beef stock (chicken)
28	Al Ferdaws Food Industries Materials Co. (Fimcobase)	Land No. 4, Block No. 1, 2nd Industrial Zone, New Borg El-Arab City, Alexandria, Egypt	045/2013	1/2/2013	2 years	Gum base (blocks, sheets, pearls)
29	Green Land Group for Food Industries Co.	Piece No.155, 156, 161, 162, Industrial Zone B3, 10th of Ramadan City, Egypt	046/2013	1/2/2013	2 years	Dairy (12 products)
30	Orion Food Industries Co.	Plot No. 6/1, 6/2, 6/3, 4/2, 4/3, 6th Industrial Zone, 6th October City, Egypt	047/2013	1/3/2013	2 years	Gardenio (Natural mango aseptic pulp, natural guava aseptic pulp)
31	Egyptian Canning Company (Americana)	Lot No. 38, 3rd Industrial Zone, 6th October City, Egypt	048/2013	1/4/2013	2 years	Green olives, black olives, mixed pickles, artichoke (12 products)

No.	Company	Address	Registration No.	Date	Duration	Products
32	Agrana Nile Fruits Processing	Units 4 and 5, Block 13 019, 1st Industrial Zone, Al Obour City, Egypt	049/2013052/2013	1/4/20131/6/2013	2 years	Fruit preparation (6 products) Fruit preparation (20 products)
33	Halawni Brother Co.	Piece No 160, 163a, 163b, 164, 165, Industrial Zone B3, 10th of Ramadan City, Egypt	051/2013	1/5/2013	2 years	Halawa, tahina, biscuits, jam, luncheon beef, smoked roast beef, drinks (33 products)
34	El-Marwa Food Industries Co.	Lot No. 43, 4th Industrial Zone, 6th October City, Egypt	054/2013	1/6/2013	2 years	Natural puree (mango, guava, apricot, peach, apple), aseptic pulp, frozen concentrated juice (11 products)
35	Modern Concentrates Industries	Lot no. 42, 4th Industrial Zone, 6th October City, Egypt	055/2013	1/6/2013	2 years	Frozen concentrated juice (orange, grapefruit, lemon) (3 products)
36	Integrated Food Franchising	Piece no. 19A, 2nd Industrial Zone, Badr City, Cairo, Egypt	056/2013	1/7/2013	2 years	Pizza, sandwich, sauce meals, frozen ready meals (20 products)
37	Delta Aromatic for Manufacturing Fruit Concentrates and Flavours	Plot no. M22, M20, M18, North Area Expansion, Industrial Zone, 6th October City, Egypt	057/2013	1/6/2013	2 years	Dry dill (tips, steam), dry hibiscus (flowers, crashed, flower fine cut) (4 products)
38	Al-Ain Food Beverages	7th Industrial Zone, Area Plot7198, Sadat City, Al Menofiya Governorate, Egypt	058/2013	1/7/2013	2 years	Tomato concentrate pulp (guava, strawberry) individually quick frozen food, strawberry (8 products)
39	International Co. for Food Industries	Piece 1, 16, Block 13023, Industrial Zone A, Elobour City, Egypt	059/2013	1/9/2013	2 years	Dry dill (tips, steam) dry hibiscus (flowers, crashed, flower fine cut) (14 products)
40	Caravan Global Industries (CGI)	District 6, 3rd Industrial Zone, 6th October City, Egypt	061/2013	1/11/2013	2 years	Coffee mix, cappuccino, hot chocolate (9 products)

(Continued)

Table 23.2 (Continued)

No.	Company	Address	Certificate no.	Date of issue	Validity	Product(s) certified
41	El Seba Spices and Herbs Co.	Lot No1/G, 6th Industrial Zone, 6th October City, Egypt	063/2014	1/2/2014	2 years	Spices and herbs (24 products)
42	Angel Yeast Egypt Co.	Area of Medium Industries, New BeniSuef City, BeniSuef Governorate, Egypt	064/2014	1/3/2014	2 years	Instant dry yeast (8 products)
43	Classic Food Industries Co.	2nd Industrial Zone, 6th October City, Egypt	065/2014	1/5/2014	2 years	Vanilla butter cookies (23 products)
44	El Jawhara for Drying, Packing Dates and Developing Agricultural Crops	Siwa, Marsa Matrouh Governorate, Egypt	066/2014	1/5/2014	2 years	Dried unpressed dates (2 products)
45	Negmet Siwa Factory	Siwa, Marsa Matrouh Governorate, Egypt	067/2014	1/5/2014	2 years	Dried unpressed dates (2 products)
46	Dried and Packed Dehydrated Pressed Dates Co.	Ghormy, Siwa, Marsa Matrouh Governorate, Egypt	068/2014	1/5/2014	2 years	Dried unpressed dates (2 products)
47	El-Wady Company for the Development of Food Industries	El-Kordy Zone, Siwa, Marsa Matrouh Governorate, Egypt	069/2014	1/5/2014	2 years	Dried unpressed dates (2 products)
48	Konoz Siwa Factory	Taba Zone, Siwa, Marsa Matrouh Governorate, Egypt	070/2014	1/5/2014	2 years	Dried unpressed dates (2 products)
49	Port Said Co. for Food Industry (Riyada)	Industrial Zone South Port Said, Port Said, Egypt	071/2014	1/6/2014	2 years	Dried unpressed dates (2 products)
50	Advanced Global Industries (AGI) Co.	1st Industrial Zone, 6th October City, Egypt	072/2014	1/7/2014	2 years	Full cream milk, skimmed milk
51	Industries (Safe for Pharmaceuticals Pharma)	2nd Industrial Zone, New Borg El Arab City, Cairo, Egypt	073/2014	15/7/2014	2 years	Soft gelatine capsules

enough knowledge about candy and chocolate ingredients. Additives such as the E-numbers 472 (fatty acid derivatives and modified fats), E472a (acetic acid esters of mono- and diglycerides of fatty acids), E472b (lactic acid esters of mono- and diglycerides of fatty acids), E472c (citric acid esters of mono- and diglycerides of fatty acids), E472d (tartaric acid esters of mono- and diglycerides of fatty acids), E472e (mono- and diacetyltartaric acid esters of mono- and diglycerides of fatty acids), E472f (mixed acetic and tartaric acid esters of mono- and diglycerides of fatty acids), E473 (sucrose esters of fatty acids), E485 (gelatin), and E471 (mono- and diglycerides of fatty acids) are also likely to be haram. (The numbering scheme follows that of the International Numbering System [INS] as determined by the Codex Alimentarius committee, although only a subset of INS additives is approved for use in the European Union.) But will Muslims shoppers go around with a long list of E-numbers checking every individual product? In addition, certain pharmaceutical and healthcare products contain gelatin or other products derived from non-halal sources.

The best one can do is to believe the information that the producer has put on the package or in enclosed leaflet, but even that frequently requires quite a leap of faith. The distances in space and culture, and the decomposition and recomposition processes, provide countless opportunities for fraud and deception. This holds all the more for halal food, as quite a few of its standards are process standards rather than product standards. The latter can in principle be tested at the final product, the former not, making it even more difficult to detect fraud.

Notwithstanding these detection problems, stories about fraud with halal products abound. Several thousand tons of meat was sold wrongly as halal to Muslims in France during 2009 and in the same period it was claimed that 60% of all halal products sold there were impure. Modern methods of food preparation and processing pose special problems. Take the case of chicken filets. When Muslims buy chicken, they may be under the impression that it is halal. They think they are buying halal meat, but with modern food manipulation they cannot be sure. Muslims are not alone in their distrust of food. According to the Dutch Food Inspection Consumer Monitor 2004, only one in every three Dutch people consider the food on the shelves of their stores sufficiently safe. The lowest scores were for chicken (trusted by only 33%) and ready-made meals (36%). There was more trust in bread (82%) and cheese (80%).

Food adulteration has been practiced from time immemorial. Bread has been diluted with plaster, sand, bone meal, and even poisonous lead-white; milk with ditchwater. In 1978 Egypt dumped eight containers of beef lanchos from Dutch meat producer Zwanenberg into the Red Sea. The lanchos were a mixture of beef fat and powdered pig bones, and were, notwithstanding the name, 99% pork. Before 1977, Egypt did not have the right equipment to analyze the composition. Saudi Arabia did, and sold it in 1978 to Egypt, and the fraud was detected. During that time, the Netherlands was blacklisted for a while by several Middle East countries.

Currently, the Egyptians and Saudis use expensive laboratory methods to protect their citizens from haram food. Cases like those above highlighted the importance of halal certification for exports and led to the first private attempts at halal certification, encouraged by most Middle East countries, including Egypt, which are major meat consumers.

23.7 Conditions, Regulations, and Certification of Halal Food Imported to Egypt

All imported products are subjected to strict inspection to ensure they meet Egyptian standards. The import inspection is conducted for safety, hygiene, and environmental reasons. Imports into Egypt must comply with very strict labelling and marking requirements that are covered by number of regulations. All imported products must meet the relevant Egyptian standard(s), as conformity to the standards will be assessed during the customs clearance procedure. The majority of Egyptian standards relating to products in the food sector are periodically reviewed to ensure their relevance to the current requirements of the country by the Egyptian Regulatory and Enforcement body. The majority of Egyptian standards conform to international standards. If a local product standard for a specific imported item does not exist, Egyptian authorities may apply the standard for that product used in the country of origin.

The general requirements and standards for food and agricultural imports into Egypt are inspected and controlled by the following governmental organizations:

1. **Regulatory body**

   The Egyptian Organization for Standardization and Quality Control (EOSQ) is located in the MTI (http://www.eos.org.eg/public/en-us/). The EOSQ has sole responsibility for establishing, adopting, and publishing food standards, technical regulations, and codes of practice by consulting with other MTI departments. However, verification and assessment are carried out by other ministries, including the Ministry of Health and Population and the Ministry of Agriculture. The assessment of imported goods is made by the GOEIC of MTI. In cases where no mandatory Egyptian standards exist, the Codex standards are acceptable. In the absence of an Egyptian or international standard, authorities often refer to the Analysis Certificate accompanying the product. The EOSQ is also responsible for issuing quality and conformity marks. The conformity marks are mandatory for certain goods that may affect health and safety. The quality mark is issued by the EOSQ upon request by a producer and is valid for two years. However, goods carrying the mark are still subject to random testing.

2. **Enforcement body**

   The Ministry of Health and Population (http://www.mohp.gov.eg/) and the MTI (http://www.mti.gov.eg/) are responsible for applying standards as well as for registration and approval of all specialty and dietary foods.

3. **GOEIC** are responsible for inspection and certification of imported goods. GOEIC is part of the MTI and is made up of representatives from the Ministry of Agriculture, Health, and GOEIC. Each ministry performs its analysis on the same product sample and issues its results to GOEIC. If one ministry rejects the product, GOEIC, in turn, also rejects the product.

The Egyptian Nutrition Institute and the Drug Planning and Policy Center of the Ministry of Health and Population are are responsible for the registration and approval of all nutritional supplements and dietary foods.

23.8 Control of Halal Slaughtering of Animals for Human Consumption

The Islamic method of slaughtering (the halal method) is the least painful method of slaughter and is not a traumatic experience for the animals. In order for the slaughtering to be lawful, several measures must be taken by the slaughterer. This is to ensure the highest benefit to both the animal and the consumer.

There are two methods of halal slaughter. The first of these is *dabh*, which involves severing the trachea, oesophagus, and jugular veins of the animal. This method is used for sheep, goats, cows, and buffalo. The other method, *nahr*, involves cutting the blood vessels at the base of the neck. This method is used for camels and also for cows and buffaloes. Egyptian slaughtering houses, local butchers, and almost all Egyptian people follow these strict requirements for the slaughtering of animals, according to the Islamic (Shariah) law. The Prophet Mohammed (pbuh) said, 'Allah has ordained kindness (or excellence) in everything. If slaughtering is to be done, do it in the manner, and when you slaughter, do it in the best manner by first sharpening the knife and putting the animals at ease'. The legal purification of the flesh of animals requires that the following conditions be met. The laws are translated into practice as follows:

1) The slaughterer must be a mature and pious Muslim of sound mind who understands fully the fundamentals and conditions relating to halal slaughter. According to another opinion in Shariah law, the slaughterer maybe from the people of the book (Christian or Jew) and not necessarily be a Muslim.

2) The halal animals and birds species are slaughtered by an authorized Muslim slaughterer in accordance with the Islamic law. The means by which the animals and birds are rendered unconscious (whilst ensuring that they are not killed prior to the slaughtering ritual) should be slaughter by a sharp knife, which is capable of making the animal bleed by severing the blood vessels, respiratory tract, and oesophagus. Animals to be slaughtered should be checked by a qualified veterinarian following the standard inspection methodologies.

3) The name of Allah should be mention while slaughtering. No name other than Allah's should be mentioned over the animal at the time of slaughter. This is clear from the Holy Quran texts and ahadith, where Allah Ta'ala says: 'So eat of that [meat] upon which the name of Allah has been mentioned, if you are believers in His verses' (6 : 118). Allah Ta'ala also says: 'And do not eat of that upon which the name of Allah has not been mentioned, for indeed, it is grave disobedience. And indeed do the devils inspire their allies [among men] to dispute with you. And if you were to obey them, indeed, you would be associates [of others with Him]' (6 : 121).

4) Certain specifications also require checks by the halal inspector. These include the type and frequency of inspection of the stunning apparatus, to be carried out by the works electrician, and mechanical aspects, the frequency of checking on the operating skill of the stunner to ensure that the animals are slaughtered properly according to Shariah law, Muslim slaughterer and how often an animal will be removed from the process before sticking in order to guarantee animal welfare and to ensure it has been properly stunned. Stunning should meet the conditions acceptable for slaughtering

halal animals. Specifications must also include cleanliness as halal slaughter laws are based on cleanliness, sanitation, and purity. All utensils must be clean and free of contamination from any unlawful or harmful substances.

23.9 Compliance with Animal Welfare in Halal Slaughter

The swift cutting of the vessels of the neck during halal slaughter causes ischemia of the brain and makes the animal insensitive to pain. This method results in the rapid gush of blood which drains out from the animal's body. If the spinal cord is cut, the nerve fibres to the heart might be damaged, leading to cardiac arrest and stagnation of blood in the blood vessels. Bleeding ensures that the meat is of good quality.

23.10 Halal Certification

Halal certification allows access to growing export markets that require certification. Meat must be certified as halal by an accredited halal certifying body. The certifying body must be recognized in the export market(s) in question. No standard certification exists and certification requirements vary by country. Egypt requires government approval of halal certifying agencies, therefore the exporting country should check with the Egypt government to determine which organizations are recognized as halal certifiers for specific products. All certificates should be countersigned by the country of origin's chamber of commerce and notarized by the Egyptian Embassy or Consulate in the country of origin. The certification requirements for imported food products differ according to the product. For example, in Egypt veterinary and halal certificates are required for meat, poultry, and dairy products. If the product is further processed in another country during transit, the appropriate certification may be executed in the country where the additional processing is done.

Egyptian authorities claim that all product standards and requirements applied to imported food are identical to those applicable to domestically produced products. In fact, Egyptian authorities are stricter in enforcing product standards on imported food products than on locally produced food products.

The following criteria must be completed to meet the certification requirements for the export of meat and meat products to Egypt:

- The exporter must obtain a halal certificate or Certificate of Islamic Slaughter from a member of an Islamic centre or Islamic organization, certifying that the animals were slaughtered according to Islamic religious requirements. This certificate must accompany all shipments of products labelled 'halal'. The Egyptian Embassy or Consulate in the exporting country must authenticate the certificate.
- Poultry can only be imported to Egypt as whole birds with a halal certification. The import of poultry parts is forbidden.
- Halal labelling is the responsibility of the exporter. Halal-certified products must be labelled with a statement indicating that the product has been slaughtered in accordance with Islamic principles.

- Fresh/frozen unprocessed products labelled as halal must be accompanied by an appropriate halal certificate.
- Raw materials used in processed products labelled as halal must be accompanied by an appropriate halal certificate.
- General labelling must be in Arabic and English and include storage temperatures, production and expiration dates, and metric net weight.
- Additionally, for fresh/frozen products the country of origin, producer's name, name of the slaughter plant, date of production, importer's name and address, and the name of the approved Islamic organization must be included on a label inserted inside the bag or wrapping.

All imported products must meet the relevant Egyptian standard(s), as conformity with the standards will be assessed during the customs clearance procedure. Of Egypt's current 5000 standards, compliance with 543 is mandatory. EOSQ reports that it has begun to harmonize its mandatory standards with international standards and that about 80% of its mandatory standards are based on standards issued by international institutions. According to Ministerial Decree No. 180/1996, if a relative mandatory Egyptian standard is absent, importers may choose a standard from seven international systems: international (ISO/IEC), European (EN), American, Japanese, British, German, and Codex standards. Most of these specifications are optional except for those related to general health, public security, and consumer protection and in the absence of EN standards, British (BS), German (DIN), and French (NF) standards may be applied, American standards (ANS), Japanese standards (JAS), and Codex standards. If a local product standard for a specific imported item does not exist, Egyptian authorities may apply the standard for that product used in the country of origin.

The majority of Egyptian standards relate to products in the food, engineering, textiles, and clothing sectors, including the importation of meat and live animals. These standards are periodically reviewed to ensure their relevance to the current requirements of the country. The modification to standards and regulations are announced in the Official Gazette. Others are implemented without official notification. Once a ministerial rule (e.g. decree, law, etc.) is issued, importers are strongly advised to periodically check with the various government clearing authorities to see if it has been amended (e.g. by internal memoranda). Laws often change and invariably take effect as soon as they are announced.

Egyptian Halal Food Standard Number 4249/2008 deals with the requirements and provisions for labelling halal foods.

23.11 Halal Slaughter Facilities and Products Registration

Facilities that produce and process halal meat need to be registered with an Egyptian regulatory authority prior to export. All approved slaughter facilities are subject to re-examination every three years to renew their eligibility to export to Egypt. The examinations evaluate both food safety and halal practices. In addition, meat product should be produced according to the Islamic rules and in accordance with Egyptian standard specifications. As for poultry and its products, approval should be issued through the Animal and Poultry Wealth Development department at the Ministry of Agriculture.

In 2015 the Egyptian government contemplated requiring representatives from Cairo University and the Islamic institution Al Azhar to accompany veterinarians to exporter facilities on audits. For the import beef meat from the USA, beef plant audits are implemented by authorized Egyptian veterinarians in cooperation with the General Organization for Veterinary Services (GOVS) and the US Meat Export Federation (USMEF). To date, around 57 beef slaughter facilities have been audited and approved for export to Egypt.

The Egyptian National Nutrition Institute (NNI) and the Ministry of Health and Population's Drug Planning and Policy Center are responsible for the registration and approval of all nutritional supplements and dietary foods. Special dietary foods have a different composition to the traditional, conventional food. They include calorie-modified foods, baby and infant foods, energy foods, special health foods, including diabetic and weight-control foods, vitamin and mineral supplements, medicinal herbs, and bottled water. Any food making a nutritional claim also falls under the NNI's remit. The registration process for special dietary foods involves two technical committees:

1) The NNI internal Technical Review Committee (TRC) comprises NNI researchers, including biochemists, food technologists, pharmacists, medical doctors, and nutritionists.
2) The Higher Committee for Nutrition (HCN) is used as an advisory committee to the NNI and is responsible for the final approval of all special dietary foods. The HCN comprises representatives from the Ministry of Health and Population, universities (departments of pharmacy and food technology), representatives of research institutes (e.g. food technology research institutes), and the Egyptian Academy of Sciences.

To register a product, the interested party must submit an application form that includes the product name, manufacturer, country of origin, importer (if imported), ingredients, specifications, the manufacturing process, a certificate of analysis, a health certificate, and a certificate of free sale. Samples are submitted and analyzed to ensure the product meets the nutritional claims, and to test for the presence of heavy metals, pesticide residues, and microbial contaminants. Labels are reviewed and claims verified. The results are forwarded to the NNI TRC for review. In the absence of an Egyptian or international standard, authorities often refer to the analysis certificate accompanying the product. If no suitable standard exists for a product, a committee may be formed to develop a new standard. Halal certificates are required if any product contains gelatin. The applicant must disclose the source of the gelatin as well as any pharmaceuticals and cosmetic products obtained from various animal tissues, such as those from the glands, e.g. the pancreas, thyroid, adrenal, and pineal, as well as organs such as the liver, stomach, lungs, and blood. The sources of any fine chemicals obtained from animals, such as bovine serum albumin, which are classified under protein, must also be disclosed. Bovine serum albumin is sometimes used as a component of moisturizing creams and lotions. The Egyptian Ministry of Health and Population prohibits the importation of these as natural ingredient products or as ingredients and premixes to be prepared and packed by local pharmacies. Only halal-certified products can be used to produce pharmaceutical products and food supplements. The government is supportive of the production of agricultural biotech products and has drawn up regulations on the review and approval of biotech seed.

23.12 Egypt Opportunities

Halal food is subject to special requirements. All products (except pork meat products, products containing alcohol and intoxicants, blood and blood products, and foods containing ingredients such as gelatine and enzymes) can be certified as halal food. The products certified must bear the inscription 'halal food' on the label. Egyptian standard guidelines (SN-4249/2008) indicate the requirements and provisions for labelling halal foods. A halal unit has been established at the EOSQ, supporting the development of an increased halal-certified export base. Their halal dietary guidelines are not limited to meat products, but also include other processed foods such as confectionery, snacks, beverages, and chocolates which could contain non-halal ingredients like pork-based gelatin. The halal dietary guidelines indicate that the use of alcohol during the production process renders products non-halal.

Over 50 halal-certified Egyptian companies produced a variety of halal products (Table 23.2) and there is growing concern for food safety and quality in halal markets so significant opportunities exist for Egyptian companies wishing to enter the halal food market. Consumers seek a wide range of halal-certified products in variety of food categories (not just meat), including food supplements, pharmaceuticals, cosmetic products, leather goods, and services. Strong demand for new or differentiated halal products that are not already in the market also exists. Consumers are also demanding genuine halal products, as there have been a number of incidents globally where food marketed as halal has failed to meet halal requirements.

The market for halal products worldwide provides great opportunities for Egyptian companies because there is enormous export potential in value-added halal foods that have high-quality standards that meet the demand for halal supplies from European hypermarkets and supermarket chains. Moreover, certifying and branding products made by Egyptian companies is an important marketing tool and enables Egyptian companies to capture new markets. Halal products are also growing in popularity among non-Muslim consumers due to humane animal treatment concerns and the perception that halal products are healthier and safer.

23.13 Halal Food Testing

Food fraud is a collective term used to encompass the deliberate and intentional substitution, addition, tampering, or misrepresentation of food, food ingredients, or food packaging, or false or misleading statements made about a product for economic gain (Spink and Moyer 2011a). Fraud includes adulteration, tampering, overrun, theft, diversion, simulation, and counterfeiting (Spink and Moyer 2011b). Food fraud is not a new phenomenon, but has gained attention in recent years due to the number of cases and the costs associated with the fraud. Recent food fraud cases include horse meat scandals and cases related to the intentional mislabelling of fish in stores. Some of the common foods involved in adulteration include olive oil, milk, honey, saffron, coffee, tea, fish, and meat. The adulteration of meat products with undeclared animal species can result in recalls of the products and reduce consumer confidence in well-established product brands. Laboratory testing is used to determine the species composition of processed meat products and other food ingredients to assess their authenticity.

The determination of food authenticity, particularly meat species identification, has great importance in the halal food system, food quality control, and safety. The most suitable technique for any particular sample is often determined by the nature of the sample itself, for instance whether it is raw or cooked, whole muscle or comminuted (Hargin 1996). Technological advances have facilitated research applications to detect foreign DNA in food products. Sensitive and reliable methods for the detection of halal product adulteration are of paramount important for implementation of halal food labelling, regulations, and product quality control. Halal product analyses, determination of food authenticity, and detection of adulteration are major concerns not only to consumers, but also to industries and policy makers at all levels of the production process. Egyptian research institutes and universities have sophisticated equipment and accredited laboratories that already conduct fraud analysis, authenticity, and detection of adulteration research.

Halal food laboratory detection can be implemented in two ways, namely the chemical and molecular methods. Alcohol content can be identified by chemical methods and molecular methods can be used to detect DNA in a product. The analytical techniques currently used for detecting pork adulteration rely on either protein or DNA analysis. Enzyme linked immunosorbent assay (ELISA) is the most commonly used protein-based technique and a number of commercial immunoassays and ELISA kits are available. Other protein-based methods include electrophoresis, Fourier transform infrared (FTIR) spectroscopy, near-infrared spectroscopy, electronic nose, and chromatography. However, protein-based techniques have numerous limitations. In addition to being limited when assaying heat-treated products (due to denaturation of proteins during thermal processing), analyses of immunoassays that rely on the use of antibodies raised against a specific protein are often hindered by cross-reactions occurring among closely related species (Fajardo et al. 2010). Due to the advantages DNA-based techniques have over protein-based techniques, there has been a shift from protein to DNA analysis. The degeneration of DNA offers the advantage of differentiating among different animal species solely using DNA analysis. In addition to DNA being a stable molecule that allows analysis of processed and heat-treated products, it is present in the majority of cells and the information content of DNA is not only greater than that of protein but can also be extracted from all kinds of tissues (Ballin 2010).

Most solid food samples, such as fresh meat, processed meat products, chicken, sausage, and biscuits, can be tested using standard methods to ensure the absence of foreign DNA, particularly pig DNA in halal food products. The products that are crushed last are treated with chemicals to separate waste organelles, fat, and pure DNA. Finally, genomic isolation analysis can be carried out to identify the origin, quality, quantity, and purity of the total DNA. Polymerase chain reaction (PCR) has the capacity to amplify a few copies of DNA and its detection limit is much lower than that observed with protein-based assays. The basis of PCR amplification is the hybridization of specific oligonucleotides to a target DNA and synthesis of million copies flanked by these primers. Detection of species with PCR involves choosing adequate genetic markers, either nuclear or mitochondrial genes, to develop the assay. Techniques that are commonly used to detect species for halal authentication include PCR amplification of mitochondrial DNA, PCR-restriction fragment length polymorphism (RFLP) analysis, PCR using species-specific primers, real-time PCR, random amplified polymorphic DNA (RAPD) analysis, and PCR sequencing. However, other techniques are in use for detecting pork, lard, and other unlawful meat adulteration. The most critical test parameters for evaluating halal products are as follows:

1) determination of pork fat, lard, and lard derivatives via gas chromatography (GC), chromatography-mass spectrometry (GC–MS), FTIR, PCR, ELISA
2) detection of the presence of alcohol via GC, GC–MS, FTIR
3) characterization and comparison of gelatin from animal- and plant-based origin via FTIR, PCR, ELISA
4) protein expression to predict quality of halal meat via immunochemistry, ELISA, fluorimetry
5) detection of pig's blood in processed and raw products via PCR, ELISA, FTIR, DNA
6) fatty acid profiling of meat products via GC–MS
7) detection of high-performance liquid chromatography (HPLC) and FTIR
8) characterization of thermal behaviour of gelatins using other complementary techniques such as rheometry and differential scanning calorimetry (DSC).

23.14 The Egyptian Governmental Agencies in Charge of Halal Food

- Ministry of Agriculture and Land Reclamation, General Authority of Veterinary Services.
  Website: www.agr-egypt.gov.eg
- Ministry of Health and Population, The Central Laboratories of the Ministry of Health and Population
  Website: www.mohp.gov.eg
- Ministry of Trade and Industry, General Authority of Import and Export Control
  Website: www.goeic.gov.eg

References

Ballin, N.Z. (2010). Authentication of meat and meat products. *Meat Science* 86 (3): 577–587.

Fajardo, V., González, I., Rojas, M. et al. (2010). A review of current PCR-based methodologies for the authentication of meats from game animal species. *Trends in Food Science & Technology* 21 (8): 408–421.

GIEI (2014–2015). The Global Islamic Economy Indicator Report 2014–2015.

GIEI (2018–2019). The Global Islamic Economy Indicator Report 2018–2019.

Hargin, K.D. (1996). Authenticity issues in meat and meat products. *Meat Science* 43 (Supp. 1): 277–289.

Hayati, T.A., Habibah, M.A., Anuar, K., and Jamaludin, K.R. (2008). Quality assurance in halal food manufacturing in Malaysia: a preliminary study. *Proceedings of International Conference on Mechanical & Manufacturing Engineering (ICME2008)*, Johor Bahru, Malaysia (21–23 May 2008).

Qureshi, S.S., Jamal, M., Qureshi, M.S. et al. (2012). A review of halal food with special reference to meat and its trade potential. *Journal of Animal and Plant Sciences* 22 (Sup 2): 79–83.

Spink, J. and Moyer, D.C. (2011a). Defining the public health threat of food fraud. *Journal of Food Science* 76 (9): R157–R162.

Spink, J. and Moyer, D.C. (2011b). Defining the Public Health Threat of Food Fraud. *Journal of Food Science* 76 (9): R157–R163.

van Waarden, F. and van Dalen, R. (2010). The market for halal certificates: competitive private regulation. *Proceedings of the Third Biennial Conference of the ECPR Standing Groups on Regulation and Governance*, University College, Dublin (17–19 June 2010).

Zakaria, Z. (2008). Tapping into the world Halal market: some discussions on Malaysian laws and standards. *Shariah Journal* 16: 603–616.

24

Halal Food in the USA

Joe M. Regenstein[1] and Umar Moghul[2]

[1] *Department of Food Science, College of Agriculture and Life Sciences, Cornell University, Ithaca, NY, USA*
[2] *Gateway Islamic Advisory LLP, New York, USA*

24.1   Halal in the USA

The US business community is becoming aware of the size, growth, and dynamism of global Muslim communities. As might be expected, it has begun to recognize the opportunity to serve the domestic American community with goods and services, and the opportunities and uniqueness of its needs in many areas for products that fulfil the principles of their faith. Thus, a significant expansion of the number of available halal food products in the USA has occurred and for that the design and development of a halal products infrastructure has also begun very recently.

The growing American Muslim population, estimated to be some seven to eight million, comprises different ethnicities as well as a significant African American population. Its understanding and practice of religion is not uniform, and so there is diversity in its acceptance of different foods. After 11 September 2001, the political complexities for the Muslim community in the USA increased; the acceptance of Muslims as part of American society remains a challenge – at least in some quarters.

The development of modern religious supervision of food products in an industrial context started in the 1920s and really blossomed after the Second World War. As such, an infrastructure and general familiarity with religious supervision has existed in the US food industry for some time. However, rather than this being the result of any government mandates, it is the religious communities employing their purchasing power as consumers that has driven the agenda. The expansion of halal foods markets in the USA takes place within this context and, in a sense, under the shadow of kosher food products. This is not to say that the path of kosher will be the most suitable in all respects for halal foods, but many of the issues and experiences will be shared.

The Halal Food Handbook, First Edition. Edited by Yunes Ramadan Al-Teinaz, Stuart Spear, and Ibrahim H. A. Abd El-Rahim.

24.2   Religion, Food, and Government

Religious freedom is a strong component of the US legal system, protected in particular by the First Amendment (which came into force in 1791) to the US Constitution (which came into force two years earlier in 1789). The First Amendment frames a great deal of how the USA views religion, both legally and otherwise. The relevant part reads: 'Congress shall make no law respecting an establishment of religion, or prohibiting the free exercise thereof'. It presents two primary legal norms: an Establishment Clause, which disallows the government from favouring any one religion over another religion, and a Free Exercise Clause, which prohibits governmental interference with the practice of any religion, except in limited instances. The legal right to religious freedom has commercial implications. One of the more dramatic instances is the right to specify faith as a credential in hiring – something that would otherwise be deemed unlawful discrimination in hiring in ways that would otherwise be unlawful, so long as such hiring can be shown to be necessary for religious purposes. For example, a plant slaughtering halal may insist that all slaughterers are Muslim, but it cannot insist that all its employees be Muslim.

By keeping American governmental agencies largely out of the religious standards 'business' and encouraging entrepreneurship within religious certification, a supervision mechanism has dynamically developed to fulfil the various requirements of diverse consumers and the different food businesses. The benefits of this less well-defined and more adaptable system, in the view of the authors, outweighs the benefits of a more centralized and standardized system. Diversity, although frustrating for some at times, is part of the strength of the modern religious certification system that facilitates religious choice. With the strong emphasis on religious freedom and independence in the USA, this approach better respects religious pluralism, both amongst religions and within them. This has meant that both Jewish and Muslim communities have had to accept and work through the issues that arise due to diverse religious interpretations within their respective faiths leading to varying interpretations and 'official' rulings. Neither faith, however, is without extensive intellectual history developing and addressing intra-religious legal diversity. As might be expected, this has sometimes left the food industry and consumers wishing for the simplicity of uniformity, and, at times, to efforts to standardize. But, in fact, as will hopefully be shown, standardization either by the government or by Muslim certifiers through means other than a natural evolution of dialogue and consensus or by increased governmental involvement can neither be accomplished nor is it necessarily desirable.

This is not to say that the US governmental involvement is entirely inappropriate, for it provides (i) consumer protection laws and enforcement mechanisms, and, in this regard, some states have enacted laws specific to halal, (ii) enforcement of intellectual property rights, such as trademark symbols used by halal certifying agencies, and (iii) laws aimed at ensuring animal welfare, a concern shared by Islam and Judaism in the use of animals for food production generally and in their slaughter in particular. Much of the foregoing takes place under the protection of religious freedom pursuant to the First Amendment. We take up these three topics in greater detail below to help frame the status and functioning of halal food certification in the USA and within each topic we analyse certain current challenges and opportunities.

Many of the original state laws protecting religiously certified foods worked with the assumption that it was somehow possible for the American government to determine what was religiously acceptable. Thus, these laws tried to define a substantive standard *of* religion. The earliest such is known as the Orthodox Hebrew Standard, established by New York State. Other states followed in a similar fashion with some so specific as to cite Hebrew documents as the determinative authority for kosher! Similar approaches were used by the first few states that created halal laws. Unfortunately, this approach to determining religious standards leads to two problems: (i) any committee created by the government that involves religious leaders making such a legal determination of religion was eventually decided to be unconstitutional and (ii) without such religious authorities' inclusion, the problem arises of what religious standards to authoritatively use following initial determinations or in ongoing determinations.

24.3 Consumer Protection

After a long history of various cases disputing constitutionality, the State of New Jersey's Supreme Court declared in 1996 (Perretti v. Ran-Dav's County Kosher Inc., 289 N.J. Super 618, 674 A. 2d 647 (Superior Ct. Appellate Div. 1996) that the New Jersey kosher law was unconstitutional. The court suggested that the legitimate role of the state is to create a framework for two consumer rights that are broadly applicable, even for the religious certification of commercial products: 'Truth in Labeling' and the 'Consumer's Right to Know'. This approach calls for food stores and producers to disclose practices and positions on issues deemed of major religious consumer concern.

A major challenge to religiously sanctioned foods was, and continues to be, the need to address the issue of fraud. Once a food is identified as, for example, halal, it has additional market value, which some have exploited by intentionally misrepresenting non-halal food as halal. If the fraud is widespread, i.e. where a large part of the halal food industry sees that fraud is essentially a normal halal market practice, then there it is very possible that other players will resist tighter controls by the certifying agencies or other regulatory bodies. Consumers in such markets, on the other hand, may very well seek stronger enforcement. Accordingly, religious communities need to address this issue by working with those stakeholders that demand transparency and provide goods that properly meet consumer needs.

State inspectors (such as those of New York, New Jersey, and a few other states and local governments) have limited religious inspection regulations and may check whether products and procedures in practice are consistent with what has been publicly disclosed. If there is fraud, authorities may take action. Similarly, states without formal religious food protection laws can take action if fraud can be proven. As mentioned earlier, a handful of states have enacted halal food consumer protection laws: California, Illinois, Michigan, Minnesota, and New Jersey.

Interestingly, consumer protection laws are, generally speaking, not without their precedent in the classical Islamic legal literature and history. The administrative institution of the hisbah, employed in many Muslim societies, monitored the implementation of various laws. Officers of this institution regulated consumer protection matters and established professional rules for butchers, milk and spice sellers, and flour merchants, amongst many other areas, received complaints, and had limited enforcement rights. Many of these rules had significant implications for health and safety, and sustainable food practices. Because halal is a relative newcomer to the market place in the USA, much of today's relevant legal precedent in US federal and state laws comes from the efforts of various kosher market stakeholders.

24.4 Certification: Agencies and Standards

The US kosher market developed a product marking system in the first part of the twentieth century, namely the use of a symbol that may be trademarked and that belonged, by law, to the certifying party. Using this system, a food business enters into a written agreement with a religious certification body to develop a product that will be acceptable according to the agency's interpretations. In return, upon review and successful assessment and payment of fees, the agency grants the producing company a right to mark the product with its trademarked symbol. The mark signifies a determination of the food's permissibility, in the case of halal for both following the specific halal laws applicable to the product, but also, depending on the agency or instance, with the company being in compliance with Islamic ethics to the extent defined by the halal certifier. This highlights the importance of transparency on the part of the agency so that other stakeholders know the exact scope of agency review and the significance of its mark.

This system has two major advantages. First, by owning the trademark, the certifying agency has a presumably strong legal basis for enforcement. This right also typically exists, at least in principle, in many other countries so long as the trademark has been registered there, and thus, certainly serves the franchise. Second, consumers benefit from the transparency and consequent greater reliability. The certifying agency can become known to the consumer in terms of, amongst other things, its religious interpretations, methods, and standards, and can, furthermore, inform and educate consumers and other stakeholders about faith and food standards. With the advent of the Internet, most of the major certifying agencies in the USA have a website which gives some idea of their standards and certainly provides contact information. Consumers are thereby enabled with a source with which they can engage in dialogue and feedback.

It is important to pause here and note a few relevant features of Islamic ethics and jurisprudence. First is that this certification determination is effectively an ethical-legal opinion (or fatwa, in Arabic) issued in accordance with Islamic jurisprudence and which would be expected (by many consumers) to be issued by competent and qualified Muslim scholars of Islamic jurisprudence and related disciplines. The certifying agency should identify those scholars and other advisors on whom it relies. Indeed, Muslims are advised to seek out those certifiers who have, or are known to have, a good reputation for knowledge and piety. A very recent study of what is looked for (by those seeking knowledge or guidance) in determining the validity of, and confidence in, a fatwa shows that the leading factor is that it be issued by a known, reputable authority. Other factors, almost equal in importance to the first, are that the scriptural or textual evidence underlying the determination be cited and explained by the jurist in reaching his or her conclusion. Last but not least of the factors cited by survey respondents is that the fatwa include some official stamp or marking to evidence its authenticity (Furber 2013). Normally, a fatwa is the response to a petition; in this case, the petition would be to verify whether a product is halal and that would entail an assessment of its composition and production process. It may also be that a fatwa is given without a particular petition in the case when an authority wishes to address a general matter of concern.

Islamic law assesses actions and then classifies them into five categories (generally speaking): obligatory (fard), recommended (mustahab), disliked (makruh), prohibited (haram), and permissible (mubah). This is for known rulings but there is also a category of doubtful (mashbuh) for emerging, unknown issues. An assessment of 'doubtfulness' generates questions and priorities to help establish subjects of further research to help assess assigning the substance in question into one of the five above categories. We mention this because certification bodies should make readily available information to explain to Muslim consumers at what level the food (or other product) has been assessed within the framework of Islamic jurisprudence and by whom. Just as importantly it would seem (both for the agency and consumers) is that the rationale of the determination be set forth in relatively detailed fashion. Islamic law is certainly not monolithic, and those instances in which there is a difference of opinion amongst Muslim jurists (such as with cheese, for example) should be duly noted, in our view with the agency specifying which view it has adopted. (In some cases, an agency may accept more than one ruling, in which case it must clearly indicate which ruling applies to the product being certified.) Finally, it is also worth mentioning that we approach questions herein from the perspective of a determining, certifying authority once it has been asked for, and decided to provide, an Islamic ethical-legal assessment. The certifying agency's responsibility is greater, under Islamic ethics, than that of the layperson consumer.

The agreement entered into between the food business and the certification agency often permits the latter access to confidential information allowing the agency to apply its principles across the business processes, ask questions comprehensively, and provide more detailed training of its own people and company personnel. The agreement typically provides more comprehensive access to company facilities than might normally be acceptable – even more so than government inspectors. This is important to assure that things are being done properly at the facility, and will hopefully enable the certification agency to conduct its review and issue its determination in an informed manner. Note that a government inspector must enter through the main door, announce him or herself, and allow the company opportunity to identify appropriate personnel to accompany and guide the inspector. A thoughtful religious certifying agency should contractually establish a right to enter unannounced (with personnel trained to meet the company's health and safety requirements), which offers a more robust approach, though it may not wish to always use this privilege.

Another important task of the certifying body is to review product labels prior to printing. This is important to the food company as well as the agency. The former wants to be able to market its products successfully with the certification mark. The latter seeks to support this success with transparency and clarity to serve its constituency and fulfil its responsibilities in issuing its determination. Practically speaking, for the agency to perform this function and maintain a good relationship with food companies, it must be prepared to respond promptly.

Costs for a certification program are usually borne by the company and taken from its advertising budget. From its perspective, the primary reason to be under religious supervision is a business one – to increase sales. So the decision to take on religious certification must fulfil the same criteria as any other component of the advertising budget and the selection of any other vendor. In most cases, these businesses usually get a high 'bang for the buck', i.e. religious certification when done correctly is extremely cost effective as it allows access to multiple additional markets for its products. Thus, the cost of religious certification is not a 'tax' as is alleged at times, but a marketing opportunity to increase incremental sales with a net return to the company similar to say organic certification. Actual pricing for the certification and ongoing supervision are of course negotiated and largely based on the complexity of the facility and products as well as the required number of inspections.

These certification systems have become a fee-for-service. Coupled with moral and spiritual incentives, this has led, as it should, to a greater professionalism amongst religious foods certifiers to perform to high standards and maintain their reputation, client base, and consumer confidence. As such, agencies must work efficiently with the food industry by providing, for instance, timely services. It also means that the value of the certification to the food industry is only as good as its market acceptance. We note that this method allows a food company to select a religious certification body that meets its capabilities or willingness, and its target markets, either because the products it produces may only meet the standards of certain agencies and/or it fits the company's business setup, plans, strategies, and cost requirements.

24.5 Markings

Markings may take – and have taken – a number of different forms. The most obvious is to simply write 'kosher' or 'halal' (in the relevant language) upon the packaging. However, this method would likely mean all consumers, including those who might object to their foods meeting religious requirements, would know that the company was working with a particular faith community which to a minority of other consumers may be objectionable. If the information was set forth in a 'foreign' language, some consumers may be particularly put off! Accordingly, another approach developed is to simply place a letter such as a 'k' or 'h' on the product. This technique is currently employed by multiple businesses, but it is a very weak approach from both the company's point of view and certainly from the consumers' point of view. Why? In the USA, at least, one cannot trademark a letter of the alphabet, regardless of the language. So a single letter 'H' may be used by any company without actually submitting to an Islamic ethical-legal review and certification process, and even if such a process were successfully completed, consumers would probably find it rather difficult to associate the particular organization or individuals responsible for certification with a letter used by many and owned by none. In an effort to proceed without such obstacles, these trademark symbols were designed to be subtle. So the 'certification mark' was born.

The key to making these markings successful depends on consumers and the religious community. When a religious certifying agency clearly has a standard that a community considers unacceptable, that community and its leadership ought to make known its disagreement, but this needs to be done intelligently and carefully (because of, amongst other things, the litigiousness of modern US society). The focus of any claim against a religious certifying agency should generally be upon a particular product, indicating that it does not meet the community's standard. This makes it a protected personal/community statement. The community should not assert that the product is 'not halal', for that unproductively challenges the certifier's authority and may allow the agency, product manufacturers, and/or sellers to file suit for damages on the basis of libel. In the highly unlikely event an agency takes on a position categorically unacceptable in Islamic ethical and legal discourse, that error can be pointed out on its own merits. If, as is more likely, an agency asserts a standard which it does not meet or is controversial, that legitimately raises the issue of trust.

A certifying agency must also be careful about the misuse of its trademarked symbol. In some instances, the agency's client may incorrectly understand which products are permitted to carry the agency's seal. In such a case, the agency can negotiate with the company and agree on how to ameliorate the situation. The agency may need to decide if (i) publicly announcing the problem (e.g. on both the company and the certifying agency's website) and letting various alerting services know about the problem, (ii) requiring a recall in which product returns are accepted by the seller with full refund to customers, and/or (iii) requiring product relabelling is appropriate. Obviously, if the certifying agency suspects that the company is intentionally being misleading, then it must take stronger action to protect its consumers, i.e. terminate the contract. A more egregious concern arises when the product is manufactured by a company and uses the symbol of an agency with which the manufacturer has no licencing relationship. That is one of the times when the state agency, either one specifically dealing with religious food or the general consumer protection agency, can be extremely helpful; they can more readily employ legal action and also have a wider range of non-legal options.

24.5.1 Inspections

Religious certifying agencies generally visit any facility much more often than US government inspectors. The US Food and Drug Administration (FDA) under the recently passed Food Safety and Modernization Act hopes to visit all high-risk food processing facilities at least once every three years! However, in fairness to US food plants, the state and often local health departments certainly visit food plants more often. Based on the experience of kosher in the USA, a good certifying agency should visit each plant under its supervision at least once every six months, even if the plant is very straightforward in its operations and only producing halal products. Many plants are visited more often, as often as once a month, especially those producing both halal and non-halal products. Best practices call for agencies to dispatch a senior inspector once a year from the 'home office' of the certifying agency even if other inspections may be done by those in the field. For intermediate inspections the use of local Muslims with appropriate training can be beneficial as it permits more inspections at reasonable cost, and sends the message to companies that inspectors are readily available.

Another concern as agencies grow larger is consistency. All inspectors of a given agency should be consistently applying its standards as representatives of the agency. As agencies mature, they will likely need to consider the education of their own inspectors to help ensure this consistency, amongst other things. Food businesses, especially those processing meat products and most foodservice operations, are generally under either continuous inspection while operating or subject to daily visits by kosher or halal authorities. Ultimately, the agency will have to determine inspection frequency based on its standards, objectives, the product in question, compliance risk, and its responsibility scope.

24.6 Plants: Halal and Non-halal

In the case of a plant that intends to only make halal products, all of the equipment should be proper for the production or preparation of halal foods. Only if newly purchased used equipment is brought in would there be any issues as its history of use would not be known. A comprehensive list covering not only direct ingredients but any materials with which food comes into contact as well as packaging materials must be provided by the company to the certification body. Of course, the food business should be maintaining and updating such a list for reasons other than religious certification. The responsible religious certifying agencies will regularly assess that the lists are accurate. Confirming the accuracy of such lists is an important task of the agency's inspector. There may be instances where the decision as to whether something is halal relies on representations of fact made by the food business and on the agency's inspectors' findings and reports.

In many cases plants will prepare both halal and non-halal products. This raises the issue of the status of the equipment. How does one ensure that the equipment is sufficiently free of the non-halal so as to maintain the halal status of foods that will be prepared using such equipment? The certifying agency must determine if the cleaning procedure is acceptable, provide feedback if not, and receive proof, such as time/temperature recordings on critical equipment, from the plant that the agreed upon cleaning protocol was in fact implemented. Again, this is where it is essential for the certifying agency to understand the plant and its processes in detail, so that it can communicate what is required to meet certification requirements.

Such plants (producing both halal and non-halal products) present another interesting issue: pipe layout. The engineering department of most plants maintains official layout diagrams. These are helpful in determining if the halal product processing line is completely separate from a non-halal line (and thus whether there is any mixing of the latter with the former, rendering the food prohibited or disliked for consumption under Islamic norms). However, pipe diagrams are often incorrect as changes in the actual layout may be made from time to time that are not communicated to those responsible for the diagrams. Therefore, the certifying agency's inspectors should check the actual pipe layout, rather than relying solely on a written diagram and consider whether periodic checks are appropriate. A careful certifying agency will provide training in this matter (of pipe layout) as a particular specialty, and those certified with this expertise should be amongst those inspecting the plant. However, regular inspectors should also be enabled to do some spot checks of potentially critical points in the piping.

A food producer may very well have multiple production facilities which are rarely identical to one another. In some instances, a producer may only be interested in having one of its plants deemed halal. So, only product from that very plant is to be deemed halal; product similar to it but from another of that producer's facilities is not necessarily so. On occasion, as does sometimes happen, a company may bring in the same product type from another facility or country because of a supply shortage or because it makes economic sense to do so. A consumer may use this particular brand's product regularly, and, for instance, purchase it while travelling, not noticing that it is missing the certifying agency's mark of acceptance. (Islamic law teaches that an individual is not liable before God for a mistake.) Certification agencies should develop policies for these events, preferably before they materialize. Their agreements should carefully set out the scope of certification and any halal determination. That scope ought to be disclosed to consumers so they can note the potential for a discrepancy that can be better and more transparently assessed. However, the authors strongly recommend that within a single country a particular SKU (stocking unit, a product with a unique universal product code, i.e. the bar code on packages that permit them to be scanned at checkout) be deemed either halal or not halal, including those that might be questionable. A company's branding policy may also help consumers in this regard.

The above discussion deals with plants that make products that are normally 'packaged'. Ingredients and processes for such rarely change and producers usually maintain detailed operations records. But what about the certification of local food production facilities like a bakery or restaurant that do not tend to follow a fairly set routine, but rather are much more reactive and are known to be often changing their practices? These require more intensive inspection. The agency has to decide if continuous inspection is needed or if daily or weekly inspection will suffice. The certifier may require a halal compliance officer be appointed by the company who has ultimate responsibility to ensure the relevant religious standard is continually met. Whatever approach is taken by the certifier, it will ultimately have to perform a risk assessment that an unlawful and/or disliked product may from time to time be produced or that one which it has been deemed lawful could, unbeknownst to the agency, be changed so as to become unlawful. Then it will have to balance these factors with the cost and other difficulties or hardships of regular inspection (for inspectors, the inspected, and consumers, amongst others) as well as objectives of Shariah law. It may also have to consider not certifying such a facility due to an uncontrollable level of uncertainty and risk.

24.7 Packaging

Packaging materials are actually quite complex and the nature of these materials is an issue with which some religious certifying agencies are currently concerned both for kosher and halal products. While many packaging components include ingredients that are part of the package matrix, others components may be migratory and actually come in contact and eventually be eaten with the foodstuff. Some materials are placed on the surface of the packaging specifically for that purpose, i.e. to be able to migrate to the surface of the food. Thus, the presence of substances of usual concern, such as stearates (which might be of animal or plant origin) need to be considered by the certifying body. Halal certification agency scholars will have to understand whether there has been any transfer, of what, and if and to what extent a disliked or prohibited element remains with the final food product, relying on the appropriate technical expertise and working with the food manufacturer in determining the final Islamic legal assessment of the product. Halal certifying agency scholars will of course be mindful of the well-established Islamic legal ethic of not questioning excessively and thereby not creating hardship both upon themselves as well as others.

24.8 Ingredients: Alcohol

For halal food markets, perhaps the most unique non-meat ingredient issue is alcohol, referred to more specifically as ethyl alcohol or ethanol. From the scientific point of view, there are many different types of alcohols, two of which, methanol (wood alcohol) and isopropyl alcohol (rubbing alcohol), are routinely found in the home and in industrial goods, and which are not relevant to our discussion. We limit our discussion to the drinking of alcohol and the eating of things in which it is found. Beverage ethyl alcohol, i.e. wine, beer, and hard cider as well as distilled spirits like whiskey, gin, etc., is prohibited by Islam (for consumption by Muslims) under the category of khamr (intoxicant), but what does it mean for food ingredients to be free of alcohol?

Standards for alcohol in foods amongst halal certifying agencies vary. The largest US halal certifying agency, the Islamic Food and Nutrition Council of America (IFANCA), espouses the view that a standard is needed to articulate and apply classical Islamic ethical-legal views to contemporary foods. Why? With the increased sophistication of analytical instrumentation, it is now possible to find naturally occurring alcohol in many products even in their freshest form, such as many fruits. So it is necessary to allow for some such residual amount of alcohol in foods since such foods, such as the aforementioned fruits, are permitted in Islam as evidenced by Islam's foundational texts, including the Prophetic hadith about the fermentation of date-water. Accordingly, IFANCA has set a level of 0.1% in any finished food product, i.e. the product as purchased by the consumer, which includes any alcohol used as a processing aid, although such alcohol cannot be any beverage/drinking alcohol.

Many flavours and some other important food chemicals are extracted with ethyl alcohol to get the best possible capture of the flavours in the product. When sold to consumers flavour extracts such as vanilla, lemon, and maple must, by US law, comprise at least 35% ethyl alcohol. These extracts are, by the majority of Sunni Muslim scholars, impermissible. A research project done within the US Department of Agriculture (USDA) has shown that if alcohol is added to water, and that water is heated to a boil, the alcohol does not all boil away even after two and a half hours of boiling. Rena Cutrufelli of the USDA prepared a table of alcohol content in cooked foods (Larsen 1995). IFANCA has, therefore, also established a level of 0.5% ethyl alcohol for any ingredients used in food products. That does make it possible to 'dry' products like vanilla flavour and then use them in the finished product. It must be recognized that some of the volatiles captured in the ethanol may be lost during this drying process.

We raise this issue for two reasons. First, so that readers pay attention to the presence of alcohol and appreciate that most Muslim jurists take a thoughtful, nuanced approach to the issue. Second, because this has commercial relevance. Halal certifying agencies, as with kosher ones, do not all apply the same principles; this is important to both food producers and the agencies. Manufacturers should keep in mind whether their food products are able to freely move between facilities operating under different religious certifying bodies.

24.9   The Issue of Multiple Agencies

One of the challenges that the religious supervision agencies have is how to make the system work in the real world. In modern food production the supply chain can be composed of many complicated steps. For example, a particular flavour compound is extracted in a field someplace in Asia. It is shipped via boat to the USA, possibly in bulk. A flavour house in America uses this flavour as part of creating an end-user flavour, sending it to a company that assembles this flavour with other components to provide a pre-formulated mixture that goes to the actual product manufacture. If the product manufacturer has sought to market the product as halal, each of the various steps described above will likely require certification agency review. Given the geographic distance and various parties involved, multiple certification bodies may be needed that will have to coordinate amongst themselves to complete their work.

Such instances demonstrate the importance of communication amongst agencies, but this is certainly not the only reason they should do so. In fact, such instances would be made easier if agencies were already in a consultative relationship with one another. Consultation allows sharing of principles, findings, and experiences to, hopefully, thereby arrive at more informed, correct solutions. Agencies should also have developed a sense of which other religious certification agencies' judgments it is willing to accept.

24.10   Selecting an Agency

In practice certifying agencies seem to breakdown de facto into three categories. The main category is the de facto normative standard, which represents views broadly accepted by a given religious community. But to remain 'normative', as such, the agency needs to be acceptable to their 'competitors' so products can be used between agencies. The second category includes those certifying agencies that are 'stricter', or less allowing if you will, and exert greater practical effort in product supervision. In some cases, these certifying agencies will work with the 'normative' ones, but bring in their own inspectors and may even require continuous inspection in an industrial plant (where continuous inspection is not undertaken by the normative agency). They are limited in their reach; products under their supervision become more expensive than general products in the supermarket that have a supervision mark because of the greater effort required and the generally smaller market volume. Normally, with the exception of meats, products under 'normative' agency supervision are at the same price as competitive products without any supervision. However, with these 'stricter' agencies there is likely to be an upcharge for what they review and certify. The third category of supervision agencies includes those agencies that are more 'lenient', or allowing if you will, than the 'normative' agency. These often supervise companies with major problems in becoming mainstream certified.

They also tend to be focused on retail products like meat, where the consumer can make their own personal choice.

This categorization of certifying agencies is important for the food industry, and particularly for ingredient manufacturers. Any company that wants to sell halal ingredients or products to another halal company needs to be certain that their supervision agency, and thus their halal-certified products, is widely accepted. Otherwise, the market in which they may sell their ingredients will be severely limited. So, as a practical matter, it is generally necessary as a supplier to have at least a 'normative' mainstream religious supervision agency.

One of the developments on the kosher side that is worth noting is the development of the Association of Kashrus Organizations (AKO). This group, representing mostly the normative mainstream and some of the stricter agencies, has grown into a functional trade association in North America (mainly made up of certifying agencies in the USA and Canada but with a few members from beyond). They take on some of the general research projects that are needed to determine what is happening in the food industry and to try to establish guidelines that will help the religious supervision agencies do a better job. But they are careful to still respect differences between religious supervision agencies and to function legally in the USA, where trade associations need to be sure not to overstep areas that can be seen as restraint of trade. A recent book entitled *Kosher: Private Regulation in the Age of Industrial Foods* by Timothy Lytton (2013) describes the history of this organization and how the modern kosher supervision of foods has finally come to a more professional standing and a better balance between competition and cooperation with respect to the kosher supervision of foods in North America.

One further benefit of the many different 'shades' of supervision agencies is their ability to do the research necessary to understand the food industry and to monitor changes. Some of this need in the kosher market is being served by AKO, but another part of this process is that the larger, stronger agencies will assist the smaller, local agencies by sharing intelligence. The larger agencies actually do not necessarily want local production facilities clientele, e.g. bakery, butcher, etc.; instead, they often supervise the local agencies, where being local is much more efficient. In return, the local agencies will often provide the larger agencies with extra trained inspectors in their local area who can be contracted to do inspections. So both sides benefit by these relationships. A key issue in this case is to be sure the local inspector is clear about the standards of the agency for which they are doing the inspection. This is not trivial since some of the local inspectors may be contracted by multiple certification agencies, sometimes within one day.

24.11   The Religious Slaughter of Animals

The religious slaughter of animals is an issue of contention in many parts of the world, although less so in the USA. Before beginning to focus particularly on the religious slaughter of animals, it is important to frame the issue of slaughter in the American legal system generally. As an outgrowth of consumer reaction to the book *The Jungle* written by Upton Sinclair (1906), the US Congress passed various laws to create new approaches to food regulation. At this time most traditional red meat animals (mammals) and birds used for food and liquid eggs are under the jurisdiction of the Food Safety and Inspection Service (FSIS) of the US Department of Agriculture. All other food products are under the jurisdiction of the FDA within the Department of Health and Human Services.

The handling of meat (meaning for our purposes, the species that are under FSIS jurisdiction) in the USA is subject to very intensive inspection (continuous for slaughter, at least once a day for processing) and a great deal more direct control (e.g. label approval) than all other food products. An animal covered by FSIS regulations cannot be moved into the market place without having been inspected. Some states have systems in place for state inspection – until very recently such slaughtered products could only be sold within the state of inspection. Under some recently developed procedures, state-inspected meat can sometimes be sold outside the state by basically demonstrating that the plant's safety programs and the state's inspection is equivalent to that of a directly inspected federal plant. The other exceptions to federal inspection are custom slaughter and the personal exemption – an individual may slaughter an animal he or she owns or purchases prior to slaughter (or the person can have the animal slaughtered at a custom exempt plant) following more limited state specific rules for one's own personal purposes but one cannot sell any of the meat or by-products obtained thereby. The Muslim tradition of keeping a third of a sacrificed animal's meat for one's immediate family, a third for the extended family and community, and a third for charity during the Muslim holiday of Eid al-Adha is thus possible under this rule.

Because of the continuous involvement of FSIS, the practical issues that arise with the religious slaughter of animals and how that process interacts with government inspectors had to be negotiated. All labels are approved by USDA, which means that at some level it is approving the religious slaughter of animals by affixing the mandatory USDA 'wholesomeness' seal on products. As a result, the FSIS does at least try as part of its label review to determine whether the claim of kosher or halal has some broad validity by requiring food companies to provide documentation to substantiate its claims. The FSIS, given US tradition, does not try to judge the 'quality' of the certification provided.

In the late 1950s the US government began considering the humane slaughter of animals, including by religious methods. In 1958 a Humane Slaughter Act was debated and eventually passed and became law. It has two portions that are directly of interest to the religious slaughter of animals. The first section is a 'finding' of the US government that the following methods of slaughter are humane:

7 U.S.C.A. § 1902. Humane methods:

No method of slaughtering or handling in connection with slaughtering shall be deemed to comply with the public policy of the United States unless it is humane. Either of the following two methods of slaughtering and handling are hereby *found to be humane*:

a) in the case of cattle, calves, horses, mules, sheep, swine, and other livestock, all animals [not poultry] are rendered insensible to pain [unconscious] by a single blow or gunshot or an electrical, chemical or other means that is *rapid and effective*, before being shackled, hoisted, thrown, cast, or cut; or

b) by slaughtering *in accordance with* the ritual requirements of the Jewish faith or any other religious faith that prescribes a method of slaughter whereby the animal suffers loss of consciousness by anaemia of the brain caused by the simultaneous and instantaneous severance of the carotid arteries with a sharp instrument and handling in connection with such slaughtering.

CREDIT(S): (Pub. L. 85–765, § 2 Aug. 27, 1958, 72 Stat. 862; Pub. L. 95–445, § 5[a], Oct. 10, 1978, 92 Stat. 1069.)

The second section of the bill provides for an 'exemption' for the religious slaughter of animals. The exemption covers both the pre-slaughter handling of the animal and the actual slaughter.

7 U.S.C.A. § 1906. Exemption of ritual slaughter

Nothing in this chapter shall be construed to prohibit, abridge, or in any way hinder the religious freedom of any person or group. Notwithstanding any other provision of this chapter, in order to protect freedom of religion, *ritual slaughter and the handling or other preparation of livestock for ritual slaughter are exempted from the terms of this chapter.* For the purposes of this section the term 'ritual slaughter' means slaughter in accordance with section 1902(b) of this title.

CREDIT(S): (Pub. L. 85–765, § 6, Aug. 27, 1958, 72 Stat. 864.)

For our purposes, there are two important provisions to discuss. The first is that the two sections seem to contradict each other. Is the religious slaughter of animals humane or is it exempt from such considerations? Historian Roger Horowitz, in his recently published book *Kosher USA: How Coke Became Kosher and Other Tales of Modern Food* (2016) has reviewed the history of this bill and shows that its ambiguity was intentional, i.e. that to get acceptance by various stakeholders these conflicting statements were necessary. Most people when they discuss this bill in terms of the religious slaughter of animals tend to focus on the one of these two approaches that suits their rhetorical need.

The second issue is that of the exemption for pre-slaughter handling of animals in case of religious slaughter. The bulk of the handling of the animal is covered by requirements imposed on all slaughter houses, but the immediate handling prior to the religious slaughter of animals is exempted. It is the first author's contention that this exemption with the benefit of hindsight is unfortunate because progress in improving animal handling for religious slaughter in the USA would probably be much further along without this exemption. As long as the handling even just prior to slaughter does not interfere with the religious slaughter of the animal, there is no reason to exempt it from meeting good animal handling practices.

In recent years animal welfare has become a more important consideration in the animal foods industry. The concern for animals should not be foreign to halal market stakeholders given the strong ethical and legal concern for animal welfare expressed in Islam. It is incumbent on slaughter houses in the USA, including those doing the religious slaughter of animals, to upgrade their activities to meet the dictates of applicable laws. For those holding to a religious persuasion, these religious teachings should be a significant motivation. The issues of animal slaughter have been addressed for many years by the North American Meat Institute (NAMI, formerly the American Meat Institute). Working with the pre-eminent animal welfare scientist Dr Temple Grandin, Professor of Animal Science at Colorado State University and owner of Grandin Livestock Handling, NAMI (2017) has developed a set of animal welfare guidelines. Based on Dr Grandin's work, the key to a good set of guidelines is to have measurable parameters that are routinely monitored. The NAMI guidelines include a section on the religious slaughter of animals. These are found below (courtesy of NAMI, with its permission).

24.12   Religious Slaughter (Kosher and Halal)

Cattle, calves, sheep, or other animals that are ritually slaughtered without prior stunning should be restrained in a comfortable upright position. For both humane and safety reasons, plants should install modern upright restraining equipment whenever possible. Shackling and hoisting, shackling and dragging, trip floor boxes and leg clamping boxes should never be used. In a very limited number of glatt kosher plants in the United States and more commonly in South America and Europe, restrainers that position animals on their backs are used. For information about these systems and evaluating animal welfare, refer to www.grandin.com (Ritual Slaughter Section). The throat cut should be made immediately after the head is restrained (within 10 seconds). Small animals such as sheep and goats can be held manually by a person during ritual slaughter. Plants that conduct ritual slaughter should use the same scoring procedures except for stunning scoring, which should be omitted in plants that conduct ritual slaughter without stunning.

Cattle vocalization percentages should be 5 % or less of the cattle in the crowd pen, lead up chute and restraint device. A slightly higher vocalization percentage is acceptable because the animal must be held longer in the restraint device compared to conventional slaughter. A 5 % or less vocalization score can be reasonably achieved. Scoring criteria for electric prod use and slipping on the floor should be the same as for conventional slaughter.

Animals must be completely insensible before any other slaughter procedure is performed (shackling, hoisting, cutting, etc.) If the animal does not become insensible within 60 seconds, it should be stunned with a captive bolt gun or other apparatus and designated as non-Kosher or non-Halal.

Upright Pen – This device consists of a narrow stall with an opening in the front for the animal's head. After the animal enters the box, it is nudged forward with a pusher gate and a belly lift comes up under the brisket.

The head is restrained by a chin lift that holds it still for the throat cut.

Vertical travel of the belly lift should be restricted to 28 in. (71.1 cm) so that it does not lift the animal off the floor. The rear pusher gate should be equipped with either a separate pressure regulator or special pilot-operated check valves to allow the operator to control the amount of pressure exerted on the animal. Pilot operated check valves enable the operator to stop the air cylinders that control the apparatus at mid-stroke positions.

The pen should be operated from the rear towards the front.

Head restraint is the last step. The operator should avoid sudden jerking of the controls. Many cattle will stand still if the box is slowly closed up around them and less pressure will be required to hold them.

Ritual slaughter should be performed immediately after the head is restrained (**within 10 seconds of restraint**).

An ASPCA pen can be easily installed in one weekend with minimum disruption of plant operations. It has a maximum capacity of 100 cattle per hour and it works best at 75 head per hour **or less**. A small version of this pen could be easily built for calf plants.

Conveyor Restrainer Systems – Either V restrainer or center track restrainer systems can be used for holding cattle, sheep, or calves in an upright position during Shehita or Halal slaughter (Figures 24.1 and 24.2). The restrainer is stopped for each animal and a

Figure 24.1 Restrainer system for religious slaughter of calves and sheep: upright pen.

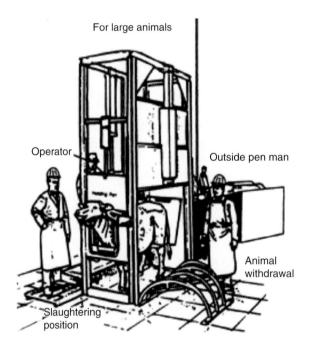

Figure 24.2 Upright pen for religious slaughter of cattle.

head holder positions the head for the ritual slaughter official. For cattle, a head holder similar to the front of the ASPCA pen can be used on the center track conveyor restrainer. A bi-parting chin lift is attached to two horizontal sliding doors.

   Small Restrainer Systems – For small locker plants that ritually slaughter a few calves or sheep per week, an inexpensive rack constructed from pipe can be used to hold the animal in a manner similar to the center track restrainer (Figure 24.3). Animals must be allowed to bleed out and become completely insensible before any other slaughter procedure is performed (shackling, hoisting, cutting, etc.).

With the increased interest in animal welfare, these NAMI guidelines are being required by more and more end users, both in the fast food/restaurant industry and by supermarkets and other retail outlets. FSIS has also begun to take animal welfare more seriously, employing animal welfare veterinary inspectors and enforcing higher animal welfare standards. This has meant that the larger slaughter houses are meeting these higher thresholds. The generally smaller religious slaughter houses have a choice: To only sell to only the religious communities or to fulfil these guidelines and potentially sell to larger markets. In practice this has meant that larger animals are being slaughtered religiously in some sort of restraining pen. In the USA at this time it is the authors' understanding that all large animals such as cattle and bison are handled with some sort of restraint system. Small animals can also be handled by high speed upright handling systems such as those developed by Dr Grandin. Many sheep facilities around the world are now using the 'V-restrainer' developed by Dr Grandin. However, the use of shackle and hoisting (i.e. the handling of live sheep and goats by chaining one leg to hoist it off the ground) is still done in the USA. This is one area where acceptance of proper equipment and making the required investment would improve animal welfare and would demonstrate the commitment of the religious communities to animal welfare by combining religious principles with contemporary methods and technologies. It should be noted that the first author is involved in improving the religious slaughter of animals, focusing at present on the use of a razor sharp knives of the appropriate length (twice the width of the neck with a straight blade) free of nicks. It is also worth noting that regulations and mechanisms which serve the achievement of both the particulars and general objectives of the Shari'ah are themselves to be upheld by Islamic thought.

One area where a difference of opinion amongst Muslim jurists persists relates to the lawfulness of animals being made unconscious prior to slaughter   whether this slaughter is the traditional horizontal cut across the neck, known in Arabic as zabihah, severing the oesophagus, trachea, and two major blood vessels on both sides of the neck, or the method involving a stab at the base of the neck and then a horizontal cut that is ideally used for camels and allowed for cattle, known in Arabic as nahr, cutting the foregoing passages. The issue – for those Muslim jurists who might consider such to be permitted – is how to be sufficiently certain that all of the animals rendered unconscious are still alive at the time of slaughter and that blood loss following slaughter is not impeded. In the USA both a traditional slaughter and pre-slaughter intervention are used for halal meat but a further possible option of a post-cut intervention, where the animal being alive at the point of slaughter is guaranteed, has not to our knowledge been sufficiently explored in Muslim jurisprudence. Although there have been surveys in other countries, there are no surveys of American Muslims indicating whether and to what extent they accept the pre-slaughter intervention or simply assume that the traditional method is being used for meat and poultry labelled as halal. It would be helpful if a framework of labelling halal meat were developed by halal certifying agencies to indicate whether animals had been subject to a pre-slaughter intervention so Muslim consumers may make an informed decision.

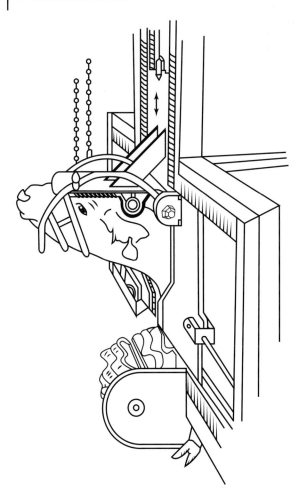

Figure 24.3 Centre track restrainer being used for ritual slaughter.

24.13   The Commercial Side of Halal Foods

Halal certification agencies in the USA are of two types. In recent years, the development of broader-based halal certification agencies has occurred. Many others are mainly focused on certifying meats that are intended for export and therefore must meet the needs of the receiving country. The largest and most dominant agency in the USA is IFANCA. However, the Islamic Services of America in Des Moines, IA, and a few other certifying agencies are also involved in certification of halal products for the domestic market.

   At the retail, finished product level, many more products are now appearing with halal certification. Some of these products are dual certified, i.e. have both kosher and halal certification. With respect to meat products, this has meant at times that kosher meat is being used for halal production in keeping with the concept that the slaughter of the 'people of the book' is lawful for Muslim consumption. The most interesting area for halal in the USA has been in the meats area. Initially, butchers provided uncooked halal meat; later, the thrust was to provide halal versions of traditional US meat products, such as hamburgers, hot dogs, processed luncheon meats, 'southern fried' chicken, and chicken nuggets. Some of these products are available for fresh distribution and others for frozen distribution to mimic the non-halal market. Over time a few companies branched out to make more traditional dishes, based on recipes from Muslim communities, generally frozen to permit national distribution, that were made with halal meat and poultry. Some are also trying to reach non-Muslim consumers. Another exciting area of activity has been the work of producing gelatin from halal sources to avoid the use of questionable gelatin from sources such as non-halal slaughtered beef or pork. Not all Muslim scholars share the same views regarding gelatin, it is interesting to note.

The future of halal in the USA is very positive. Already the USA is a major exporter of halal meats to the Muslim world, and as Muslim populations in America increase and further integrate, their potential will be further appreciated. As the halal industry matures and halal is itself better understood, more and more businesses will seek to create opportunities to serve this global population.

References

Furber, M. (2013). Elements of a fatwa and their contribution to confidence in its validity. *Tabah Analytic Brief* 14: 5.

Grandin, T. (2017). *Recommended Animal Handling Guidelines & Audit Guide: A Systematic Approach to Animal Welfare*. Washington, DC: North American Meat Institute.

Horowitz, R. (2016). *Kosher USA: How Coke Became Kosher and Other Tales of Modern Food*. New York: Columbia University Press.

Larsen J. 1995. *Ask the Dietitian*. Hopkins, MN: Hopkins Technology, LLC. http://www.dietitian.com/alcohol.html (accessed 18 May 2015).

Lytton, T.D. (2013). *Kosher: Private Regulation in the Age of Industrial Foods*. Cambridge, MA: Harvard University Press.

25

Halal Food in Italy

Beniamino Cenci Goga

University of Perugia, Perugia, Italy

25.1 Conventional and Religious Slaughter: Animal Protection

Nowadays, in European society, animal welfare and protection are undisputed values, widespread and shared amongst citizens and, moreover, subject to regulation. The general criterion is to give a double value to the human/animal relationship. A live animal is considered a sentient being and, like a human, deserves protection from mistreatment. At the same time it is seen as being different from a human, therefore killing it is considered possible and justified by the profit obtained. In short, a human being's needs are prevalent. Therefore, the reason for the slaughtering of animals is that they are a source of nourishment and sustenance in general for people. There are also other circumstances that make animal killing possible, such as the necessity to avoid the spread of infectious disease, management issues of overcrowding in some kinds of livestock farms, and the control of animal populations, whether synanthropic, wild or domestic. The essential condition, for the sake of common sensitivity, is to reasonably avoid any possible cause of anxiety, pain, and suffering. Although it could seem a contradiction, this condition is required even though the animal is still being slaughtered or killed. The number of laws on animal protection is growing in parallel with people's increased interest in these issues. As specifically regards protection during slaughter, the continuous changes in the laws on this issue follow different directions in every European country. In Italy regulations were introduced with the first laws on meat hygiene, and they were amended in order to incorporate European Community guidelines in them. Even amongst lawmakers there is a growing interest in animals considered as sentient beings. In 1974 the European Community introduced measures for protection during slaughter, with the obligation of pre-stunning. The usable methods and elements of protection during religious slaughter, already allowed in some Member States, were explained within this context. Unfortunately, there have been few pronouncements about other procedures that regulation now puts under the general definition of 'related operations', i.e. operations such as handling, lairaging, restraining, stunning, and

The Halal Food Handbook, First Edition. Edited by Yunes Ramadan Al-Teinaz, Stuart Spear, and Ibrahim H. A. Abd El-Rahim.
© 2020 John Wiley & Sons Ltd. Published 2020 by John Wiley & Sons Ltd.

bleeding of animals taking place in the context and at the location where they are to be killed. The only procedure they considered at that time was animal restraint, which, 'if [...] it is necessary, it is supposed to be done just before stunning'. In 1979 the European Convention took into account, as elements for protection, the side regarding '[...] the movement, lairaging, restraint, stunning and slaughter of domestic solipeds, ruminants, pigs, rabbits and poultry' (Anonymous 1979). Although there were no detailed rules, it was acknowledged that animal welfare conditions are guaranteed when the entire approach is adequately considered for any type of livestock. Moreover, the concept of the need to avoid any unnecessary suffering to animals, both during their life at the farm and during slaughter, was reasserted. Furthermore, evaluation criteria were established for staff competence when approaching livestock from admission of animal to the abattoir to slaughter. The Member States where religious slaughter was allowed were required by the competent authority (the central authority of a Member State competent to carry out veterinary checks or any authority to which it has delegated that competence) to verify that the slaughterers were recognized by religious authorities, or to verify directly their qualifications and therefore issue the qualification title. In 1993 the European law concerning protection during slaughter was broadened, with detailed prescriptions regarding all aspects and all stages, from the entering of the slaughterhouse to the slaughter (Anonymous 1993). The possibility of continuing with religious slaughter was maintained in those Member States where it was not forbidden. In 1997, the European Union (EU) reasserted its own interest in animal welfare: 'In formulating and implementing the Community's agriculture, transport, internal market and research policies, the Community and the Member States shall pay full regard to the welfare requirements of animals, while respecting the legislative or administrative provisions and customs of the Member States relating in particular to religious rites, cultural traditions and regional heritage...' (Anonymous 1997). Protection becomes, in this context, a group of actions directed towards avoiding, as far as possible, suffering for the animals. In this perspective, the obligation of pre-stunning is upheld in the law and kept up to date according to the most recent scientific findings. The only dispensation provided allows the exclusion of this practice during religious slaughter. The main religions with dietary and ritual rules connected with animal sacrifice are Islam and Judaism. In these religions, ritual slaughter is integrated with additional more or less detailed food hygiene regulations and, in all likelihood, is aimed at preserving health for Muslims and Jews. According to these precepts, only flesh obtained from animals killed intact and fully conscious is permissible for the people of these religions. For this reason, religious slaughter highlights the need to adopt the greatest degree of animal wardship and protection in view of the guarantee of freedom of worship. A cue for reflection is offered by the sacredness that the ceremony gives to animal life. Briefly, the inclusion of slaughter in a religious prescription can be interpreted as a way of reminding humans that the act of killing cannot be done without concern. According to some authors, the slaughter prescriptions found in the Torah and in the Koran represent the first sources of rules about animal protection during slaughter and hygienic meat production. Both shechita and dhabiha provide for a clean neck cut through the trachea, oesophagus, carotids, and jugulars without detachment of the head. In this way the bleeding is complete, as prescribed, and according to people who practice this technique it makes it possible to avoid suffering and pain (Gibson et al. 2009b).

In the precepts about slaughter, in both slaughter techniques it is expected that knife blades be carefully sharpened. This way the sacrificial cut is clean, complete, deep, and rapid. The latest updating of the law still provides for the respecting of the religious precept, making it possible, through dispensation, to avoid stunning methods before the cutting of the soft neck tissues. There is still the obligation to prevent any avoidable suffering, lesions, anxiety, and distress in the animals. This is obtained in the current procedure by adopting systems and methods of animal care and handling based on the study of their behaviour and reaction, considering the breeding methods used during in their life. These systems and methods are identical in both conventional and religious slaughter. The difference between them is the method of restraint for slaughter and the device used for this aim. Therefore, for *related operations*, the control and monitoring instruments used by operators, as well as the methods for the evaluation of official control compliance, are the same for both slaughter systems.

Here we should quote Temple Grandin (1996), who says about religious slaughter:

'I will never forget having nightmares after visiting the now defunct Spencer Foods plant in Spencer, Iowa fifteen years ago. Employees wearing football helmets attached a nose tong to the nose of a writhing beast suspended by a chain wrapped around one back leg. Each terrified animal was forced with an electric prod to run into a small stall which had a slick floor on a forty-five degree angle. This caused the animal to slip and fall so that workers could attach the chain to its rear leg [in order to raise it into the air]. As I watched this nightmare, I thought, "This should not be happening in a civilized society." In my diary I wrote, "If hell exists, I am in it." I vowed that I would replace the plant from hell with a kinder and gentler system'.

The first common step for food business operators (FBO) and for the competent authority (CA) for official control is to define the aims established in the regulations. The FBO will plan their self-control programme in order to achieve these aims. The CA will provide an objective and uniform evaluation system regarding the adequacy of the measures described and practiced by the FBO. Operators should be aware of the real meaning of *self control*. It should be recalled that this term means the *processing of procedures and control plans* carried out by the FBO. The revision of the food law, the steps and fundamental goals of which have been set out in the White Book on Food Safety, brought about a reassessment of control systems. In this new view, the responsibility to guarantee food product safety is assigned to the FBO. The official control must ensure, by the application of control methods and techniques, that all FBOs carry out correctly what is required by food law. The aim is to allow every operator to study and adjust to his own particular activity the necessary number of procedures in order to guarantee the achievement of the goals established by law. The FBO must carry out the method that is established by this definition through the *hazard analysis and critical control points* (HACCP) system. This system must be applied by any FBO except primary producers. For them, it is still compulsory to adopt the recording system, already provided by other laws, and the possibility of adopting and applying in the practice the so-called 'manual of correct praxis' provided by the category's associations.

25.2 Restraining Animals

Any manual operations to be carried out on the animal to be slaughtered require as an inexorable condition that it be restrained. Today this is done in special cages in which the animal is restrained for the purpose of stunning or, more rarely, ritual cutting. The cages, also called restraints or boxes, traps, crates, restrainers, or pens, are defined by the operation that will be carried out in them. The law does not contain provisions regarding the structural characteristics of these devices, other than the general requirements for restraint as a 'related operation'. The requirements for which compliance must be assessed and demonstrated are therefore those of performance, formulated for the achieving of protection objectives. It is up to the manufacturers of these devices to provide the measures necessary for ensuring the best animal welfare conditions and to provide users with all instructions for proper use and proper maintenance. Animal restraint that is properly done makes it possible to minimize the disturbing elements that at this point of the path towards the slaughter area are often difficult to eliminate. The success of stunning is largely influenced by the ability to properly apply the chosen instrument for this purpose; likewise, the ritual cut will be made in an appropriate manner if the animal's position exposes the cutting area in the best way. This presupposes that the animal is calm when it arrives in the restrainer, a condition that is helped by the elimination of 'disturbing elements'. At this point in the procedure, special consideration must be given to abrupt changes in lighting, noise caused by the operations being carried out and by the movement of the cage closing systems, and the presence of the operator. Of no less importance are the place and the way in which the floor of the corridor through which the animals are led joins with the floor of the cage. As already indicated, the unevenness of the floor is a major cause of animal hesitation, starting or baulking. When this crucial stretch of the route is optimized, the animal held in the restraint must be 'guided' into assuming almost spontaneously the correct head position, which is a necessary prerequisite for the success of the subsequent operations. For poultry, the restraint method used depends on the type of slaughter and the stunning method chosen. In the opinion which the European Commission requested from the Scientific Committee on Animal Health and Welfare (SCAHAW), under the authority of the European Food Safety Authority (EFSA), the need to develop restraint systems suitable for the application of both mechanical and electrical stunning methods is indicated as a high-priority research objective.

25.2.1 Restraint in Conventional Slaughter

The cages used in conventional slaughter are intended to hold the animal still for the time necessary for the correct application of the stunning method. The technical construction specifications may be different depending on the production capacity of the slaughterhouse. Typically, the cage has four compact partitions, alongside which is a raised platform on which the operator stands. The manual operations necessary for stunning are carried out through the opening in the upper part. The entrance has a guillotine gate, operated by pneumatic or hydraulic lifting mechanisms. It is important that the opening and closing of the gate be timed so as not to obstruct the entry of the animals and to avoid hitting the back or limbs still not well placed inside the cage. After stunning, the animal is released to the outside through the opening of one of the side partitions. If the cage is equipped with systems that facilitate the correct positioning of the animal's head, the application of the stunning device will be more effective. In general, those systems that do this passively by the way they are designed are preferable, rather than the use of active restraint devices. In any case, the presence of such devices must not disturb the animals, nor prolong the time they stay in the cage, nor hinder the removal of the animal after stunning. For poultry, the restraint system depends on the stunning method chosen. Generally, at slaughterhouses where a water bath is used for the application of an electric current, the birds are hung from metal supports before stunning. This system has been judged to be highly painful and stressful for the animal. Holding the animal upside down causes anxiety and distress because the position is unnatural. Added to this is the pain caused by compression of the leg shackle on the periosteum of the metatarsus and the activation of nociceptors that are found in the dermis of the limbs. Another source of suffering for the animal is the compression that the abdominal organs exert on the heart due to the absence of the diaphragm muscle. Some methodological characteristics may also bring about attempts to escape, expressed with beating of the wings. Sudden and loud noises, passing into brightly lit areas, and sharp bends in the path are factors that could frighten the animal, which will try to break free and, wriggling, feel greater pain from compression. A system to make this method less stressful and painful consists of providing support at the breast by means of a plastic or rubber installation that follows the entire path of the shackle line with the leg shackles (Gentle and Tilston 2000). If the stunning method used calls for the use of gaseous mixtures, the animals can remain in the containers used for the journey, or they can be sent individually towards a conveyor and transport system. These conditions are considered more respectful of animal welfare.

25.2.2 Restraint in Religious Slaughter

In religious slaughter the animal must be killed by cutting the large vessels of the neck, without the animal having been previously stunned. This requires the use of restraint systems that do not cause suffering for the animal and do not expose workers to occupational hazards. In some slaughter facilities, rotating cages are used. Once the animal is restrained, these devices rotate entirely to hold the animal with its back on the ground. Losing its quadrupedal stance is a condition that triggers reactions of fear and anxiety in the animal, thus it follows that rotating cages should not be considered adequate for the purposes of guaranteeing protection for the animals. In order for religious slaughter to be done correctly, the restraint cage should allow the animals to remain in a quadrupedal stance, and it must also be equipped with devices to hold and raise the head upwards until the moment when the cut is made. To ensure that this happens with greater simplicity and suitability, the rear of the cage must be equipped with a system that makes it possible to block and prod the animal towards the front part equipped with a head restraint (Grandin 2010). Birds can be restrained by hanging from the legs or by putting them into a truncated cone, the funnel for bleeding, that holds the bird up to the greater part of its breast and lets the head stick out to expose the neck for the ritual cut. The animals may also be restrained manually for the cut.

25.2.3 Correct Procedure

In both conventional slaughter and religious slaughter, great care must be taken to ensure that animals are not subjected to the distress of being abruptly, persistently, and violently forced to enter the cage. This special equipment must always be made so as not to cause injuries, wounds or bruises and not to cause anxiety in the animal that must enter it. It is for this reason that the floor must be well connected to that of the corridor. The size should be appropriate to the animals for which it is used. This ensures that the animals do not have room to move too far forward, sideways or backwards in order to avoid the manual operations or to cause injury or hinder the correct execution of the operations. The guillotine gate that closes the rear should not be closed until the animal is correctly positioned in the cage, restraining it with both hind limbs. Special attention must be given to the choice of the partition movement mechanisms and other equipment. Preference should be given to those that allow for smooth, silent movement. No less important than the aspects regarding the design and construction of the cage is how the device is managed. Agitation, anxiety, and suffering can be reduced considerably if the restraint in the cage is limited to a few seconds, i.e. the time that is necessary to proceed with stunning or the cutting of the neck. Access to the cage should therefore be allowed only when both the people and the equipment are ready to carry out the stunning and cutting operations. In the event of interruptions in the slaughter line, the structure must be equipped so that any animals restrained can exit safely and quickly. For sheep, goats, pigs, and calves, automatic advancement and restraint systems, such as double rail and V restrainers, have proven to be very useful in improving the handling of animals and the accuracy of stunning.

25.3 Animal Welfare for Farm Animals

Before attempting to see whether any guidelines can be suggested as to when it is ethically acceptable to slaughter animals while still conscious, we need to turn the question of how breeding and transportation may alter the essence of an animal. The absence of any specific legal provisions for animal welfare in the context of EU policies has been criticized in the past, in particular in the context of the high-profile campaigns concerning the transport of animals and the role of the EC legislation in this regard, and although the EC Treaty (the so-called Treaty of Rome) was not actually amended to include such provision, a somewhat ambivalent Protocol has been annexed to it. This states that the EC and the Member States 'shall pay full regard to the welfare recruitments of animals', while apparently also respecting the law and customs of Member States 'relating to particular religious rites, cultural traditions and regional heritage'. The Treaty of Rome, which established the European Economic Community (EEC) and the European Atomic Energy Community (EAEC), was signed by France, Germany, Italy, Belgium, the Netherlands and Luxembourg on 25 March 1957. While still remaining separate entities the EEC and EACE were merged during the 1960s into one organizational structure, called the European Community, which was the precursor of today's EU. Animal welfare is concerned with the humane treatment of animals under our care. It is the conviction that all animals should be treated in such a way that they do not suffer unnecessarily, while animal rights is the viewpoint that animals have rights and are worthy of ethical consideration in how humans interact with them. Examples of animal welfare positions are the EU Directives on transport (Anonymous 1991, 1995, 2005). Animal rights is the concept that sentient animals, because they are capable of valuing their own life, should be entitled to possess their own flesh, and therefore are deserving of rights to protect their autonomy. The animal rights view rejects the concept that animals are commodities or property that exist to serve humans. Many animal rights and animal welfare advocates make a clear distinction between the two philosophies. Animal rights advocates argue that the animal welfare position (advocating for the betterment of the condition of animals, but without abolishing animal use) is logically inconsistent and ethically unacceptable. Most animal welfarists argue that the animal rights view is extreme, and they do not advocate the elimination of all animal use, rather they advocate a more humane and compassioned use of animals for food or companionship. However, there are also some animal rights groups which support animal welfare measures in the short term to alleviate animal suffering until all animal use is ended. Ethicist David Sztybel distinguishes six different types of animal welfare views (Sztybel 2004): (i) animal exploiters' animal welfare: the reassurance from those who use animals that they already treat animals well, (ii) common-sense animal welfare: the average person's concern to avoid cruelty and be kind to animals, (iii) humane animal welfare: a more principled opposition to cruelty to animals, which does not reject most animal-using practices (except perhaps the use of animals for fur and sport), (iv) animal liberationist animal welfare: a philosophy championed by Peter Singer (Singer 2004; Singer and Mason 2006), which strives to minimize suffering but accepts some animal use for the greater good, such as the use of animals in some medical research, (v) new welfarism: a term coined by Gary Francione (2018) to refer to the belief that measures to improve the lot of animals used by humans will lead to the abolition of animal use, and (vi) animal welfare/animal rights views which do not distinguish the two. Animal welfare principles are codified by positive law in many nations, but animal rights are recognized in none. The concept of 'animal rights' is often confused with animal welfare, which is the philosophy that takes animal suffering into account. Animal welfarists range in their degree from organizations that profit from the exploitation of animals whilst claiming to be concerned with their welfare, to radical animal welfarists such as Peter Singer. While Singer is often associated with the animal rights movement, and many animal rightists agree with much of Singer's work, his philosophy does not rely on a concept of rights. Rather, Singer's radical animal welfarist philosophy is based on ethical utilitarianism that takes the capacity of sentient animals to suffer into account, and thus promotes the concept of animal liberation and veganism as logical outcomes of this consideration (Francione 2018). The animal rights philosophy does not necessarily maintain that human and non-human animals have equal moral standing. However, animal rightists do believe that because animals are capable of valuing their own life, regardless of whether humans have use for animals or not, then they should be afforded the right to live and to decide upon their life. This means that, according to a rights view, any human or human institution that commoditizes animals for food, entertainment, clothing, scientific testing, or any other purpose infringes upon the rights of the animal to possess its own being, and thus the property status of animals, which is used to maintain the use of animals for human ends, is unethical because it ignores the rights of animals.

25.4 Do Animal Have Rights?

It is often held by philosophers that the notion of 'rights', whether or not applied to animals, is problematic. In fact the word is used in different contexts and it has been generally linked with the notion of legal rights. Moreover, rights are commonly taken to be correlative with duties while many philosophers follow Kant in rejecting the notion of duties to animals. We can by-pass the question and focus on the extent to which our use of animals is often 'speciesist' (Reiss and Straughan 1996). It was the Australian philosopher Peter Singer who, in 1975, produced the first really sustained argument that most humans are guilty of 'speciesism'. This passage, taken from a book by Reiss and Straughan (1996), is enlightening:

> '... humans belong to a different biological species from, say, chimpanzees, dogs, farm animals and laboratory mice; we do not have the right to treat such species merely as we choose and for our own ends. Think of the conditions we normally require before humans are permitted to be used as research subjects. We require that two conditions be met: first, that the participating individual gives their informed consent; secondly, that there is no intent to do harm to that individual. The second of these conditions is inviolate. The first can only be overturned when patients are unable to give their consent, for instance because they are babies or in a coma, when it can be given on their behalf. Further, most of us are not, generally, persuaded by the utilitarian argument that these conditions can be overturned if a number of other people would benefit. This, of course, is why most people hold that, in everyday language, humans have rights. For example, nowadays few of us would be persuaded that subjecting even a few people to slavery is acceptable whatever the beneficial consequences that might result for the rest of us', or: '... the problem is, why do most people hold that it is not permissible to subjugate people into slavery or to experiment on them without their consent when we regularly do these things to nonhumans, including even our closest evolutionary relatives, namely chimpanzees and other mammals? We can also note that if one adopts solely the criterion of suffering to decide whether or not an organism should be used for human ends, not only would the use of most animals for medical research cease, but a case could be made, abhorrent as it sounds, for mentally handicapped new-born human infants to be used for such research, on the grounds that such infants are arguably closer to self-conscious sentient people than are the laboratory animals presently used' (Anonymous 1993; Reiss and Straughan 1996).

Several authors have raised the concept that '... being of the human species may be a sufficient condition of being awarded enhanced moral status. Possessing the nature of a rational self-conscious creature may be sufficient for being awarded this status even though this nature may be impaired or underdeveloped in the individual case' (Grayson 2000). There is still considerable controversy both over whether animals have rights and whether animals are qualitatively different from humans in some way that allows us to treat them differently.

It has been argued that, from a technical point of view and from an animal welfare perspective, an acceptable level of animal welfare during religious slaughter can be maintained if scores at the critical control points for slaughtering, animal insensibility, slipping and falling, vocalization and electric prod are in the acceptable range (Grandin and Regenstein 1994). Scoring performance on these variables is simple and easy to do under commercial plant conditions. In conclusion, managers must be committed to good animal welfare. Plants which have managers who insist on good handling and stunning practices, have managers that insist that employees handle and stun animals correctly. You manage the things that you measure. To maintain good handling and stunning practices requires continuous measurement and monitoring and management.

25.5 Religious Slaughter in Italy

Italy has recently become the destination of a massive immigration by people from less developed areas of the world with different, social, religious, and cultural backgrounds. These populations have brought with them different ideas and lifestyles to which they have had to make some considerable and sometimes controversial adjustments in order to even legally integrate them with local national customs. Italy is now, in many aspects, a multi-ethnic nation: the Caritas report of 2013 estimated that over 8% of the population living in our country consists of immigrants, the majority of whom are Muslims (Anonymous 1991). Approximately 2.2% of the current population in Italy is, in fact, Muslim, whereas there are fewer people of other faiths: Buddhists make up 0.17, Hindus 0.18, and Jews 0.06% of the population, respectively (Caritas/Migrantes 2018). These numbers are continually rising and the influence of these minorities is, therefore, gradually on the increase, as new commercial markets open up.

One of the more complicated issues we face in order to facilitate the integration of these populations is that of ritual slaughter, a practice which sees the freedom of faith clashing with the protection of animal welfare. The terms 'religious' or 'ritual' slaughter indicate the practice of slaughtering animals according to the precepts of Islamic and Jewish religions. In particular, both religions require animals to be alive at the time of killing. However, all Jews and some Muslims refuse any kind of stunning: only in this way can the meat be considered 'halal' (for Muslims) or 'kosher' (for Jews) and fit for consumption. The Council Regulation (EC) No. 1099/2009, in force since 1 January 2013, regulates the protection of animals during slaughter. The killing of animals is defined by the regulation as 'any intentional process that causes the death of animals' and must 'save animals from avoidable pain, distress or suffering'. The same regulation also specifies that 'animals are slaughtered only after stunning and that loss of consciousness and sensibility shall be maintained until the death of the animal' and it adds that, for the slaughter of cattle, the 'inversion or any unnatural position of the animal, such as an immobilization system, is forbidden'. The same articles of this regulation, however, allow some derogations both for stunning and the systems of immobilization during religious slaughter, provided it is carried out in approved slaughterhouses. This anomaly in the current legal situation is also contributed to by the lack of a central, religious figure which can provide unambiguous guidelines on this question, especially for the Muslim communities. Jews or Muslims are, in fact, organized into fairly numerous groups, each with a local, religious authority represented by the Imam (for Muslims) or the Rabbi (for Jews). Each religious authority may, therefore, give his community different instructions, based on his own interpretation of the sacred rules, his fundamentalism and his personal opinion.

The aim of the following chapter is to provide an overview of the methods currently used in Italy during religious slaughter, paying particular attention to the types of restraint methods and to the use of stunning, in order to provide concrete answers to the issues of animal welfare.

25.5.1 Data from the European Project Dialrel (www.dialrel.eu)

The Italian coordination of the Dialrel project made a request to the Ministry of Health to receive the complete list of Italian slaughterhouses authorized to perform ritual slaughter. for a complete list of Italian slaughterhouses authorized to perform ritual slaughter from the Ministry of Health. A questionnaire was then sent to the veterinary services of all 136 abattoirs in order to obtain preliminary information regarding the type of slaughter (halal or kosher), the number of animals slaughtered annually, the restraint systems used and the stunning method applied, if applicable.

We performed a simple, random sampling in five plants selected amongst the 136 slaughterhouses, using the formula $x = Z(c/100)^2 r(100r)$, where r is the fraction of response and $Z(c/100)$ is the critical level value for the confidence level c, with a 20% margin of error and a 95% confidence level, to obtain a statistically significant sample of Italian slaughterhouses (Bruce et al. 2008). A team consisting of at least two auditors visited these five slaughterhouses over the following six months. In each plant, a simple, random selection to sample a reasonable number of animals was made and analysis carried out on a total of 313 animals. A two-part questionnaire was completed for each animal. The first part collected information regarding the type of slaughter (halal and kosher), the restraint system, the stunning method applied, if applicable, and the general, pre-, and post-sticking management of the animal. The quality of bleeding and the structures severed during cutting (carotid, jugular, trachea, and oesophagus) were determined. The second part of the questionnaire was focused on the evaluation of the signs of pain and stress shown by the animals during restraint, cutting, and pre-agonic phases (before death). To standardize the information obtained, we designed three guidelines, one for each class of stock slaughtered: small ruminants, cattle, and poultry. We selected and sampled two slaughterhouses for cattle, two for small ruminants, and one for poultry from amongst the 136 plants. The information in Table 25.1, concerning the slaughterhouses under examination, is divided according to the species, and the methods of slaughter and restraint.

Table 25.1 Data concerning the five slaughterhouses under examination

Species	Method	Restraint	No. of animals observed	No. of visits to the slaughterhouse
Bovine	Halal	Upright	30	5
Bovine	Halal	Mechanically turned on its side	14	2
Ovine[a]	Halal	Suspended before sticking	79	1
Ovine[a]	Kosher	Suspended before sticking	114	1
Ovine	Halal	Manually turned on its side	6	1
Poultry	Kosher	Suspended before sticking	70	1

[a] At the same slaughterhouse.

We sent a preliminary questionnaire to all 136 slaughterhouses authorized for ritual slaughter in Italy. Only 29 questionnaires (18% of the total) were returned: 25 for halal slaughter and three for kosher slaughter; one questionnaire was received without data for reasons of privacy. No information is available for 107 slaughterhouses (82% of the total) and therefore the data obtained may not be representative of the situation in Italy. As regards halal practice, nine questionnaires were analysed for cattle, 12 for small ruminants, and four for poultry. As regards kosher practice, one questionnaire was analysed for each category (cattle, small ruminants, and poultry). Data obtained from the 29 questionnaires concerning the procedure are shown in Table 25.2 and for the restraint system in Table 25.3. No abattoir authorized to perform religious halal or kosher slaughter used pre- or post-cut stunning, with the exception of one for the halal slaughtering of small ruminants: 100% of halal or kosher bovine, 100% of halal or kosher poultry, and 94.1% of halal or kosher small ruminants were slaughtered without stunning. Just 5.9% of halal-slaughtered small ruminants were killed by head-only electrical stunning.

Of the 136 Italian abattoirs authorized for ritual slaughter, we selected five to perform on-site visits during which a total of 313 animals (44 bovines, 70 chickens, and 199 small ruminants) were examined. A two-part questionnaire was completed for each individual animal: the first part contained generic data concerning the slaughterhouse (Tables 25.4 and 25.5) and the second part contained information on each individual animal slaughtered (Tables 25.4–25.12) (Figures 25.1 and 25.2). All the results shown are classified according to the animal species slaughtered and the procedure performed.

Table 25.2 Type of slaughter

Species	Conventional slaughter[a]	Halal slaughter	Kosher slaughter
Bovine	95.30%	4.27%	0.43%
Small ruminants	90.37%	5.47%	4.16%
Poultry	98.69%	1.31%	0.00%

[a] Stunning before sticking.

Table 25.3 Restraint methods

Species	Procedure	Upright	Turned on their side	Turned on their back	Shackled before sticking
Bovine	Halal	100.00%	0.00%	0.00%	0.00%
	Kosher	100.00%	0.00%	0.00%	0.00%
Small ruminants	No stun halal	0.00%	69.07%	30.30%	0.00%
	Pre-cut stun halal	0.00%	100.00%	0.00%	0.00%
	Kosher	0.00%	0.00%	0.00%	100.00%
Poultry	Halal	0.00%	0.00%	0.00%	100.00%
	Kosher	0.00%	0.00%	0.00%	100.00%

Table 25.4 Speed of halal or kosher slaughter line

	Halal			Kosher	
	Upright or mechanically turned bovine	Manually turned sheep/goats	Suspended sheep/goats	Suspended sheep/goats	Suspended poultry
Speed	13 animals/hour	20 animals/hour	162.3 animals/hour	234.4 animals/hour	792 animals/hour

Table 25.5 Average time required to turn the animals 90°

	Mechanically turned bovine	Mechanically turned sheep/goats
Time (s)	10.3 ± 1.4	5.5 ± 2.9

Table 25.6 Average time from beginning of restraint to sticking

Procedure	Species	Restraint	Time (s)
Halal	Bovine	Upright	97.5 ± 56.6
	Bovine	Mechanically turned	115.8 ± 86.8
	Small ruminants	Manually turned	21.3 ± 12.1
	Small ruminants	Suspended	57.2 ± 19.1
Kosher	Small ruminants	Suspended	228.8 ± 59
	Poultry	Suspended	26.2 ± 11.6

Table 25.7 Average time from sticking to the next handling

Procedure	Species	Restraint	Time (s)
Halal	Bovine	Upright	93.3 ± 23.9
	Bovine	Mechanically turned	114.1 ± 21.6
	Small ruminants	Manually turned	105.8 ± 59.6
	Small ruminants	Suspended	379.3 ± 47.2
Kosher	Small ruminants	Suspended	677.3 ± 176.3
	Poultry	Suspended	136.7 ± 13.1

The results obtained in this study provide only a general picture of the current situation concerning shechita and halal slaughter in Italy. In fact, it was impossible to analyse all the aspects related to these practices. In particular, it was difficult to form a univocal judgement on the controversial issue of animal welfare because it proved very difficult to find

Table 25.8 Average number of cuts performed with knife during incision

Procedure	Species	Restraint	No. of cuts
Halal	Bovine	Upright	25.2±9.4
	Bovine	Mechanically turned	7.4±2.5
	Small ruminants	Manually turned	3±0.9
	Small ruminants	Suspended	2.9±1.1
Kosher	Small ruminants	Suspended	1.25±1.1
	Poultry	Suspended	1±0

Table 25.9 Percentage of animals that showed excitement during incision

Procedure	Species	Restraint	Percentage
Halal	Bovine	Upright	42.90
Halal	Bovine	Mechanically turned	42.90
Halal	Small ruminants	Manually turned	33.30
Halal	Small ruminants	Suspended	34.20
Kosher	Small ruminants	Suspended	31.60
Kosher	Poultry	Suspended	7.10

Table 25.10 Average time(s) of struggling[a] from restraint to sticking

Procedure	Species	Restraint	Time (s)
Halal	Bovine	Upright	14.7±16.5
	Bovine	Mechanically turned	31.3±33.5
	Small ruminants	Manually turned	5.3±2.36
	Small ruminants	Suspended	13.1±7.6
kosher	Small ruminants	Suspended	23.6±11.7
	Poultry	Suspended	6.1±4.3

[a] Average time of struggling was calculated by summing each single time of coordinated movements shown by each animal and calculating the mathematical average for the number of observed animals for this parameter.

objective and measurable parameters that could give an idea of the real state of the animals during the slaughter. In order to protect animal welfare, particular importance was given to certain parameters, such as the speed of the slaughter line, the average time between restraint and cutting, the number of cuts, the average time of struggling, and the time taken to lose reflexes and the symptoms of conscience (corneal reflex and rhythmic breathing). Taking into account these limitations, this study highlighted the following results.

Table 25.11 Average time of struggling from sticking to the next handling

Procedure	Species	Restraint	Time (s)
Halal	Bovine	Upright	12.4±14.2
	Bovine	Mechanically turned	8.6±5.5
	Small ruminants	Manually turned	5.8±7.1
	Small ruminants	Suspended	4.1±4.3
Kosher	Small ruminants	Suspended	4.2±3.2
	Poultry	Suspended	21±4.3

Table 25.12 Average time of the loss of corneal reflex and rhythmic breathing[a]

Reflex lost	Species	Restraints	Percentage of animals	Time (s)	Time (s)	Time (s)
Corneal	Bovine	Upright	0	–	–	–
		Mechanically turned	7.10%	79	79	79
	Small ruminants	Manually turned	33.30%	50	50	38–62
Rhythmic breathing	Bovine	upright	90.00%	85.1	85	37–167
		Mechanically turned	92.20%	99.8	96	45–166
	Small ruminants	Manually turned	83.30%	29	27	26–33

[a] It is often very difficult to observe these parameters in practical situations, therefore the data are only relevant for a reduced percentage of animals in some abattoirs.

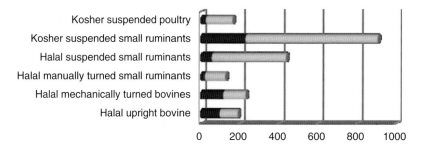

Figure 25.1 Average time(s) from restraint to sticking (black; see Table 25.6) and from sticking to next handling (grey, see Table 25.7) as recorded during spot check visits.

The parameter for the speed of the slaughter line in the slaughterhouse is closely related to the farmers' and slaughterers' economic profit. From this point of view, the practice of suspending the animals before sticking is much more productive. However, this practice is illegal for small ruminants (Anonymous 2009; Cenci Goga et al. 2009) and should not, therefore, be used. In line with these principles, despite the speed of the slaughter line of kosher, suspended, small ruminants being faster, some rabbinical committees have declared this practice to be unsuitable from the perspective of animal welfare (Dorff and Roth 2002).

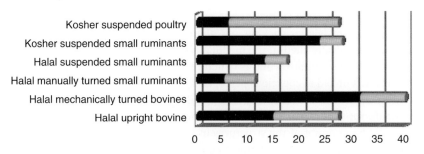

Figure 25.2 Average time(s) of struggling from restraint to sticking (black; see Table 25.10) and from sticking to the next handling (grey; see Table 25.11).

The average time between the restraint of the animal and cut corresponds, in practice, to the time taken by the operators to place the animal in a suitable position for incision to provide an optimum cut. This time must be as short as possible from the point of view of animal welfare. It is obvious that this time depends mainly on the restraint system used, although it can also vary depending on the organization of the slaughter line and the position of the operator performing the incision. In fact, the average time difference between halal and kosher slaughter of suspended small ruminants, estimated at 172 seconds, is explained by the operator's position. The latter stands further down the slaughter line from the hoist point of the sheep/goats in kosher slaughter, whereas the operator in the halal method is closer to this host point (Cenci Goga 2009; Cenci Goga and Catanese 2009; Cenci Goga et al. 2009; 2010). The animal's perception of pain at the time of sticking has been the subject of a controversial debate within the scientific community (Levinger 1995; Rosen 2004). However, recent studies using electroencephalography have shown that animals feel pain during sticking (Gibson et al. 2007, 2009a–d). The number of cuts, together with the direction of the incision influence the degree of pain in slaughtered animals and depends on both the restraint system and on the operator's skill (Catanese and Cenci Goga 2010). The cut determines a sudden nociceptor discharge at central levels which ends in four seconds. The cut nociceptors then lose their sensitive ability, whereas those left intact and exposed from the wound may still be stimulated, especially as a result of the mechanical effects of traction: their activation depends on cutting manipulation (Gregory 2004).

The average time of struggling is the period taken by the animal to perform voluntary coordinated movement. The average time for struggling before sticking is scientifically shorter and corresponds to the time between the restraint and cutting of the animal. Conversely, the average time of struggling after sticking is much more important in order to assess animal welfare. This parameter can then be used as a clinical indicator of the presence in the animal of consciousness and of its perception of pain, caused especially by the presence of blood in the trachea (Catanese and Cenci Goga 2010; Gregory et al. 2009, 2010; von Holleben et al. 2010). The gap between the average time after sticking for bovine and small ruminants could be attributed to the presence of a pathway for cerebral blood circulation and anastomosis (between branches of the vertebral and carotids arteries), which can be found in bovines, but not in ovine (Gregory 1998). Moreover, a false aneurysm is often established at the cut point in cattle, which can occlude the carotid and prolong the state of consciousness of the animal (Gregory 2004). The loss of corneal reflex and rhythmic breathing were used in this work to provide concrete answers regarding the perception of pain in animals after sticking and, therefore, their state of consciousness. These reflexes are anatomically sited in the medulla oblongata and therefore their disappearance following the hypoxia/anoxia of this structure indicates a loss of sensibility in the cranial encephalic areas linked to conscious perception. In other words, the disappearance of these reflexes indicates with certainty that the animal is unconscious and cannot, therefore, feel pain. Unfortunately, a positive reflex does not indicate consciousness, since it merely indicates the functioning of the structures of the medulla oblongata, which only coordinates involuntary vital reflexes (Catanese and Cenci Goga 2010).

The data we processed and examined demonstrate that it could take some varying times for a ritually slaughtered animal to lose consciousness after sticking and therefore pain may be felt. Nevertheless, religious practices and animal welfare could possibly be reconciled if future scientific research were to propose acceptable alternative methods of stunning to limit animal consciousness without cancelling it completely. It is also essential to have a CA at the slaughterhouse, e.g. a vet, who would ensure animal welfare in all the other phases of slaughtering. At the same time, research needs to identify new parameters which will enable us to truly and objectively assess the state of animal consciousness during slaughter (Velarde et al. 2014). Demand for ritual slaughter and halal meat is rapidly expanding in an economic market which may represent new industrial opportunities (D'Amico et al. 2017). This practice should not, therefore, be demonized by the public. On the contrary, it should be regulated more precisely. Slaughterhouses should be encouraged to modernize their lines and put in place measures to ensure correct handling and neck cutting procedures that reduce the suffering of the animal and minimize animal welfare problems. All these elements could improve not only animal welfare and operator safety, but also the economic profits of slaughterhouses and of all the related industries. Finally, they may alleviate the social tensions between the different communities.

25.6 Ritual Slaughter in Italy: Critical Aspects and Proposals

In an attempt to explore the acceptability of certain modifications to the procedure for halal slaughter and to stimulate the debate in the field of religious slaughter, we have selected the Reggio Emilia province for several logistic and demographic reasons.

First, the Reggio Emilia Public Health Department (PHD) has implemented an action plan specifically addressed to the Muslim population living in Emilia Romagna Region (ERR), in collaboration with the ERR Veterinary Service and with National Centre for Animal Welfare. Two establishments authorized to perform ritual slaughter currently operate in the Reggio Emilia province, one in the Rio Saliceto municipality and the other in Correggio.

Second, a large proportion of the people coming from non-EU member countries and living in Italy belong to the Muslim faith, currently the second religion in Italy and addressing, therefore, a considerable number of immigrants and their descendants, second and third generation, born and raised in Italy (Caritas/Migrantes 2018). The foreign population living in the Reggio Emilia province is quite substantial, amounting to 52 672 non-EU citizens (10.1% of the total resident population) and 6752 EU citizens (1.3%) (31.12.2008); Reggio Emilia ranks, therefore, as the second province of the ERR and the fourth Italian province, following Brescia, Prato, and Piacenza, for the proportion of foreign citizens in the resident population (Gregory et al. 2009).

Finally, people coming from 141 nations live in the Reggio Emilia province; the nations with the highest numbers coming in are Morocco (25.4% of all foreign citizens), Albania (15.7% of all foreign citizens), India (11.4% of all foreign citizens), China (9% of all foreign citizens), and Romania (7.5% of all foreign citizens) (Gibson et al. 2009c).

Usually there is a direct correlation between health condition and access to the health services; immigrants' access to health is sporadic and unpredictable due also to the scarce knowledge of preventive medicine in their native countries. The PHD is therefore actively engaged in diffusing its prevention activities in order to increase the access rate to the health services (Drewnowski and Specter 2004).

The analysis of the condition of immigrants' health has detected some major problems: a growing incidence of scabies and tuberculosis, little access to the health services (provided free to the community), and non-compliance not only with safety requirements in the work place, but also with the hygiene requirements during slaughter and all the stages of food production, from the early preparation stages to distribution. The PHD policy for these major problems is based on consumer education and on the continuous application of preventive measures.

One difficulty in applying the modern concepts of food safety, especially as far as prevention is concerned, is due to the strong food traditions of most immigrants, in which the religious component is a major characteristic; this has brought about an increase in the production and selling of traditional foods (also called ethnic foods). As far as Muslims are concerned, a big role is played by Islamic centres; they provide a wide, articulated support network to Muslim citizens (e.g. with the regular implementation of Arab language and Islamic culture courses), but they also interact with the local communities to sustain and develop food services delivered according to the Islamic law, particularly of meats produced according to the rules of Islamic ritual slaughter (Campagnolo and Perocco 2002).

25.6.1 Demonstrative Stage on Site for Stunning

In 2007 a process of consultation with representatives of the Islamic religious communities at the regional level was implemented. The aim was to reach a formal agreement concerning the slaughtering practices to be adopted in the ERR abattoirs in order to comply with food safety and the religious requirements of the Islamic and Jewish communities, and to limit the trauma deriving from the incision of the neck vessels in a conscious animal as far as possible.

A presentation slaughter was performed in a slaughterhouse of the Modena province: ritual incision of the neck vessels was preceded by stunning to explore the feasibility that lessening animal suffering could, in any event, conform to religious prescriptions, as it does in other countries.

These 'educational' slaughters were organized by ERR and implemented by the veterinary services operating in adjoining provinces (Bologna, Modena, and Reggio Emilia). The religious authorities of the Bologna, Modena, and Reggio Emilia provinces were invited to this event; in particular, the following representatives of the Reggio Emilia province were present: the Correggio Islamic Culture Arab Association, the Novellara Culture Arab Association, the Reggio Emilia Moroccan Immigrants Association, the Reggio Emilia Islamic Culture House, the Reggio Emilia Mosque Islamic Centre, the Montecchio Egyptian Community, and the Reggio Emilia Islamic Community Cultural Centre.

In Italy there are several Islamic associations (the most important are Unione delle Comunità Islamiche d'Italia (UCOII) and Centro Culturale Islamico d'Italia (CCII), but there are not many prominent Muslim scholars, so many members of Islamic communities rely on foreign scholars. As for the requirements about religious slaughter, we followed the prescription suggested by the religious authorities that participated to the project.

Two alternative methods to classical ritual slaughter without prior stunning were illustrated in order not only to limit animal suffering during killing, but also to comply with Islamic ritual requirements.

25.6.1.1 Electro-narcosis Prior to Slaughter

Electro-narcosis applied only to the head using electrical tongs is a reversible event; it does not result in any anatomical alteration and does not modify the integrity of the animal. Immediately after the ritual incision of the neck, the thoracic vessels are incised to quicken bleeding.

This method requires two operators: the first performs the electro-narcosis and the other must perform slaughter as quickly as possible. Loss of sensibility in cattle lasts between 31 and 90 seconds from the electro-narcosis, after which the animals regain consciousness. This fairly ample interval implies that the animal could regain consciousness and therefore suffer before slaughter has been completed so it was agreed to also carry out the incision of the sternal vessels in order to quicken the bleeding.

25.6.1.2 Stunning by Captive Bolt Immediately after Incision of the Neck Vessels

This method complies with ritual slaughter rules in so far as the animal's wholeness is not affected prior to slaughter.

Two operators must be present, the first one performing the slaughter and the other stunning the animal as quickly as possible. Animal suffering would be limited to the neck incision only, the extent of which should, in any case, be negligible when stunning follows the incision immediately.

The demonstrative stage showed that electro-narcosis prior to sticking produces instantaneous loss of consciousness without adversely affecting the integrity of the animal and is completely reversible, i.e. if sticking is not performed the animal regains the vital functions within 1–1.5 minutes. After this demonstrative stage on site, the local Solidarity Office organized a series of meetings to discuss the issues regarding ritual slaughter. Participation at the meetings was considerable: in addition to the head of the Reggio Emilia Solidarity Office, PHD representatives were also present (the head of the PHD and the head of the veterinary service), together with representatives of the animal protection associations, the Islamic immigrants associations, the ERR Veterinary Services, and the National Centre for Animal Welfare. All the participants acknowledged the need not only to respect the ritual practices and to promote their inclusion within the legal frame concerning food production, but also to safeguard the health of the Islamic people when handling animals during slaughter and of their families when consuming the products of the ritual slaughter, as well as the need to safeguard animal welfare during transport and slaughter. The alternative slaughter techniques previously mentioned were discussed further with the parties involved, who showed interest in these proposals.

25.6.1.3 Local Initiatives Regarding Ritual Slaughter and Results

Ritual sheep slaughter occurs most frequently in the occasion of the Sacrifice of Abraham (Id al-Adha), which falls on about the 70th day after the end of Ramadan. Sheep are slaughtered in such numbers as to raise attention and concern not only amongst the local health and political institutions, but also amongst the animal protection organizations in countries where religious slaughter is not predominant. During recent years the Id al-Kabir celebration has sometimes given rise to strong disapproval within the local non-Muslim community. This is due to matters of public order, especially when considering past frequent irregularities detected by different control authorities. Critics were influenced by some animal activist movements against ritual slaughter practice, as this was considered to conflict with the basic animal welfare principles not only established by EU and by national legislation, but also largely acknowledged by large sectors of Italian society.

The Reggio Emilia Veterinary Service has developed several initiatives regarding ritual slaughter: provincial guidelines, multilingual brochures, official meetings with the leaders of the local religious communities. All these initiatives have made Muslim leaders increasingly aware of their legal duties. They have become thoroughly acquainted with the current food safety legislation and have induced Muslim community to comply with legal requirements when slaughtering animals. The purchase and sale of animals, farming, and transport have also been affected. The veterinary service personnel have been available to perform their inspection and control duties during the celebration day, even when this falls on a bank holiday.

The slaughter trend has continuously increased since 2004, with a marked peak registered in 2008 with special reference to the stunning of animals prior to slaughter, particularly cattle.

The following factors have played a major role in the increasing awareness of the Reggio Emilia Islamic population regarding the need to use authorized slaughterhouses:

- the acknowledgement of an increased need for food safety for their family, especially where women and young children are concerned
- the purchase of healthy animals from registered farms under veterinary control
- the perception of the potential risk of illness due to contact with sick animals during ritual slaughter performed at home
- attention to the sensitivity of their Italian neighbours, given that Islamic immigrants wish to maintain friendly relations
- awareness of the risk of environmental contamination as a result of the disposal of post-slaughter waste
- awareness of the different sensitivity of the Italians concerning slaughter without stunning, which results in an often quite harsh, direct clash with the local and national animal protection movements.

Currently, the supply of sheep and goat meat complying with Islamic law requirements is guaranteed in the Reggio Emilia province throughout the year, thanks to the two authorized abattoirs operating locally. Another five slaughterhouses have said they are willing to host ritual slaughter during the three days of the religious feast; they request, however, stunning by electro-narcosis or captive bolt.

This experiment was carried out for the first time in sheep and goats in the Reggio Emilia province in 2007, and a marked increase in the number of animals stunned prior to slaughter was recorded the following year. Moreover, in the same year stunning prior to slaughter was applied for the first time to cattle.

Seven out of 13 abattoirs currently operating in the Reggio Emilia territory have declared their availability to guarantee slaughter service to the Islamic community, although not all under the same conditions. This concurs with the current policy to have all the local abattoirs working during public holidays, so that all Islamic citizens could have access to the nearest abattoir and therefore the only two slaughterhouses currently operating would not be overworked. No other alternative exists: the current legislation does actually agree to the possibility of not stunning the animals, but does not allow ritual slaughter outside the slaughterhouses.

In 2010 Italy registered the Halal Italy trademark in order to certify compliance with Islamic religious requirements for food and agricultural products, cosmetics, and pharmaceuticals. In the guidelines on the slaughtering of animals for halal meat produced by the Certification Ethics Committee of COREIS (Comunità Religiosa Islamica, the Islamic religious community) stunning is taken into consideration: 'the animals must be alive at the moment of sticking, even if is stunned' (Cenci Goga 2009). This last point is very important because it represents openness towards methods of stunning that do not lead to death and do not injure the animal prior to bleeding.

The potential risk of an animal being conscious after sticking is a welfare issue (Cenci-Goga et al. 2010), enhancing the necessity to encourage the use of stunning for religious slaughter where religious communities can accept it. New stunning methods are regularly developed and proposed on the market in order to face the new challenges of the farming and meat industry, and scientific and technical progress is regularly made with regard to the handling and restraining of animals at slaughterhouses (Anonymous 2009). Current technical-scientific advances allow a reappraisal of some practices used during religious slaughter without compromising its deep and essential meaning, through to the identification of techniques that limit animal vigilance without causing any lesion that may impair its integrity. All this is done with respect for the religious principles of the Jewish and Muslim community and with respect for animal welfare, minimizing as much as possible the risk of causing needless suffering to the animals.

25.7 Halal Certification in Italy

There are three different types of halal certification. Individual products can be certified, meaning the production process and ingredients in that particular product are halal. So a consumer could buy halal yoghurt, for example, from a store that also sells non-halal yoghurt. Production facilities can be certified, so that any products produced according to the certification standards can claim to be halal. For example, in an abattoir that is certified to produce halal meat, the meat will be halal no matter what cuts or final shape the meat takes. However, it may not even get labelled as halal when it reaches the market. Retail premises can also be certified so that all food prepared and sold from that business is halal. The halal certification process varies depending on who is performing the service. This is where uncertainty creeps in. Muslim consumers are largely unable to find out exactly what process has been followed in the certification process and what standards have been set by the certification provider.

Halal certification is needed for two key reasons. First, certification helps local Muslims decide which products to buy. Modern food processing and globalized markets make it hard for Muslims to know how their food was produced and where it has come from. To get around this uncertainty, consumers who want to buy halal food need a system that checks whether products meet the requirements of being halal. In this sense, halal certification is similar to any type of food certification and audit system. Whether it be halal, kosher, gluten-free or organic, food certification services help consumers to make informed decisions about the food they eat. The second reason has to do with trade. With the global halal food trade estimated at €500 billion annually, with 2 billion de facto consumers, Muslim markets provide a lucrative opportunity for companies. If companies want to export their products to those markets, they need to have halal certification.

Certified halal products in Italy can come from two sources: domestic products that are produced locally and certified by local businesses, or imported products that have been certified overseas. Numerous halal certifiers operate in Italy (Table 25.13).

Access to halal market is facilitated by halal certification, which can be provided by halal Italia. In 2010 the Health Minister, along with the Foreign Affairs Minister and the Minister of Economy and Agriculture Minister signed an agreement to promote voluntary certification for Italian premises and to certify the conformity to religious recommendations for food, cosmetics, and drugs (Anonymous 2010). This was done under the control and supervision of COREIS.

Table 25.13 Some organizations operating in Italy

Halal Italy	http://www.halalitaly.org/certificazione-halal-italy.html
Halal Italia	http://www.halalitalia.org/index.php?p=chi_siamo_it
Tüv Italia	http://www.tuv.it/it-it/settori/prodotti-di-consumo-e-retail/alimentare-salute-bellezza/halal
Halal Global	http://www.halalglobal.it
Whad	http://m.whad-it.com/1/chi_siamo_1674327.html
Halalint	http://www.halalint.org/index.php/it/servizi/certificazione/questo-prodotto-e-halal.html
Ihsan	http://www.ihsansrl.com/?pag=chi-siamo
Rina services	https://www.rina.org/it/halal
Bureau veritas italia	https://www.bureauveritas.it/services+sheet/certificazione-halal
Simply halal	https://www.sistemieconsulenze.it/halal

There are several halal certification organizations operating in Italy, but Italian government regulation applies only to providers that certify meat for export. While much of this meat may end up in the domestic market, certification providers that service only the Italian market do not come under any government regulation.

While some halal certification providers are associated with, or part of, larger Islamic organizations, such as COREIS, others are stand-alone businesses that provide local certification services.

With so much uncertainty about what constitutes halal, how products are certified and who is doing the certification, consumers who wish to buy halal food can find it a difficult task.

For non-Muslim consumers, however, halal food is little different to any other food available. It only matters whether or not food is halal if a person has the religious conviction and desire to eat only halal food. Although improvements could be made, halal certification is one way Muslims are able to do this.

25.7.1 Halal Italia

Halal Italia was founded in July 2010 on the initiative of six Muslim professionals with different qualifications, levels of aptitudes, and linguistic competence.

Halal Italia is a voluntary Italian certification organization for the most outstanding Italian products in conformity with the rules as to what is permissible (halal) under Islamic law in the agricultural and food, cosmetics, health, pharmaceutical, finance, and insurance sectors.

The promotion of high-quality Italian products is guaranteed by rigorous compliance with international halal standards and European regulations regarding the production processes that require certification. Halal Italia also supplies training, commercial assistance, and marketing services for the Italian and foreign market to firms interested in developing new products and exploring new opportunities. Thanks to its collaboration with the Ethical Committee for Halal Certification of COREIS, it has been possible to ensure a high level of reliability in the certification service: the Ethical Committee is the independent religious authority that issues the certificates, while the staff of Halal Italia carry out inspections on the premises of participating firms.

The Ethical Committee has prepared an important tool: a booklet containing the rules and regulations for halal certification. This contains an analysis of the Islamic doctrine regarding food and body care that has allowed Halal Italia to implement the guidelines for the certification of production processes and staff training.

When published, the forthcoming rules and regulations for pharmaceuticals will open up new prospects for the export and domestic markets in a sector where there is a high demand and where the ethical value of the certification constitutes a precious plus point.

25.7.2 COREIS

For over 20 years, COREIS has represented the religious values of Muslims in harmony with the cultural identity of Italy and Europe. Since 2006, Yahya Pallavicini, Imam of the Al-Wahid Mosque in Milan, has been a member of the Council for Italian Islam. Set up by the Ministry of the Interior, this is the only official body catering for the needs of Muslims in Italy.

The fruitful collaboration with, first, the Minister of the Interior Giuseppe Pisanu, then with Minister Giuliano Amato and with Minister Roberto Maroni, is indicative of the trust that is placed in COREIS at an institutional level. In recognition of 'its important contribution to the development of an intercultural and interreligious dialogue over more than twenty years', in December 2008 the mayor of Milan awarded the Ambrogino prize for civic merit to COREIS. In 2009 COREIS started its first pilot project for halal certification in Italy in collaboration with the region of Lombardy and PROMOS (the special agency of the Milan Chamber of Commerce for international activities). It registered the Halal Italia mark, the first official one in Italy, and set up an Ethical Committee for Halal Certification consisting of Muslim theologians and jurists and technical experts.

The Ethical Committee of COREIS has the task of drawing up the rules and regulations for halal certification for the various production sectors and ensuring their application, taking into account the common ground between the four Sunni schools of Islamic law as well as the Shi'ite one. This activity was officially recognized in an agreement signed on 30 June 2010 at the Ministry of Foreign Affairs in Rome between the Italian Minister of Foreign Affairs Franco Frattini and the Ministers of Economic Development, Agricultural, Food and Forestry Policies, and Health in the presence of the ambassadors of the Organization of the Islamic Conference (OIC) countries.

References

Anonymous (1979). *European Convention for the Protection of Animals for Slaughter.* Strasbourg: Council of Europe.

Anonymous (1991). Council Directive 91/628/EEC of 19 November 1991 on the protection of animals during transport and amending Directives 90/425/EEC and 91/496/EEC. *Official Journal L* 340: 17–27.

Anonymous (1993). Directive 93/119/EC on the protection of animals at the time of slaughter or killing. *European Community Official Journal* 340: 21–34.

Anonymous (1993). *Working Party on Research on the Mentally Incapacitated: The Ethical Conduct of Research on the Mentally Incapacitated.* Medical Research Council.

Anonymous (1995). Council Directive 95/29/EC of 29 June 1995 amending Directive 90/628/EEC concerning the protection of animals during transport. *Official Journal L* 148: 52–63.

Anonymous (1997). *Treaty of Amsterdam.* Amsterdam: Office for Official Publications of the European Communities, 2 October 1997.

Anonymous (2005). Council Regulation (EC) No 1/2005 of 22 December 2004 on the protection of animals during transport and related operations and amending Directives 64/432/EEC and 93/119/EC and Regulation (EC) No 1255/97. *Official Journal of the European Union L* 3, 5 January 2005 p 1–44.

Anonymous (2009). Council Regulation (EC) No 1099/2009 of 24 September 2009 on the protection of animals at the time of killing. *Official Journal of the European Union L* 52 (303): 1–30. (Vol. 1099/2009).

Anonymous (2010). Convenzione interministeriale a sostegno dell'iniziativa Halal Italia. http://www.halalitalia.org/index.php?p=convenzione (accessed December 2018).

Bruce, N., Pope, D., and Stanistreet, D. (2008). *Quantitative Methods for Health Research: A Practical Interactive Guide to Epidemiology and Statistics*. Wiley.

Campagnolo, M.T. and Perocco, F. (2002). *La comunicazione degli immigrati e le tecniche dei nuovi media: l'associazionismo in rete. Riflessioni sulla comunità islamica digitale*. Venezia: Università di Venezia.

Caritas/Migrantes. (2018). Immigrazione dossier statistico XXIII rapporto. www.caritasitaliana.it (accessed December 2018).

Catanese, B. and Cenci Goga, B. (2010). Religious slaughter: data from Italy. *Dialrel Final Workshop. Religious Slaughter: Improving Knowledge and Expertise Through Dialogue and Debate on Welfare, Legislation and Socio-Economic Aspects*, Istanbul, Turkey (15–16 March 2010).

Cenci Goga, B.T. (2009). La macellazione rituale: benessere animale e aspetti giuridici. *Ingegneria alimentare,* 6: 55–58.

Cenci Goga, B. and Catanese, B. (2009). Religious slaughter of poultry in Italy: animal welfare issues. *Hygiena Alimentorum* XXX: 59–162.

Cenci Goga, B.T., Mattiacci, C., De Angelis, G. et al. (2009). La macellazione religiosa in Italia. *Atti della Società Italiana delle Scienze Veterinarie* LXIII: 356–358.

Cenci-Goga, B.T., Mattiacci, C., De Angelis, G. et al. (2010). Religious slaughter in Italy. *Veterinary Research Communications* 34 (Suppl 1): S139–S143.

D'Amico, P., Vitelli, N., Cenci Goga, B. et al. (2017). Meat from cattle slaughtered without stunning sold in the conventional market without appropriate labelling: a case study in Italy. *Meat Science* 134: 1–6.

Dorff, E.N. and Roth, J. (2002). *Shackling and Hoisting: The Committee on Jewish Law and Standards of the Conservative Movement*. New York: The Rabbinical Assembly.

Drewnowski, A. and Specter, S.E. (2004). Poverty and obesity: the role of energy density and energy costs. *American Journal of Clinical Nutrition* 79: 6–16.

Francione, G.L. (2018). A short essay on the meaning of "new welfarism". http://www.abolitionistapproach.com/a-short-essay-on-the-meaning-of-new-welfarism (accessed December 2018).

Gentle, M.J. and Tilston, V.L. (2000). Nociceptors in the legs of poultry: implications for potential pain in preslaughter shackling. *Animal Welfare* 9: 227–236.

Gibson, T.J., Johnson, C.B., Murrell, J.C. et al. (2009a). Components of electroencephalographic responses to slaughter in halothane-anaesthetised calves: effects of cutting neck tissues compared with major blood vessels. *New Zealand Veterinary Journal* 57 (2): 84–89.

Gibson, T.J., Johnson, C.B., Murrell, J.C. et al. (2009b). Electroencephalographic responses of halothane – anaesthetised calves to slaughter by ventral-neck incision without prior stunning. *New Zealand Veterinary Journal* 57 (2): 77–83.

Gibson, T.J., Johnson, C.B., Murrell, J.C. et al. (2009c). Amelioration of electroencephalographic responses to slaughter by non-penetrative captive-bolt stunning after ventral-neck incision in halothane – anaesthetised calves. *New Zealand Veterinary Journal* 57 (2): 96–101.

Gibson, T.J., Johnson, C.B., Murrell, J.C. et al. (2009d). Electroencephalographic responses to concussive non-penetrative captive-bolt stunning in halothane – anaesthetised calves. *New Zealand Veterinary Journal* 57 (2): 90–95.

Gibson, T.J., Johnson, C.B., Stafford, K.J. et al. (2007). Validation of the acute electroencephalographic responses of calves to noxious stimulus with scoop dehorning. *New Zealand Veterinary Journal* 55: 152–157.

Grandin, T. (1996). *Thinking in Pictures and Other Reports from My Life with Autism*. New York: Vintage Books, Random House.

Grandin, T. (2010). ASPCA Pen. http://www.grandin.com/ritual/aspca.html (accessed December 2018).

Grandin, T. and Regenstein, J.M. (1994). Religious slaughter and animal welfare: a discussion for meat scientists. *Meat Focus International* 3: 115–123.

Grayson, L. (2000). *Animals in Research: For and against*. The British Library.

Gregory, N.G. (1998). *Animal Welfare and Meat Science*. Wallingford: CABI Publishing.

Gregory, N.G. (2004). *Physiology and Behaviour of Animal Suffering*. Blackwell.

Gregory, N.G., Fielding, H.R., Wenzlawowicz, M.V., and Holleben, K.V. (2010). Time to collapse following slaughter without stunning in cattle. *Meat Science* 85: 66–69.

Gregory, N.G., Wenzlawowicz, M.V., and Holleben, K.V. (2009). Blood in the respiratory tract during slaughter with and without stunning in cattle. *Meat Science* 82: 13–16.

von Holleben, K. V., Wenzlawowicz, M. V., Gregory, N. et al. (2010). Report on Good and Adverse Practices. Animal Welfare Concerns in Relation to Slaughter Practices from the Viewpoint of Veterinary Sciences. *Dialrel Deliverable n. 1.3, Cardiff University*.

Levinger, I.M. (1995). *Shechita in the Light of the Year 2000*. Jerusalem: Maskil L. David.

Reiss, M.J. and Straughan, R. (1996). *Improving Nature?* Cambridge: University Press.

Rosen, S.D. (2004). Physiological insights into Shechita. *Veterinary Record* 154: 759–765.

Singer, P. A. D. (2004). Subject: Heavy Petting.

Singer, P. and Mason, J. (2006). *The Ethics of What We Eat: Why our Food Choices Matter*. Rodale Books, Pennsylvania State University.

Sztybel, D. (2004). Subject: Philosophy and Animal Rights Page. http://davidsztybel.info/, retrieved September 2019.

Velarde, A., Rodriguez, P., Dalmau, A. et al. (2014). Religious slaughter: evaluation of current practices in selected countries. *Meat Science* 96 (1): 278–287.

Index

Note: Page numbers in *italics* indicate figures; page numbers in **bold** indicate tables.

The Halal Food Handbook, First Edition. Edited by Yunes Ramadan Al-Teinaz, Stuart Spear,
and Ibrahim H. A. Abd El-Rahim.
© 2020 John Wiley & Sons Ltd. Published 2020 by John Wiley & Sons Ltd.